SCHAUM'S OUTLINE OF

THEORY AND PROBLEMS

OF

ELECTRONIC DEVICES AND CIRCUITS

SCHAUM'S OUTLINE OF

THEORY AND PROBLEMS

OF

ELECTRONIC DEVICES AND CIRCUITS

•

JIMMIE J. CATHEY, Ph.D.
Professor of Electrical Engineering
University of Kentucky

SCHAUM'S OUTLINE SERIES
McGRAW-HILL, INC.
New York St. Louis San Francisco Auckland Bogotá Caracas
Hamburg Lisbon London Madrid Mexico Milan Montreal
New Delhi Paris San Juan São Paulo Singapore
Sydney Tokyo Toronto

JIMMIE J. CATHEY earned the Ph.D. from Texas A&M University and has 13 years of industrial experience in the design and development of electric drive systems. Since 1980, he has taught at the University of Kentucky, and his research and teaching interests are power electronics, electric machines, and robotics. He is a Registered Professional Engineer.

Schaum's Outline of Theory and Problems of
ELECTRONIC DEVICES AND CIRCUITS

3 4 5 6 7 8 9 10 11 12 13 14 15 16 17 18 19 20 SHP SHP 9 9 8 7 6 5 4 3 2

ISBN 0-07-010274-0

Sponsoring Editor, David Beckwith
Production Supervisor, Fred Schulte
Editing Supervisor, Marthe Grice
Cover design by Amy E. Becker.

Library of Congress Cataloging-in-Publication Data

Cathey, Jimmie J.
Schaum's outline of theory and problems of electronic devices and circuits.

(Schaum's outline series)
Includes index.
1. Electronic circuits. 2. Electronic control.
I. Title. II. Title: Theory and problems of electronic devices and circuits.
TK7867.C34 1989 621.3815'3 88-23090
ISBN 0-07-010274-0

To Phillip and Julia

with encouragement to stand morally strong in the Way

Preface

The subject matter of electronics may be divided into two broad categories: the application of physical properties of materials in the development of electronic control devices, and the utilization of electronic control devices in circuit applications. The emphasis in this book is on the latter category, beginning with the terminal characteristics of electronic control devices. Other topics are dealt with only as necessary to an understanding of these terminal characteristics.

This book is designed to supplement the text for a first course in electronic circuits for engineers. It will also serve as a refresher for those who have previously taken a course in electronic circuits. Engineering students enrolled in a nonmajors' survey course on electronic circuits will find that portions of Chapters 1 to 6, 10, and 11 offer a valuable supplement to their study. Each chapter contains a brief review of pertinent topics along with governing equations and laws, with examples inserted to immediately clarify and emphasize principles as introduced. As in other Schaum's Outlines, primary emphasis is on the solution of problems; to this end, over 640 solved problems are presented.

The author is indebted to his wife, Mary Ann, for her untiring labor in typing the manuscript. Thanks are also due the editor, Ed Millman, for helpful suggestions and careful checking of the material. For any lapses, the author accepts blame and offers apology.

JIMMIE J. CATHEY

Contents

Chapter 1

Circuit Analysis: Port Point of View

1.1 INTRODUCTION

Electronic devices are described by their nonlinear terminal voltage-current characteristics. Circuits containing electronic devices are analyzed and designed either by utilizing graphs of experimentally measured characteristics or by linearizing the voltage-current characteristics of the devices. Depending upon applicability, the latter approach involves the formulation of either small-perturbation equations valid about an operating point or a piecewise-linear equation set. The linearized equation set describes the circuit in terms of its interconnected passive elements and independent or controlled voltage and current sources; formulation and solution require knowledge of the circuit analysis and circuit reduction principles reviewed in this chapter.

1.2 CIRCUIT ELEMENTS

The time-stationary (or constant-value) elements of Fig. 1-1(*a*) to (*c*) (the resistor, inductor, and capacitor, respectively) are called *passive elements*, since none of them can continuously supply energy to a circuit. For voltage v and current i, we have the following relationships: For the resistor,

$$v = Ri \qquad \text{or} \qquad i = Gv \tag{1.1}$$

where R is its *resistance* in ohms (Ω), and $G \equiv 1/R$ is its *conductance* in siemens (S). Equation (*1.1*) is known as *Ohm's law*. For the inductor,

$$v = L\frac{di}{dt} \qquad \text{or} \qquad i = \frac{1}{L}\int_{-\infty}^{t} v\,d\tau \tag{1.2}$$

where L is its *inductance* in henrys (H). For the capacitor,

$$v = \frac{1}{C}\int_{-\infty}^{t} i\,d\tau \qquad \text{or} \qquad i = C\frac{dv}{dt} \tag{1.3}$$

where C is its *capacitance* in farads (F). If R, L, and C are independent of voltage and current (as well as of time), these elements are said to be linear: Multiplication of the current through each by a constant will result in the multiplication of its terminal voltage by that same constant. (See Problems 1.1 and 1.3.)

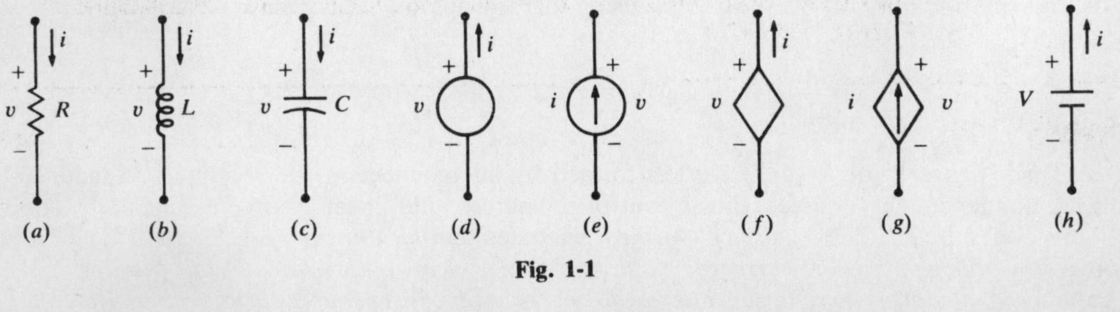

Fig. 1-1

The elements of Fig. 1-1(*d*) to (*h*) are called *active elements* because each is capable of continuously supplying energy to a network. The *ideal voltage source* in Fig. 1-1(*d*) provides a terminal voltage v that is independent of the current i through it. The *ideal current source* in Fig.

1-1(e) provides a current i that is independent of the voltage across its terminals. However, the *controlled* (or *dependent*) *voltage source* in Fig. 1-1(f) has a terminal voltage that depends upon the voltage across or current through some other element of the network. Similarly, the *controlled* (or *dependent*) *current source* in Fig. 1-1(g) provides a current whose magnitude depends on either the voltage across or current through some other element of the network. If the dependency relation for the voltage or current of a controlled source is of the first degree, then the source is called a *linear* controlled (or dependent) source. The *battery* or *dc voltage source* in Fig. 1-1(h) is a special kind of independent voltage source.

1.3 CIRCUIT LAWS

Along with the three voltage-current relationships (*1.1*) to (*1.3*), Kirchhoff's laws are sufficient to formulate the simultaneous equations necessary to solve for all currents and voltages of a network. (We use the term *network* to mean any arrangement of circuit elements.)

Kirchhoff's voltage law (KVL) states that *the algebraic sum of all voltages around any closed loop of a circuit is zero*; it is expressed mathematically as

$$\sum_{k=1}^{n} v_k = 0 \tag{1.4}$$

where n is the total number of passive- and active-element voltages around the loop under consideration.

Kirchhoff's current law (KCL) states that *the algebraic sum of all currents entering every node* (junction of elements) *must be zero*; that is,

$$\sum_{k=1}^{m} i_k = 0 \tag{1.5}$$

where m is the total number of currents flowing into the node under consideration.

1.4 STEADY-STATE CIRCUITS

At some (sufficiently long) time after a circuit containing linear elements is energized, the voltages and currents become independent of initial conditions and the time variation of circuit quantities becomes identical to that of the independent sources; the circuit is then said to be operating in the *steady state*. If all nondependent sources in a network are independent of time, the steady state of the network is referred to as the *dc steady state*. On the other hand, if the magnitude of each nondependent source can be written as $K \sin(\omega t + \phi)$, where K is a constant, then the resulting steady state is known as the *sinusoidal steady state*, and well-known frequency-domain, or phasor, methods are applicable in its analysis. In general, electronic circuit analysis is a combination of dc and sinusoidal steady-state analysis, using the principle of superposition discussed in the next section.

1.5 NETWORK THEOREMS

A *linear network* (or *linear circuit*) is formed by interconnecting the terminals of independent (that is, nondependent) sources, linear controlled sources, and linear passive elements to form one or more closed paths. The *superposition theorem* states that *in a linear network containing multiple sources, the voltage across or current through any passive element may be found as the algebraic sum of the individual voltages or currents due to each of the independent sources acting alone, with all other independent sources deactivated.*

An ideal voltage source is deactivated by replacing it with a short circuit. An ideal current source is deactivated by replacing it with an open circuit. In general, controlled sources remain active when the superposition theorem is applied.

Example 1.1 Is the network of Fig. 1-2 a linear circuit?

The definition of a linear circuit is satisfied if the controlled source is a linear controlled source; that is, if α is a constant.

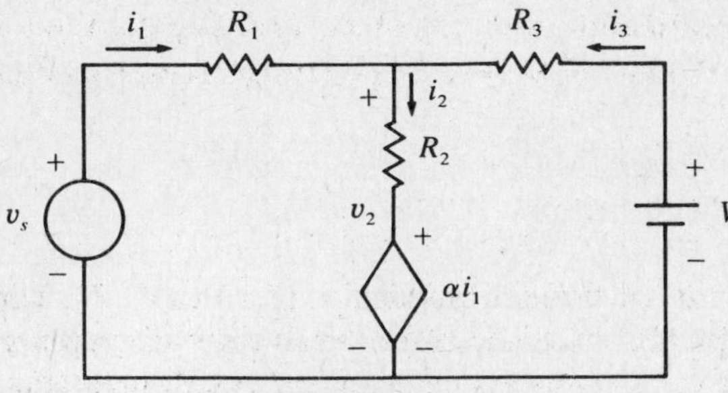

Fig. 1-2

Example 1.2 For the circuit of Fig. 1-2, $v_s = 10 \sin \omega t$ V, $V_b = 10$ V, $R_1 = R_2 = R_3 = 1\ \Omega$, and $\alpha = 0$. Find current i_2 by use of the superposition theorem.

We first deactivate V_b by shorting, and use a single prime to denote a response due to v_s alone. Using the method of node voltages with unknown v_2' and summing currents at the upper node, we have

$$\frac{v_s - v_2'}{R_1} = \frac{v_2'}{R_2} + \frac{v_2'}{R_3}$$

Substituting given values and solving for v_2', we obtain

$$v_2' = \tfrac{1}{3} v_s = \tfrac{10}{3} \sin \omega t$$

Then, by Ohm's law,

$$i_2' = \frac{v_2'}{R_2} = \tfrac{10}{3} \sin \omega t \qquad \text{A}$$

Now, deactivating v_s and using a double prime to denote a response due to V_b alone, we have

$$i_3'' = \frac{V_b}{R_3 + R_1 \| R_2}$$

where

$$R_1 \| R_2 \equiv \frac{R_1 R_2}{R_1 + R_2}$$

so that

$$i_3'' = \frac{10}{1 + 1/2} = \frac{20}{3} \text{ A}$$

Then, by current division,

$$i_2'' = \frac{R_1}{R_1 + R_2} i_3'' = \frac{1}{2} i_3'' = \frac{1}{2} \frac{20}{3} = \frac{10}{3} \text{ A}$$

Finally, by the superposition theorem,

$$i_2 = i_2' + i_2'' = \frac{10}{3}(1 + \sin \omega t) \qquad \text{A}$$

Terminals in a network are usually considered in pairs. A *port* is a terminal pair across which a voltage can be identified and such that the current into one terminal is the same as the current out of the other terminal. In Fig. 1-3, if $i_1 \equiv i_2$, then terminals 1 and 2 form a port. Moreover, as viewed to the left from terminals 1,2, network A is a one-port network. Likewise, viewed to the right from terminals 1,2, network B is a one-port network.

Thévenin's theorem states that *an arbitrary linear, one-port network such as network A in Fig. 1-3(a) can be replaced at terminals 1,2 with an equivalent series-connected voltage source V_{Th} and impedance Z_{Th} (=$R_{Th} + jX_{Th}$) as shown in Fig. 1-3(b). V_{Th} is the open-circuit voltage of network A at*

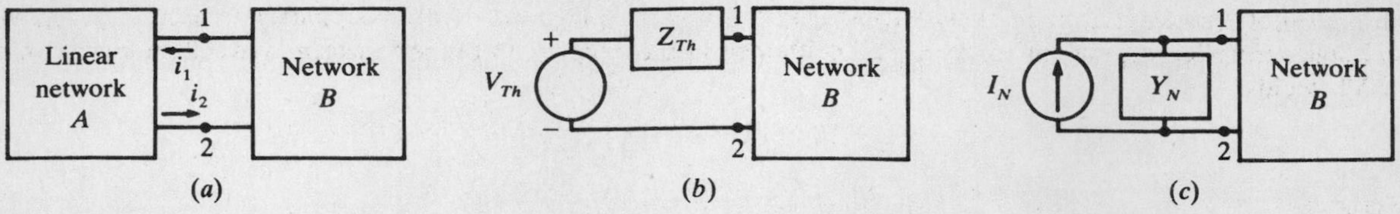

Fig. 1-3

terminals 1,2 and Z_{Th} is the ratio of open-circuit voltage to short-circuit current of network A determined at terminals 1,2 with network B disconnected. If network A or B contains a controlled source, its controlling variable must be in that same network. Alternatively, Z_{Th} is the equivalent impedance looking into network *A* through terminals 1,2 with all independent sources deactivated. If network *A* contains a controlled source, Z_{Th} is found as the *driving-point impedance*. (See Example 1.4.)

Example 1.3 In the circuit of Fig. 1-4, $V_A = 4$ V, $I_A = 2$ A, $R_1 = 2\ \Omega$, and $R_2 = 3\ \Omega$. Find the Thévenin equivalent voltage V_{Th} and impedance Z_{Th} for the network to the left of terminals 1,2.

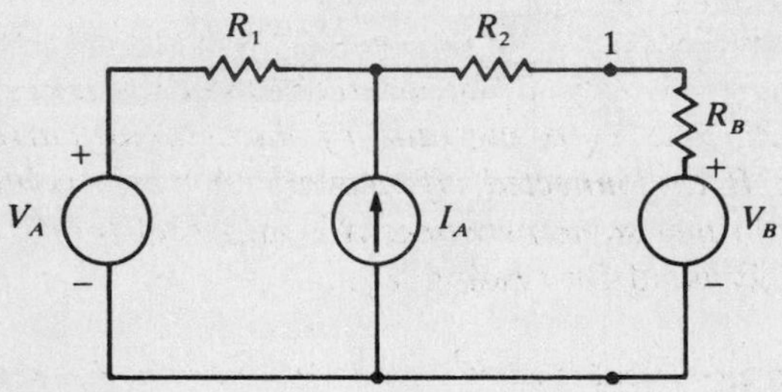

Fig. 1-4

With terminals 1,2 open-circuited, no current flows through R_2; thus, by KVL,

$$V_{Th} = V_{12} = V_A + I_A R_1 = 4 + (2)(2) = 8\ \text{V}$$

The Thévenin impedance Z_{Th} is found as the equivalent impedance for the circuit to the left of terminals 1,2 with the independent sources deactivated (that is, with V_A replaced by a short circuit, and I_A replaced by an open circuit):

$$Z_{Th} = R_{Th} = R_1 + R_2 = 2 + 3 = 5\ \Omega$$

Example 1.4 In the circuit of Fig. 1-5(*a*), $V_A = 4$ V, $\alpha = 0.25$ A/V, $R_1 = 2\ \Omega$, and $R_2 = 3\ \Omega$. Find the Thévenin equivalent voltage and impedance for the network to the left of terminals 1,2.

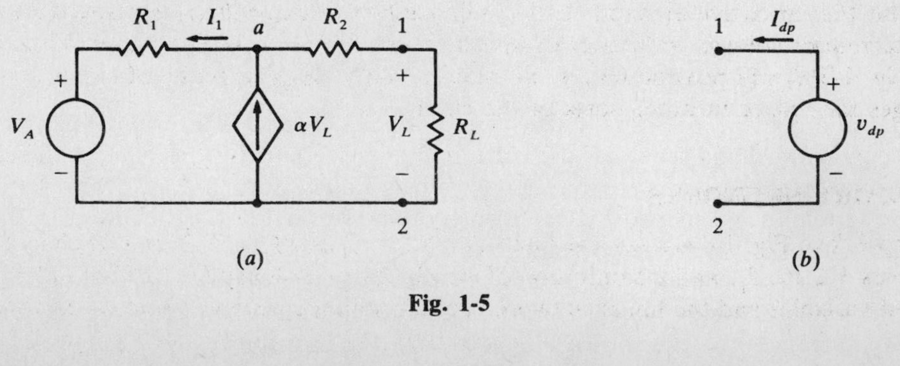

Fig. 1-5

With terminals 1,2 open-circuited, no current flows through R_2. But the control variable V_L for the voltage-controlled dependent source is still contained in the network to the left of terminals 1,2. Application of KVL yields

$$V_{Th} = V_L = V_A + \alpha V_{Th} R_1$$

so that

$$V_{Th} = \frac{V_A}{1 - \alpha R_1} = \frac{4}{1 - (0.25)(2)} = 8 \text{ V}$$

Since the network to the left of terminals 1,2 contains a controlled source, Z_{Th} is found as the driving-point impedance V_{dp}/I_{dp}, with the network to the right of terminals 1,2 in Fig. 1-5(*a*) replaced by the driving-point source of Fig. 1-5(*b*) and V_A deactivated (short-circuited). After these changes, KCL applied at node *a* gives

$$I_1 = \alpha V_{dp} + I_{dp} \tag{1.6}$$

Application of KVL around the outer loop of this circuit (with V_A still deactivated) yields

$$V_{dp} = I_{dp} R_2 + I_1 R_1 \tag{1.7}$$

Substitution of (*1.6*) into (*1.7*) allows solution for Z_{Th} as

$$Z_{Th} = \frac{V_{dp}}{I_{dp}} = \frac{R_1 + R_2}{1 - \alpha R_1} = \frac{2 + 3}{1 - (0.25)(2)} = 10 \ \Omega$$

Norton's theorem states that *an arbitrary linear, one-port network such as network A in Fig. 1-3(a) can be replaced at terminals 1,2 by an equivalent parallel-connected current source I_N and admittance Y_N as shown in Fig. 1-3(c). I_N is the short-circuit current that flows from terminal 1 to terminal 2 due to network A, and Y_N is the ratio of short-circuit current to open-circuit voltage at terminals 1,2 with network B disconnected. If network A or B contains a controlled source, its controlling variable must be in that same network.* It is apparent that $Y_N = 1/Z_{Th}$; thus, any method for determining Z_{Th} is equally valid for finding Y_N.

Example 1.5 Find the Norton equivalent current I_N and admittance Y_N for the circuit of Fig. 1-4 with values as given in Example 1.3.

The Norton current is found as the short-circuit current from terminal 1 to terminal 2 by superposition; it is

$$I_N = I_{12} = \text{current due to } V_A + \text{current due to } I_A = \frac{V_A}{R_1 + R_2} + \frac{R_1 I_A}{R_1 + R_2}$$

$$= \frac{4}{2 + 3} + \frac{(2)(2)}{2 + 3} = 1.6 \text{ A}$$

The Norton admittance is found from the result of Example 1.3 as

$$Y_N = \frac{1}{Z_{Th}} = \frac{1}{5} = 0.2 \text{ S}$$

We shall sometimes double-subscript voltages and currents to show the terminals that are of interest. Thus, V_{13} is the voltage across terminals 1 and 3, where terminal 1 is at a higher potential than terminal 3. Similarly, I_{13} is the current that flows *from* terminal 1 *to* terminal 3. As an example, V_L in Fig. 1-5(*a*) could be labeled V_{12} (but not V_{21}).

Note also that an active element (either independent or controlled) is restricted to its assigned, or stated, current or voltage, no matter what is involved in the rest of the circuit. Thus the controlled source in Fig. 1-5(*a*) will provide αV_L A no matter what voltage is required to do so and no matter what changes take place in other parts of the circuit.

1.6 TWO-PORT NETWORKS

The network of Fig. 1-6 is a *two-port* network if $I_1 = I_1'$ and $I_2 = I_2'$. It can be characterized by the four variables V_1, V_2, I_1, and I_2, only two of which can be independent. If V_1 and V_2 are taken as independent variables and the linear network contains no independent sources, the independent and

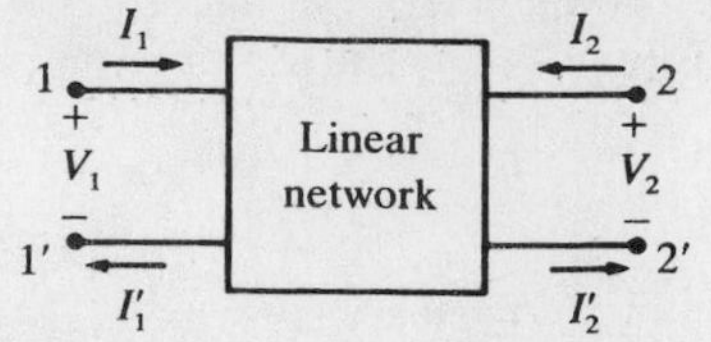

Fig. 1-6

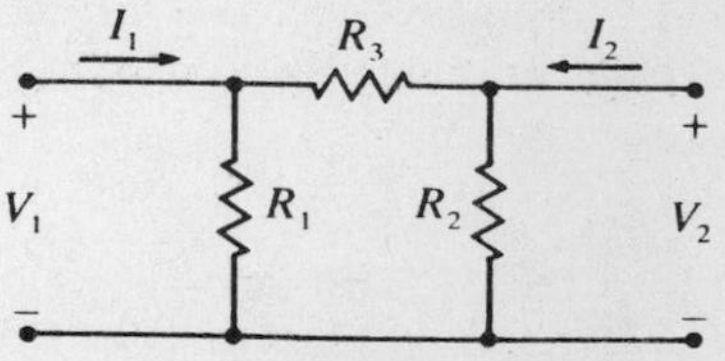

Fig. 1-7

dependent variables are related by the *open-circuit impedance parameters* (or, simply, the *z parameters*) z_{11}, z_{12}, z_{21}, and z_{22} through the equation set

$$V_1 = z_{11}I_1 + z_{12}I_2 \tag{1.8}$$

$$V_2 = z_{21}I_1 + z_{22}I_2 \tag{1.9}$$

Each of the z parameters can be evaluated by setting the proper current to zero (or, equivalently, by open-circuiting an appropriate port of the network). They are:

$$z_{11} = \left.\frac{V_1}{I_1}\right|_{I_2=0} \tag{1.10}$$

$$z_{12} = \left.\frac{V_1}{I_2}\right|_{I_1=0} \tag{1.11}$$

$$z_{21} = \left.\frac{V_2}{I_1}\right|_{I_2=0} \tag{1.12}$$

$$z_{22} = \left.\frac{V_2}{I_2}\right|_{I_1=0} \tag{1.13}$$

In a similar manner, if V_1 and I_2 are taken as the independent variables, a characterization of the two-port network via the *hybrid parameters* (or, simply, the *h-parameters*) results:

$$V_1 = h_{11}I_1 + h_{12}V_2 \tag{1.14}$$

$$I_2 = h_{21}I_1 + h_{22}V_2 \tag{1.15}$$

Two of the h parameters are determined by short-circuiting port 2, while the remaining two parameters are found, by open-circuiting port 1:

$$h_{11} = \left.\frac{V_1}{I_1}\right|_{V_2=0} \tag{1.16}$$

$$h_{12} = \left.\frac{V_1}{V_2}\right|_{I_1=0} \tag{1.17}$$

$$h_{21} = \left.\frac{I_2}{I_1}\right|_{V_2=0} \tag{1.18}$$

$$h_{22} = \left.\frac{I_2}{V_2}\right|_{I_1=0} \tag{1.19}$$

Example 1.6 Find the z parameters for the two-port network of Fig. 1-7.

With port 2 (on the right) open-circuited, $I_2 = 0$ and the use of (*1.10*) gives

$$z_{11} = \left.\frac{V_1}{I_1}\right|_{I_2=0} = R_1 \| (R_2 + R_3) = \frac{R_1(R_2 + R_3)}{R_1 + R_2 + R_3}$$

Also, the current I_{R2} flowing downward through R_2 is, by current division,

$$I_{R2} = \frac{R_1}{R_1 + R_2 + R_3} I_1$$

But, by Ohm's law,

$$V_2 = I_{R2}R_2 = \frac{R_1R_2}{R_1 + R_2 + R_3} I_1$$

Hence, by (*1.12*),

$$z_{21} = \frac{V_2}{I_1}\bigg|_{I_2=0} = \frac{R_1R_2}{R_1 + R_2 + R_3}$$

Similarly, with port 1 open-circuited, $I_1 = 0$ and (*1.13*) leads to

$$z_{22} = \frac{V_2}{I_2}\bigg|_{I_1=0} = R_2 \| (R_1 + R_3) = \frac{R_2(R_1 + R_3)}{R_1 + R_2 + R_3}$$

The use of current division to find the current downward through R_1 yields

$$I_{R1} = \frac{R_2}{R_1 + R_2 + R_3} I_2$$

and Ohm's law gives

$$V_1 = R_1 I_{R1} = \frac{R_1R_2}{R_1 + R_2 + R_3} I_2$$

Thus, by (*1.11*),

$$z_{12} = \frac{V_1}{I_2}\bigg|_{I_1=0} = \frac{R_1R_2}{R_1 + R_2 + R_3}$$

Example 1.7 Find the h parameters for the two-port network of Fig. 1-7.
With port 2 short-circuited, $V_2 = 0$ and, by (*1.16*),

$$h_{11} = \frac{V_1}{I_1}\bigg|_{V_2=0} = R_1 \| R_3 = \frac{R_1R_3}{R_1 + R_3}$$

By current division,

$$I_2 = -\frac{R_1}{R_1 + R_3} I_1$$

so that, by (*1.18*),

$$h_{21} = \frac{I_2}{I_1}\bigg|_{V_2=0} = -\frac{R_1}{R_1 + R_3}$$

If port 1 is open-circuited, voltage division and (*1.17*) lead to

$$V_1 = \frac{R_1}{R_1 + R_3} V_2$$

and

$$h_{12} = \frac{V_1}{V_2}\bigg|_{I_1=0} = \frac{R_1}{R_1 + R_3}$$

Finally, h_{22} is the admittance looking into port 2, as given by (1.19):

$$h_{22} = \frac{I_2}{V_2}\bigg|_{I_1=0} = \frac{1}{R_2 \| (R_1 + R_3)} = \frac{R_1 + R_2 + R_3}{R_2(R_1 + R_3)}$$

1.7 INSTANTANEOUS, AVERAGE, AND RMS VALUES

The *instantaneous value* of a quantity is the value of that quantity at a specific time. Often we will be interested in the average value of a time-varying quantity. But obviously, the average value of a sinusoidal function over one period is zero. For sinusoids, then, another concept, that of the *root-mean-square* (or *rms*) value, is more useful: For any time-varying function $f(t)$ with period T, the *average* value over one period is given by

$$F_0 = \frac{1}{T}\int_{t_0}^{t_0+T} f(t)\, dt \tag{1.20}$$

and the corresponding *rms* value is defined as

$$F = \sqrt{\frac{1}{T}\int_{t_0}^{t_0+T} f^2(t)\, dt} \tag{1.21}$$

where, of course, F_0 and F are independent of t_0. The motive for introducing rms values can be gathered from Example 1.9.

Example 1.8 Since the average value of a sinusoidal function of time is zero, the *half-cycle* average value, which is nonzero, is often useful. Find the half-cycle average value of the current through a resistance R connected directly across a periodic (ac) voltage source $v(t) = V_m \sin \omega t$.

By Ohm's law,

$$i(t) = \frac{v(t)}{R} = \frac{V_m}{R} \sin \omega t$$

and from (*1.20*), applied over the half cycle from $t_0 = 0$ to $T/2 = \pi$,

$$I_0 = \frac{1}{\pi}\int_0^{\pi} \frac{V_m}{R} \sin \omega t\, d(\omega t) = \frac{1}{\pi}\frac{V_m}{R}[-\cos \omega t]_{\omega t=0}^{\pi} = \frac{2}{\pi}\frac{V_m}{R} \tag{1.22}$$

Example 1.9 Consider a resistance R connected directly across a dc voltage source V_{dc}. The power absorbed by R is

$$P_{dc} = \frac{V_{dc}^2}{R} \tag{1.23}$$

Now replace V_{dc} with an ac voltage source, $v(t) = V_m \sin \omega t$. The *instantaneous power* is now given by

$$p(t) = \frac{v^2(t)}{R} = \frac{V_m^2}{R} \sin^2 \omega t \tag{1.24}$$

Hence, the *average power* over one period is, by (*1.20*),

$$P_0 = \frac{1}{2\pi}\int_0^{2\pi} \frac{V_m^2}{R} \sin^2 \omega t\, d(\omega t) = \frac{V_m^2}{2R} \tag{1.25}$$

Comparing (*1.23*) and (*1.25*), we see that, insofar as power dissipation is concerned, an ac source of amplitude V_m is equivalent to a dc source of magnitude

$$\frac{V_m}{\sqrt{2}} = \sqrt{\frac{1}{T}\int_0^T v^2(t)\, dt} \equiv V \tag{1.26}$$

For this reason, the rms value of a sinusoid, $V = V_m/\sqrt{2}$, is also called its *effective* value.

From this point on, unless an explicit statement is made to the contrary, all currents and voltages in the frequency domain (phasors) will reflect rms rather than maximum values. Thus, the time-domain voltage $v(t) = V_m \cos(\omega t + \phi)$ will be indicated in the frequency domain as $\bar{V} = V\underline{|\phi}$, where $V = V_m/\sqrt{2}$.

Solved Problems

1.1 Prove that the inductor element of Fig. 1-1(b) is a linear element by showing that (*1.2*) satisfies the converse of the superposition theorem.

Let i_1 and i_2 be two currents that flow through the inductors. Then by (*1.2*) the voltages across the inductor for these currents are, respectively,

$$v_1 = L\frac{di_1}{dt} \quad \text{and} \quad v_2 = L\frac{di_2}{dt} \qquad (1)$$

Now suppose $i = k_1 i_1 + k_2 i_2$, where k_1 and k_2 are distinct arbitrary constants. Then by (*1.2*) and (*1*),

$$v = L\frac{d}{dt}(k_1 i_1 + k_2 i_2) = k_1 L\frac{di_1}{dt} + k_2 L\frac{di_2}{dt} = k_1 v_1 + k_2 v_2 \qquad (2)$$

Since (*2*) holds for any pair of constants (k_1, k_2), superposition is satisfied and the element is linear.

1.2 If $R_1 = 5\ \Omega$, $R_2 = 10\ \Omega$, $V_s = 10$ V, and $I_s = 3$ A in the circuit of Fig. 1-8, find the current i by using the superposition theorem.

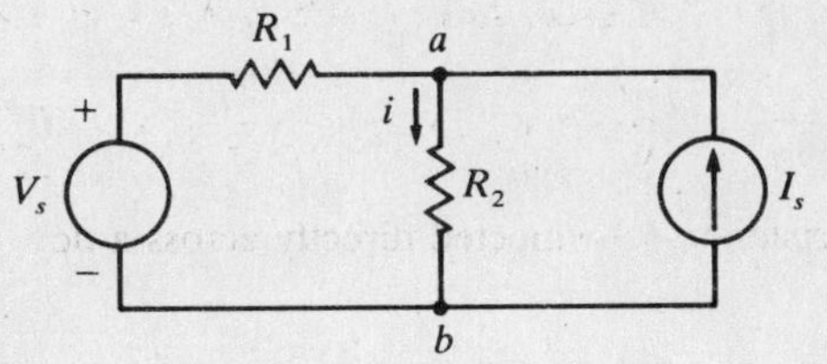

Fig. 1-8

With I_s deactivated (open-circuited), KVL and Ohm's law give the component of i due to V_s as

$$i' = \frac{V_s}{R_1 + R_2} = \frac{10}{5 + 10} = 0.667 \text{ A}$$

With V_s deactivated (short-circuited), current division determines the component of i due to I_s:

$$i'' = \frac{R_1}{R_1 + R_2} I_s = \frac{5}{5 + 10} 3 = 1 \text{ A}$$

By superposition, the total current is

$$i = i' + i'' = 0.667 + 1 = 1.667 \text{ A}$$

1.3 In Fig. 1-8, assume all circuit values as in Problem 1.2 except that $R_2 = 0.25i\ \Omega$. Determine the current i using the method of node voltages.

By (*1.1*), the voltage-current relationship for R_2 is

$$v_{ab} = R_2 i = (0.25i)(i) = 0.25i^2$$

so that

$$i = 2\sqrt{v_{ab}} \qquad (1)$$

Applying the method of node voltages at a and using (*1*), we get

$$\frac{v_{ab} - V_s}{R_1} + 2\sqrt{v_{ab}} - I_s = 0$$

Rearrangement and substitution of given values lead to

$$v_{ab} + 10\sqrt{v_{ab}} - 25 = 0$$

Letting $x^2 = v_{ab}$ and applying the quadratic formula, we obtain

$$x = \frac{-10 \pm \sqrt{(10)^2 - 4(-25)}}{2} = 2.071 \qquad \text{or} \qquad -12.071$$

The negative root is extraneous, since the resulting value of v_{ab} would not satisfy KVL; thus,

$$v_{ab} = (2.071)^2 = 4.289 \text{ V} \qquad \text{and} \qquad i = 2 \times 2.071 = 4.142 \text{ A}$$

Notice that, because the resistance R_2 is a function of current, the circuit is not linear and the superposition theorem cannot be applied.

1.4 For the circuit of Fig. 1-9, find v_{ab} if (*a*) $k = 0$ and (*b*) $k = 0.01$. Do not use network theorems to simplify the circuit prior to solution.

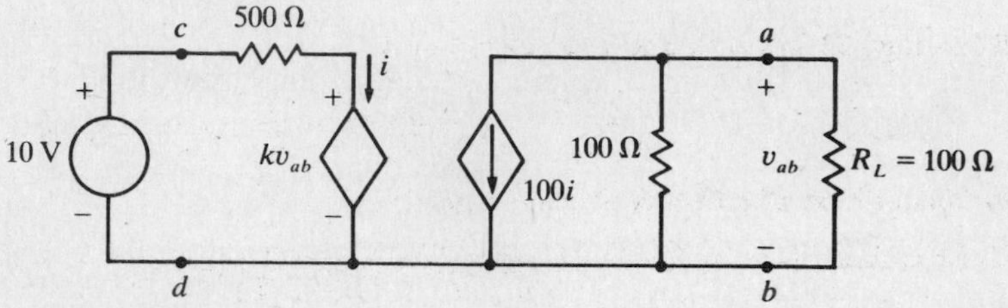

Fig. 1-9

(*a*) For $k = 0$, the current i can be determined immediately with Ohm's law:

$$i = \frac{10}{500} = 0.02 \text{ A}$$

Since the output of the controlled current source flows through the parallel combination of two 100-Ω resistors, we have

$$v_{ab} = -(100i)(100 \| 100) = -100 \times 0.02 \frac{(100)(100)}{100 + 100} = -100 \text{ V} \tag{1}$$

(*b*) With $k \neq 0$, it is necessary to solve two simultaneous equations with unknowns i and v_{ab}. Around the left loop, KVL yields

$$0.01v_{ab} + 500i = 10 \tag{2}$$

With i unknown, (*1*) becomes

$$v_{ab} + 5000i = 0 \tag{3}$$

Solving (*2*) and (*3*) simultaneously by Cramer's rule leads to

$$v_{ab} = \frac{\begin{vmatrix} 10 & 500 \\ 0 & 5000 \end{vmatrix}}{\begin{vmatrix} 0.01 & 500 \\ 1 & 5000 \end{vmatrix}} = \frac{50{,}000}{-450} = -111.1 \text{ V}$$

1.5 For the circuit of Fig. 1-10, find i_L by the method of node voltages if (*a*) $\alpha = 0.9$ and (*b*) $\alpha = 0$.

(*a*) With v_2 and v_{ab} as unknowns and summing currents at node c, we obtain

$$\frac{v_2 - v_s}{R_1} + \frac{v_2}{R_2} + \frac{v_2 - v_{ab}}{R_3} + \alpha i = 0 \tag{1}$$

But

$$i = \frac{v_s - v_2}{R_1} \tag{2}$$

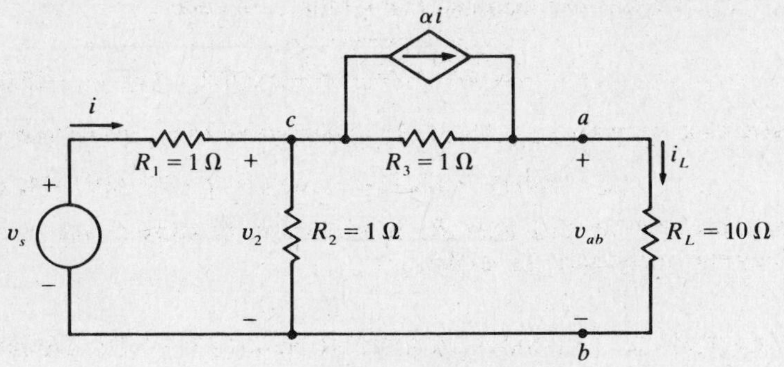

Fig. 1-10

Substituting (*2*) into (*1*) and rearranging give

$$\left(\frac{1-\alpha}{R_1} + \frac{1}{R_2} + \frac{1}{R_3}\right)v_2 - \frac{1}{R_3}\,v_{ab} = \frac{1-\alpha}{R_1}\,v_s \qquad (3)$$

Now, summation of currents at node *a* gives

$$\frac{v_{ab} - v_2}{R_3} - \alpha i + \frac{v_{ab}}{R_L} = 0 \qquad (4)$$

Substituting (*2*) into (*4*) and rearranging yield

$$-\left(\frac{1}{R_3} - \frac{\alpha}{R_1}\right)v_2 + \left(\frac{1}{R_3} + \frac{1}{R_L}\right)v_{ab} = \frac{\alpha}{R_1}\,v_s \qquad (5)$$

Substitution of given values into (*3*) and (*5*) and application of Cramer's rule finally yield

$$v_{ab} = \frac{\begin{vmatrix} 2.1 & 0.1v_s \\ -0.1 & 0.9v_s \end{vmatrix}}{\begin{vmatrix} 2.1 & -1 \\ -0.1 & 1.1 \end{vmatrix}} = \frac{1.9v_s}{2.21} = 0.8597v_s$$

and by Ohm's law,

$$i_L = \frac{v_{ab}}{R_L} = \frac{0.8597v_s}{10} = 0.08597v_s \quad \text{A}$$

(*b*) With the given values (including $\alpha = 0$) substituted into (*3*) and (*5*), Cramer's rule is used to find

$$v_{ab} = \frac{\begin{vmatrix} 3 & v_s \\ -1 & 0 \end{vmatrix}}{\begin{vmatrix} 3 & -1 \\ -1 & 1.1 \end{vmatrix}} = \frac{v_s}{2.3} = 0.4348v_s$$

Then i_L is again found with Ohm's law:

$$i_L = \frac{v_{ab}}{R_L} = \frac{0.4348v_s}{10} = 0.04348v_s \quad \text{A}$$

1.6 If $V_1 = 10$ V, $V_2 = 15$ V, $R_1 = 4\ \Omega$, and $R_2 = 6\ \Omega$ in the circuit of Fig. 1-11, find the Thévenin equivalent for the network to the left of terminals *a*,*b*.

With terminals *a*,*b* open-circuited, only loop current *I* flows. Then, by KVL,

$$V_1 - IR_1 = V_2 + IR_2$$

so that

$$I = \frac{V_1 - V_2}{R_1 + R_2} = \frac{10 - 15}{4 + 6} = -0.5 \text{ A}$$

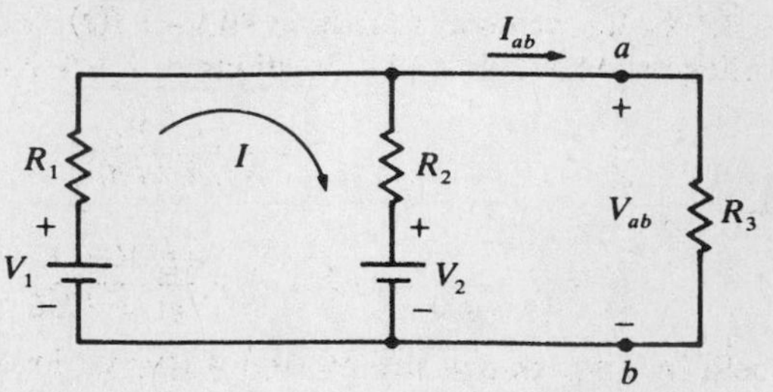

Fig. 1-11

The Thévenin equivalent voltage is then

$$V_{Th} = V_{ab} = V_1 - IR_1 = 10 - (-0.5)(4) = 12 \text{ V}$$

Deactivating (shorting) the independent voltage sources V_1 and V_2 gives the Thévenin impedance to the left of terminals a,b as

$$Z_{Th} = R_{Th} = R_1 \| R_2 = \frac{R_1 R_2}{R_1 + R_2} = \frac{(4)(6)}{4+6} = 2.4 \ \Omega$$

V_{Th} and Z_{Th} are connected as in Fig. 1-3(*b*) to produce the Thévenin equivalent circuit.

1.7 For the circuit and values of Problem 1.6, find the Norton equivalent for the network to the left of terminals a,b.

With terminals a,b shorted, the component of current I_{ab} due to V_1 alone is

$$I'_{ab} = \frac{V_1}{R_1} = \frac{10}{4} = 2.5 \text{ A}$$

Similarly, the component due to V_2 alone is

$$I''_{ab} = \frac{V_2}{R_2} = \frac{15}{6} = 2.5 \text{ A}$$

Then, by superposition,

$$I_N = I_{ab} = I'_{ab} + I''_{ab} = 2.5 + 2.5 = 5 \text{ A}$$

Now, with R_{Th} as found in Problem 1.6,

$$Y_N = \frac{1}{R_{Th}} = \frac{1}{2.4} = 0.4167 \text{ A}$$

I_N and Y_N are connected as in Fig. 1-3(*c*) to produce the Norton equivalent circuit.

1.8 For the circuit and values of Problems 1.6 and 1.7, find the Thévenin impedance as the ratio of open-circuit voltage to short-circuit current to illustrate the equivalence of the results.

The open-circuit voltage is V_{Th} as found in Problem 1.6, and the short-circuit current is I_N from Problem 1.7. Thus,

$$Z_{Th} = \frac{V_{Th}}{I_N} = \frac{12}{5} = 2.4 \ \Omega$$

which checks with the result of Problem 1.6.

1.9 Thévenin's and Norton's theorems are applicable to other than dc steady-state circuits. For the "frequency-domain" circuit of Fig. 1-12 (where s is frequency), find (*a*) the Thévenin equivalent and (*b*) the Norton equivalent of the circuit to the right of terminals a,b.

(*a*) With terminals *a*,*b* open-circuited, only loop current $I(s)$ flows; by KVL and Ohm's law, with all currents and voltages understood to be functions of s, we have

$$I = \frac{V_2 - V_1}{sL + 1/sC}$$

Now KVL gives

$$V_{Th} = V_{ab} = V_1 + sLI = V_1 + \frac{sL(V_2 - V_1)}{sL + 1/sC} = \frac{V_1 + s^2LCV_2}{s^2LC + 1}$$

With the independent sources deactivated, the Thévenin impedance can be determined as

$$Z_{Th} = sL \parallel \frac{1}{sC} = \frac{sL(1/sC)}{sL + 1/sC} = \frac{sL}{s^2LC + 1}$$

(*b*) The Norton current can be found as

$$I_N = \frac{V_{Th}}{Z_{Th}} = \frac{\dfrac{V_1 + s^2LCV_2}{s^2LC + 1}}{\dfrac{sL}{s^2LC + 1}} = \frac{V_1 + s^2LCV_2}{sL}$$

and the Norton admittance as

$$Y_N = \frac{1}{Z_{Th}} = \frac{s^2LC + 1}{sL}$$

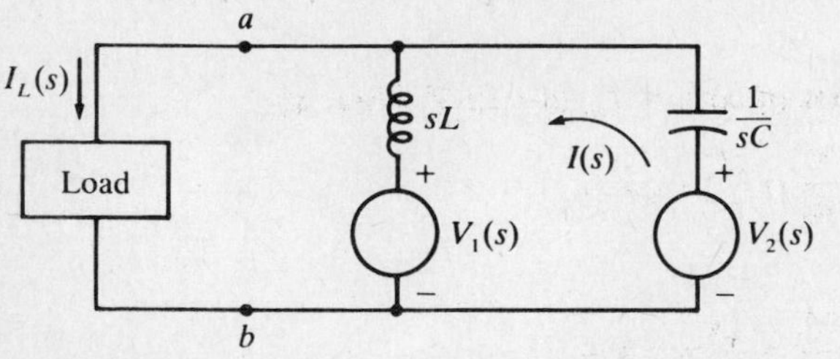

Fig. 1-12

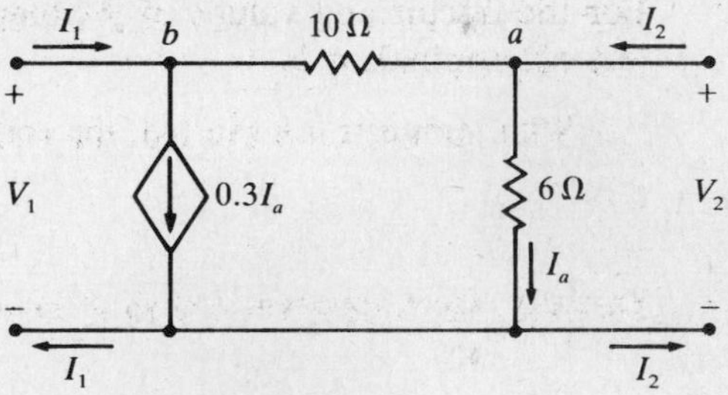

Fig. 1-13

1.10 Determine the z parameters for the two-port network of Fig. 1-13.

For $I_2 = 0$, by Ohm's law,

$$I_a = \frac{V_1}{10 + 6} = \frac{V_1}{16}$$

Also, at node b, KCL gives

$$I_1 = 0.3I_a + I_a = 1.3I_a = 1.3\,\frac{V_1}{16} \tag{1}$$

Thus, by (*1.10*),

$$z_{11} = \frac{V_1}{I_1}\bigg|_{I_2=0} = \frac{16}{1.3} = 12.308\ \Omega$$

Further, again by Ohm's law,

$$I_a = \frac{V_2}{6} \tag{2}$$

Substitution of (*2*) into (*1*) yields

$$I_1 = 1.3\,\frac{V_2}{6}$$

so that, by (*1.12*),

$$z_{21} = \frac{V_2}{I_1}\bigg|_{I_2=0} = \frac{6}{1.3} = 4.615\ \Omega$$

Now with $I_1 = 0$, applying KCL at node a gives us

$$I_2 = I_a + 0.3I_a = 1.3I_a \qquad (3)$$

The application of KVL then leads to

$$V_1 = V_2 - (10)(0.3I_a) = 6I_a - 3I_a = 3I_a = \frac{3I_2}{1.3}$$

so that, by (*1.11*),

$$z_{12} = \frac{V_1}{I_2}\bigg|_{I_1=0} = \frac{3}{1.3} = 2.308\ \Omega$$

Now, substitution of (*2*) in (*3*) gives

$$I_2 = 1.3I_a = 1.3\,\frac{V_2}{6}$$

Hence, from (*1.13*),

$$z_{22} = \frac{V_2}{I_2}\bigg|_{I_1=0} = \frac{6}{1.3} = 4.615\ \Omega$$

1.11 Determine the h parameters for the two-port network of Fig. 1-13.

For $V_2 = 0$, $I_a \equiv 0$; thus, $I_1 = V_1/10$ and, by (*1.16*),

$$h_{11} = \frac{V_1}{I_1}\bigg|_{V_2=0} = 10\ \Omega$$

Further, $I_2 = -I_1$ and, by (*1.18*),

$$h_{21} = \frac{I_2}{I_1}\bigg|_{V_2=0} = -1$$

Now, $I_a = V_2/6$. With $I_1 = 0$, KVL yields

$$V_1 = V_2 - 10(0.3I_a) = V_2 - 10(0.3)\,\frac{V_2}{6} = \tfrac{1}{2}V_2$$

and, from (*1.17*),

$$h_{12} = \frac{V_1}{V_2}\bigg|_{I_1=0} = 0.5$$

Finally, applying KCL at node a gives

$$I_2 = I_a + 0.3I_a = 1.3\,\frac{V_2}{6}$$

so that, by (*1.19*),

$$h_{22} = \frac{I_2}{V_2}\bigg|_{I_1=0} = \frac{1.3}{6} = 0.2167\ \text{S}$$

1.12 Use (*1.8*), (*1.9*), and (*1.16*) to (*1.19*) to find the h parameters in terms of the z parameters.

Setting $V_2 = 0$ in (*1.9*) gives

$$0 = z_{21}I_1 + z_{22}I_2 \qquad \text{or} \qquad I_2 = -\frac{z_{21}}{z_{22}}\,I_1 \qquad (1)$$

from which we get

$$h_{21} = \frac{I_2}{I_1}\bigg|_{V_2=0} = -\frac{z_{21}}{z_{22}}$$

Back substitution of (*1*) into (*1.8*) and use of (*1.16*) give

$$h_{11} = \frac{V_1}{I_1}\bigg|_{V_2=0} = z_{11} - \frac{z_{12}z_{21}}{z_{22}}$$

Now, with $I_1 = 0$, (*1.8*) and (*1.9*) become

$$V_1 = z_{12}I_2 \qquad \text{and} \qquad V_2 = z_{22}I_2$$

so that, from (*1.17*),

$$h_{12} = \frac{V_1}{V_2}\bigg|_{I_1=0} = \frac{z_{12}}{z_{22}}$$

and, from (*1.19*),

$$h_{22} = \frac{I_2}{V_2}\bigg|_{I_1=0} = \frac{I_2}{z_{22}I_2} = \frac{1}{z_{22}}$$

1.13 The h parameters of the two-port network of Fig. 1-14 are $h_{11} = 100\ \Omega$, $h_{12} = 0.0025$, $h_{21} = 20$, and $h_{22} = 1$ mS. Find the voltage-gain ratio V_2/V_1.

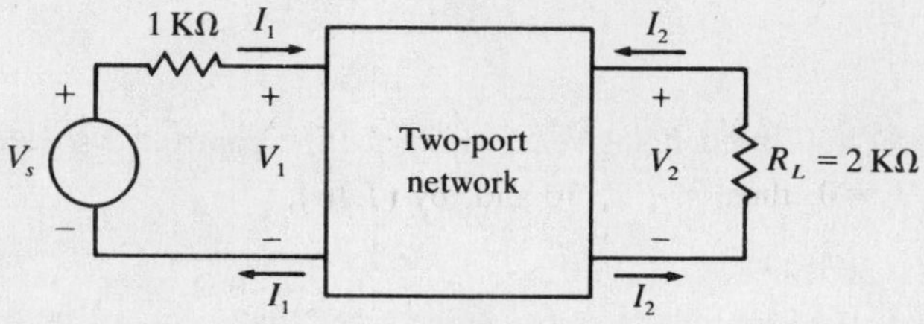

Fig. 1-14

By Ohm's law, $I_2 = -V_2/R_L$, so that (*1.15*) may be written

$$-\frac{V_2}{R_L} = I_2 = h_{21}I_1 + h_{22}V_2$$

Solving for I_1 and substitution into (*1.14*) give

$$V_1 = h_{11}I_1 + h_{12}V_2 = \frac{-(1/R_L + h_{22})}{h_{21}}V_2h_{22} + h_{12}V_2$$

which can be solved for the voltage gain ratio:

$$\frac{V_2}{V_1} = \frac{1}{h_{12} - (h_{11}/h_{21})(1/R_L + h_{22})} = \frac{1}{0.0025 - (100/20)(1/2000 + 0.001)} = -200$$

1.14 Determine the Thévenin equivalent voltage and impedance looking right into port 1 of the circuit of Fig. 1-14.

The Thévenin voltage is V_1 of (*1.8*) with port 1 open-circuited:

$$V_{Th} = V_1|_{I_1=0} = z_{12}I_2 \tag{1}$$

Now, by Ohm's law,

$$V_2 = -R_LI_2 \tag{2}$$

But, with $I_1 = 0$, (*1.9*) reduces to

$$V_2 = z_{22}I_2 \tag{3}$$

Subtracting (*2*) from (*3*) leads to

$$(z_{22} + R_L)I_2 = 0 \tag{4}$$

Since, in general, $z_{22} + R_L \neq 0$, we conclude from (*4*) that $I_2 = 0$ and, from (*1*), $V_{Th} = 0$.

Substituting (2) into (1.8) and (1.9) gives

$$V_1 = z_{11}I_1 + z_{12}I_2 = z_{11}I_1 - \frac{z_{12}}{R_L} V_2 \tag{5}$$

and

$$V_2 = z_{21}I_1 + z_{22}I_2 = z_{21}I_1 - \frac{z_{22}}{R_L} V_2 \tag{6}$$

V_1 is found by solving for V_2 and substituting the result into (5):

$$V_1 = z_{11}I_1 - \frac{z_{12}z_{21}}{z_{22} + R_L} I_1$$

Then Z_{Th} is calculated as the driving-point impedance V_1/I_1:

$$Z_{Th} = \frac{V_{dp}}{I_{dp}} = \frac{V_1}{I_1} = z_{11} - \frac{z_{12}z_{21}}{z_{22} + R_L}$$

1.15 Find the Thévenin equivalent voltage and impedance looking into port 1 of the circuit of Fig. 1-14 if R_L is replaced with a current-controlled voltage source such that $V_2 = \beta I_1$, where β is a constant.

As in Problem 1.14,

$$V_{Th} = V_1|_{I_1=0} = z_{22}I_2$$

But if $I_1 = 0$, (1.9) and the defining relationship for the controlled source lead to

$$V_2 = \beta I_1 = 0 = z_{22}I_2$$

from which $I_2 = 0$ and, hence, $V_{Th} = 0$.

Now we let $V_1 = V_{dp}$, so that $I_1 = I_{dp}$, and we determine Z_{Th} as the driving-point impedance. From (1.8), (1.9), and the defining relationship for the controlled source, we have

$$V_1 = V_{dp} = z_{11}I_{dp} + z_{12}I_2 \tag{1}$$

$$V_2 = \beta I_{dp} = z_{21}I_{dp} + z_{22}I_2 \tag{2}$$

Solving (2) for I_2 and substituting the result into (1) yields

$$V_{dp} = z_{11}I_{dp} + z_{12}\frac{\beta - z_{21}}{z_{22}} I_{dp}$$

from which Thévenin impedance is found to be

$$Z_{Th} = \frac{V_{dp}}{I_{dp}} = \frac{z_{11}z_{22} + z_{12}(\beta - z_{21})}{z_{22}}$$

1.16 The periodic current waveform of Fig. 1-15 is composed of segments of a sinusoid. Find (a) the average value of the current and (b) the rms (effective) value of the current.

(a) Because $i(t) = 0$ for $0 \le \omega t < \alpha$, the average value of the current is, according to (1.20),

$$I_0 = \frac{1}{\pi}\int_\alpha^\pi I_m \sin \omega t \, d(\omega t) = \frac{I_m}{\pi}[-\cos \omega t]_{\omega t=\alpha}^{\pi} = \frac{I_m}{\pi}(1 + \cos \alpha)$$

(b) By (1.21) and the identity $\sin^2 x = \frac{1}{2}(1 - \cos 2x)$,

$$I^2 = \frac{1}{\pi}\int_\alpha^\pi I_m^2 \sin^2(\omega t)\, d(\omega t) = \frac{I_m^2}{2\pi}\int_\alpha^\pi (1 - \cos 2\omega t)\, d(\omega t)$$

$$= \frac{I_m^2}{2\pi}\left[\omega t - \frac{1}{2}\sin 2\omega t\right]_{\omega t=\alpha}^{\pi} = \frac{I_m^2}{2\pi}\left(\pi - \alpha + \frac{1}{2}\sin 2\alpha\right)$$

so that

$$I = I_m\sqrt{\frac{\pi - \alpha + \frac{1}{2}\sin 2\alpha}{2\pi}}$$

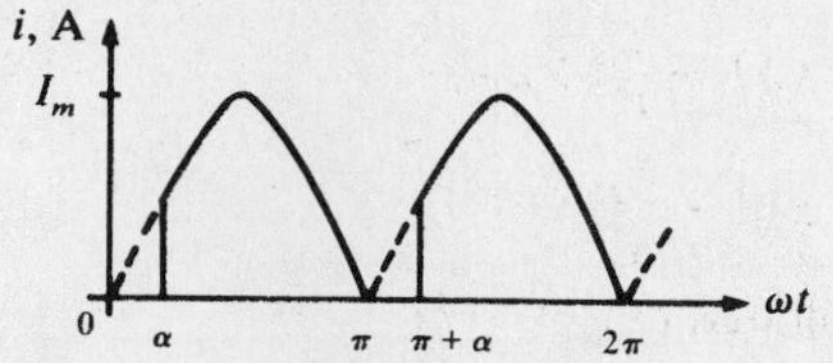

Fig. 1-15

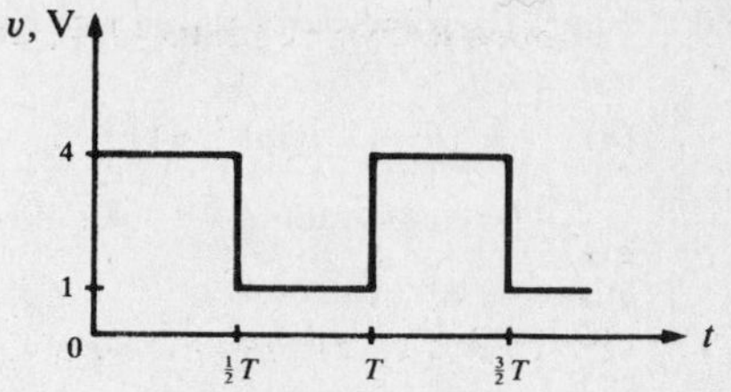

Fig. 1-16

1.17 Assume that the periodic waveform of Fig. 1-16 is a current (rather than a voltage). Find (*a*) the average value of the current and (*b*) the rms value of the current.

(*a*) The integral in (*1.20*) is simply the area under the $f(t)$ curve for one period. We can, then, find the average current as

$$I_0 = \frac{1}{T}\left(4 \times \frac{T}{2} + 1 \times \frac{T}{2}\right) = 2.5 \text{ A}$$

(*b*) Similarly, the integral in (*1.21*) is no more than the area under the $f^2(t)$ curve. Hence,

$$I = \left[\frac{1}{T}\left(4^2\,\frac{T}{2} + 1^2\,\frac{T}{2}\right)\right]^{1/2} = 4.25 \text{ A}$$

1.18 Calculate the average and rms values of the current $i(t) = 4 + 10 \sin \omega t$ A.

Since $i(t)$ has period 2π, (*1.20*) gives

$$I_0 = \frac{1}{2\pi}\int_0^{2\pi} (4 + 10 \sin \omega t)\, d(\omega t) = \frac{1}{2\pi}\,[4\omega t - 10 \cos \omega t]_{\omega t=0}^{2\pi} = 4 \text{ A}$$

This result was to be expected, since the average value of a sinusoid over one cycle is zero.

Equation (*1.21*) and the identity $\sin^2 x = \frac{1}{2}(1 - \cos 2x)$ provide the rms value of $i(t)$:

$$I^2 = \frac{1}{2\pi}\int_0^{2\pi} (4 + 10 \sin \omega t)^2\, d(\omega t) = \frac{1}{2\pi}\int_0^{2\pi} (16 + 80 \sin \omega t + 50 - 50 \cos 2\omega t)\, d(\omega t)$$

$$= \frac{1}{2\pi}\left[66\omega t - 80 \cos \omega t - \frac{50}{2} \sin 2\omega t\right]_{\omega t=0}^{2\pi} = 66$$

so that $I = \sqrt{66} = 8.125$ A

1.19 Find the rms (or effective) value of a current consisting of the sum of two sinusoidally varying functions with frequencies whose ratio is an integer.

Without loss of generality, we may write

$$i(t) = I_1 \cos \omega t + I_2 \cos k\omega t$$

where k is an integer. Applying (*1.21*) and recalling that $\cos^2 x = \frac{1}{2}(1 + \cos 2x)$ and $\cos x \cos y = \frac{1}{2}[\cos(x + y) + \cos(x - y)]$, we obtain

$$I^2 = \frac{1}{2\pi}\int_0^{2\pi} (I_1 \cos \omega t + I_2 \cos k\omega t)^2\, d(\omega t)$$

$$= \frac{1}{2\pi}\int_0^{2\pi} \left\{\frac{I_1^2}{2}(1 + \cos 2\omega t) + \frac{I_2^2}{2}(1 + \cos 2k\omega t) + I_1 I_2[\cos (k + 1)\omega t + \cos (k - 1)\omega t]\right\} d(\omega t)$$

Performing the indicated integration and evaluating at the limits result in

$$I = \sqrt{\frac{I_1^2}{2} + \frac{I_2^2}{2}}$$

1.20 Find the average value of the power delivered to a one-port network with *passive sign convention* (that is the current is directed from the positive to the negative terminal) if $v(t) = V_m \cos \omega t$ and $i(t) = I_m \cos(\omega t + \theta)$.

The instantaneous power flow into the port is given by

$$p(t) = v(t)i(t) = V_m I_m \cos \omega t \cos(\omega t + \theta)$$
$$= \tfrac{1}{2} V_m I_m [\cos(2\omega t + \theta) + \cos \theta]$$

By (*1.20*),

$$P_0 = \frac{1}{2\pi}\int_0^{2\pi} p(t)\, dt = \frac{V_m}{4\pi} I_m \int_0^{2\pi} [\cos(2\omega t + \theta) + \cos \theta]\, d(\omega t)$$

After the integration is performed and its limits evaluated, the result is

$$P_0 = \frac{V_m I_m}{2} \cos \theta = \frac{V_m}{\sqrt{2}} \frac{I_m}{\sqrt{2}} \cos \theta = VI \cos \theta$$

Supplementary Problems

1.21 Prove that the capacitor element of Fig. 1-1(*c*) is a linear element by showing that it satisfies the converse of the superposition theorem. (*Hint:* See Problem 1.1.)

1.22 Use the superposition theorem to find the current i in Fig. 1-8 if $R_1 = 5\ \Omega$, $R_2 = 10\ \Omega$, $V_s = 10 \cos 2t$ V, and $I_s = 3 \cos(3t + \pi/4)$ A. *Ans.* $i = 0.667 \cos 2t + \cos(3t + \pi/4)$ A

1.23 In Fig. 1-17, (*a*) find the Thévenin equivalent voltage and impedance for the network to the left of terminals a,b, and (*b*) use the Thévenin equivalent circuit to determine the current I_L.
Ans. (*a*) $V_{Th} = V_1 - I_2 R_2$, $Z_{Th} = R_1 + R_2$; (*b*) $I_L = (V_1 - I_2 R_2)/(R_1 + R_2 + R_L)$

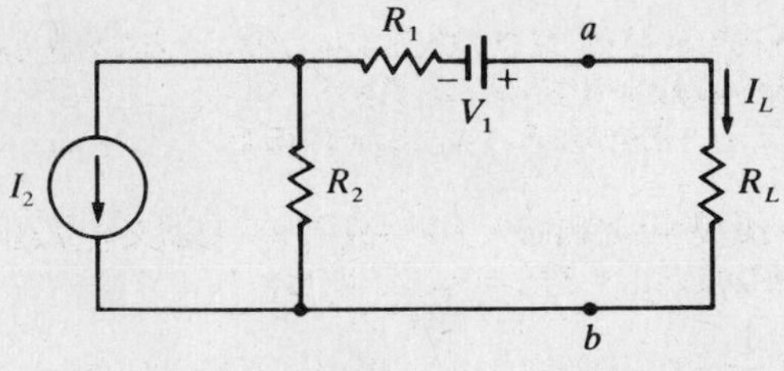

Fig. 1-17

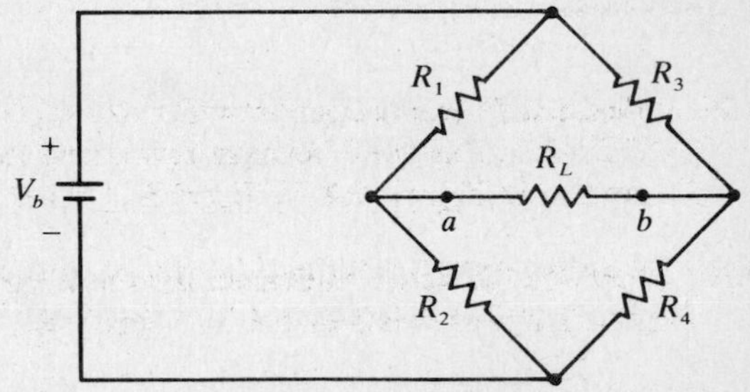

Fig. 1-18

1.24 In the circuit of Fig. 1-12, $V_1 = 10 \cos 2t$ V, $V_2 = 20 \cos 2t$ V, $L = 1$ H, $C = 1$ F, and the load is a 1-Ω resistor. (*a*) Determine the Thévenin equivalent for the network to the right of terminals a,b. (*b*) Use the Thévenin equivalent to find the load current $\bar{I}_L$. (*Hint:* The results of Problem 1.9 can be used here, with $s = j2$.) *Ans.* (*a*) $\bar{V}_{Th} = 23.333\angle 0°$ V, $Z_{Th} = -j0.667\ \Omega$; (*b*) $\bar{I}_L = 19.4\angle 33.69°$ A.

1.25 In Fig. 1-18, find the Thévenin equivalent for the bridge circuit as seen through the load resistor R_L. *Ans.* $V_{Th} = V_b(R_2R_3 - R_1R_4)/(R_1 + R_2)(R_3 + R_4)$, $Z_{Th} = R_1R_3/(R_1 + R_3) + R_2R_4/(R_2 + R_4)$

1.26 Suppose the bridge circuit in Fig. 1-18 is balanced by letting $R_1 = R_2 = R_3 = R_4 = R$. Find the elements of the Norton equivalent circuit. *Ans.* $I_N = 0$, $Y_N = 1/R$

1.27 For the circuit of Fig. 1-19, (*a*) determine the Thévenin equivalent of the circuit to the left of terminals *a*,*b*, and (*b*) use the Thévenin equivalent to find the load current i_L.
Ans. (*a*) $V_{Th} = 120$ V, $Z_{Th} = 20\ \Omega$; (*b*) $i_L = 4$ A

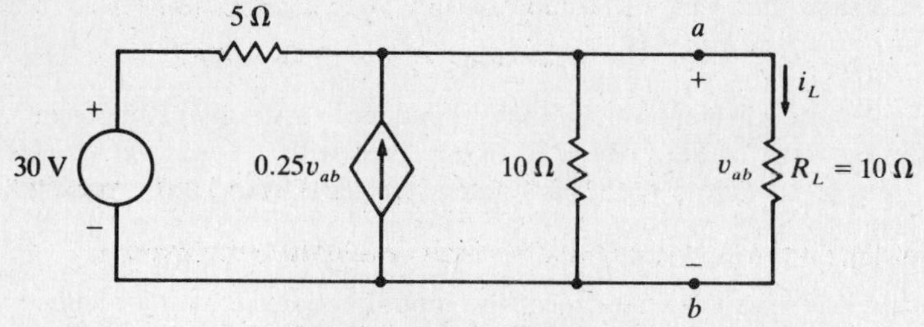

Fig. 1-19

1.28 In the circuit of Fig. 1-20, let $R_1 = R_2 = R_C = 1\ \Omega$ and find the Thévenin equivalent for the circuit to the right of terminals *a*,*b* (*a*) if $v_C = 0.5i_1$ and (*b*) if $v_C = 0.5i_2$.
Ans. (*a*) $V_{Th} = 0$, $Z_{Th} = R_{Th} = 1.75\ \Omega$; (*b*) $V_{Th} = 0$, $Z_{Th} = R_{Th} = 1.667\ \Omega$

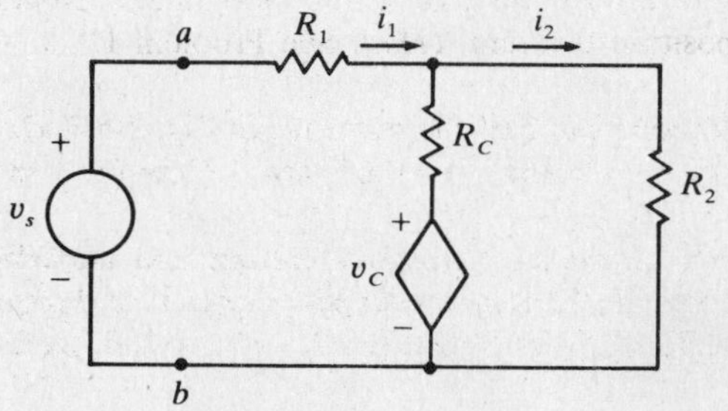

Fig. 1-20

1.29 Find the Thévenin equivalent for the network to the left of terminals *a*,*b* in Fig. 1-9 (*a*) if $k = 0$, and (*b*) if $k = 0.1$. Use the Thévenin equivalent to verify the results of Problem 1.4.
Ans. (*a*) $V_{Th} = -200$ V, $Z_{Th} = R_{Th} = 100\ \Omega$; (*b*) $V_{Th} = -250$ V, $Z_{Th} = R_{Th} = 125\ \Omega$

1.30 Find the Thévenin equivalent for the circuit to the left of terminals *a*,*b* in Fig. 1-10, and use it to verify the results of Problem 1.5. *Ans.* $V_{Th} = \frac{1}{2}(1 + \alpha)v_s$, $Z_{Th} = R_{Th} = \frac{1}{2}(3 - \alpha)\ \Omega$

1.31 An alternative solution to Problem 1.3 involves finding a Thévenin equivalent circuit which, when connected across the nonlinear $R_2 = 0.25i$, allows a quadratic equation in current i to be written via KVL. Find the elements of the Thévenin circuit and the resulting current.
Ans. $V_{Th} = 25$ V, $Z_{Th} = R_{Th} = 5\ \Omega$, $i = 4.142$ A.

1.32 Use (*1.10*) to (*1.15*) to find expressions for the z parameters in terms of the h parameters.
Ans. $z_{11} = h_{11} - h_{12}h_{21}/h_{22}$, $z_{12} = h_{12}/h_{22}$, $z_{21} = -h_{21}/h_{22}$, $z_{22} = 1/h_{22}$

1.33 For the two-port network of Fig. 1-14, (*a*) find the voltage-gain ratio V_2/V_1 in terms of the z parameters, and then (*b*) evaluate the ratio, using the h-parameter values given in Problem 1.13 and the results of Problem 1.32. *Ans.* (*a*) $z_{21}R_L/(z_{11}R_L + z_{11}z_{22} - z_{12}z_{21})$; (*b*) -200

1.34 Find the current-gain ratio I_2/I_1 for the two-port network of Fig. 1-14 in terms of the h parameters.
Ans. $h_{21}/(1 + h_{22}R_L)$

1.35 Find the current-gain ratio I_2/I_1 for the two-port network of Fig. 1-14 in terms of the z parameters. *Ans.* $-z_{21}/(z_{22}+R_L)$

1.36 Determine the Thévenin equivalent voltage and impedance, in terms of the z parameters, looking right into port 1 of the two-port network of Fig. 1-14 if R_L is replaced with an independent dc voltage source V_d, connected such that $V_2 = V_d$.
Ans. $V_{Th} = z_{12}V_d/z_{22}$, $Z_{Th} = (z_{11}z_{22} - z_{12}z_{21})/z_{22}$

1.37 Find the Thévenin equivalent voltage and impedance, in terms of the h parameters, looking right into port 1 of the network of Fig. 1-14 if R_L is replaced with a voltage-controlled current source such that $I_2 = -\alpha V_1$, where $\alpha > 0$ and the h parameters are understood to be positive.
Ans. $V_{Th} = 0$, $Z_{Th} = (h_{11}h_{22} - h_{12}h_{21})/(h_{22} + \alpha h_{12})$

1.38 Determine the driving-point impedance (the input impedance with all independent sources deactivated) of the two-port network of Fig. 1-14. *Ans.* $(z_{11}R_L + z_{11}z_{22} - z_{12}z_{21})/(z_{22} + R_L)$

1.39 Evaluate the z parameters of the network of Fig. 1-10.
Ans. $z_{11} = 2\ \Omega$, $z_{12} = 1\ \Omega$, $z_{21} = \alpha + 1\ \Omega$, $z_{22} = 2\ \Omega$

1.40 Find the current i_1 in Fig. 1-2 if $\alpha = 2$, $R_1 = R_2 = R_3 = 1\ \Omega$, $V_b = 10$ V, and $v_s = 10 \sin \omega t$ V.
Ans. -2 A

1.41 For a one-port network with passive sign convention (see Problem 1.20), $v = V_m \cos \omega t$ V and $i = I_1 + I_2 \cos(\omega t + \theta)$ A. Find (*a*) the instantaneous power flowing to the network and (*b*) the average power to the network. *Ans.* (*a*) $V_m I_1 \cos \omega t + \frac{1}{2} V_m I_2[\cos(2\omega t + \theta) + \cos \theta]$; (*b*) $\frac{1}{2} V_m I_2 \cos \theta$

Chapter 2

Semiconductor Diodes

2.1 INTRODUCTION

Diodes are among the oldest and most widely used of electronic devices. A *diode* may be defined as a near-unidirectional conductor whose state of conductivity is determined by the polarity of its terminal voltage. The subject of this chapter is the *semiconductor diode*, formed by the metallurgical junction of p-type and n-type materials. (A p-type material is a group-IV element *doped* with a small quantity of a group-V material; n-type material is a group-IV base element doped with a group-III material.)

2.2 THE IDEAL DIODE

The symbol for the *common*, or *rectifier*, *diode* is shown in Fig. 2-1. The device has two terminals, labeled *anode* (p-type) and *cathode* (n-type), which makes understandable the choice of *di*ode as its name. When the terminal voltage is nonnegative ($v_D \geq 0$), the diode is said to be *forward-biased* or "on"; the positive current that flows ($i_D \geq 0$) is called *forward current*. When $v_D < 0$, the diode is said to be *reverse-biased* or "off," and the corresponding small negative current is referred to as *reverse current*.

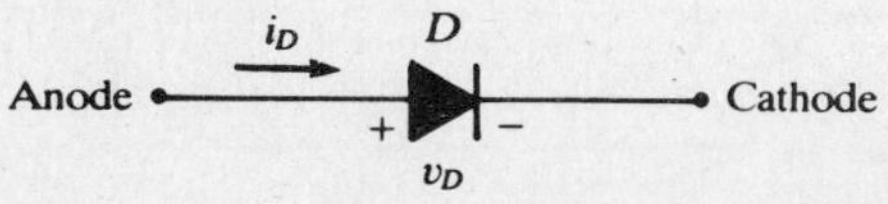

Fig. 2-1

The ideal diode is a perfect two-state device that exhibits zero impedance when forward-biased and infinite impedance when reverse-biased (Fig. 2-2). Note that since either current or voltage is zero at any instant, no power is dissipated by an ideal diode. In many circuit applications, diode forward voltage drops and reverse currents are small compared to other circuit variables; then, sufficiently accurate results are obtained if the actual diode is modeled as ideal.

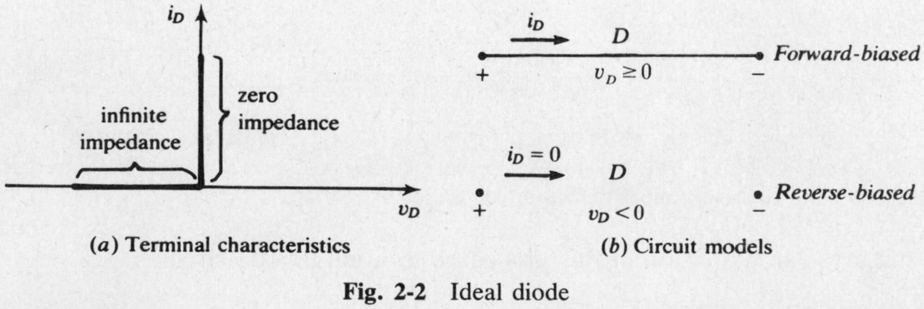

Fig. 2-2 Ideal diode

The *ideal diode analysis procedure* is as follows:

Step 1: Assume forward bias, and replace the ideal diode with a short circuit.

Step 2: Evaluate the diode current i_D, using any linear circuit-analysis technique.

Step 3: If $i_D \geq 0$, the diode is actually forward-biased, the analysis is valid, and step 4 is to be omitted.

Step 4: If $i_D < 0$, the analysis so far is invalid. Replace the diode with an open circuit, forcing $i_D = 0$, and solve for the desired circuit quantities using any method of circuit analysis. Voltage v_D must be found to have a negative value.

Example 2.1 Find voltage v_L in the circuit of Fig. 2-3(a), where D is an ideal diode.

The analysis is simplified if a Thévenin equivalent is found for the circuit to the left of terminals a,b; the result is

$$v_{Th} = \frac{R_1}{R_1 + R_S} v_S \quad \text{and} \quad Z_{Th} = R_{Th} = R_1 \| R_S = \frac{R_1 R_S}{R_1 + R_S}$$

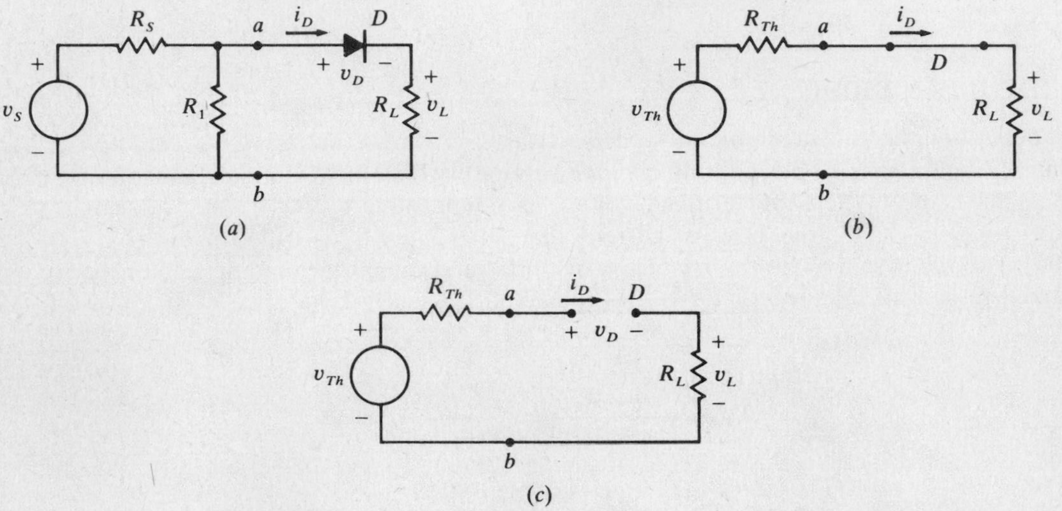

Fig. 2-3

Step 1: After replacing the network to the left of terminals a,b with the Thévenin equivalent, assume forward bias and replace diode D with a short circuit, as in Fig. 2-3(b).

Step 2: By Ohm's law,

$$i_D = \frac{v_{Th}}{R_{Th} + R_L}$$

Step 3: If $v_S \geq 0$, then $i_D \geq 0$ and

$$v_L = i_D R_L = \frac{R_L}{R_L + R_{Th}} v_{Th}$$

Step 4: If $v_S < 0$, then $i_D < 0$ and the result of step 3 is invalid. Diode D must be replaced by an open circuit as illustrated in Fig. 2-3(c), and the analysis performed again. Since now $i_D = 0$, $v_L = i_D R_L = 0$. Since $v_D = v_S < 0$, the reverse bias of the diode is verified.

(See Problem 2.4 for an extension of this procedure to a multidiode circuit.)

2.3 DIODE TERMINAL CHARACTERISTICS

Use of the Fermi-Dirac probability function to predict charge neutralization gives the *static* (non-time-varying) equation for diode junction current:

$$i_D = I_o(e^{v_D/\eta V_T} - 1) \quad \text{A} \tag{2.1}$$

where $V_T \equiv kT/q$, V
$v_D \equiv$ diode terminal voltage, V
$I_o \equiv$ temperature-dependent saturation current, A
$T \equiv$ absolute temperature of *p-n* junction, K
$k \equiv$ Boltzmann's constant (1.38×10^{-23} J/K)
$q \equiv$ electron charge (1.6×10^{-19} C)
$\eta \equiv$ empirical constant, 1 for Ge and 2 for Si

Example 2.2 Find the value of V_T in (*2.1*) at 20°C.

Recalling that absolute zero is −273°C, we write

$$V_T = \frac{kT}{q} = \frac{(1.38 \times 10^{-23})(273 + 20)}{1.6 \times 10^{-19}} = 25.27 \text{ mV}$$

While (*2.1*) serves as a useful model of the junction diode insofar as dynamic resistance is concerned, Fig. 2-4 shows it to have regions of inaccuracy:

1. The actual (measured) forward voltage drop is greater than that predicted by (*2.1*) (due to ohmic resistance of metal contacts and semiconductor material).
2. The actual reverse current for $-V_R \le v_D < 0$ is greater than predicted (due to leakage current I_S along the surface of the semiconductor material).
3. The actual reverse current increases to significantly larger values than predicted for $v_D < -V_R$ (due to a complex phenomenon called *avalanche breakdown*).

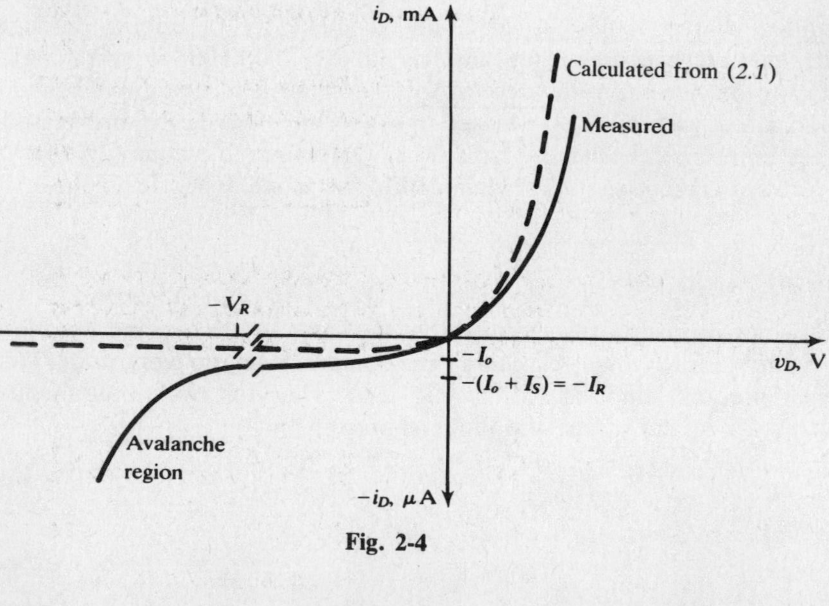

Fig. 2-4

In commercially available diodes, proper doping (impurity addition) of the base material results in distinct static terminal characteristics. A comparison of Ge- and Si-base diode characteristics is shown in Fig. 2-5. If $-V_R < v_D < -0.1$ V, both diode types exhibit a near-constant reverse current I_R. Typically, $1\ \mu\text{A} < I_R < 500\ \mu\text{A}$ for Ge, while $10^{-3}\ \mu\text{A} < I_R < 1\ \mu\text{A}$ for Si, for signal-level diodes (forward current ratings of less than 1 A). For a forward bias, the onset of low-resistance conduction is between 0.2 and 0.3 V for Ge, and between 0.6 and 0.7 V for Si.

For both Si and Ge diodes, the saturation current I_o doubles for an increase in temperature of 10°C; in other words, the ratio of saturation current at temperature T_2 to that at temperature T_1 is

$$\frac{(I_o)_2}{(I_o)_1} = 2^{(T_2 - T_1)/10} \qquad (2.2)$$

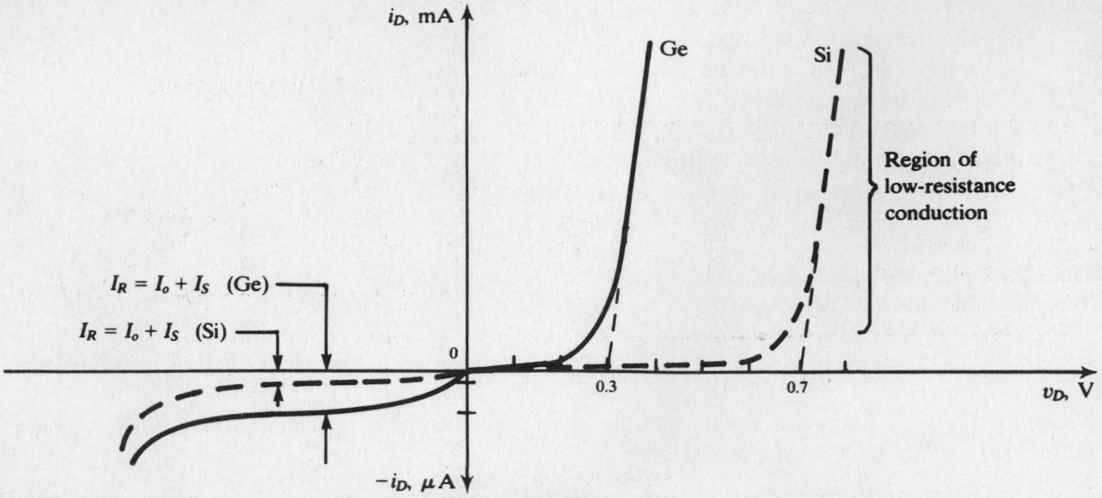

Fig. 2-5

Example 2.3 Find the percentage increase in the reverse saturation current of a diode if the temperature is increased from 25°C to 50°C.

By (2.2),

$$\frac{(I_o)_2}{(I_o)_1} = 2^{(50-25)/10} \times 100\% = 565.7\%$$

Static terminal characteristics are generally adequate for describing diode operation at low frequency. However, if high-frequency analysis (above 100 kHz) or switching analysis is to be performed, it may be necessary to account for the small *depletion capacitance* (typically several picofarads) associated with a reverse-biased *p-n* junction; for a forward-biased *p-n* junction, a somewhat larger *diffusion capacitance* (typically several hundred picofarads) that is directly proportional to the forward current should be included in the model. (See Problem 2.25.)

2.4 GRAPHICAL ANALYSIS

A graphical solution necessarily assumed that the diode is resistive and therefore instantaneously characterized by its static i_D-versus-v_D curve. The balance of the network under study must be linear so that a Thévenin equivalent exists for it (Fig. 2-6). Then the two simultaneous equations to be solved graphically for i_D and v_D are the diode characteristic

$$i_D = f_1(v_D) \tag{2.3}$$

and the *load line*

$$i_D = f_2(v_D) = -\frac{1}{R_{Th}} v_D + \frac{v_{Th}}{R_{Th}} \tag{2.4}$$

Example 2.4 In the circuit of Fig. 2-3(*a*), $v_s = 6$ V and $R_1 = R_S = R_L = 500\ \Omega$. Determine i_D and v_D graphically, using the diode characteristic in Fig. 2-7.

The circuit may be reduced to that of Fig. 2-6, with

$$v_{Th} = \frac{R_1}{R_1 + R_S} v_S = \frac{500}{500 + 500} 6 = 3 \text{ V}$$

and

$$R_{Th} = R_1 \| R_S + R_L = \frac{(500)(500)}{500 + 500} + 500 = 750\ \Omega$$

Then, with these values the load line (*2.4*) must be superimposed on the diode characteristic, as in Fig. 2-7. The desired solution, $i_D = 3$ mA and $v_D = 0.75$ V, is given by the point of intersection of the two plots.

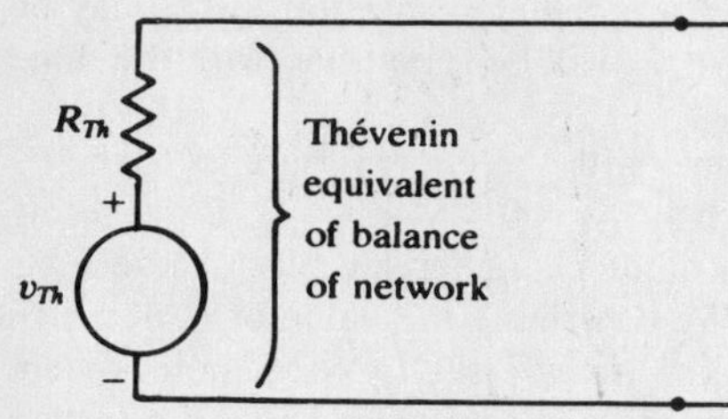

Fig. 2-6

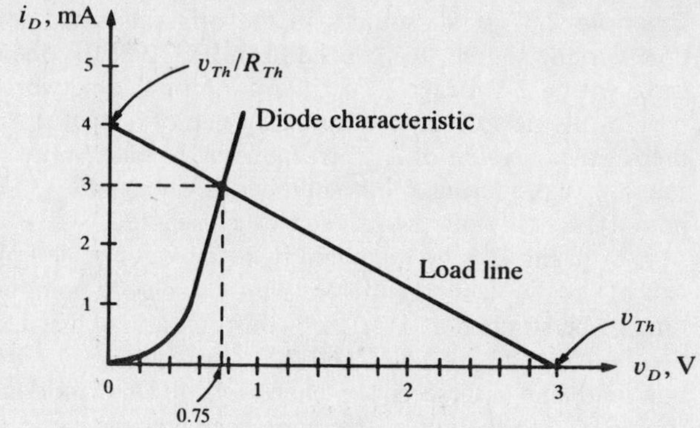

Fig. 2-7

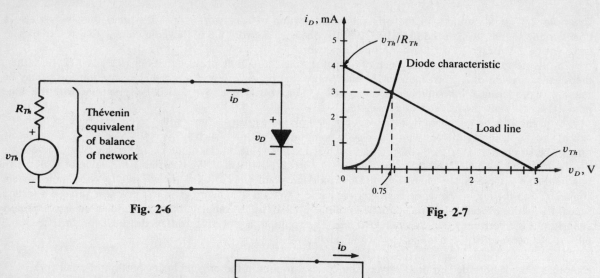

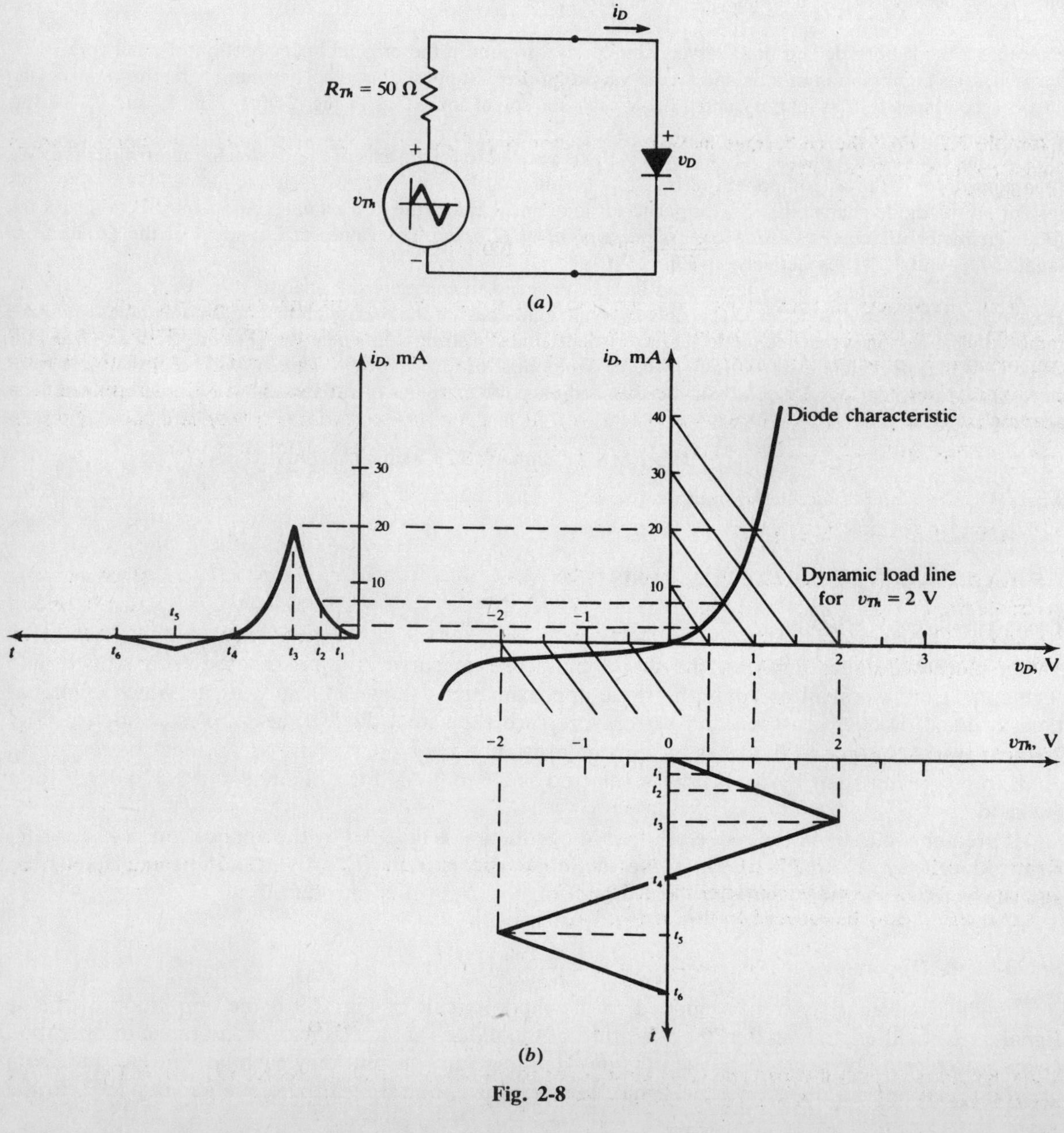

Fig. 2-8

Example 2.5 If all sources in the original linear portion of a network vary with time, then v_{Th} is also a time-varying source. In reduced form [Fig. 2-8(*a*)], one such network has a Thévenin voltage that is a triangular wave with a 2-V peak. Find i_D and v_D for this network.

In this case there is no unique value of i_D that satisfies the simultaneous equations (*2.3*) and (*2.4*); rather, there exists a value of i_D corresponding to each value that v_{Th} takes on. An acceptable solution for i_D may be found by considering a finite number of values of v_{Th}. Since v_{Th} is repetitive, i_D will be repetitive (with the same period), so only one cycle need be considered.

As in Fig. 2-8(*b*), we begin by laying out a scaled plot of v_{Th} versus time, with the v_{Th} axis parallel to the v_D axis of the diode characteristic. We then select a point on the v_{Th} plot, such as $v_{Th} = 0.5$ V at $t = t_1$. Considering time to be stopped at $t = t_1$, we construct a load line for this value on the diode characteristic plot; it intersects the v_D axis at $v_{Th} = 0.5$ V, and the i_D axis at $v_{Th}/R_{Th} = 0.5/50 = 10$ mA. We determine the value of i_D at which this load line intersects the characteristic, and plot the point (t_1, i_D) on a time-versus-i_D coordinate system constructed to the left of the diode characteristic curve. We then let time progress to some new value, $t = t_2$, and repeat the entire process. And we continue until one cycle of v_{Th} is completed. Since the load line is continually changing, it is referred to as a *dynamic load line*. The solution, a plot of i_D, differs drastically in form from the plot of v_{Th} because of the nonlinearity of the diode.

Example 2.6 If both dc and time-varying sources are present in the original linear portion of a network, then v_{Th} is a series combination of a dc and a time-varying source. Suppose that the Thévenin source for a particular network combines a 0.7-V battery and a 0.1-V-peak sinusoidal source, as in Fig. 2-9(*a*). Find i_D and v_D for the network.

We lay out a scaled plot of v_{Th}, with the v_{Th} axis parallel to the v_D axis of the diode characteristic curve. We then consider v_{th}, the ac component of v_{Th}, to be momentarily at zero ($t = 0$), and we plot a load line for this instant on the diode characteristic. This particular load line is called the *dc load line*, and its intersection with the diode characteristic curve is called the *quiescent point* or *Q point*. The values of i_D and v_D at the Q point are labeled I_{DQ} and V_{DQ}, respectively, in Fig. 2-9(*b*).

In general, a number of dynamic load lines are needed to complete the analysis of i_D over a cycle of v_{th}. However, for the network under study, only dynamic load lines for the maximum and minimum values of v_{th} are required. The reason is that the diode characteristic is almost a straight line near the Q point [from *a* to *b* in Fig. 2-9(*b*)], so that negligible distortion of i_d, the ac component of i_D, will occur. Thus, i_d will be of the same form as v_{th} (i.e., sinusoidal), and it can easily be sketched once the extremes of variation have been determined. The solution for i_D is thus

$$i_D = I_{DQ} + i_d = I_{DQ} + I_{dm} \sin \omega t = 36 + 8 \sin \omega t \qquad \text{mA}$$

where I_{dm} is the amplitude of the sinusoidal term.

2.5 EQUIVALENT-CIRCUIT ANALYSIS

Piecewise-Linear Techniques

In piecewise-linear analysis, the diode characteristic curve is approximated with straight-line segments. Here we shall use only the three approximations shown in Fig. 2-10, in which combinations of ideal diodes, resistors, and batteries replace the actual diode. The simplest model, in Fig. 2-10(*a*), treats the actual diode as an infinite resistance for $v_D < V_F$, and as an ideal battery if v_D tends to be greater than V_F. V_F is usually selected as 0.6 to 0.7 V for a Si diode and 0.2 to 0.3 V for a Ge diode.

If greater accuracy in the range of forward conduction is dictated by the application, a resistor R_F is introduced, as in Fig. 2-10(*b*). If the diode reverse current ($i_D < 0$) cannot be neglected, the additional refinement (R_R plus an ideal diode) of Fig. 2-10(*c*) is introduced.

Small-Signal Techniques

Small-signal analysis can be applied to the diode circuit of Fig. 2-9 if the amplitude of the ac signal v_{th} is small enough so that the curvature of the diode characteristic over the range of operation (from *b* to *a*) may be neglected. Then the diode voltage and current may each be written as the sum of a dc signal and an *undistorted* ac signal. Furthermore, the ratio of the diode ac voltage v_d to the

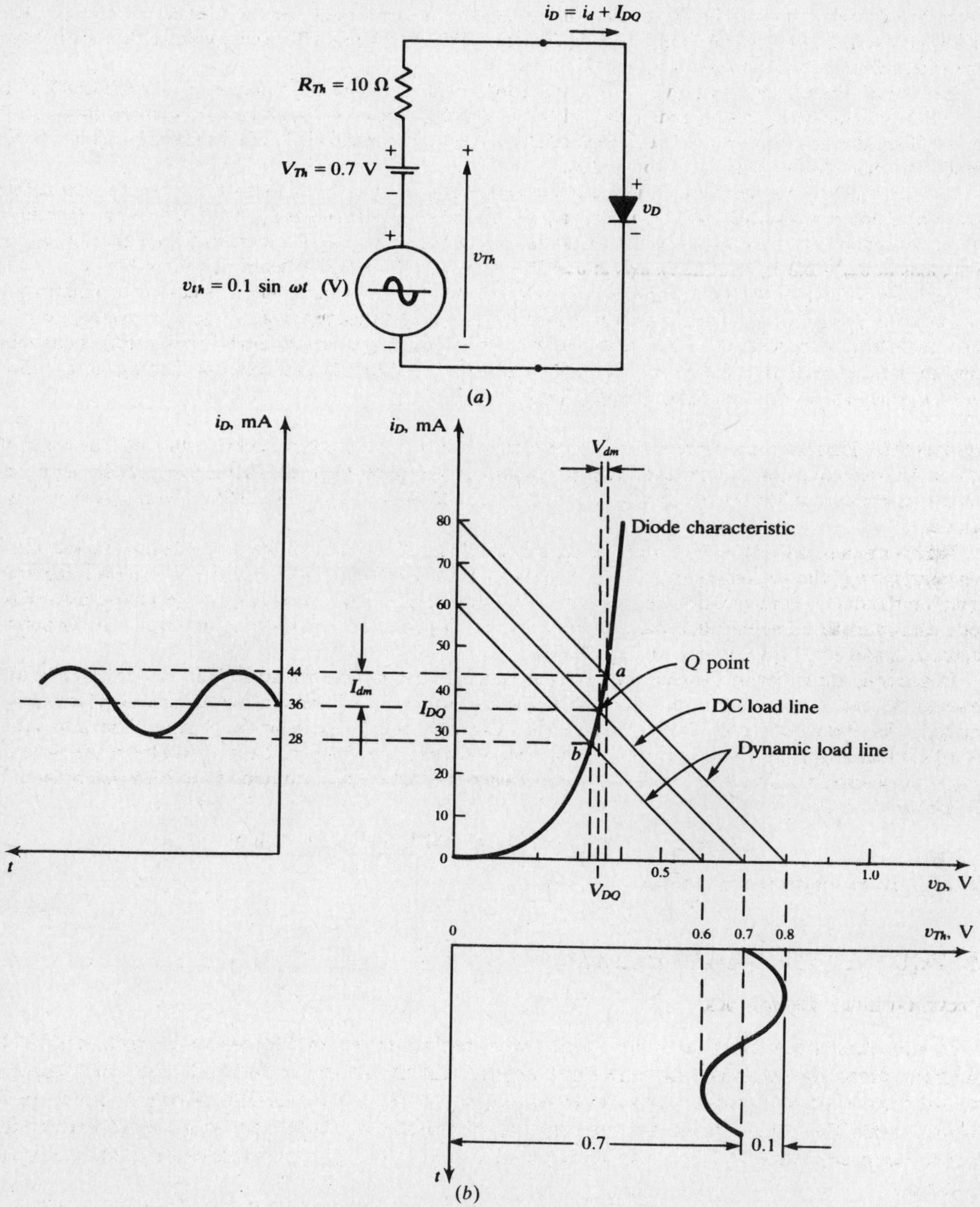

Fig. 2-9

diode ac current i_d will be constant and equal to

$$\frac{v_d}{i_d} = \frac{2V_{dm}}{2I_{dm}} = \frac{v_D|_a - v_D|_b}{i_D|_a - i_D|_b} = \left.\frac{\Delta v_D}{\Delta i_D}\right|_Q = \left.\frac{dv_D}{di_D}\right|_Q \equiv r_d \tag{2.5}$$

where r_d is known as the *dynamic resistance* of the diode. It follows (from a linear circuit argument) that the ac signal components may be determined by analysis of the "small-signal" circuit of Fig. 2-11; if the frequency of the ac signal is large, a capacitor can be placed in parallel with r_d to model

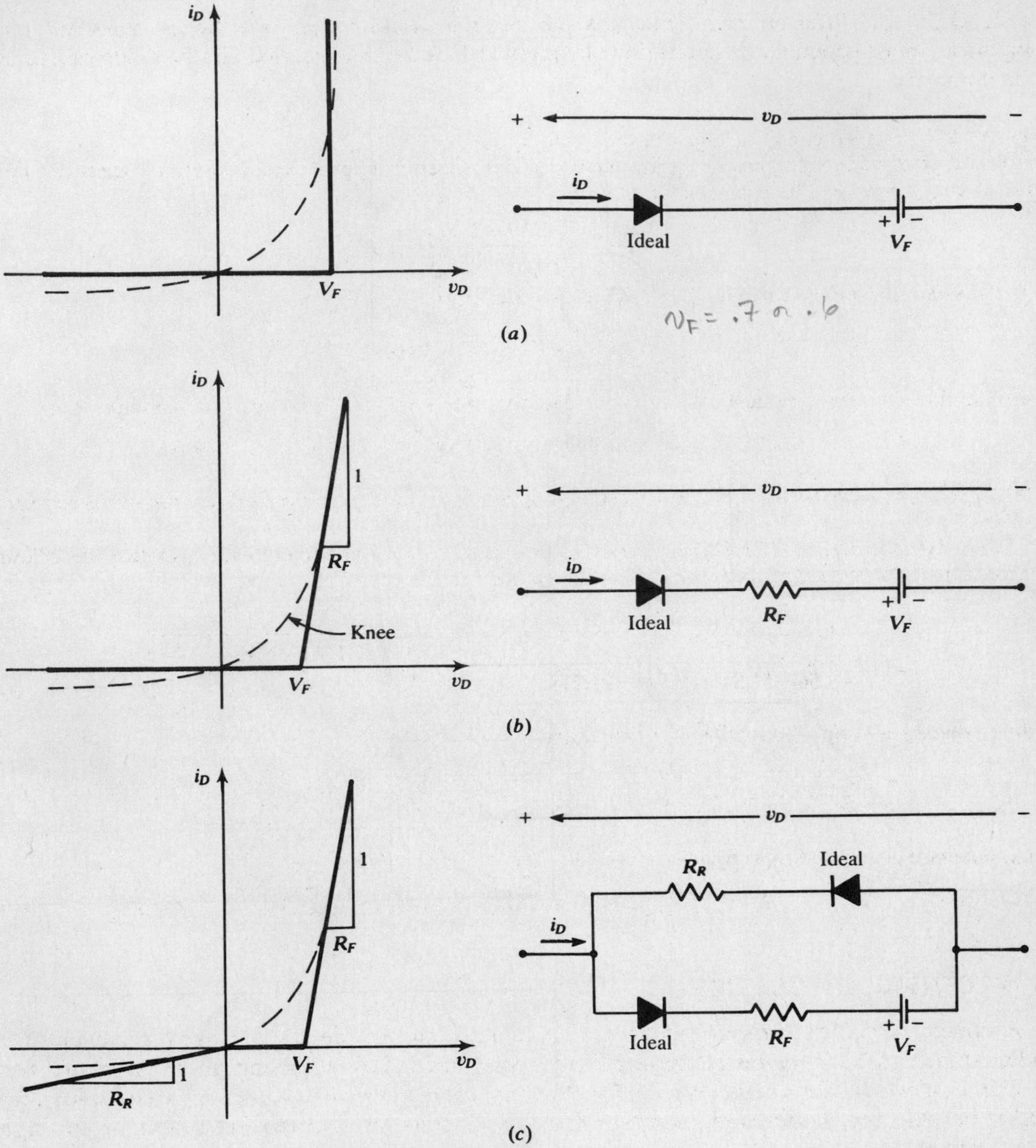

Fig. 2-10

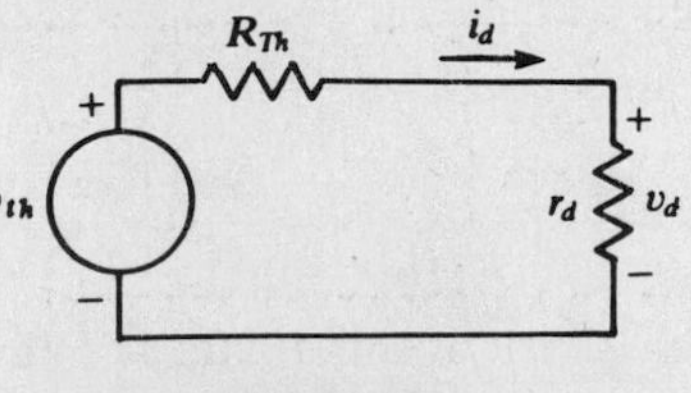

Fig. 2-11

the depletion or diffusion capacitance as discussed in section 2.3. The dc or quiescent signal components must generally be determined by graphical methods since, overall, the diode characteristic is nonlinear.

Example 2.7 For the circuit of Fig. 2-9, determine i_D.

The Q-point current I_{DQ} has been determined as 36 mA (see Example 2.6). The dynamic resistance of the diode at the Q point can be evaluated graphically:

$$r_d = \frac{\Delta v_D}{\Delta i_D} = \frac{0.37 - 0.33}{0.044 - 0.028} = 2.5\ \Omega$$

Now the small-signal circuit of Fig. 2-11 can be analyzed to find i_d:

$$i_d = \frac{v_{th}}{R_{Th} + r_d} = \frac{0.1 \sin \omega t}{10 + 2.5} = 0.008 \sin \omega t \qquad \text{A}$$

The total diode current is obtained by superposition and checks well with that found in Example 2.6:

$$i_D = I_{DQ} + i_d = 36 + 8 \sin \omega t \qquad \text{mA}$$

Example 2.8 For the circuit of Fig. 2-9, determine i_D if $\omega = 10^8$ rad/s and the diffusion capacitance is known to be 5000 pF.

From Example 2.7, $r_d = 2.5\ \Omega$. The diffusion capacitance C_d acts in parallel with r_d to give the following equivalent impedance for the diode, as seen by the ac signal:

$$Z_d = r_d \,\|\, jx_d = r_d \,\|\left(-j\frac{1}{\omega C_d}\right) = \frac{r_d}{1 + j\omega C_d r_d} = \frac{2.5}{1 + j(10^8)(5000 \times 10^{-12})(2.5)}$$

$$= 1.56\underline{/-51.34^\circ} = 0.974 - j1.218$$

In the frequency domain, the small-signal circuit (Fig. 2-11) yields

$$\bar{I}_d = \frac{\bar{V}_{th}}{R_{th} + Z_d} = \frac{0.1\underline{/-90^\circ}}{10 + 0.974 - j1.218} = \frac{0.1\underline{/-90^\circ}}{11.041\underline{/-6.33^\circ}} = 0.0091\underline{/-83.67^\circ} \qquad \text{A}$$

In the time domain, with I_{DQ} as found in Example 2.6, we have

$$i_D = I_{DQ} + i_d = 36 + 9.1 \cos (10^8 t - 83.67^\circ) \qquad \text{mA}$$

2.6 RECTIFIER APPLICATIONS

Rectifier circuits are two-port networks that capitalize on the nearly one-way conduction of the diode: An ac voltage is impressed upon the input port, and a dc voltage appears at the output port.

The simplest rectifier circuit (Fig. 2-12) contains a single diode. It is commonly called a *half-wave rectifier* because the diode conducts over either the positive or the negative halves of the input-voltage waveform.

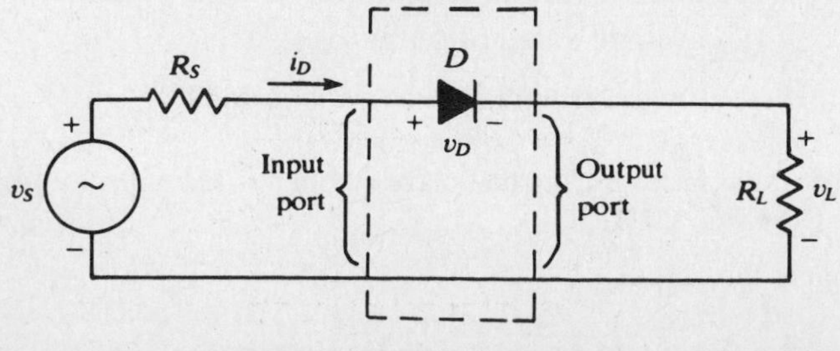

Fig. 2-12

Example 2.9 In Fig. 2-12, $v_S = V_m \sin \omega t$ and the diode is ideal. Calculate the average value of v_L.
Only one cycle of v_S need be considered. For the positive half-cycle, $i_D > 0$ and, by voltage division,

$$v_L = \frac{R_L}{R_L + R_S}(V_m \sin \omega t) \equiv V_{Lm} \sin \omega t$$

For the negative half-cycle, the diode is reverse-biased, $i_D = 0$, and $v_L = 0$. Hence,

$$V_{L0} = \frac{1}{2\pi}\int_0^{2\pi} v_L(\omega t)\, d(\omega t) = \frac{1}{2\pi}\int_0^{\pi} V_{Lm} \sin \omega t\, d(\omega t) = \frac{V_{Lm}}{\pi}$$

Although the half-wave rectifier gives a dc output, current flows through R_L only half the time, and the average value of the output voltage is only $1/\pi = 0.318$ times the peak value of the sinusoidal input voltage. The output voltage can be improved by use of a *full-wave rectifier* (see Problems 2.27 and 2.48).

When rectifiers are used as dc power supplies, it is desirable that the average value of the output voltage remain nearly constant as the load varies. The degree of constancy is measured as the *voltage regulation*,

$$\text{Reg} \equiv \frac{(\text{no-load } V_{L0}) - (\text{full-load } V_{L0})}{\text{full-load } V_{L0}} \tag{2.6}$$

which is usually expressed as a percentage. Note that 0 percent regulation implies a constant output voltage.

Example 2.10 Find the voltage regulation of the half-wave rectifier of Fig. 2-12.
From Example 2.9, we know that

$$\text{Full-load } V_{L0} = \frac{V_{Lm}}{\pi} = \frac{R_L}{\pi(R_L + R_S)} V_m \tag{2.7}$$

Realizing that $R_L \rightarrow \infty$ for no load, we may write

$$\text{No-load } V_{L0} = \lim_{R_L \rightarrow \infty} \left| \frac{R_L}{\pi(R_L + R_S)} V_m \right| = \frac{V_m}{\pi}$$

Thus, the voltage regulation is

$$\text{Reg} = \frac{\dfrac{V_m}{\pi} - \dfrac{R_L}{\pi(R_L + R_S)} V_m}{\dfrac{R_L}{\pi(R_L + R_S)} V_m} = \frac{R_S}{R_L} = \frac{100 R_S}{R_L} \%$$

2.7 WAVEFORM FILTERING

The output of a rectifier alone does not usually suffice as a power supply, due to its variation in time. The situation is improved by placing a *filter* between the rectifier and the load. The filter acts to suppress the harmonics from the rectified waveform and to preserve the dc component. A measure of goodness for rectified waveforms, both filtered and unfiltered, is the *ripple factor*,

$$F_r \equiv \frac{\text{maximum variation in output voltage}}{\text{average value of output voltage}} = \frac{\Delta v_L}{V_{L0}} \tag{2.8}$$

A small value, say $F_r \leq 0.05$, is usually attainable and practical.

Example 2.11 Calculate the ripple factor for the half-wave rectifier of Example 2.9 (*a*) without a filter and (*b*) with a shunt capacitor filter as in Fig. 2-13(*a*).

(*a*) For the circuit of Example 2.9,

$$F_r = \frac{\Delta v_L}{V_{L0}} = \frac{V_{Lm}}{V_{Lm}/\pi} = \pi \approx 3.14$$

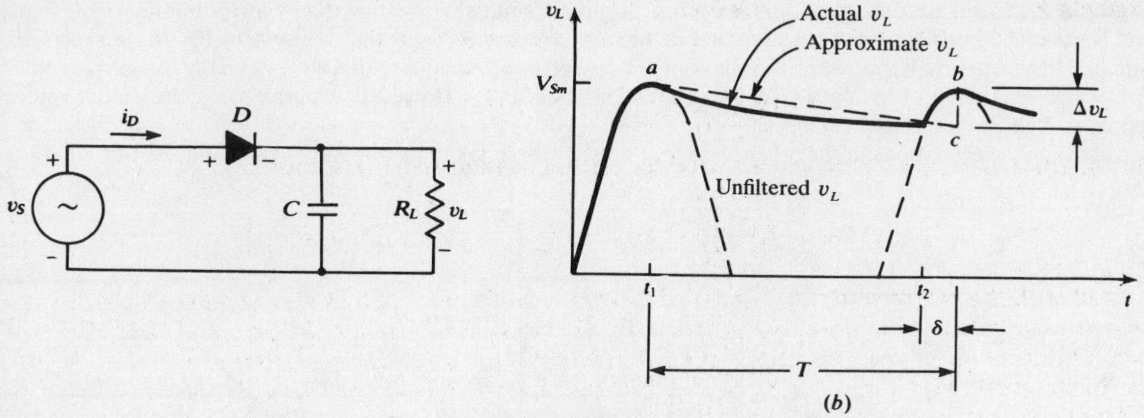

Fig. 2-13

(*b*) The capacitor in Fig. 2-13 stores energy while the diode allows current to flow, and delivers energy to the load when current flow is blocked. The actual load voltage v_L that results with the filter inserted is sketched in Fig. 2-13(*b*), for which we assume that $v_S = V_{Sm} \sin \omega t$ and D is an ideal diode. For $0 < t \leq t_1$, D is forward-biased and capacitor C charges to the value V_{Sm}. For $t_1 < t \leq t_2$, v_S is less than v_L, reverse-biasing D and causing it to act as an open circuit. During this interval the capacitor is discharging through the load R_L, giving

$$v_L = V_{Sm}\, e^{-(t-t_1)/R_L C} \qquad (t_1 < t \leq t_2) \tag{2.9}$$

Over the interval $t_2 < t \leq t_2 + \delta$, v_S forward-biases diode D and again charges the capacitor to V_{Sm}. Then v_S falls below the value of v_L and another discharge cycle identical to the first occurs.

Obviously, if the time constant $R_L C$ is large enough compared to T to result in a decay like that indicated in Fig. 2-13(*b*), a major reduction in Δv_L and a major increase in V_{L0} will have been achieved, relative to the unfiltered rectifier. The introduction of two quite reasonable approximations leads to simple formulas for Δv_L and V_{L0}, and hence for F_r, that are sufficiently accurate for design and analysis work:

1. If Δv_L is to be small, then $\delta \to 0$ in Fig. 2-13(*b*) and $t_2 - t_1 \approx T$.
2. If Δv_L is small enough, then (*2.9*) can be represented over the interval $t_1 < t \leq t_2$ by a straight line with a slope of magnitude $V_{Sm}/R_L C$.

The dashed line labeled "Approximate v_L" in Fig. 2-13(*b*) implements these two approximations. From right triangle *abc*,

$$\frac{\Delta v_L}{T} = \frac{V_{Sm}}{R_L C} \qquad \text{or} \qquad \Delta v_L = \frac{V_{Sm}}{f R_L C}$$

where f is the frequency of v_S. Since, under this approximation,

$$V_{L0} = V_{Sm} - \tfrac{1}{2}\Delta v_L$$

and $R_L C/T = f R_L C$ is presumed large,

$$F_r = \frac{\Delta v_L}{V_{L0}} = \frac{2}{2 f R_L C - 1} \approx \frac{1}{f R_L C} \tag{2.10}$$

2.8 CLIPPING AND CLAMPING OPERATIONS

Diode *clipping circuits* separate an input signal at a particular dc level and pass to the output, without distortion, the desired upper or lower portion of the original waveform. They are used to eliminate amplitude noise or to fabricate new waveforms from an existing signal.

Example 2.12 Figure 2-14(*a*) shows a *positive* clipping circuit, which removes any portion of the input signal v_i that is greater than V_b and passes as the output signal v_o any portion of v_i that is less than V_b. As you can see, v_D is negative when $v_i < V_b$, causing the ideal diode to act as an open circuit. With no path for current to flow through R, the value of v_i appears at the output terminals as v_o. However, when $v_i \geq V_b$, the diode conducts, acting as a short circuit and forcing $v_o = V_b$. Figure 2-14(*b*), the *transfer graph* or *transfer characteristic* for the circuit, shows the relationship between the input voltage, here taken as $v_i = 2V_b \sin \omega t$, and the output voltage.

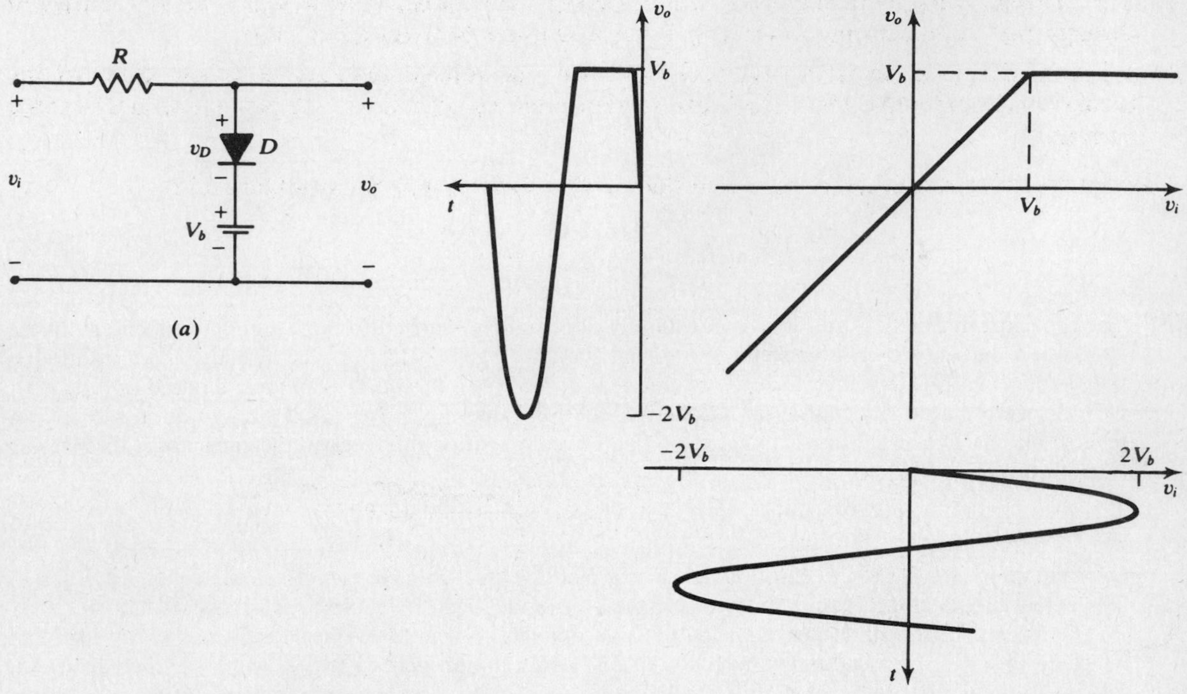

Fig. 2-14

Clamping is a process of setting the positive or negative peaks of an input ac waveform to a specific dc level, regardless of any variation in those peaks.

Example 2.13 An ideal clamping circuit is shown in Fig. 2-15(*b*), and a triangular ac input waveform in Fig. 2-15(*a*). If the capacitor C is initially uncharged, the ideal diode D is forward-biased for $0 < t \leq T/4$, and it acts as a short circuit while the capacitor charges to $v_C = V_p$. At $t = T/4$, D open-circuits, breaking the only possible discharge path for the capacitor. Thus, the value $v_C = V_p$ is preserved; since v_i can never exceed V_p, D remains reverse-biased for all $t > T/4$, giving $v_o = v_D = v_i - V_p$. The function v_o is sketched in Fig. 2-15(*c*); all positive peaks are clamped at zero, and the average value is shifted from 0 to $-V_p$.

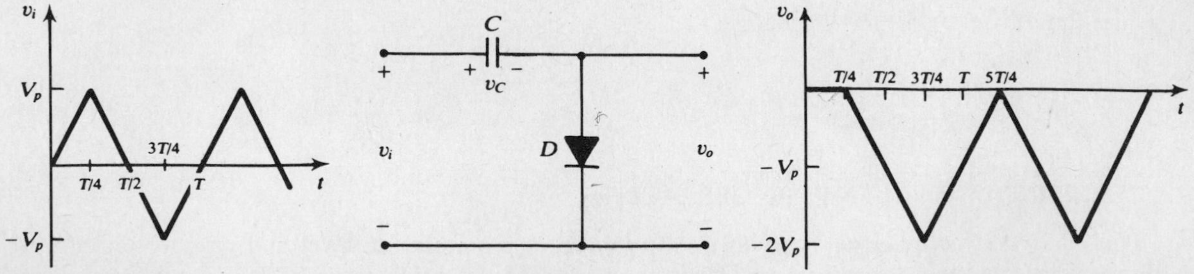

Fig. 2-15

2.9 THE ZENER DIODE

The *Zener diode* or *reference diode*, whose symbol is shown in Fig. 2-16(*a*), finds primary usage as a voltage regulator or reference. The forward conduction characteristic of a Zener diode is much the same as that of a rectifier diode; however, it usually operates with a reverse bias, for which its characteristic is radically different. Note, in Fig. 2-16(*b*), that:

1. The reverse voltage breakdown is rather sharp. The breakdown voltage can be controlled through the manufacturing process so it has a reasonably predictable value.
2. When a Zener diode is in reverse breakdown, its voltage remains extremely close to the breakdown value while the current varies from rated current (I_Z) to 10 percent or less of rated current.

A Zener regulator should be designed so that $i_Z \geq 0.1 I_Z$ to assure the constancy of v_Z.

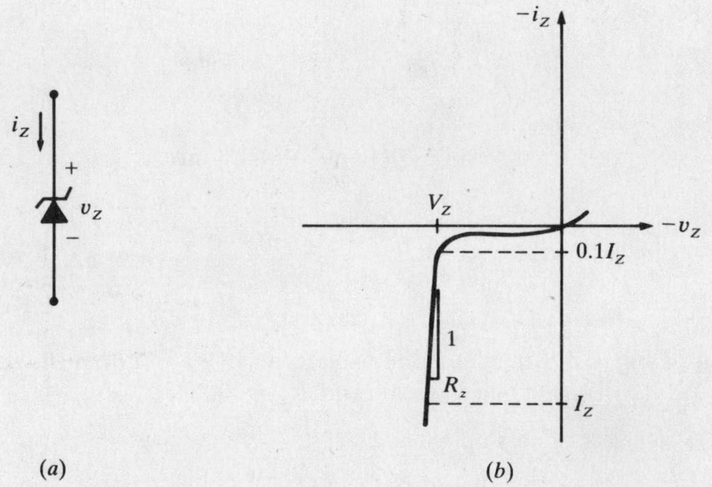

Fig. 2-16

Example 2.14 Find the voltage v_Z across the Zener diode of Fig. 2-16(*a*) if $i_Z = 10$ mA and it is known that $V_Z = 5.6$ V, $I_Z = 25$ mA, and $R_Z = 10\ \Omega$.

Since $0.1 I_Z \leq i_Z \leq I_Z$, operation is along the safe and predictable region of Zener operation. Consequently,

$$v_Z \approx V_Z + i_Z R_Z = 5.6 + (10 \times 10^{-3})(10) = 5.7 \text{ V}$$

R_Z is frequently neglected in the design of Zener regulators. Problem 2.29 illustrates the design technique.

Solved Problems

2.1 At a junction temperature of 25°C, over what range of forward voltage drop v_D can (*2.1*) be approximated as $i_D \approx I_o e^{v_D/V_T}$ with less than 1 percent error for a Ge diode?

From (*2.1*) with $\eta = 1$, the error will be less than 1 percent if $e^{v_D/V_T} > 101$. In that range,

$$v_D > V_T \ln 101 = \frac{kT}{q} \ln 101 = \frac{(1.38 \times 10^{-23})(25 + 273)}{1.6 \times 10^{-19}} 4.6151 = 0.1186 \text{ V}$$

2.2 A Ge diode described by (*2.1*) is operated at a junction temperature of 27°C. For a forward current of 10 mA, v_D is found to be 0.3 V. (*a*) If $v_D = 0.4$ V, find the forward current. (*b*) Find the reverse saturation current.

(*a*) We form the ratio

$$\frac{i_{D2}}{i_{D1}} = \frac{I_o(e^{v_{D2}/V_T} - 1)}{I_o(e^{v_{D1}/V_T} - 1)} = \frac{e^{0.4/0.02587} - 1}{e^{0.3/0.02587} - 1} = 47.73$$

Then $\quad i_{D2} = (47.73)(10 \text{ mA}) = 477.3 \text{ mA}$

(*b*) By (*2.1*),

$$I_o = \frac{i_{D1}}{e^{v_{D1}/V_T} - 1} = \frac{10 \times 10^{-3}}{e^{0.3/0.02587} - 1} = 91 \text{ nA}$$

2.3 For the circuit of Fig. 2-17(*a*), sketch the waveforms of v_L and v_D if the source voltage v_S is as given in Fig. 2-17(*b*). The diode is ideal, and $R_L = 100\ \Omega$.

If $v_S \geq 0$, D conducts, so that $v_D = 0$ and

$$v_L = \frac{R_L}{R_L + R_S} v_S = \frac{100}{100 + 10} v_S = 0.909 v_S$$

If $v_S < 0$, D blocks, so that $v_D = v_S$ and $v_L = 0$. Sketches of v_D and v_L are shown in Fig. 2-17(*c*).

2.4 Extend the ideal diode analysis procedure of Section 2.2 to the case of multiple diodes by solving for the current i_L in the circuit of Fig. 2-18(*a*). Assume D_1 and D_2 are ideal. $R_2 = R_L = 100\ \Omega$, and v_S is a 10-V square wave of period 1 ms.

Step 1: Assume both diodes are forward-biased, and replace each with a short circuit as shown in Fig. 2-18(*b*).

Step 2: Since D_1 is "on," or in the zero-impedance state, current division requires that

$$i_{D2} = -\frac{0}{R_2 + 0} i_L = 0 \tag{1}$$

Hence, by Ohm's law,

$$i_L = i_{D1} = \frac{v_S}{R_L} \tag{2}$$

Step 3: Observe that when $v_S = 10 > 0$, we have, by (*2*), $i_{D1} = 10/100 = 0.1 \text{ A} > 0$. Also, by (*1*), $i_{D2} = 0$. Thus all diode currents are greater than or equal to zero, and the analysis is valid. *However*, when $v_S = -10 < 0$, we have, by (*2*), $i_{D1} = -10/100 = -0.1 \text{ A} < 0$, and the analysis is no longer valid.

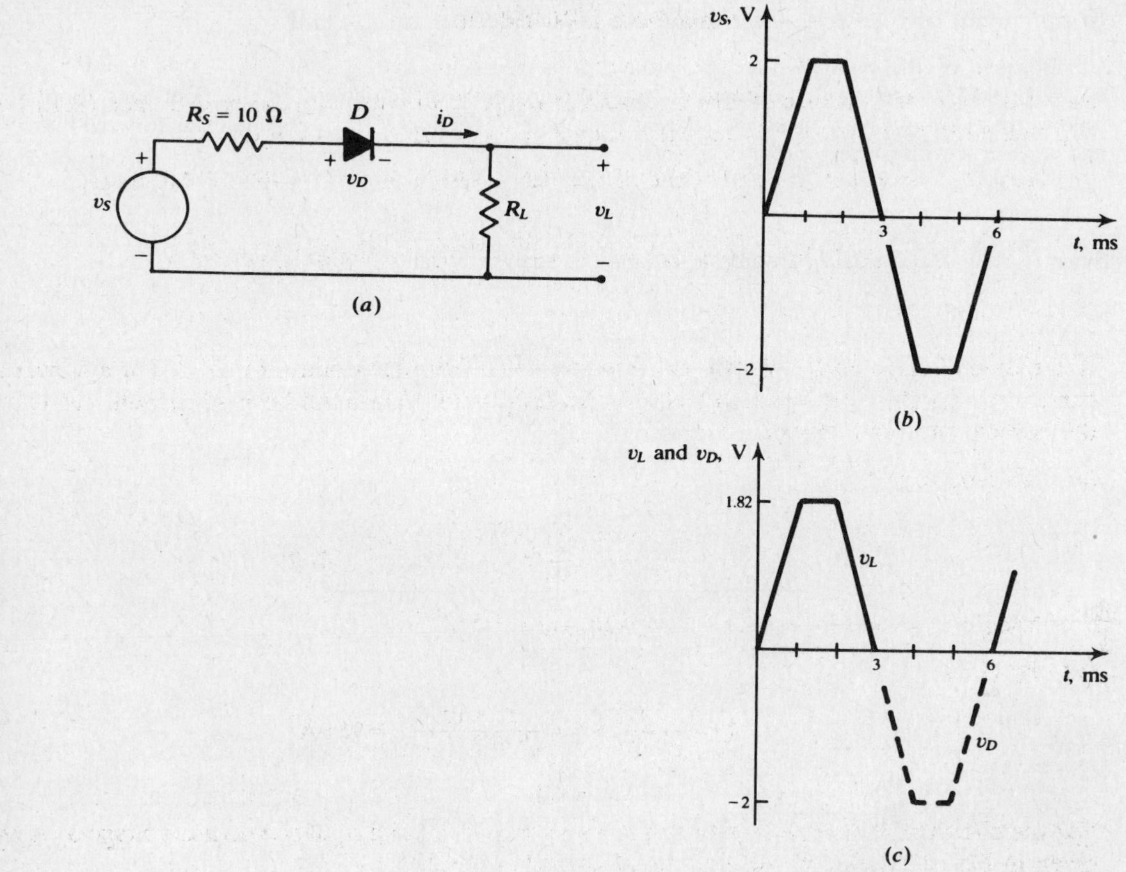

Fig. 2-17

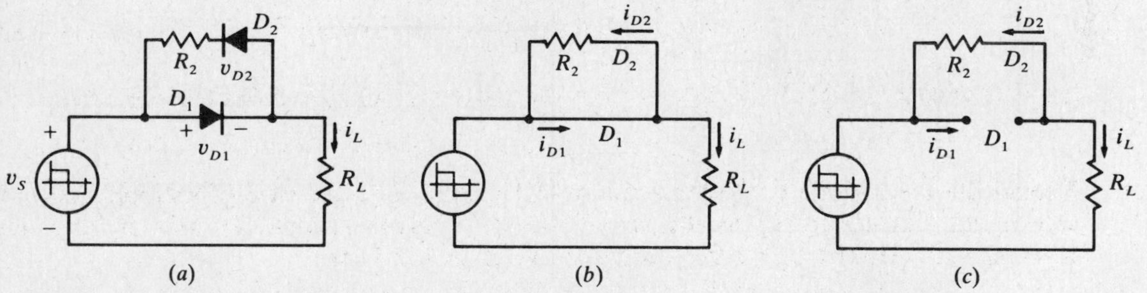

Fig. 2-18

Step 4: Replace D_1 with an open circuit as illustrated in Fig. 2-18(*c*). Now obviously $i_{D1} = 0$ and, by Ohm's law,

$$i_L = -i_{D2} = \frac{v_S}{R_2 + R_L} = \frac{-10}{100 + 100} = -0.05 \text{ A}$$

Further, voltage division requires that

$$v_{D1} = \frac{R_2}{R_2 + R_L} v_S$$

so that $v_{D1} < 0$ if $v_S < 0$, verifying that D_1 is actually reverse-biased. Note that if D_2 had been replaced with an open circuit, we would have found that $v_{D2} = -v_S = 10 \text{ V} > 0$, so D_2 would not actually have been reverse-biased.

2.5 In the circuit of Fig. 2-19, D_1 and D_2 are ideal diodes. Find i_{D1} and i_{D2}.

Because of the polarities of D_1 and D_2, it is necessary that $i_S \geq 0$. Thus, $v_{ab} \leq V_S = V_1$. But $v_{D1} = v_{ab} - V_1$; therefore, $v_D \leq 0$ and so $i_{D1} \equiv 0$, regardless of conditions in the right-hand loop. It follows that $i_{D2} = i_S$. Now using the analysis procedure of Section 2.2, we assume D_2 is forward-biased and replace it with a short circuit. By KVL,

$$i_{D2} = \frac{V_S - V_2}{500} = \frac{5-3}{500} = 4 \text{ mA}$$

Since $i_{D2} \geq 0$, D_2 is in fact forward-biased and the analysis is valid.

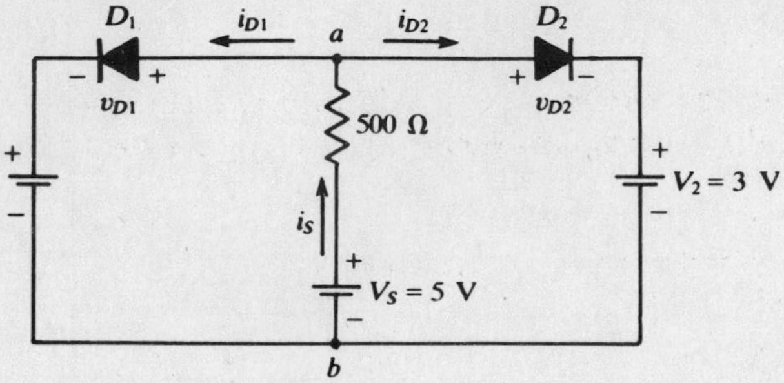

Fig. 2-19

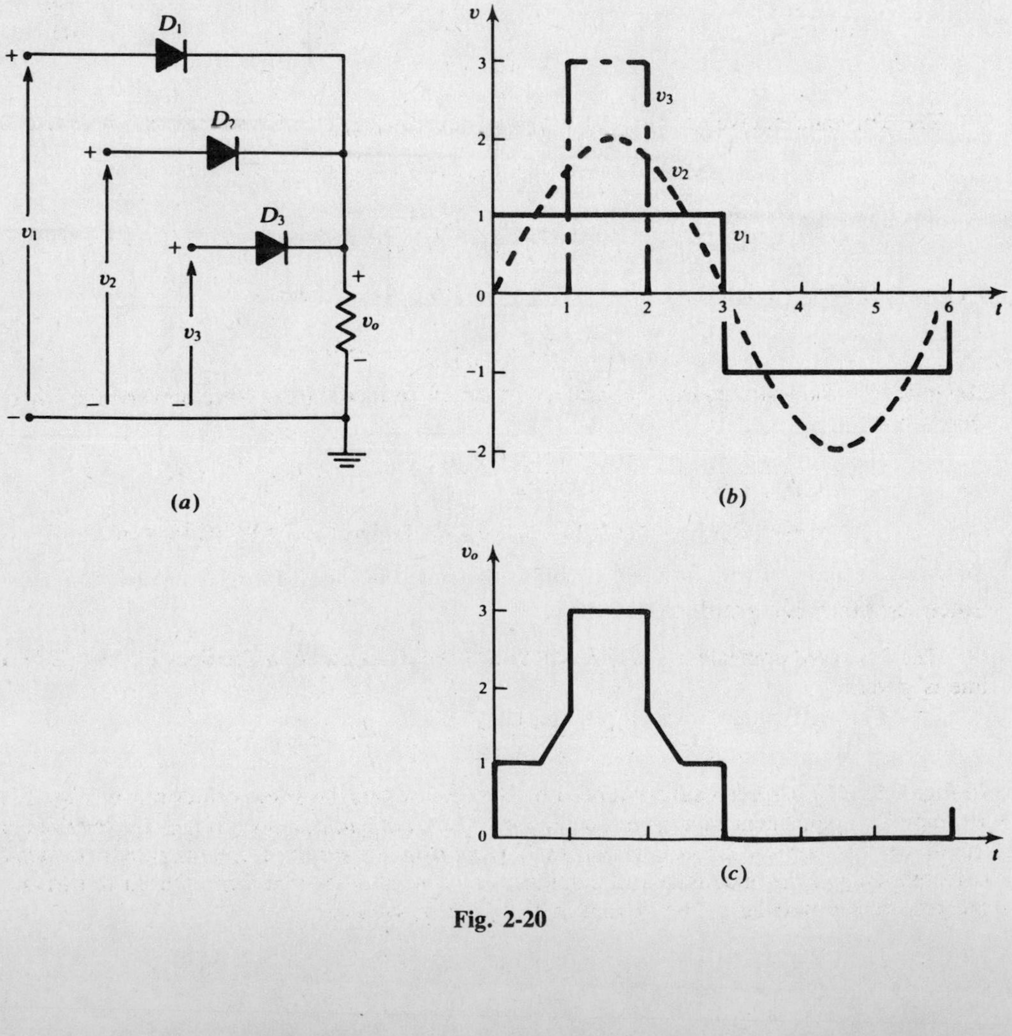

Fig. 2-20

2.6 The logic OR gate can be utilized to fabricate composite waveforms. Sketch the output v_o of the gate of Fig. 2-20(*a*) if the three signals of Fig. 2-20(*b*) are impressed on the input terminals. Assume that diodes are ideal.

For this circuit, KVL gives

$$v_1 - v_2 = v_{D1} - v_{D2} \qquad v_1 - v_3 = v_{D1} - v_{D3}$$

i.e., the diode voltages have the same ordering as the input voltages. Suppose that v_1 is positive and exceeds v_2 and v_3. Then D_1 must be forward-biased, with $v_{D1} = 0$ and, consequently, $v_{D2} < 0$ and $v_{D3} < 0$. Hence D_2 and D_3 block, while v_1 is passed as v_o. This is so in general: The logic of the OR gate is that the largest positive input signal is passed as v_o, while the remainder of the input signals are blocked. If all input signals are negative, $v_o = 0$. Application of this logic gives the sketch of v_o in Fig. 2-20(*c*).

2.7 The diode in the circuit of Fig. 2-21(*a*) has the nonlinear terminal characteristic of Fig. 2-21(*b*). Find i_D and v_D analytically, given $v_S = 0.1 \cos \omega t$ V and $V_b = 2$ V.

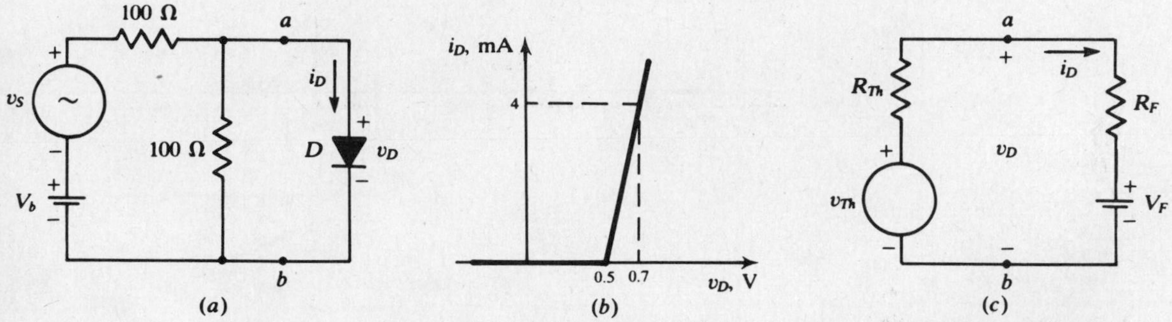

Fig. 2-21

The Thévenin equivalent circuit for the network to the left of terminals *a,b* in Fig. 2-21(*a*) has

$$V_{Th} = \frac{100}{200}(2 + 0.1 \cos \omega t) = 1 + 0.05 \cos \omega t \qquad \text{V}$$

$$R_{Th} = \frac{(100)^2}{200} = 50\ \Omega$$

The diode can be modeled as in Fig. 2-10(*b*), with $V_F = 0.5$ V and

$$R_F = \frac{0.7 - 0.5}{0.004} = 50\ \Omega$$

Together, the Thévenin equivalent circuit and the diode model form the circuit in Fig. 2-21(*c*). Now by Ohm's law,

$$i_D = \frac{V_{Th} - V_F}{R_{Th} + R_F} = \frac{(1 + 0.05 \cos \omega t) - 0.5}{50 + 50} = 5 + 0.5 \cos \omega t \qquad \text{mA}$$

$$v_D = V_F + R_F i_D = 0.5 + 50(0.005 + 0.0005 \cos \omega t) = 0.75 + 0.025 \cos \omega t \qquad \text{V}$$

2.8 Solve Problem 2.7 graphically for i_D.

The Thévenin equivalent circuit has already been determined in Problem 2.7. By (*2.4*), the dc load line is given by

$$i_D = \frac{V_{Th}}{R_{Th}} - \frac{v_D}{R_{Th}} = \frac{1}{50} - \frac{v_D}{50} = 20 - 20v_D \qquad \text{mA} \qquad (1)$$

In Fig. 2-22, (*1*) has been superimposed on the diode characteristic, replotted from Fig. 2-21(*b*). As in Example 2.6, equivalent time scales for v_{Th} and i_D are laid out adjacent to the characteristic curve. Since the diode characteristic is linear about the *Q* point over the range of operation, only dynamic load lines corresponding to the maximum and minimum of v_{Th} need be drawn. Once these two dynamic load lines are constructed parallel to the dc load line, i_D can be sketched.

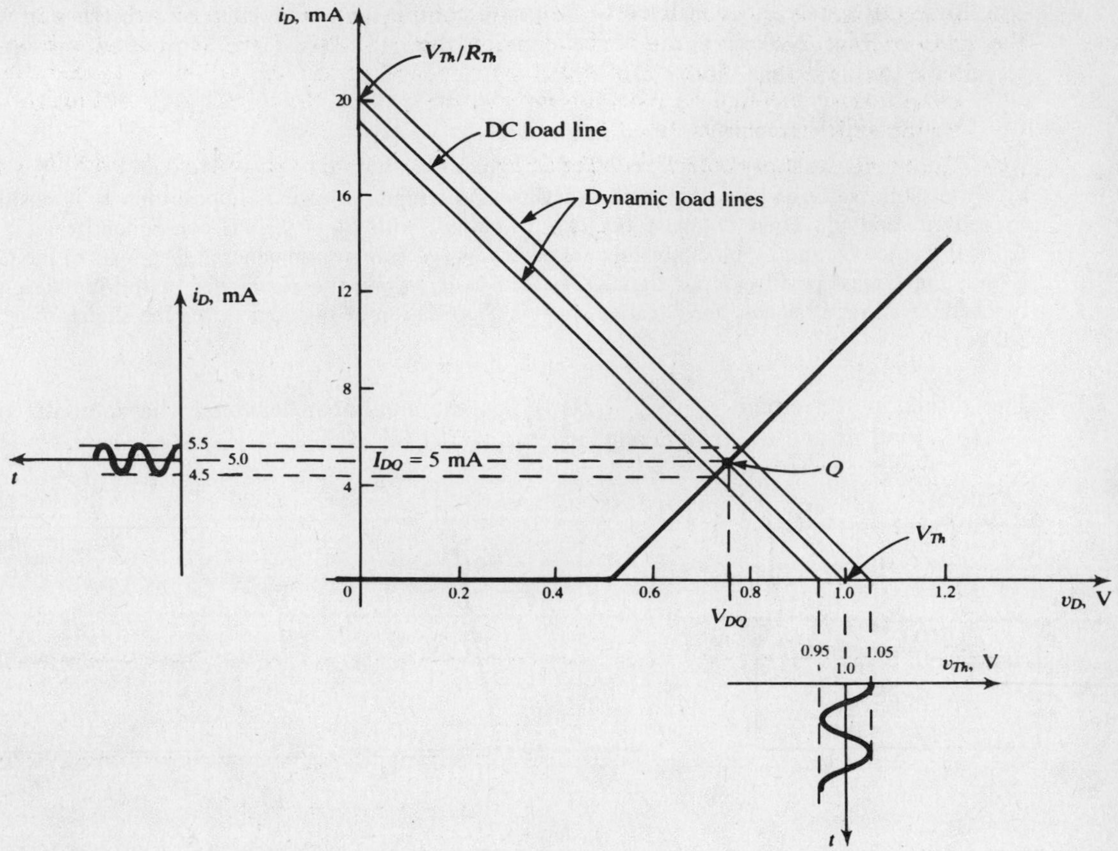

Fig. 2-22

2.9 Use the small-signal technique of Section 2.5 to find i_D and v_D in Problem 2.7.

The Thévenin equivalent circuit of Problem 2.7 is valid here. Moreover, the intersection of the dc load line and the diode characteristic in Fig. 2-22 gives $I_{DQ} = 5$ mA and $V_{DQ} = 0.75$ V. The dynamic resistance is, then, by (*2.5*),

$$r_d = \frac{\Delta v_D}{\Delta i_D} = \frac{0.7 - 0.5}{0.004} = 50\ \Omega$$

We now have all the values needed for analysis using the small-signal circuit of Fig. 2-11. By Ohm's law,

$$i_d = \frac{v_{th}}{R_{Th} + r_d} = \frac{0.05 \cos \omega t}{50 + 50} = 0.5 \cos \omega t \quad \text{mA}$$

$$v_d = r_d i_d = 50(0.0005 \cos \omega t) = 0.025 \cos \omega t \quad \text{V}$$

$$i_D = I_{DQ} + i_d = 5 + 0.5 \cos \omega t \quad \text{mA}$$

$$v_D = V_{DQ} + v_d = 0.75 + 0.025 \cos \omega t \quad \text{V}$$

2.10 A voltage source, $v_S = 0.4 + 0.2 \sin \omega t$ V, is placed directly across a diode characterized by Fig. 2-21(*b*). The source has no internal impedance and is of proper polarity to forward-bias the diode. (*a*) Sketch the resulting diode current i_D. (*b*) Determine the value of the quiescent current I_{DQ}.

(*a*) A scaled plot of v_S has been laid out adjacent to the v_D axis of the diode characteristic in Fig. 2-23. With zero resistance between the ideal voltage source and the diode, the dc load line has infinite slope and $v_D = v_S$. Thus i_D is found by a point-by-point projection of v_S onto the diode characteristic, followed by reflection through the i_D axis. Notice that i_D is extremely distorted, bearing little resemblance to v_S.

(*b*) Quiescent conditions obtain when the ac signal is zero. In this case, when $v_S = 0.4$ V, $i_D = I_{DQ} = 0$.

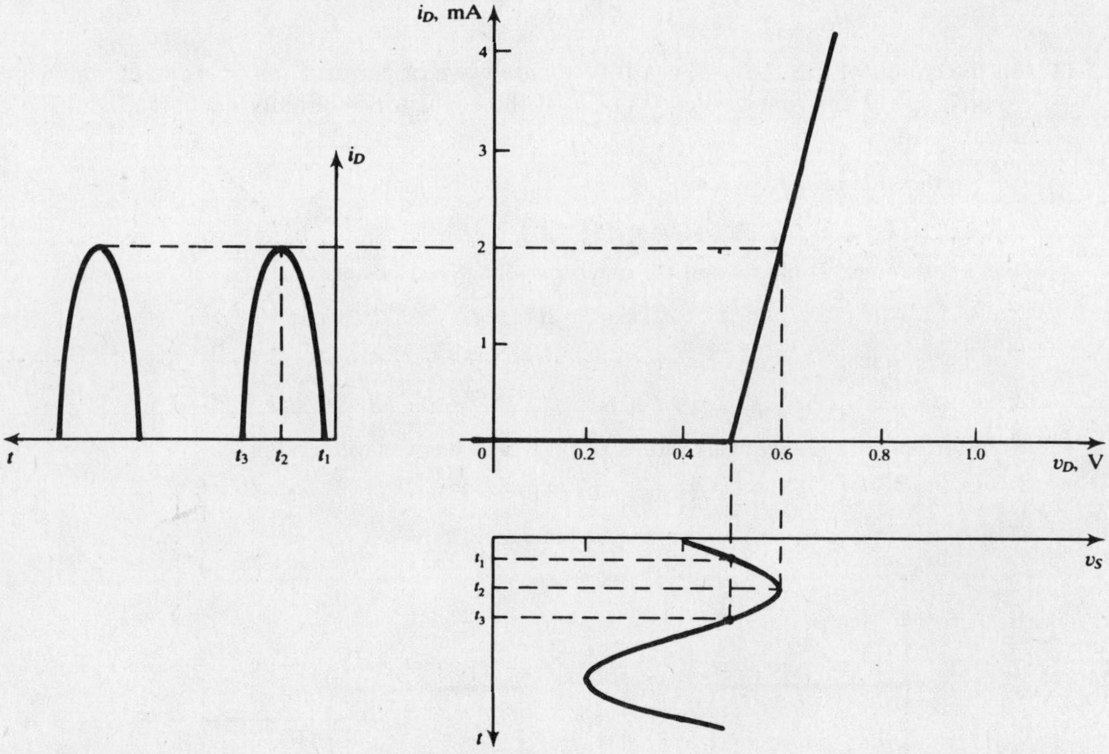

Fig. 2-23

2.11 In the circuit of Fig. 2-3(*a*), assume $R_S = R_1 = 200\ \Omega$, $R_L = 50\ \text{k}\Omega$, and $v_S = 400 \sin \omega t$ V. The diode is ideal, with reverse saturation current $I_o = 2\ \mu\text{A}$ and a peak inverse voltage (PIV) rating of $V_R = 100$ V. (*a*) Will the diode fail in avalanche breakdown? (*b*) If the diode will fail, is there a value of R_L for which failure will not occur?

(*a*) From Example 2.1,

$$v_{Th} = \frac{R_1}{R_1 + R_S} v_S = \frac{200}{200 + 200}(400 \sin \omega t) = 200 \sin \omega t \qquad \text{V}$$

$$R_{Th} = \frac{R_1 R_S}{R_1 + R_S} = \frac{(200)(200)}{200 + 200} = 100\ \Omega$$

The circuit to be analyzed is that of Fig. 2-3(*c*); the instants of concern are when $\omega t = (2n + 1)\pi/2$ for $n = 1, 2, 3, \ldots,$ at which times $v_{Th} = -200$ V and thus v_D is at its most negative value. An application of KVL yields

$$v_D = v_{Th} - i_D(R_{Th} + R_L) = -200 - (-2 \times 10^{-6})(100 + 50 \times 10^3) = -199.9 \text{ V} \qquad (1)$$

Since $v_D < -V_R = -100$ V, avalanche failure occurs.

(*b*) From (*1*), it is apparent that $v_D \geq -100$ V if

$$R_L \geq \frac{v_{Th} - v_D}{i_D} - R_{Th} = \frac{-200 - (-100)}{-2 \times 10^{-6}} - 100 = 50\ \text{M}\Omega$$

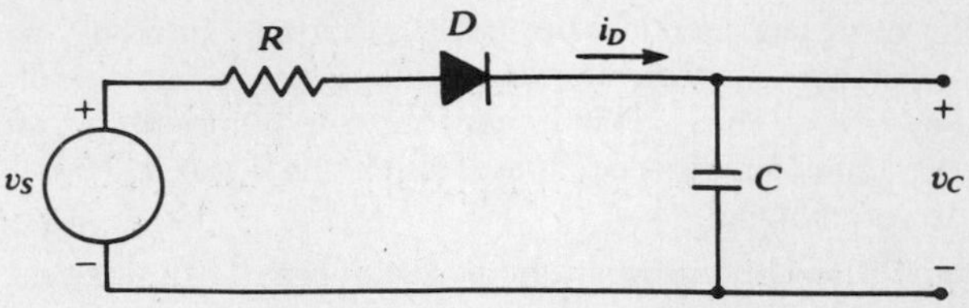

Fig. 2-24

2.12 In the circuit of Fig. 2-24, v_S is a 10-V square wave of period 4 ms, $R = 100\ \Omega$, and $C = 20\ \mu\text{F}$. Sketch v_C for the first two cycles of v_S if the capacitor is initially uncharged and the diode is ideal.

In the interval $0 \leq t < 2$ ms,

$$v_C(t) = v_S(1 - e^{-t/RC}) = 10(1 - e^{-500t}) \quad \text{V}$$

For $2 \leq t < 4$ ms, D blocks and the capacitor voltage remains at

$$v_C(2\text{ ms}) = 10(1 - e^{-500(0.002)}) = 6.32 \text{ V}$$

For $4 \leq t < 6$ ms,

$$v_C(t) = v_S - (v_S - 6.32)e^{-(t-0.004)/RC} = 10 - (10 - 6.32)e^{-500(t-0.004)} \quad \text{V}$$

And for $6 \leq t < 8$ ms, D again blocks and the capacitor voltage remains at

$$v_C(6\text{ ms}) = 10 - (10 - 6.32)e^{-500(0.002)} = 8.654 \text{ V}$$

The waveforms of v_S and v_C are sketched in Fig. 2-25.

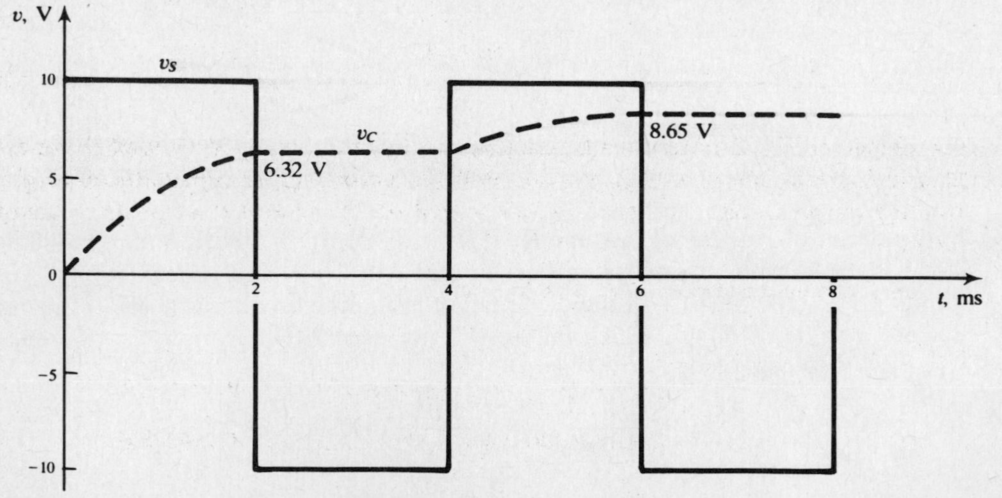

Fig. 2-25

2.13 The circuit of Fig. 2-26(*a*) is an "inexpensive" voltage regulator; all the diodes are identical and have the characteristic of Fig. 2-21(*b*). Find the regulation of v_o when V_b increases from its nominal value of 4 V to the value 6 V. Take $R = 2\ \text{k}\Omega$.

We determined in Problem 2.7 that each diode can be modeled as a battery, $V_F = 0.5$ V, and a resistor, $R_F = 50\ \Omega$, in series. Combining the diode strings between points *a* and *b* and between points *b* and *c* gives the circuit of Fig. 2-26(*b*), where

$$V_{F1} = 2V_F = 1 \text{ V} \qquad V_{F2} = 4V_F = 2 \text{ V} \qquad R_{F1} = 2R_F = 100\ \Omega \qquad R_{F2} = 4R_F = 200\ \Omega$$

By KVL,
$$I_b = \frac{V_b - V_{F1} - V_{F2}}{R + R_{F1} + R_{F2}}$$

whence
$$V_o = V_{F2} + I_b R_{F2} = V_{F2} + \frac{(V_b - V_{F1} - V_{F2})R_{F2}}{R + R_{F1} + R_{F2}}$$

For $V_{b1} = 4$ V and $V_{b2} = 6$ V,

$$V_{o1} = 2 + \frac{(4-1-2)(200)}{2000+100+200} = 2.09 \text{ V} \qquad V_{o2} = 2 + \frac{(6-1-2)(200)}{2000+100+200} = 2.26 \text{ V}$$

and (*2.6*) gives

$$\text{Reg} = \frac{V_{o2} - V_{o1}}{V_{o1}}(100\%) = 8.1\%$$

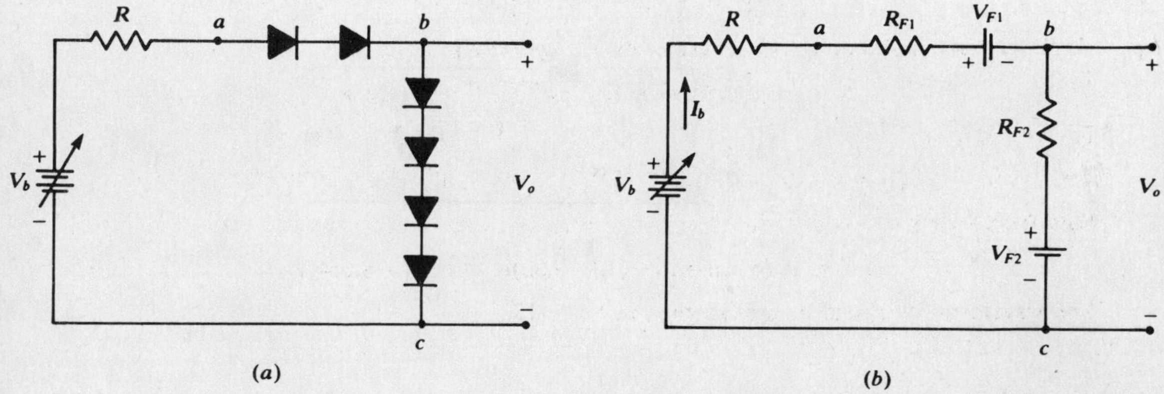

Fig. 2-26

2.14 The circuit of Fig. 2-17(*a*) is to be used as a dc power supply for a load R_L that varies from 10 Ω to 1 kΩ; v_S is a 10-V square wave. Find the percentage change in the average value of v_L over the range of load variation, and comment on the quality of regulation exhibited by this circuit.

Let T denote the period of v_S. For $R_L = 10\ \Omega$,

$$v_L = \begin{cases} \dfrac{R_L}{R_L + R_S} v_S = \dfrac{10}{10+10} 10 = 5 \text{ V} & 0 \le t < T/2 \\ 0 \text{ (diode blocks)} & T/2 \le t < T \end{cases}$$

and so
$$V_{Lo} = \frac{5(T/2) + 0(T/2)}{T} = 2.5 \text{ V}$$

For $R_L = 1\ \text{k}\Omega$,

$$v_L = \begin{cases} \dfrac{R_L}{R_L + R_S} v_S = \dfrac{1000}{1010} 10 = 9.9 \text{ V} & 0 \le t < T/2 \\ 0 \text{ (diode blocks)} & T/2 \le t < T \end{cases}$$

and so
$$V_{Lo} = \frac{9.9(T/2) + 0}{T} = 4.95 \text{ V}$$

Then, by (*2.6*) and using $R_L = 10\ \Omega$ as full load, we have

$$\text{Reg} = \frac{4.95 - 2.5}{2.5}(100\%) = 98\%$$

This large value of regulation is prohibitive for most applications. Either another circuit or a filter network would be necessary to make this power supply useful.

2.15 The circuit of Fig. 2-27 adds a dc level (a bias voltage) to a signal whose average value is zero. If v_S is a 10-V square wave of period T, $R_L = R_1 = 10\ \Omega$, and the diode is ideal, find the average value of v_L.

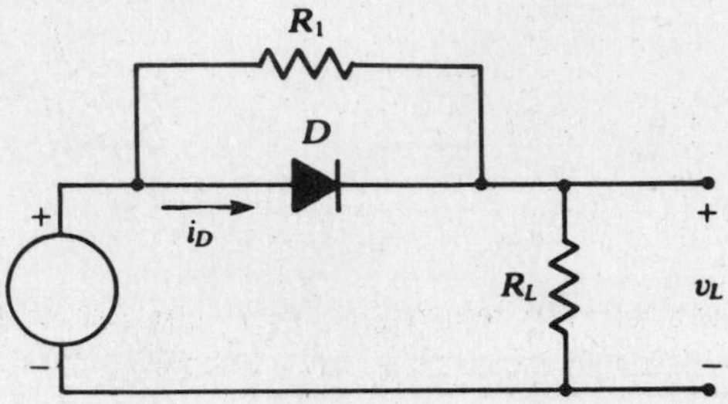

Fig. 2-27

For $v_L > 0$, D is forward-biased and $v_L = v_S = 10$ V. For $v_L < 0$, D is reverse-biased and

$$v_L = \frac{R_L}{R_L + R_1} v_S = \frac{10}{10 + 10}(-10) = -5\ \text{V}$$

Thus,
$$V_{L0} = \frac{10(T/2) + (-5)(T/2)}{T} = 2.5\ \text{V}$$

For some symmetrical input signals, this type of circuit could destroy the symmetry of the input.

2.16 Size the filter capacitor in the rectifier circuit of Fig. 2-13 so that the ripple voltage is approximately 5 percent of the average value of the output voltage. The diode is ideal, $R_L = 1\ \text{k}\Omega$, and $v_S = 90 \sin 2000t$ V. Calculate the average value of v_L for this filter.

With $F_r = 0.05$, (*2.10*) gives

$$C \approx \frac{1}{fR_L(0.05)} = \frac{1}{(2000/2\pi)(1 \times 10^3)(0.05)} = 62.83\ \mu\text{F}$$

Then, using the approximations that led to (*2.10*), we have

$$V_{L0} = V_{Sm} - \frac{1}{2}\Delta v_L = V_{Sm} - \frac{V_{Sm}}{2fR_LC} \approx V_{Sm}\left(1 - \frac{0.05}{2}\right) = (90)(0.975) = 87.75\ \text{V}$$

2.17 In the positive clipping circuit of Fig. 2-14(*a*), the diode is ideal and v_i is a 10-V triangular wave with period T. Sketch one cycle of the output voltage v_o if $V_b = 6$ V.

The diode blocks (acts as an open circuit) for $v_i < 6$ V, giving $v_o = v_i$. For $v_i \geq 6$ V, the diode is in forward conduction, clipping v_i to effect $v_o = 6$ V. The resulting output voltage waveform is sketched in Fig. 2-28.

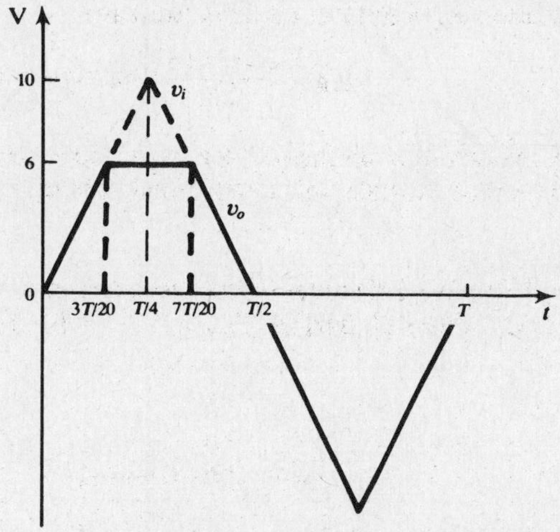

Fig. 2-28

2.18 Draw a transfer characteristic relating v_o to v_i for the positive clipping network of Problem 2.17. Also, sketch one cycle of the output waveform if $v_i = 10 \sin \omega t$ V.

The diode blocks for $v_i < 6$ V and conducts for $v_i \geq 6$ V. Thus, $v_o = v_i$ for $v_i < 6$ V, and $v_o = 6$ V for $v_i \geq 6$ V. The transfer characteristic is displayed in Fig. 2-29(*a*). For the given input signal, the output is a sine wave with the positive peak clipped at 6 V, as shown in Fig. 2-29(*b*).

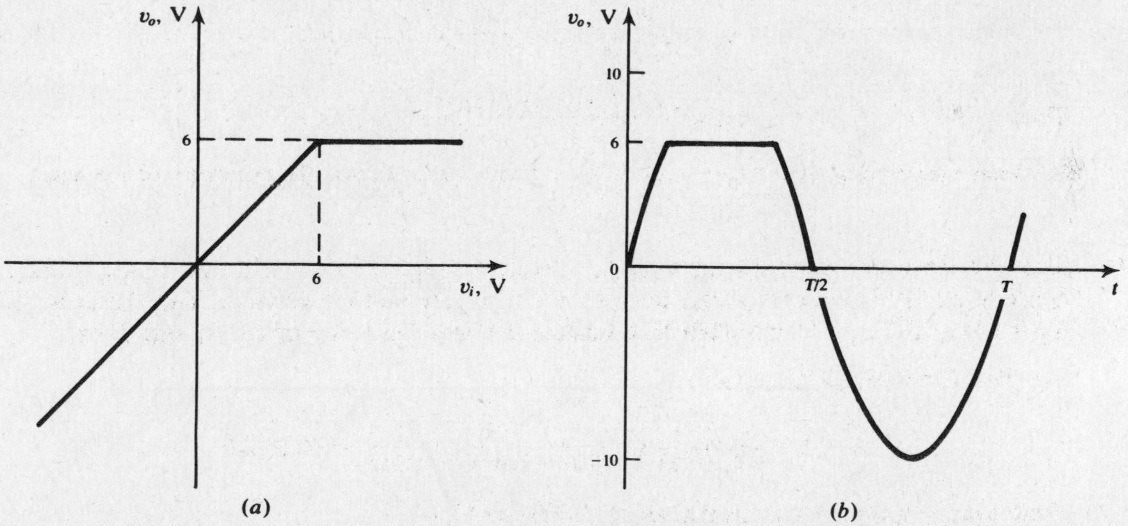

Fig. 2-29

2.19 Reverse the diode in Fig. 2-14(*a*) to create a negative clipping network. (*a*) Let $V_b = 6$ V, and draw the network transfer characteristic. (*b*) Sketch one cycle of the output waveform if $v_S = 10 \sin \omega t$ V.

(*a*) The diode conducts for $v_i \leq 6$ V and blocks for $v_i > 6$ V. Consequently, $v_o = v_i$ for $v_i > 6$ V, and $v_o = 6$ V for $v_i \leq 6$ V. The transfer characteristic is drawn in Fig. 2-30(*a*).

(*b*) With negative clipping, the output is made up of the positive peaks of $10 \sin \omega t$ above 6 V and is 6 V otherwise. Figure 2-30(*b*) displays the output waveform.

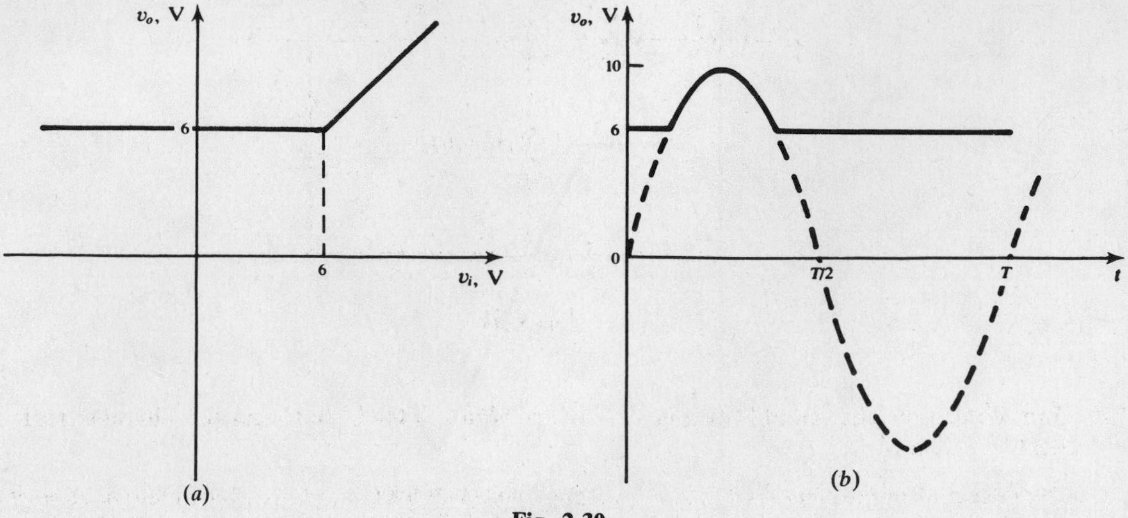

Fig. 2-30

2.20 A signal, $v_i = 10 \sin \omega t$ V, is applied to the negative clamping circuit of Fig. 2-15(*b*). Treating the diode as ideal, sketch the output waveform for $1\frac{1}{2}$ cycles of v_i. The capacitor is initially uncharged.

For $0 \le t \le T/4$, the diode is forward-biased, giving $v_o = 0$ as the capacitor charges to $v_C = +10$ V. For $t > T/4$, $v_o \le 0$, and thus the diode remains in the blocking mode, resulting in

$$v_o = -v_C + v_i = -10 + v_i = -10(1 - \sin \omega t) \qquad \text{V}$$

The output waveform is sketched in Fig. 2-31.

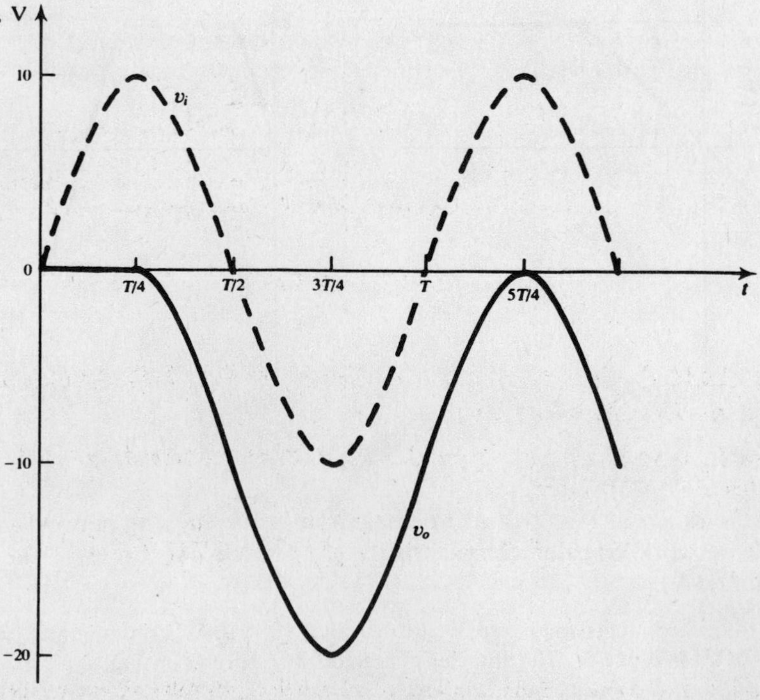

Fig. 2-31

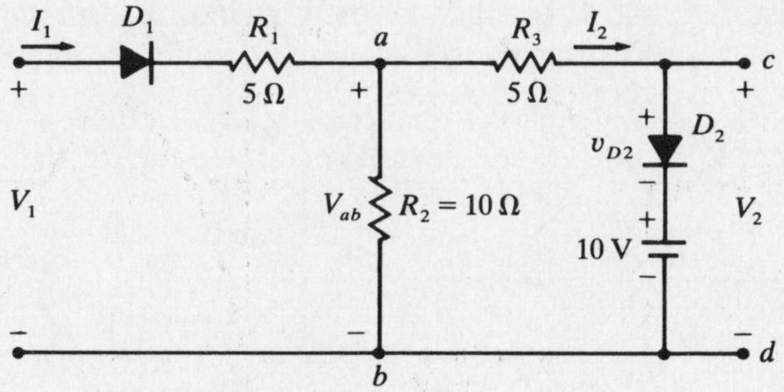

Fig. 2-32

2.21 The diodes in the circuit of Fig. 2-32 are ideal. Sketch the transfer characteristic for $-20\ \text{V} \leq V_1 \leq 20$ V.

Inspection of the circuit shows that I_2 can have no component due to the 10-V battery because of the one-way conduction property of D_2. Therefore, D_1 is "off" for $V_1 < 0$; then $v_{D2} = -10$ V and $V_2 = 0$.

Now D_1 is "on" if $V_1 \geq 0$; however, D_2 is "off" for $V_2 < 10$. The onset of conduction for D_2 occurs when $V_{ab} = 10$ V with $I_2 = 0$, or when, by voltage division,

$$V_{ab} = V_2 = 10 = \frac{R_2}{R_1 + R_2} V_1$$

Hence,

$$V_1 = \frac{R_1 + R_2}{R_2} 10 = \frac{5 + 10}{10} 10 = 15 \text{ V} \qquad (1)$$

Thus, if $V_1 \geq 15$ V, D_2 is "on" and $V_2 = 10$ V. But, for $0 \leq V_1 < 15$ V, D_2 is "off," $I_2 = 0$, and V_2 is given as a function of V_1 by (*1*). Figure 2-33 shows the composite result.

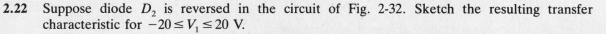

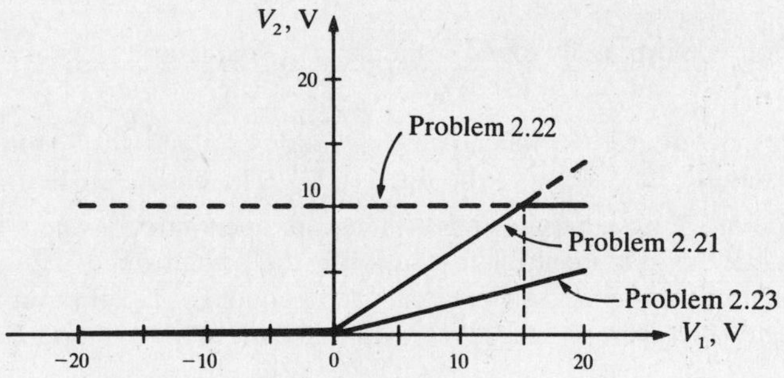

Fig. 2-33

2.22 Suppose diode D_2 is reversed in the circuit of Fig. 2-32. Sketch the resulting transfer characteristic for $-20 \leq V_1 \leq 20$ V.

Diode D_2 is now "on" and $V_2 = 10$ V until V_1 increases enough so that $V_{ab} = 10$ V, at which point $I_2 = 0$. That is, $V_2 = 10$ V until

$$V_2 = V_{ab} = 10 = \frac{R_2}{R_1 + R_2} V_1 = \frac{10}{5 + 10} V_1 = \frac{2}{3} V_1 \qquad (1)$$

or until

$$V_1 = \tfrac{3}{2} V_2 = 15 \text{ V}$$

For $V_1 > 15$ V, $I_2 = 0$ and (*1*) remains valid. The resulting transfer characteristic is shown dashed in Fig. 2-33.

2.23 Suppose a resistor $R_4 = 5\ \Omega$ is added across terminals c,d of the circuit of Fig. 2-32. Describe the changes that result in the transfer characteristic of Problem 2.21.

There is no change in the transfer characteristic for $V_1 \le 0$. However, D_2 remains "off" until $V_1 > 0$ increases to where $V_2 = 10$ V. At the onset of conduction for D_2, the current through D_2 is zero; thus,

$$I_1 = \frac{V_1}{R_1 + R_2 \| (R_3 + R_4)} = \frac{V_1}{10} \qquad \text{and} \qquad I_2 = \frac{R_2}{R_2 + R_3 + R_4} I_1 = \frac{I_1}{2}$$

Hence, by Ohm's law,

$$V_2 = I_2 R_4 = \frac{I_1 R_4}{2} = \frac{V_1 R_4}{20} = \frac{V_1}{4}$$

Thus, $V_1 = 40$ V when $V_2 = 10$ V, and it is apparent that the breakpoint of Problem 2.21 at $V_1 = 15$ V has moved to $V_1 = 40$ V. The transfer characteristic for $-20 \le V_1 \le 20$ is sketched in Fig. 2-33.

2.24 Sketch the *i-v* input characteristic of the network of Fig. 2-34(*a*) when (*a*) the switch is open and (*b*) the switch is closed.

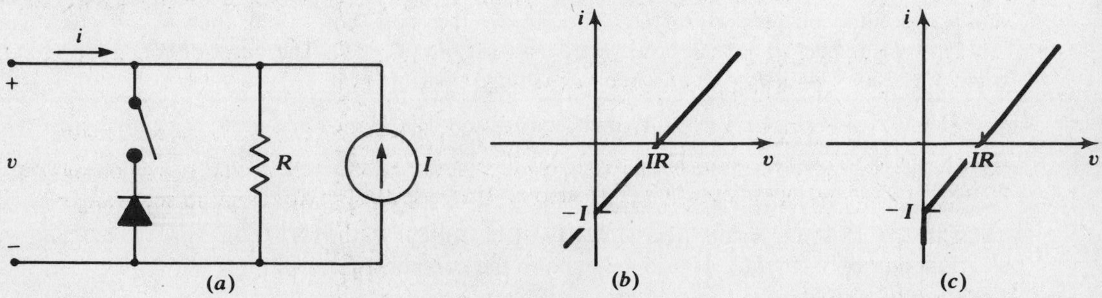

Fig. 2-34

The solution is more easily found if the current source and resistor are replaced with the Thévenin equivalents $V_{Th} = IR$ and $R_{Th} = R$.

(*a*) KVL gives $v = iR_{Th} + IR$, which is the equation of a straight line intersecting the i axis at $-I$ and the v axis at IR. The slope of the line is $1/R$. The characteristic is sketched in Fig. 2-34(*b*).

(*b*) The diode is reverse-biased and acts as an open circuit when $v > 0$. It follows that the *i-v* characteristic here is identical to that with the switch open if $v > 0$. But if $v \le 0$, the diode is forward-biased, acting as a short circuit. Consequently, v can never reach the negative values, and the current i can increase negatively without limit. The corresponding *i-v* plot is sketched in Fig. 2-34(*c*).

2.25 In the small-signal circuit of Fig. 2-35, the capacitor models the diode diffusion capacitance, so that $C = C_d = 0.02\ \mu$F, and v_{th} is known to be of frequency $\omega = 10^7$ rad/s. Also, $r_d = 2.5\ \Omega$ and $Z_{Th} = R_{Th} = 10\ \Omega$. Find the phase angle (*a*) between i_d and v_d and (*b*) between v_d and v_{th}.

(*a*) The diffusion capacitance produces a reactance

$$x_d = \frac{1}{\omega C_d} = \frac{1}{(10^7)(0.02 \times 10^{-6})} = 5\ \Omega$$

so that

$$Z_d = r_d \| (-jx_d) = \frac{(2.5)(5\angle -90^\circ)}{2.5 - j5} = 2.236\angle -26.57^\circ = 2 - j1\ \Omega$$

Thus, i_d leads v_d by a phase angle of 26.57°.

(*b*) Let Z_{eq} be the impedance looking to the right from v_{th}; then

$$Z_{eq} = Z_{Th} + Z_d = 10 + (2 - j1) = 12 - j1 = 12.04\angle -4.76^\circ\ \Omega$$

Hence, v_{th} leads v_d by an angle of $26.57^\circ - 4.76^\circ = 21.81^\circ$.

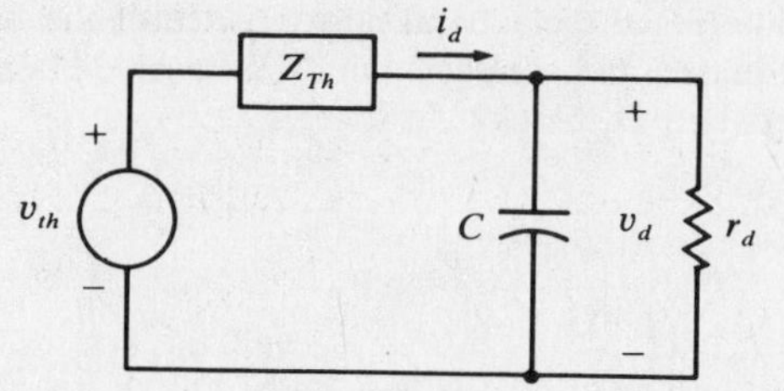

Fig. 2-35

2.26 Using ideal diodes, resistors, and batteries, synthesize a function-generator circuit that will yield the *i-v* characteristic of Fig. 2-36(*a*).

Since the *i-v* characteristic has two breakpoints, two diodes are required. Both diodes must be oriented so that no current flows for $v < -5$ V. Further, one diode must move into forward bias at the first breakpoint, $v = -5$ V, and the second diode must begin conduction at $v = +10$ V. Note also that the slope of the *i-v* plot is the reciprocal of the Thévenin equivalent resistance of the active portion of the network.

The circuit of Fig. 2-36(*b*) will produce the given *i-v* plot if $R_1 = 6$ kΩ, $R_2 = 3$ kΩ, $V_1 = 5$ V, and $V_2 = 10$ V. These values are arrived at as follows:

1. If $v < -5$ V, both v_{D1} and v_{D2} are negative, both diodes block, and no current flows.
2. If $-5 \le v < 10$ V, D_1 is forward-biased and acts as a short circuit, whereas v_{D2} is negative, causing D_2 to act as an open circuit. R_1 is found as the reciprocal of the slope in that range:

$$R_1 = \frac{10 - (-5)}{0.0025} = 6\ \text{k}\Omega$$

3. If $v \ge 10$ V, both diodes are forward-biased,

$$R_{Th} = \frac{R_1 R_2}{R_1 + R_2} = \frac{\Delta v}{\Delta i} = \frac{20 - 10}{(7.5 - 2.5) \times 10^{-3}} = 2\ \text{k}\Omega$$

and

$$R_2 = \frac{R_1 R_{Th}}{R_1 - R_{Th}} = \frac{(6 \times 10^3)(2 \times 10^3)}{4 \times 10^3} = 3\ \text{k}\Omega$$

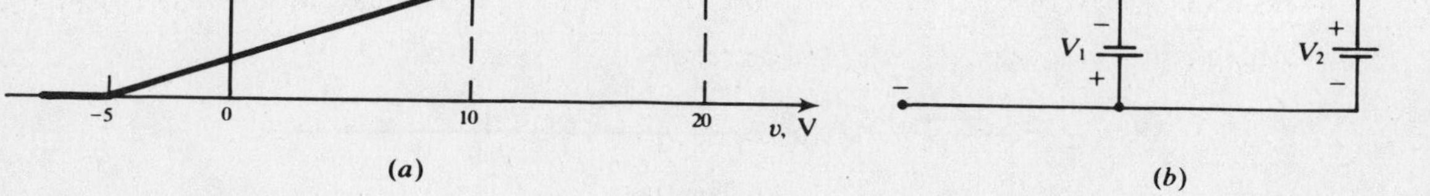

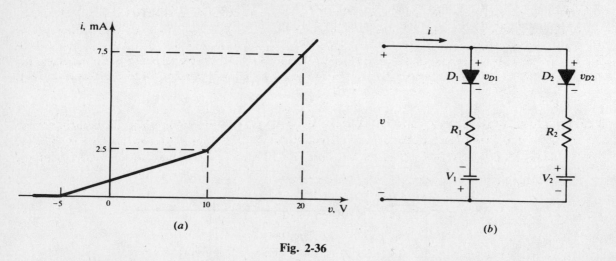

Fig. 2-36

2.27 Find v_L for the full-wave rectifier circuit of Fig. 2-37(*a*), treating the transformer and diodes as ideal.

The two voltages labeled v_2 in Fig. 2-37(*a*) are identical in magnitude and phase. The ideal transformer and the voltage source v_S can therefore be replaced with two identical voltage sources, as in

Fig. 2-37(b), without altering the electrical performance of the balance of the network. When v_S/n is positive, D_1 is forward-biased and conducts but D_2 is reverse-biased and blocks. Conversely, when v_S/n is negative, D_2 conducts and D_1 blocks. In short,

$$i_{D1} = \begin{cases} \dfrac{v_S/n}{R_L} & \dfrac{v_S}{n} \geq 0 \\ 0 & \dfrac{v_s}{n} < 0 \end{cases} \quad \text{and} \quad i_{D2} = \begin{cases} 0 & \dfrac{v_S}{n} > 0 \\ -\dfrac{v_S/n}{R_L} & \dfrac{v_S}{n} \leq 0 \end{cases}$$

By KCL,
$$i_L = i_{D1} + i_{D2} = \frac{|v_S/n|}{R_L}$$

and so $v_L = R_L i_L = |v_S/n|$.

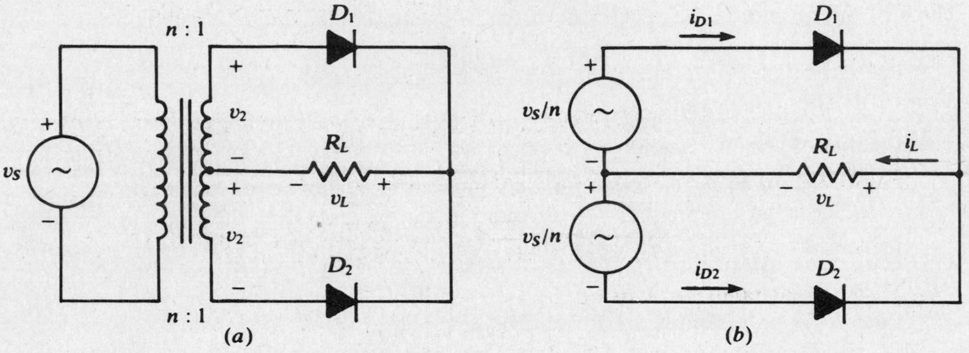

Fig. 2-37

2.28 The Zener diode in the voltage-regulator circuit of Fig. 2-38 has a constant reverse breakdown voltage $V_Z = 8.2$ V, for 75 mA $\leq i_Z \leq 1$ A. If $R_L = 9\ \Omega$, size R_S so that $v_L = V_Z$ is regulated to (maintained at) 8.2 V while V_b varies by ± 10 percent from its nominal value of 12 V.

By Ohm's law,

$$i_L = \frac{v_L}{R_L} = \frac{V_Z}{R_L} = \frac{8.2}{9} = 0.911 \text{ A}$$

Now an application of KVL gives

$$R_S = \frac{V_b - V_Z}{i_Z + i_L} \tag{1}$$

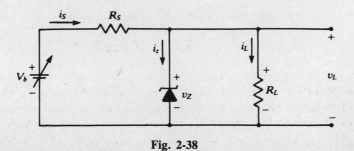

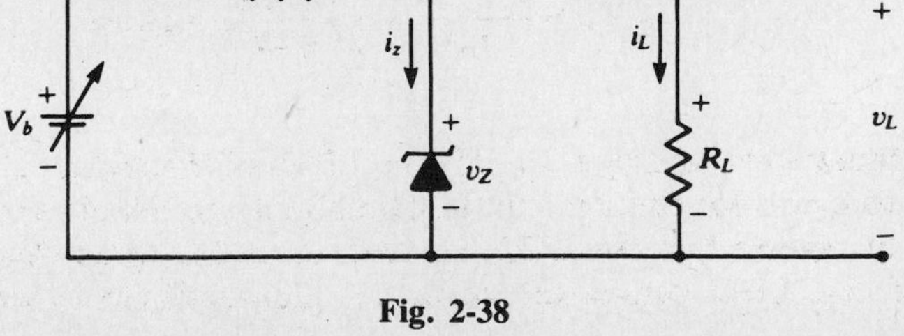

Fig. 2-38

and we use (*1*) to size R_S for maximum Zener current I_Z at the largest value of V_b:

$$R_S = \frac{(1.1)(12) - 8.2}{1 + 0.911} = 2.62\ \Omega$$

Now we check to see if $i_Z \geq 75$ mA at the lowest value of V_b:

$$i_Z = \frac{V_b - v_Z}{R_S} - i_L = \frac{(0.9)(12) - 8.2}{2.62} - 0.911 = 81.3 \text{ mA}$$

Since $i_Z > 75$ mA, $v_Z = V_Z = 8.2$ V and regulation is preserved.

2.29 A Zener diode has the specifications $V_Z = 5.2$ V and $P_{D\max} = 260$ mW. Assume $R_Z = 0$. (*a*) Find the maximum allowable current i_Z when the Zener diode is acting as a regulator. (*b*) If a single-loop circuit consists of an ideal 15-V dc source V_S, a variable resistor R, and the described Zener diode, find the range of values of R for which the Zener diode remains in constant reverse breakdown with no danger of failure.

(*a*)

$$i_{Z\max} = I_Z = \frac{P_{D\max}}{V_Z} + \frac{260 \times 10^{-3}}{5.2} = 50 \text{ mA}$$

(*b*) By KVL,

$$V_S = Ri_Z + V_Z \qquad \text{so that} \qquad R = \frac{V_s - V_Z}{i_Z}$$

From Section 2.9, we know that regulation is preserved if

$$R \leq \frac{V_S - V_Z}{0.1 I_{Z\max}} = \frac{15 - 5.2}{(0.1)(50 \times 10^{-3})} = 1.96 \text{ k}\Omega$$

Overcurrent failure is avoided if

$$R \geq \frac{V_S - V_Z}{I_{Z\max}} = \frac{15 - 5.2}{50 \times 10^{-3}} = 196\ \Omega$$

Thus, we need $196\ \Omega \leq R \leq 196$ kΩ.

2.30 A *light-emitting diode* (LED) has a greater forward voltage drop than does a common signal diode. A typical LED can be modeled as a constant forward voltage drop $v_D = 1.6$ V. Its *luminous intensity* I_v varies directly with forward current and is described by

$$I_v = 40 i_D \approx \text{millicandela (mcd)}$$

A series circuit consists of such an LED, a current-limiting resistor R, and a 5-V dc source V_S. Find the value of R such that the luminous intensity is 1 mcd.

By (*1*), we must have

$$i_D = \frac{I_v}{40} = \frac{1}{40} = 25 \text{ mA}$$

From KVL, we have

$$V_S = Ri_D + 1.6$$

so that

$$R = \frac{V_S - 1.6}{i_D} = \frac{5 - 1.6}{25 \times 10^{-3}} = 136\ \Omega$$

2.31 The reverse breakdown voltage V_R of the LED of Problem 2.30 is guaranteed by the manufacturer to be no lower than 3 V. Knowing that the 5-V dc source may be inadvertently applied so as to reverse-bias the LED, we wish to add a Zener diode to ensure that reverse breakdown of the LED can never occur. A Zener diode is available with $V_Z = 4.2$ V,

$I_Z = 30$ mA, and a forward drop of 0.6 V. Describe the proper connection of the Zener in the circuit to protect the LED, and find the value of the luminous intensity that will result if R is unchanged from Problem 2.31.

The Zener diode and LED should be connected in series so that the anode of one device connects to the cathode of the other. Then, even if the 5-V source is connected in reverse, the reverse voltage across the LED will be less than $5 - 4.2 = 0.8\ \text{V} < 3$ V. When the dc source is connected to forward-bias the LED, we will have

$$i_D = \frac{V_S - V_{FLED} - V_{FZ}}{R} = \frac{5 - 1.6 - 0.6}{136} = 20.6 \text{ mA}$$

so that

$$I_\nu = 40 i_D = (40)(20.6 \times 10^{-3}) = 0.824 \text{ mcd}$$

Supplementary Problems

2.32 A Si diode has a saturation current $I_o = 10$ nA at $T = 300°$K. (*a*) Find the forward current i_D if the forward drop v_D is 0.5 V. (*b*) This diode is rated for a maximum current of 5 A. What is its junction temperature at rated current if the forward drop is 0.7 V *Ans.* (*a*) 2.47 A; (*b*) 405.4°K

2.33 Solve Problem 2.1 for a Si diode. *Ans.* $v_D > 0.2372$ V

2.34 Laboratory data for a Si diode described by (*2.1*) show that $i_D = 2$ mA when $v_D = 0.6$ V, and $i_D = 10$ mA for $v_D = 0.7$ V. Find (*a*) the temperature for which the data were taken, and (*b*) the reverse saturation current. *Ans.* (*a*) 87.19°C; (*b*) 2.397 μA

2.35 For what voltage v_D will the reverse current of a Ge diode that is described by (*2.1*) reach 99 percent of its saturation value at a temperature of 300°K? *Ans.* $v_D = -0.1191$ V

2.36 Find the increase in temperature ΔT necessary to increase the reverse saturation current of a diode by a factor of 100. *Ans.* 66.4°C

2.37 The diode of Problem 2.32 is operating in a circuit where it has dynamic resistance $r_d = 100\ \Omega$. What must be the quiescent conditions? *Ans.* $V_{DQ} = 0.263$ V, $I_{DQ} = 0.259$ mA

2.38 The diode of Problem 2.32 has a forward current $i_D = 2 + 0.004 \sin \omega t$ mA. Find the total voltage, $v_D = V_{DQ} + v_d$, across the diode. *Ans.* $v_D = 339.5 + 0.0207 \sin \omega t$ mV

2.39 Find the power dissipated in the load resistor $R_L = 100\ \Omega$ of the circuit of Fig. 2-17(*a*) if the diode is ideal and $v_S = 10 \sin \omega t$ V. *Ans.* 206.6 mW

2.40 The logic AND gate of Fig. 2-39(*a*) has trains of input pulses arriving at the gate inputs, as indicated by Fig. 2-39(*b*). Signal v_2 is erratic, dropping below nominal logic level on occasion. Determine v_o. *Ans.* 10 V for $1 \le t \le 2$ ms, 5 V for $4 \le t \le 5$ ms, zero otherwise

2.41 The logic AND gate of Fig. 2-39(*a*) is to be used to generate a crude pulse train by letting $v_1 = 10 \sin \omega t$ V and $v_2 = 5$ V. Determine (*a*) the amplitude and (*b*) the period of the pulse train appearing as v_o. *Ans.* (*a*) 5 V; (*b*) $2\pi/\omega$

2.42 In the circuit of Fig. 2-24, v_S is a 10-V square wave with a 4-ms period. The diode is nonideal, with the characteristic of Fig. 2-21(*b*). If the capacitor is initially uncharged, determine v_C for the first cycle of v_s. *Ans.* $9.5(1 - e^{-333.3t})$ V for $0 \le t < 2$ ms and 4.62 V for $2 \le t < 4$ ms

2.43 The forward voltage across the diode of Problem 2.33 is $v_D = 0.3 + 0.060 \cos t$ V. Find the ac component of the diode current i_d. *Ans.* 2.52 cos t mA

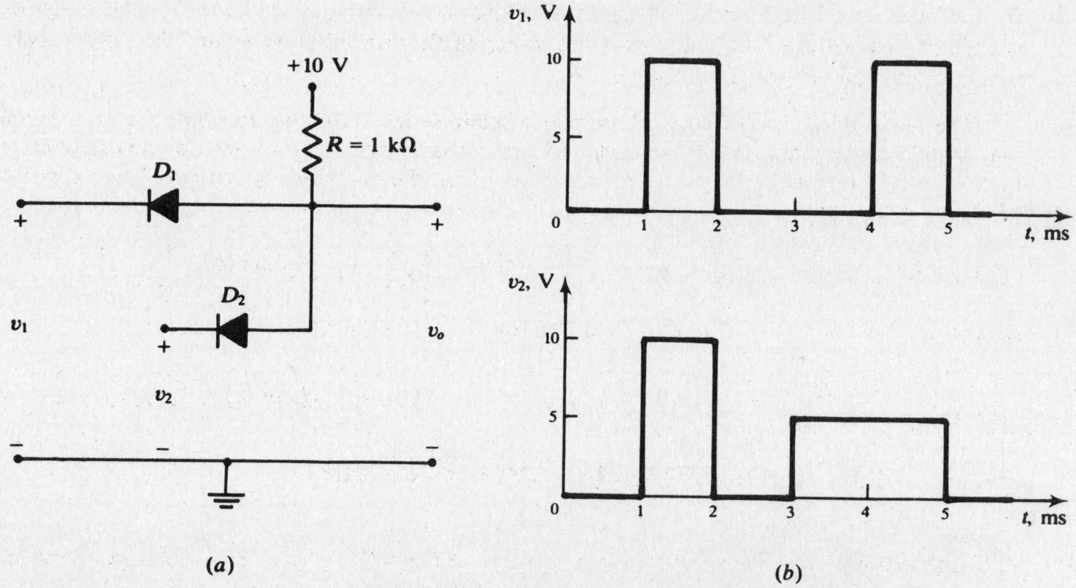

Fig. 2-39

2.44 The circuit of Fig. 2-40(*a*) is a voltage-doubler circuit, sometimes used as a low-level power supply when the load R_L is reasonably constant. It is called a "doubler" because the steady-state peak value of v_L is twice the peak value of the sinusoidal source voltage. Figure 2-40(*b*) is a sketch of the steady-state output voltage for $v_S = 10 \cos \omega t$ V. Assume ideal diodes, $\omega = 377$ rad/s, $C_1 = 200\ \mu$F, $C_2 = 10\ \mu$F, and $R_L = 1$ kΩ. (*a*) Solve, by trial and error, the transcendental equation for the decay time t_d. (*b*) Calculate the peak-to-peak value of the ripple voltage. *Ans.* (*a*) 15.2 ms; (*b*) 1.46 V

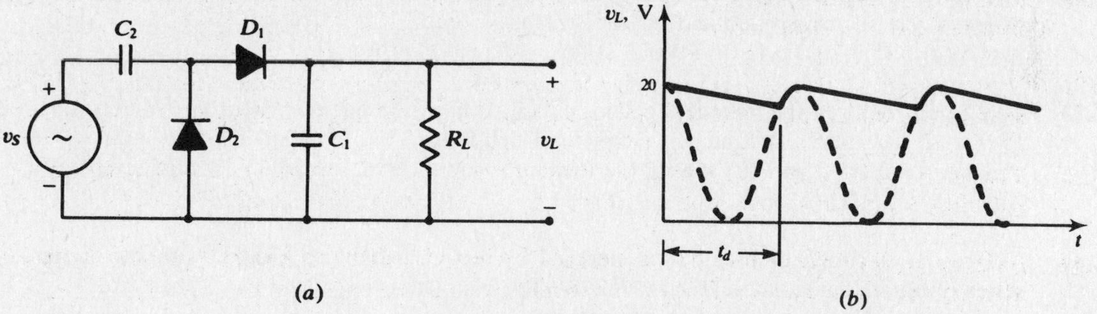

Fig. 2-40

2.45 Find the diode current during one capacitor-charging cycle in the rectifier circuit of Fig. 2-13(*a*) if $C = 47\ \mu$F, $R_L = 1$ kΩ, and $v_S = 90 \cos 2000t$ V. (*Hint:* The approximate ripple formula cannot be used, as it implicitly assumes zero capacitor charging time. Instead, solve for capacitor current and load current, and add.) *Ans.* $i_D = -8.49 \sin (2000t - 0.6°)$ A for $2.966\text{ ms} \le t < 3.142$ ms

2.46 In the circuit of Fig. 2-27, $R_1 = R_L = 10\ \Omega$. If the diode is ideal and $v_S = 10 \sin \omega t$ V, find the average value of the load voltage v_L. *Ans.* 3.18 V

2.47 Rework Problem 2.20 with the diode of Fig. 2-15(*b*) reversed and all else unchanged. (The circuit is now a positive clamping circuit.)
Ans. $v_o = 10 \sin \omega t$ V for $0 \le t < T/2$, 0 for $T/2 \le t < 3T/4$, and $10(1 - \sin \omega t)$ V for $t \ge 3T/4$

2.48 Four diodes are utilized for the full-wave bridge of Fig. 2-41. Assuming that the diodes are ideal and that $v_S = V_m \sin \omega t$, (*a*) find the output voltage v_L and (*b*) find the average value of v_L.
Ans. (*a*) $v_L = V_m|\sin \omega t|$ V; (*b*) $V_{L0} = 2V_m/\pi$

2.49 A shunt filter capacitor (see Example 2.11) is added to the full-wave rectifier of Problem 2.48. Show that the ripple factor is given by $F_r = 2/(4fR_LC - 1) \approx 1/2fR_LC$.

2.50 Devise a peak rectifier circuit by connecting a 470-μF filter capacitor across points a, b in the circuit of Fig. 2-41. $R_L = 2$ kΩ, and $v_S = \sin \omega t$ V. (*a*) Find the magnitude (peak to peak) of the ripple voltage, and (*b*) calculate the average value of v_L. *Ans.* (*a*) 3.34 V; (*b*) 98.33 V

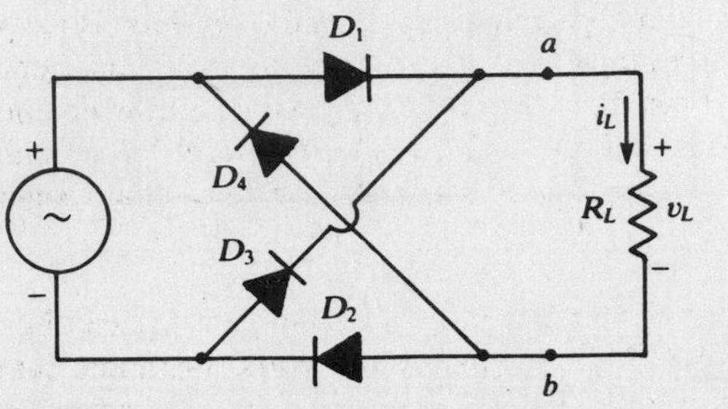

Fig. 2-41

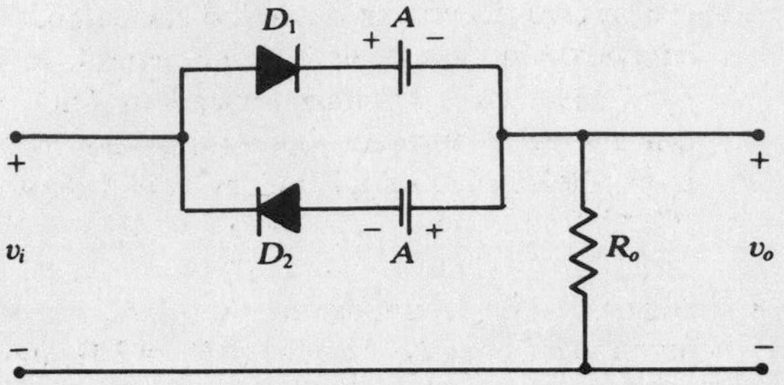

Fig. 2-42

2.51 The *level-discriminator circuit* (Fig. 2-42) has an output of zero, regardless of the polarity of the input signal, until the input reaches a threshold value. Above the threshold value, the output duplicates the input. Such a circuit can sometimes be used to eliminate the effects of low-level noise at the expense of slight distortion. Relate v_o to v_i for the circuit.
Ans. $v_o = v_i(1 - A/|v_i|)$ for $|v_i| > A$, and 0 for $|v_i| \le A$

2.52 The diode of Fig. 2-34(*a*) is reversed, but all else remains the same. Write an equation relating v and i when (*a*) the switch is open and (*b*) the switch is closed.
Ans. (*a*) $v = R(i + I)$; (*b*) $v = R(i + I)$ for $i < I$, and $v = 0$ for $i \ge I$

2.53 The Zener diode in the voltage-regulator circuit of Fig. 2-38 has $v_Z = V_Z = 18.6$ V at a minimum i_Z of 15 mA. If $V_b = 24 \pm 3$ V and R_L varies from 250 Ω to 2 kΩ, (*a*) find the maximum value of R_S to maintain regulation and (*b*) specify the minimum power rating of the Zener diode.
Ans. (*a*) 26.8 Ω; (*b*) 4.65 W

2.54 The regulator circuit of Fig. 2-38 is modified by replacing the Zener diode with two Zener diodes in series to obtain a regulated voltage of 20 V. The characteristics of the two Zeners are:

Zener 1: $V_Z = 9.2$ V for $15 \le i_Z \le 300$ mA
Zener 2: $V_Z = 10.8$ V for $12 \le i_Z \le 240$ mA

(*a*) if i_L varies from 10 mA to 90 mA and V_b varies from 22 V to 26 V, size R_S so that regulation is preserved. (*b*) Will either Zener exceed its rated current?
Ans. (*a*) 19.6 Ω; (*b*) for $V_b = 26$ V, $i_{Z1} = i_{Z2} = 296$ mA, which exceeds the rating of Zener 2

2.55 The two Zener diodes of Fig. 2-43 have negligible forward drops, and both regulate at constant V_Z for $50 \text{ mA} \le i_Z \le 500$ mA. If $R_1 = R_L = 10\ \Omega$, $V_{Z1} = 8$ V, and $V_{Z2} = 5$ V, find the average value of load voltage when v_i is a 10-V square wave. *Ans.* 0.75 V

2.56 The Zener diode of Problem 2.29 is used in a simple series circuit consisting of a variable dc voltage source V_S, the Zener diode, and a current-limiting resistor $R = 1$ kΩ. (*a*) Find the allowable range of V_S for which the Zener diode is safe and regulation is preserved. (*b*) Find an expression for the power dissipated by the Zener diode. *Ans.* (*a*) $10.2 \text{ V} \le V_S \le 55.2$ V; (*b*) $P_D = V_Z(V_S - V_Z)/R$

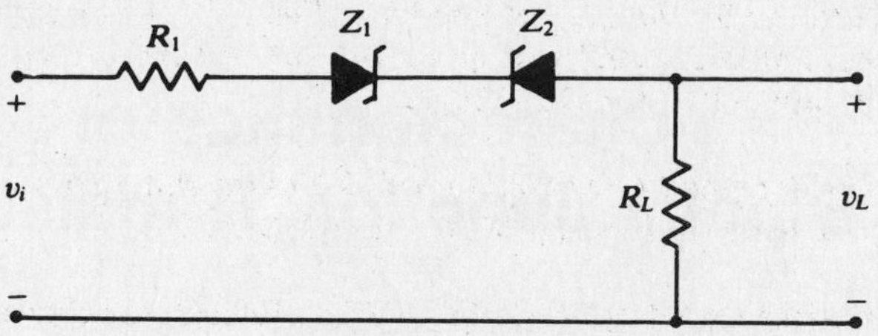

Fig. 2-43

2.57 The *varactor diode* is designed to operate reverse-biased and is manufactured by a process that increases the voltage-dependent depletion capacitance or junction capacitance C_j. A varactor diode is frequently connected in parallel with an inductor L to form a resonant circuit for which the resonant frequency, $f_R = 1/2\pi\sqrt{LC_j}$, is voltage-dependent. Such a circuit can form the basis of a *frequency modulation* (FM) transmitter. A varactor diode whose depletion capacitance is $C_j = 10^{-11}/(1 - 0.75v_D)^{1/2}$ F is connected in parallel with a 0.8-μH inductor; find the value of v_D required to establish resonance at a frequency of 100 MHz. *Ans.* $v_D = -11.966$ V

2.58 An LED with luminous intensity described by (*1*) of Problem 2.30 is modeled by the piecewise-linear function of Fig. 2-10(*b*), with $R_F = 3\ \Omega$ and $V_F = 1.5$ V. Find the maximum and minimum luminous intensities that result if the LED is used in a series circuit consisting of the LED, a current-limiting resistor $R = 125\ \Omega$, and a source $v_S = 5 + 1.13 \sin 0.1t$ V. (*Note:* Since the period of v_S exceeds 1 minute, it is logical to assume that luminous intensity follows i_D without the necessity to consider the physics of the light-emitting process.) *Ans.* $I_{v\max} = 1.798$ mcd, $I_{v\min} = 0.9204$ mcd

Chapter 3

Characteristics of Bipolar Junction Transistors

3.1 BJT CONSTRUCTION AND SYMBOLS

The *bipolar junction transistor* (BJT) is a three-element (*emitter*, *base*, and *collector*) device made up of alternating layers of *n*- and *p*-type semiconductor materials joined metallurgically. The transistor can be of *pnp* type (principal conduction by positive holes) or of *npn* type (principal conduction by negative electrons), as shown in Fig. 3-1 (where schematic symbols and positive current directions are also shown). The double-subscript notation is utilized in labeling terminal voltages, so that, for example, v_{BE} symbolizes the increase in potential from emitter terminal E to base terminal B. For reasons that will become apparent, terminal currents and voltages commonly consist of superimposed dc and ac components (usually sinusoidal signals). Table 3-1 presents the notation for terminal voltages and currents.

Table 3-1

Type of Value	Symbol		Examples
	Variable	Subscript	
total instantaneous	lowercase	uppercase	i_B, v_{BE}
dc	uppercase	uppercase	I_B, V_{BE}
quiescent-point	uppercase	uppercase plus Q	I_{BQ}, V_{BEQ}
ac instantaneous	lowercase	lowercase	i_b, v_{be}
rms	uppercase	lowercase	I_b, V_{be}
maximum (sinusoid)	uppercase	lowercase plus m	I_{bm}, V_{bem}

Example 3.1 In the *npn* transistor of Fig. 3-1(*a*), 10^8 holes/μs move from the base to the emitter region while 10^{10} electrons/μs move from the emitter to the base region. An ammeter reads the base current as $i_B = 16\ \mu$A. Determine the emitter current i_E and the collector current i_C.

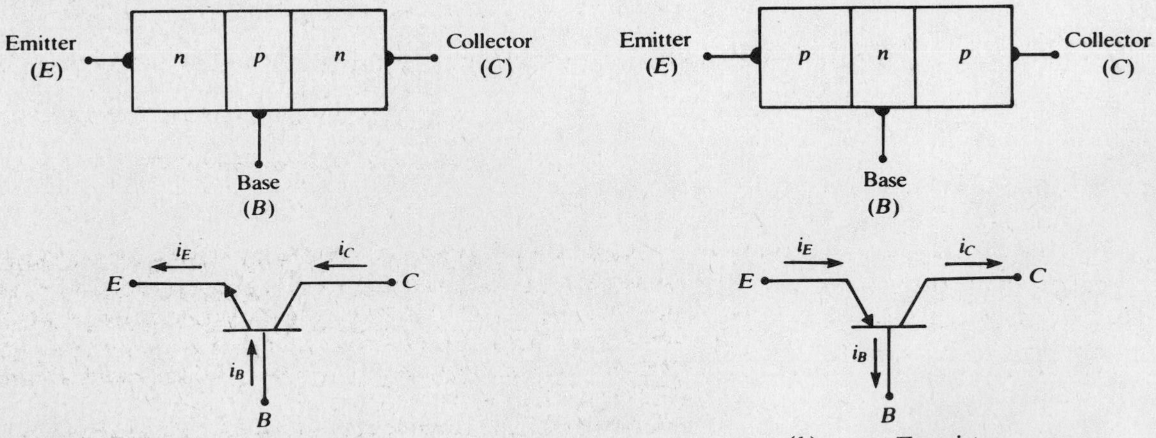

(*a*) *npn* Transistor (*b*) *pnp* Transistor

Fig. 3-1

The emitter current is found as the net rate of flow of positive charge into the emitter region:

$$i_E = (1.602 \times 10^{-19}\ \text{C/hole})(10^{14}\ \text{holes/s}) - (-1.602 \times 10^{-19}\ \text{C/electron})(10^{16}\ \text{electrons/s})$$
$$= 1.602 \times 10^{-5} + 1.602 \times 10^{-3} = 1.618\ \text{mA}$$

Further, by KCL,

$$i_C = i_E - i_B = 1.618 \times 10^{-3} - 16 \times 10^{-6} = 1.602\ \text{mA}$$

3.2 COMMON-BASE TERMINAL CHARACTERISTICS

The *common-base* (CB) connection is a two-port transistor arrangement in which the base shares a common point with the input and output terminals. The independent input variables are emitter current i_E and base-to-emitter voltage v_{EB}. The corresponding independent output variables are collector current i_C and base-to-collector voltage v_{CB}. Practical CB transistor analysis is based on two experimentally determined sets of curves:

1. *Input* or *transfer characteristics* relate i_E and v_{EB} (port input variables), with v_{CB} (port output variable) held constant. The method of laboratory measurement is indicated in Fig. 3-2(*a*), and the typical form of the resulting family of curves is depicted in Fig. 3-2(*b*).
2. *Output* or *collector characteristics* give i_C as a function of v_{CB} (port output variables) for constant values of i_E (port input variable), measured as in Fig. 3-2(*a*). Figure 3-2(*c*) shows the typical form of the resulting family of curves.

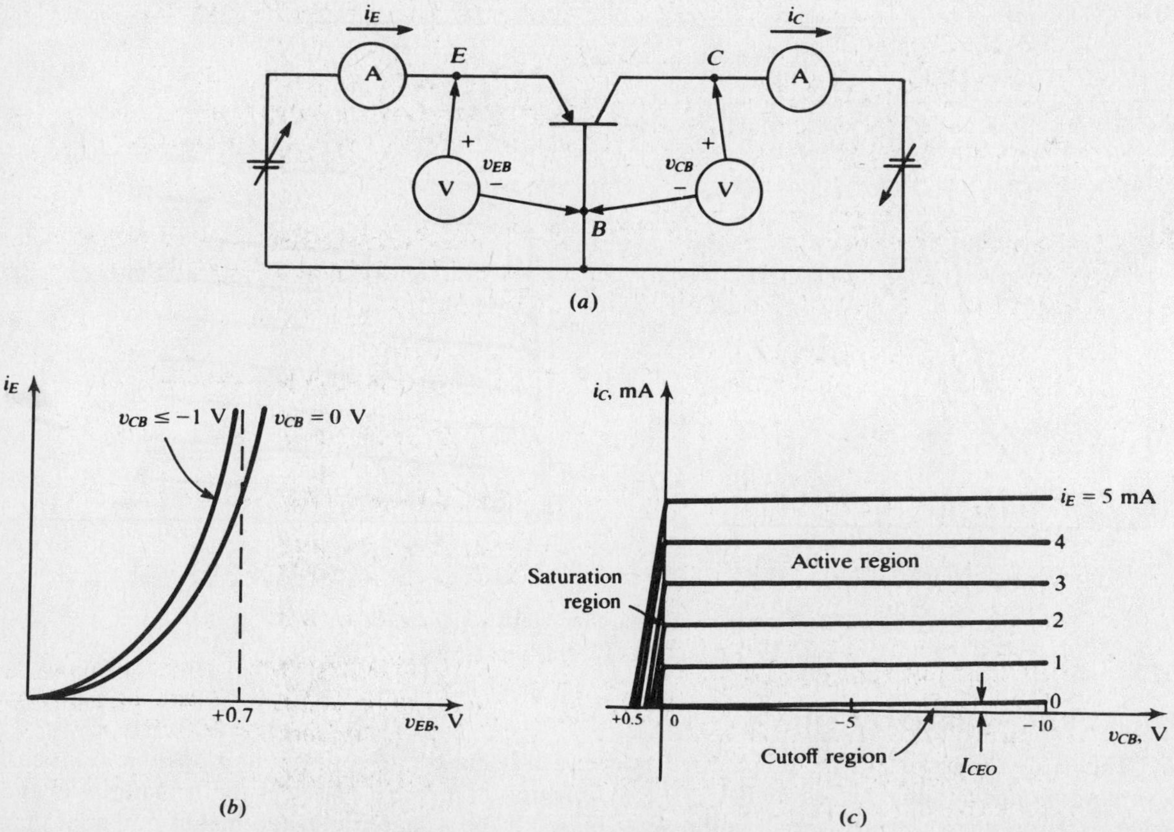

Fig. 3-2 Common-base characteristics (*pnp*, Si device)

3.3 COMMON-EMITTER TERMINAL CHARACTERISTICS

The *common-emitter* (CE) connection is a two-port transistor arrangement (widely used because of its high current amplification) in which the emitter shares a common point with the input and output terminals. The independent port input variables are base current i_B and emitter-to-base voltage v_{BE}, and the independent port output variables are collector current i_C and emitter-to-collector voltage v_{CE}. Like CB analysis, CE analysis is based on:

1. *Input* or *transfer characteristics* that relate the port input variables i_B and v_{BE}, with v_{CE} held constant. Figure 3-3(*a*) shows the measurement setup, and Fig. 3-3(*b*) the resulting input characteristics.
2. *Output* or *collector characteristics* that show the functional relationship between port output variables i_C and v_{CE} for constant i_B, measured as in Fig. 3-3(*a*). Typical collector characteristics are displayed in Fig. 3-3(*c*).

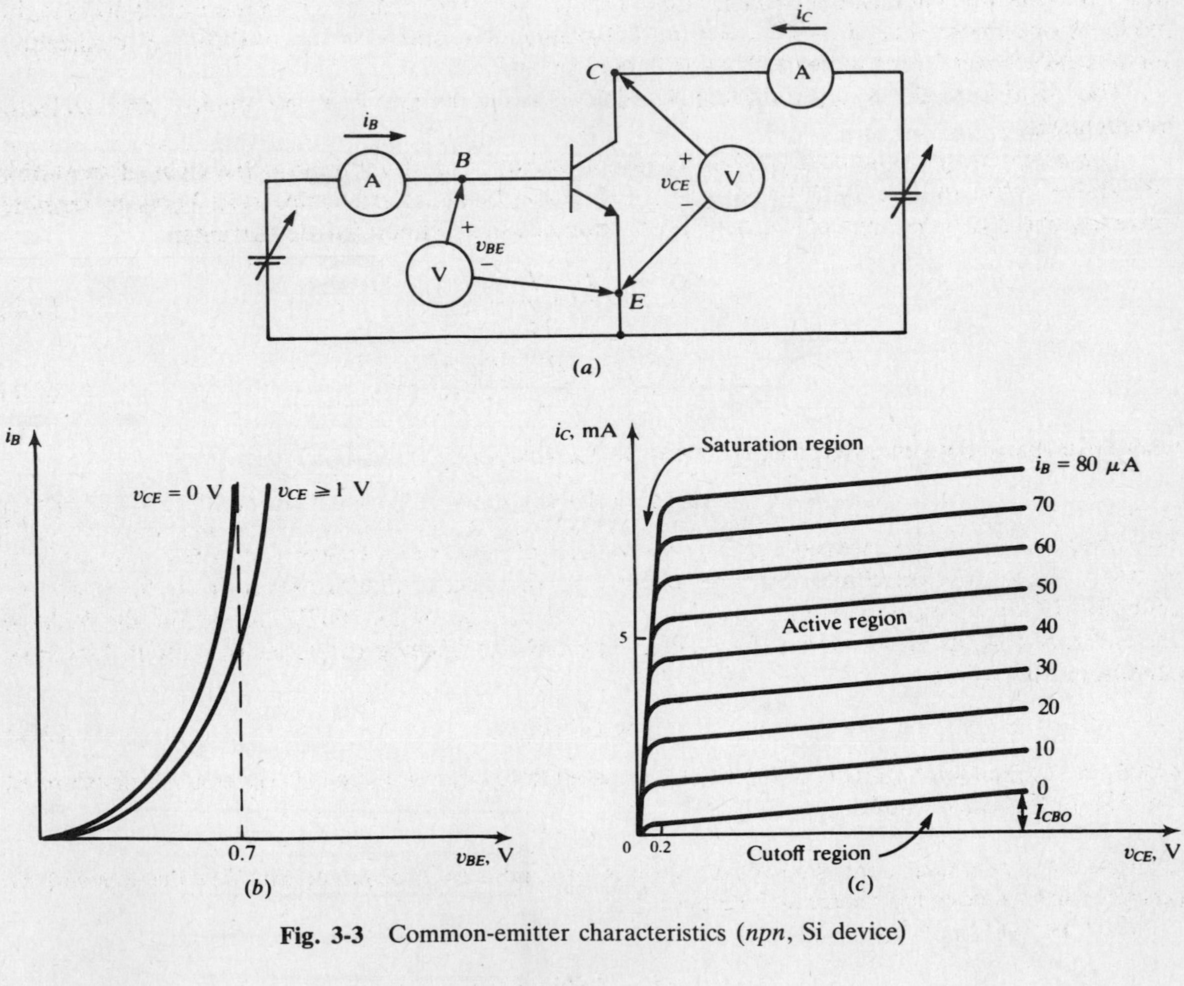

Fig. 3-3 Common-emitter characteristics (*npn*, Si device)

3.4 CURRENT RELATIONSHIPS

The two *pn* junctions of the BJT can be independently biased, to result in four possible transistor *operating modes* as summarized in Table 3-2. A junction is forward-biased if the *n* material is at a lower potential than the *p* material, and reverse-biased if the *n* material is at a higher potential than the *p* material.

Table 3-2

Emitter-Base Bias	Collector-Base Bias	Operating Mode
forward	forward	saturation
reverse	reverse	cutoff
reverse	forward	inverse
forward	reverse	linear or active

Saturation denotes operation (with $|v_{CE}| \approx 0.2$ V and $|v_{BC}| \approx 0.5$ V for Si devices) such that maximum collector current flows and the transistor acts much like a closed switch from collector to emitter terminals. [See Figures 3-2(*c*) and 3-3(*c*).]

Cutoff denotes operation near the voltage axis of the collector characteristics, where the transistor acts much like an open switch. Only leakage current (similar to I_o of the diode) flows in this mode of operation; thus, $i_C = I_{CEO} \approx 0$ for CB connection, and $i_C = I_{CBO} \approx 0$ for CE connection. Figures 3-2(*c*) and 3-3(*c*) indicate these leakage currents.

The *inverse* mode is a little-used, inefficient active mode with the emitter and collector interchanged.

The *active* or *linear* mode describes transistor operation in the region to the right of saturation and above cutoff in Figs. 3-2(*c*) and 3-3(*c*); here, near-linear relationships exist between terminal currents, and the following constants of proportionality are defined for dc currents:

$$\alpha\ (\equiv h_{FB}) \equiv \frac{I_C - I_{CBO}}{I_E} \tag{3.1}$$

$$\beta\ (\equiv h_{FE}) \equiv \frac{\alpha}{1-\alpha} \equiv \frac{I_C - I_{CEO}}{I_B} \tag{3.2}$$

where the thermally generated leakage currents are related by

$$I_{CEO} = (\beta + 1)I_{CBO} \tag{3.3}$$

The constant $\alpha < 1$ is a measure of the proportion of majority carriers (holes for *pnp* devices, electrons for *npn*) injected into the base region from the emitter that are received by the collector. Equation (*3.2*) is the dc current amplification characteristic of the BJT: Except for the leakage current, the base current is increased or amplified β times to become the collector current. Under dc conditions KCL gives

$$I_E = I_C + I_B \tag{3.4}$$

which, in conjunction with (*3.1*) through (*3.3*), completely describes the dc current relationships of the BJT in the active mode.

Example 3.2 Determine α and β for the transistor of Example 3.1 if leakage currents (flow due to holes) are negligible and the described charge flow is constant.

If we assume $I_{CBO} = I_{CEO} = 0$, then

$$\alpha = \frac{i_C}{i_E} = \frac{i_E - i_B}{i_E} = \frac{1.602 - 0.016}{1.602} = 0.99$$

and

$$\beta = \frac{i_C}{i_B} = \frac{i_E - i_B}{i_B} = \frac{1.602 - 0.016}{0.016} = 99.125$$

Example 3.3 A BJT has $\alpha = 0.99$, $i_B = I_B = 25\ \mu$A, and $I_{CBO} = 200$ nA. Find (*a*) the dc collector current, (*b*) the dc emitter current, and (*c*) the percentage error in emitter current when leakage current is neglected.

(*a*) With $\alpha = 0.99$, (*3.2*) gives

$$\beta = \frac{\alpha}{1-\alpha} = 99$$

Using (*3.3*) in (*3.2*) then gives

$$I_C = \beta I_B + (\beta + 1)I_{CBO} = 99(25 \times 10^{-6}) + (99+1)(200 \times 10^{-9}) = 2.495 \text{ mA}$$

(*b*) The dc emitter current follows from (*3.1*):

$$I_E = \frac{I_C - I_{CBO}}{\alpha} = \frac{2.495 \times 10^{-3} - 200 \times 10^{-9}}{0.99} = 2.518 \text{ mA}$$

(*c*) Neglecting the leakage current, we have

$$I_C = \beta I_B = 99(25 \times 10^{-6}) = 2.475 \text{ mA} \qquad \text{so} \qquad I_E = \frac{I_C}{\alpha} = \frac{2.475}{0.99} = 2.5 \text{ mA}$$

giving an emitter-current error of

$$\frac{2.518 - 2.5}{2.518}(100\%) = 0.71\%$$

3.5 BIAS AND DC LOAD LINES

Supply voltages and resistors *bias* a transistor; that is, they establish a specific set of dc terminal voltages and currents, thus determining a point of active-mode operation (called the *quiescent point* or *Q point*). Usually, quiescent values are unchanged by the application of an ac signal to the circuit.

With the universal bias arrangement of Fig. 3-4(*a*), only one dc power supply (V_{CC}) is needed to establish active-mode operation. Use of the Thévenin equivalent of the circuit to the left of *a*,*b* leads to the circuit of Fig. 3-4(*b*), where

$$R_B = \frac{R_1 R_2}{R_1 + R_2} \qquad V_{BB} = \frac{R_1}{R_1 + R_2} V_{CC} \tag{3.5}$$

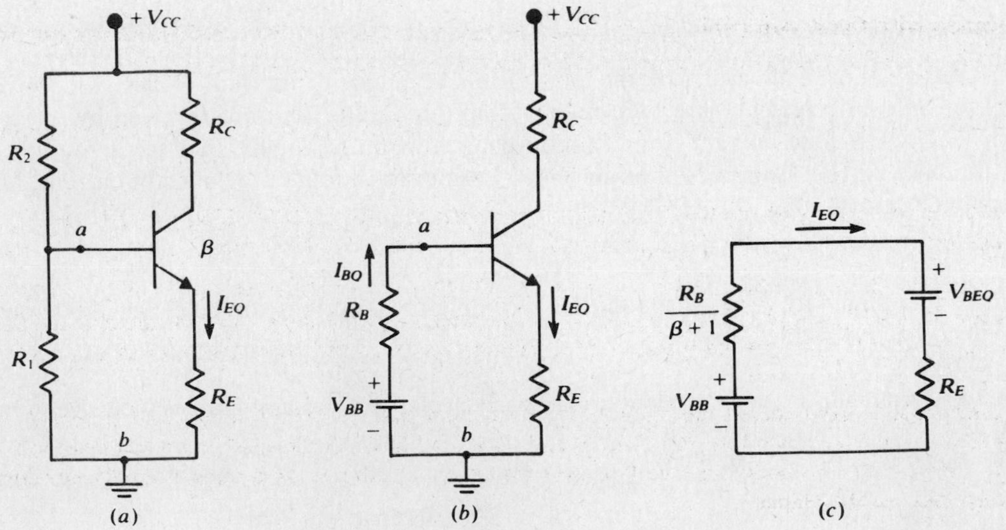

Fig. 3-4

If we neglect leakage current so that $I_{EQ} = (\beta + 1)I_{BQ}$ and assume the emitter-to-base voltage V_{BEQ} is constant (≈0.7 V and ≈0.3 V for Si and Ge, respectively), then KVL around the emitter loop of Fig. 3-4(*b*) yields

$$V_{BB} = \frac{I_{EQ}}{\beta + 1} R_B + V_{BEQ} + I_{EQ}R_E \tag{3.6}$$

which can be represented by the emitter-loop equivalent bias circuit of Fig. 3-4(c). Solving (3.6) for I_{EQ} and noting that

$$I_{EQ} = \frac{I_{CQ}}{\alpha} \approx I_{CQ}$$

we obtain

$$I_{CQ} \approx I_{EQ} = \frac{V_{BB} - V_{BEQ}}{R_B/(\beta + 1) + R_E} \tag{3.7}$$

If component values and the worst-case β value are such that

$$\frac{R_B}{\beta + 1} \approx \frac{R_B}{\beta} \ll R_E \tag{3.8}$$

then I_{EQ} (and thus I_{CQ}) is nearly constant, regardless of changes in β; the circuit then has β*-independent* bias.

From Fig. 3-3(c) it is apparent that the family of collector characteristics is described by the mathematical relationship $i_C = f(v_{CE}, i_B)$ with independent variable v_{CE} and the parameter i_B. We assume that the collector circuit can be biased so as to place the Q point anywhere in the active region. A typical setup is shown in Fig. 3-5(a), from which

$$I_{CQ} = -\frac{V_{CEQ}}{R_{dc}} + \frac{V_{CC}}{R_{dc}}$$

Thus, if the *dc load line*,

$$i_C = -\frac{v_{CE}}{R_{dc}} + \frac{V_{CC}}{R_{dc}} \tag{3.9}$$

and the specification

$$i_B = I_{BQ} \tag{3.10}$$

are combined with the relationship for the collector characteristics, the resulting system can be solved (analytically or graphically) for the collector quiescent quantities I_{CQ} and V_{CEQ}.

Example 3.4 The signal source switch of Fig. 3-5(a) is closed, and the transistor base current becomes

$$i_B = I_{BQ} + i_b = 40 + 20 \sin \omega t \quad \mu\text{A}$$

The collector characteristics of the transistor are those displayed in Fig. 3-5(b). If $V_{CC} = 12$ V and $R_{dc} = 1$ kΩ, graphically determine (a) I_{CQ} and V_{CEQ}, (b) i_c and v_{ce}, and (c) h_{FE} ($= \beta$) at the Q point.

(a) The dc load line has ordinate intercept $V_{CC}/R_{dc} = 12$ mA and abscissa intercept $V_{CC} = 12$ V and is constructed on Fig. 3-5(b). The Q point is the intersection of the load line with the characteristic curve $i_B = I_{BQ} = 40$ μA. The collector quiescent quantities may be read from the axes as $I_{CQ} = 4.9$ mA and $V_{CEQ} = 7.2$ V.

(b) A time scale is constructed perpendicular to the load line at the Q point, and a scaled sketch of $i_b = 20 \sin \omega t$ μA is drawn [see Fig. 3-5(b)] and translated through the load line to sketches of i_c and v_{ce}. As i_b swings ± 20 μA along the load line from points a to b, the ac components of collector current and voltage take on the values

$$i_c = 2.25 \sin \omega t \quad \text{mA} \qquad \text{and} \qquad v_{ce} = -2.37 \sin \omega t \quad \text{V}$$

The negative sign on v_{ce} signifies a 180° phase shift.

(c) From (3.2) with $I_{CEO} = 0$ [the $i_B = 0$ curve coincides with the v_{CE} axis in Fig. 3-5(b)],

$$h_{FE} = \frac{I_{CQ}}{I_{BQ}} = \frac{4.9 \times 10^{-3}}{40 \times 10^{-6}} = 122.5$$

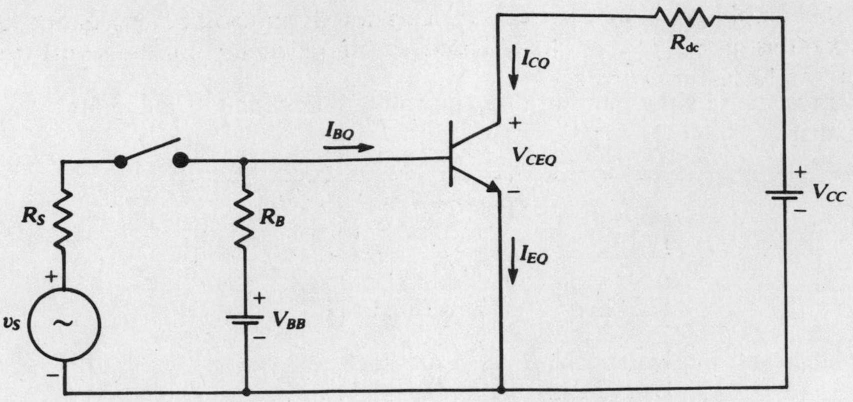

R_{dc}
I_{CQ}
I_{BQ}
V_{CEQ}
V_{CC}
R_S
R_B
I_{EQ}
v_S
V_{BB}

(a)

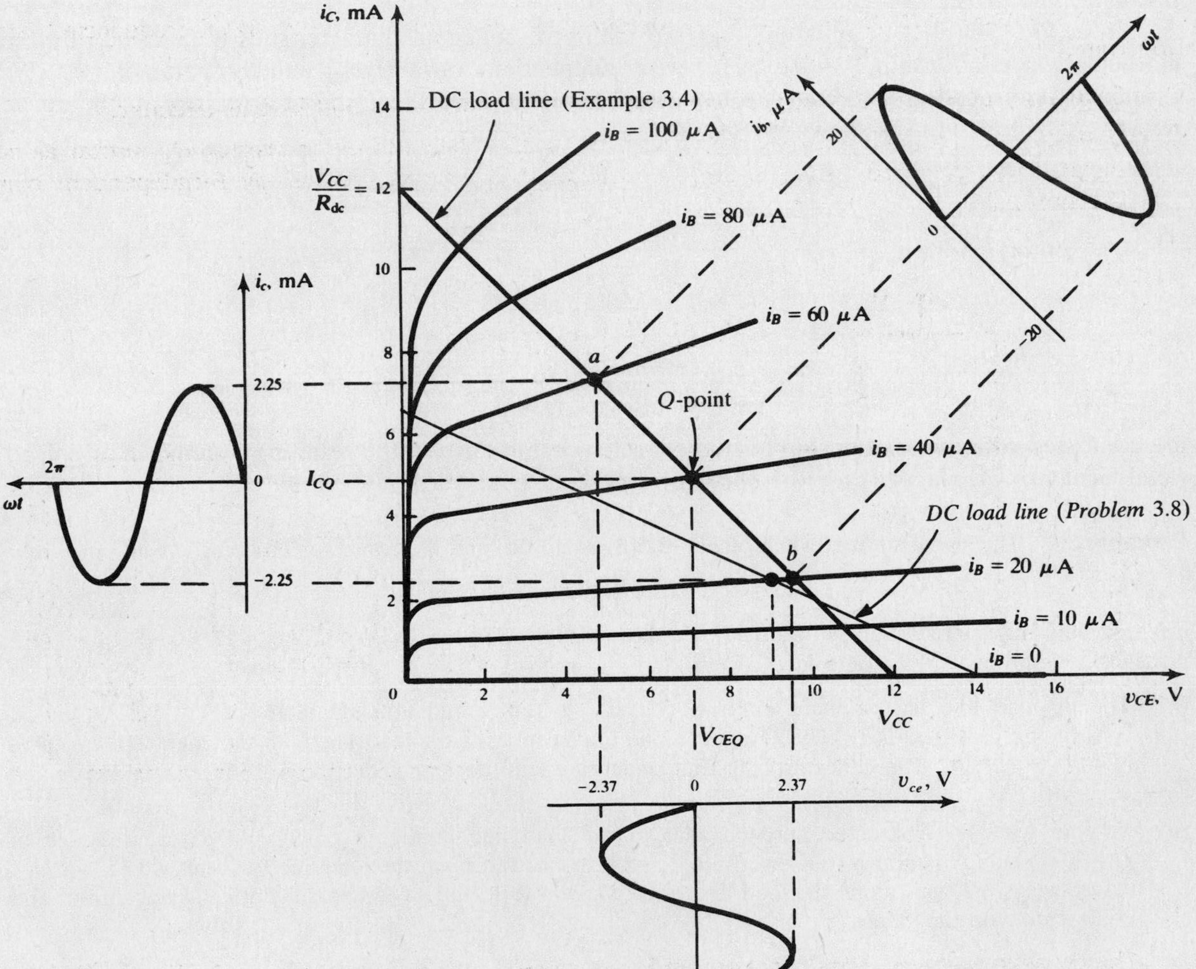

i_C, mA
DC load line (Example 3.4)
i_B = 100 μA
$\frac{V_{CC}}{R_{dc}}$ = 12
i_B = 80 μA
i_c, mA
i_B = 60 μA
a
2.25
Q-point
i_B = 40 μA
I_{CQ}
ωt
2π
DC load line (Problem 3.8)
b
i_B = 20 μA
−2.25
i_B = 10 μA
i_B = 0
v_{CE}, V
V_{CC}
V_{CEQ}
−2.37
2.37
v_{ce}, V
i_b, μA
20
−20

(b)

Fig. 3-5

It is clear that amplifiers can be biased for operation at any point along the dc load line. Table 3-3 shows the various classes of amplifiers, based on the percentage of the signal cycle over which they operate in the linear or active region.

Table 3-3

Class	Percentage of Active-Region Signal Excursion
A	100
AB	between 50 and 100
B	50
C	less than 50

3.6 CAPACITORS AND AC LOAD LINES

Two common uses of capacitors (sized to appear as short circuits to signal frequencies) are illustrated by the circuit of Fig. 3-6(*a*):

1. Coupling capacitors (C_C) confine dc quantities to the transistor and its bias circuitry.
2. Bypass capacitors (C_E) effectively remove the gain-reducing emitter resistor R_E insofar as ac signals are concerned, while allowing R_E to play its role in establishing β-independent bias (Section 3.5).

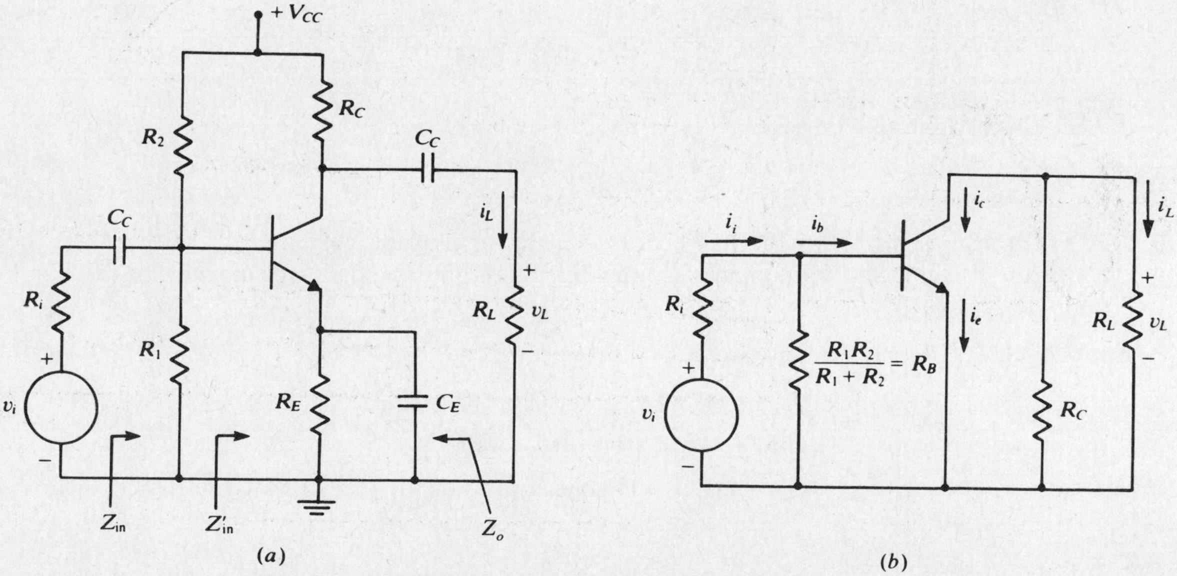

Fig. 3-6

The capacitors of Fig. 3-6(*a*) are shorted in the circuit as it appears to ac signals [Fig. 3-6(*b*)]. In Fig. 3-6(*a*), we note that the collector-circuit resistance seen by the dc bias current I_{CQ} ($\approx I_{EQ}$) is $R_{dc} = R_C + R_E$. However, from Fig. 3-6(*b*) it is apparent that the collector signal current i_c sees a collector-circuit resistance $R_{ac} = R_C R_L/(R_C + R_L)$. Since $R_{ac} \neq R_{dc}$ in general, the concept of an *ac load line* arises. By application of KVL to Fig. 3-6(*b*), the *v-i* characteristic of the external signal circuitry is found to be

$$v_{ce} = i_c R_{ac} \tag{3.11}$$

Since $i_c = i_C - I_{CQ}$ and $v_{ce} = v_{CE} - V_{CEQ}$, (3.11) can be written analogously to (3.9) as

$$i_C = -\frac{v_{CE}}{R_{ac}} + \frac{V_{CEQ}}{R_{ac}} + I_{CQ} \tag{3.12}$$

All excursions of the ac signals i_c and v_{ce} are represented by points on the ac load line, (3.12). If the value $i_C = I_{CQ}$ is substituted into (3.12), we find that $v_{CE} = V_{CEQ}$; thus, the ac load line intersects the dc load line at the Q point.

Example 3.5 Find the points at which the ac load line intersects the axes of the collector characteristic.
The i_C intercept ($i_{C\max}$) is found by setting $v_{CE} = 0$ in (3.12):

$$i_{C\max} = \frac{V_{CEQ}}{R_{ac}} + I_{CQ} \tag{3.13}$$

The v_{CE} intercept is found by setting $i_C = 0$ in (3.12):

$$v_{CE\max} = V_{CEQ} + I_{CQ} R_{ac} \tag{3.14}$$

Solved Problems

3.1 For a certain BJT, $\beta = 50$, $I_{CEO} = 3\ \mu\text{A}$, and $I_C = 1.2$ mA. Find I_B and I_E.

By (3.2),

$$I_B = \frac{I_C - I_{CEO}}{\beta} = \frac{1.2 \times 10^{-3} - 3 \times 10^{-6}}{50} = 23.94\ \mu\text{A}$$

And, directly from (3.4),

$$I_E = I_C + I_B = 1.2 \times 10^{-3} - 23.94 \times 10^{-6} = 1.224 \text{ mA}$$

3.2 A Ge transistor with $\beta = 100$ has a base-to-collector leakage current I_{CBO} of 5 μA. If the transistor is connected for common-emitter operation, find the collector current for (*a*) $I_B = 0$ and (*b*) $I_B = 40\ \mu\text{A}$.

(*a*) With $I_B = 0$, only emitter-to-collector leakage flows, and, by (3.3),

$$I_{CEO} = (\beta + 1) I_{CBO} = (100 + 1)(5 \times 10^{-6}) = 505\ \mu\text{A}$$

(*b*) If we substitute (3.3) into (3.2) and solve for I_C, we get

$$I_C = \beta I_B + (\beta + 1) I_{CBO} = (100)(40 \times 10^{-6}) + (101)(5 \times 10^{-6}) = 4.505 \text{ mA}$$

3.3 A transistor with $\alpha = 0.98$ and $I_{CBO} = 5\ \mu\text{A}$ is biased so that $I_{BQ} = 100\ \mu\text{A}$. Find I_{CQ} and I_{EQ}.

By (3.2) and (3.3),

$$\beta = \frac{\alpha}{1-\alpha} = \frac{0.98}{1 - 0.98} = 49$$

so that

$$I_{CEO} = (\beta + 1) I_{CBO} = (49 + 1)(5 \times 10^{-6}) = 0.25 \text{ mA}$$

And, from (3.2) and (3.4),

$$I_{CQ} = \beta I_{BQ} + I_{CEO} = (49)(100 \times 10^{-6}) + 0.25 \times 10^{-3} = 5.15 \text{ mA}$$

$$I_{EQ} = I_{CQ} + I_{BQ} = 5.15 \times 10^{-3} + 100 \times 10^{-6} = 5.25 \text{ mA}$$

3.4 The transistor of Fig. 3-7 has $\alpha = 0.98$ and a base current of 30 μA. Find (*a*) β, (*b*) I_{CQ}, and (*c*) I_{EQ}. Assume negligible leakage current.

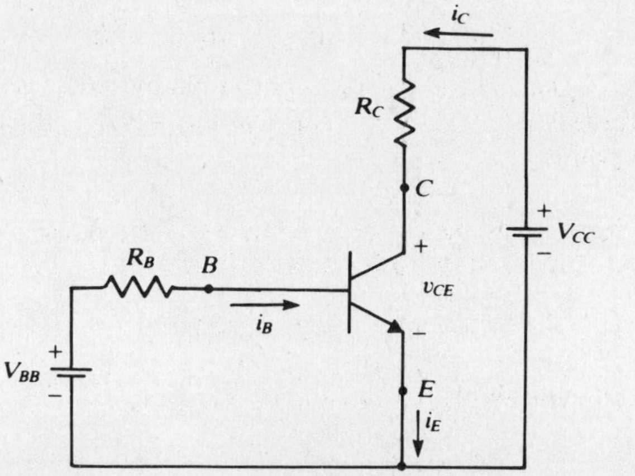

Fig. 3-7

(*a*)
$$\beta = \frac{\alpha}{1-\alpha} = \frac{0.98}{1-0.98} = 49$$

(*b*) From (*3.2*) with $I_{CEO} = 0$, we have $I_{CQ} = \beta I_{BQ} = (49)(30 \times 10^{-6}) = 1.47$ mA.

(*c*) From (*3.1*) with $I_{CBO} = 0$,

$$I_{EQ} = \frac{I_{CQ}}{\alpha} = \frac{1.47}{0.98} = 1.50 \text{ mA}$$

3.5 The transistor circuit of Fig. 3-7 is to be operated with a base current of 40 μA and $V_{BB} = 6$ V. The Si transistor ($V_{BEQ} = 0.7$ V) has negligible leakage current. Find the required value of R_B.

By KVL around the base-emitter loop,

$$V_{BB} = I_{BQ}R_B + V_{BEQ} \qquad \text{so that} \qquad R_B = \frac{V_{BB} - V_{BEQ}}{I_{BQ}} = \frac{6-0.7}{40 \times 10^{-6}} = 132.5 \text{ k}\Omega$$

3.6 In the circuit of Fig. 3-7, $\beta = 100$, $I_{BQ} = 20$ μA, $V_{CC} = 15$ V, and $R_C = 3$ kΩ. If $I_{CBO} = 0$, find (*a*) I_{EQ} and (*b*) V_{CEQ}. (*c*) Find V_{CEQ} if R_C is changed to 6 kΩ and all else remains the same.

(*a*)
$$\alpha = \frac{\beta}{\beta+1} = \frac{100}{101} = 0.9901$$

Now, using (*3.2*) and (*3.1*) with $I_{CBO} = I_{CEO} = 0$, we get

$$I_{CQ} = \beta I_{BQ} = (100)(20 \times 10^{-6}) = 2 \text{ mA}$$

and
$$I_{EQ} = \frac{I_{CQ}}{\alpha} = \frac{2 \times 10^{-3}}{0.9901} = 2.02 \text{ mA}$$

(*b*) From an application of KVL around the collector circuit,

$$V_{CEQ} = V_{CC} - I_{CQ}R_C = 15 - (2)(3) = 9 \text{ V}$$

(*c*) If I_{BQ} is unchanged, then I_{CQ} is unchanged. The solution proceeds as in part *b*:

$$V_{CEQ} = V_{CC} - I_{CQ}R_C = 15 - (2)(6) = 3 \text{ V}$$

3.7 The transistor of Fig. 3-8 is a Si device with a base current of 40 μA and $I_{CBO} = 0$. If $V_{BB} = 6$ V, $R_E = 1$ kΩ, and $\beta = 80$, find (*a*) I_{EQ} and (*b*) R_B. (*c*) If $V_{CC} = 15$ V and $R_C = 3$ kΩ, find V_{CEQ}.

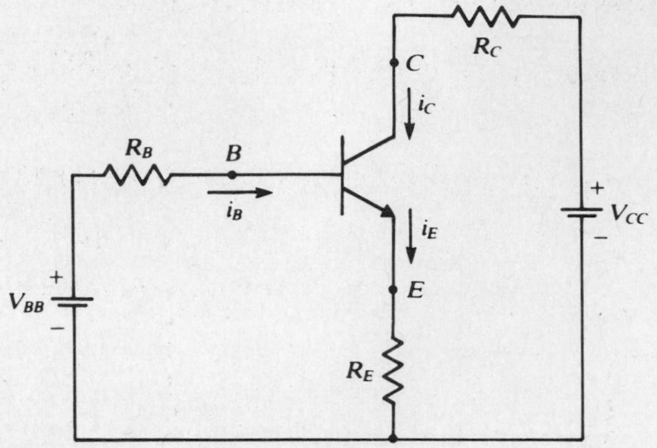

Fig. 3-8

(*a*)
$$\alpha = \frac{\beta}{\beta + 1} = \frac{80}{81} = 0.9876$$

Then combining (*3.1*) and (*3.2*) with $I_{CBO} = I_{CEO} = 0$ gives

$$I_{EQ} = \frac{I_{BQ}}{1 - \alpha} = \frac{40 \times 10^{-6}}{1 - 0.9876} = 3.226 \text{ mA}$$

(*b*) Applying KVL around the base-emitter loop gives

$$V_{BB} = I_{BQ}R_B + V_{BEQ} + I_{EQ}R_E$$

or (with V_{BEQ} equal to the usual 0.7 V for a Si device)

$$R_B = \frac{V_{BB} - V_{BEQ} - I_{EQ}R_E}{I_{BQ}} = \frac{6 - 0.7 - (3.226)(1)}{40 \times 10^{-6}} = 51.85 \text{ k}\Omega$$

(*c*) From (*3.2*) with $I_{CEO} = 0$,

$$I_{CQ} = \beta I_{BQ} = (80)(40 \times 10^{-6}) = 3.2 \text{ mA}$$

Then, by KVL around the collector circuit,

$$V_{CEQ} = V_{CC} - I_{EQ}R_E - I_{CQ}R_C = 15 - (3.226)(1) - (3.2)(3) = 2.174 \text{ V}$$

3.8 Assume that the CE collector characteristics of Fig. 3-5(*b*) apply to the transistor of Fig. 3-7. If $I_{BQ} = 20$ μA, $V_{CEQ} = 9$ V, and $V_{CC} = 14$ V, find graphically (*a*) I_{CQ}, (*b*) R_C, (*c*) I_{EQ}, and (*d*) β if leakage current is negligible.

(*a*) The *Q* point is the intersection of $i_B = I_{BQ} = 20$ μA and $v_{CE} = V_{CEQ} = 9$ V. The dc load line must pass through the *Q* point and intersect the v_{CE} axis at $V_{CC} = 14$ V. Thus, the dc load line can be drawn on Fig. 3-5(*b*), and $I_{CQ} = 2.25$ mA can be read as the i_C coordinate of the *Q* point.

(*b*) The i_C intercept of the dc load line is $V_{CC}/R_{dc} = V_{CC}/R_C$, which, from Fig. 3-5(*b*), has the value 6.5 mA; thus,

$$R_C = \frac{V_{CC}}{6.5 \times 10^{-3}} = \frac{14}{6.5 \times 10^{-3}} = 2.15 \text{ k}\Omega$$

(*c*) By (*3.4*), $I_{EQ} = I_{CQ} + I_{BQ} = 2.25 \times 10^{-3} + 20 \times 10^{-6} = 2.27$ mA.

(*d*) With $I_{CEO} = 0$, (*3.2*) yields

$$\beta = \frac{I_{CQ}}{I_{BQ}} = \frac{2.25 \times 10^{-3}}{20 \times 10^{-6}} = 112.5$$

3.9 In the *pnp* Si transistor circuit of Fig. 3-9, $R_B = 500\ \text{k}\Omega$, $R_C = 2\ \text{k}\Omega$, $R_E = 0$, $V_{CC} = 15$ V, $I_{CBO} = 20\ \mu\text{A}$, and $\beta = 70$. Find the Q-point collector current I_{CQ}.

By (*3.3*), $I_{CEO} = (\beta + 1)I_{CBO} = (70 + 1)(20 \times 10^{-6}) = 1.42$ mA. Now, application of the KVL around the loop that includes V_{CC}, R_B, $R_E\,(=0)$, and ground

$$V_{CC} = V_{BEQ} + I_{BQ}R_B \qquad \text{so that} \qquad I_{BQ} = \frac{V_{CC} - V_{BEQ}}{R_B} = \frac{15 - 0.7}{500 \times 10^3} = 28.6\ \mu\text{A}$$

Thus, by (*3.2*),

$$I_{CQ} = \beta I_{BQ} + I_{CEO} = (70)(28.6 \times 10^{-6}) + 1.42 \times 10^{-3} = 3.42\ \text{mA}$$

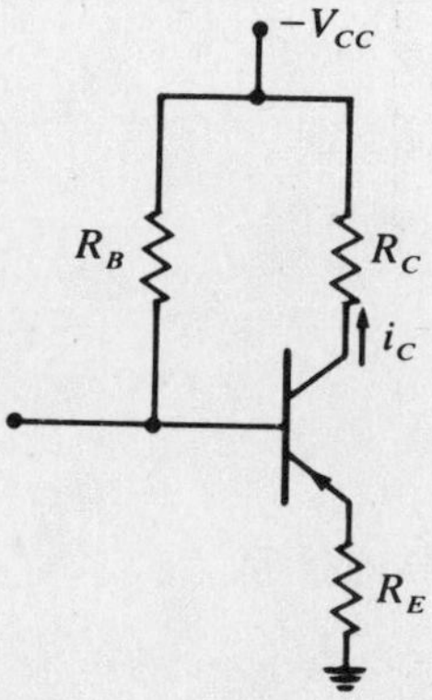

Fig. 3-9

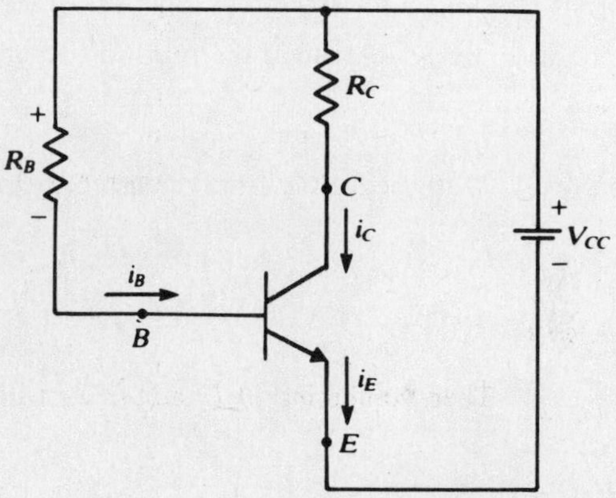

Fig. 3-10

3.10 The Si transistor of Fig. 3-10 is biased for constant base current. If $\beta = 80$, $V_{CEQ} = 8$ V, $R_C = 3\ \text{k}\Omega$, and $V_{CC} = 15$ V, find (*a*) I_{CQ} and (*b*) the required value of R_B. (*c*) Find R_B if the transistor is a Ge device.

(*a*) By KVL around the collector-emitter circuit,

$$I_{CQ} = \frac{V_{CC} - V_{CEQ}}{R_c} = \frac{15 - 8}{3 \times 10^3} = 2.333\ \text{mA}$$

(*b*) If leakage current is neglected, (*3.2*) gives

$$I_{BQ} = \frac{I_{CQ}}{\beta} = \frac{2.333 \times 10^{-3}}{80} = 29.16\ \mu\text{A}$$

Since the transistor is a Si device, $V_{BEQ} = 0.7$ V and, by KVL around the outer loop,

$$R_B = \frac{V_{CC} - V_{BEQ}}{I_{BQ}} = \frac{15 - 0.7}{29.16 \times 10^{-6}} = 490.4\ \text{k}\Omega$$

(*c*) The only difference here is that $V_{BEQ} = 0.3$ V; thus

$$R_B = \frac{15 - 0.3}{29.16 \times 10^{-6}} = 504.1\ \text{k}\Omega$$

3.11 The Si transistor of Fig. 3-11 has $\alpha = 0.99$ and $I_{CEO} = 0$. Also, $V_{EE} = 4$ V and $V_{CC} = 12$ V. (*a*) If $I_{EQ} = 1.1$ mA, find R_E. (*b*) If $V_{CEQ} = -7$ V, find R_C.

(*a*) By KVL around the emitter-base loop,

$$R_E = \frac{V_{EE} + V_{BEQ}}{I_{EQ}} = \frac{4 + (-0.7)}{1.1 \times 10^{-3}} = 3\ \Omega$$

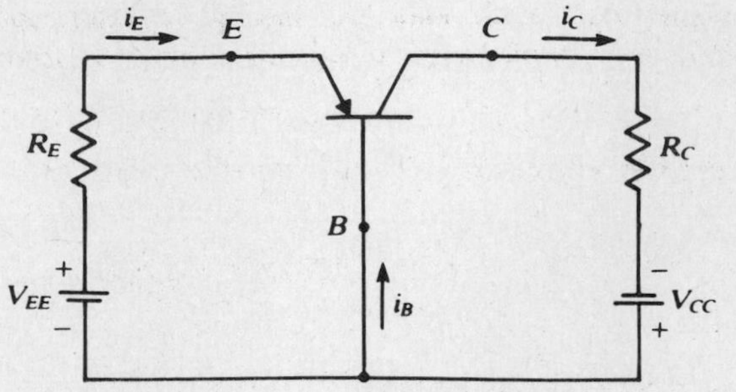

Fig. 3-11

(*b*) By KVL around the transistor terminals (which constitute a closed path),

$$V_{CBQ} = V_{CEQ} - V_{BEQ} = -7 - (-0.7) = -6.3 \text{ V}$$

With negligible leakage current, (*3.1*) gives

$$I_{CQ} = \alpha I_{EQ} = (0.99)(1.1 \times 10^{-3}) = 1.089 \text{ mA}$$

Finally, by KVL around the base-collector loop,

$$R_C = \frac{V_{CC} + V_{CBQ}}{I_{CQ}} = \frac{12 - 6.3}{1.089 \times 10^{-3}} = 5.234 \text{ k}\Omega$$

3.12 Collector characteristics for the Ge transistor of Fig. 3-11 are given in Fig. 3-12. If $V_{EE} = 2$ V, $V_{CC} = 12$ V, and $R_C = 2$ kΩ, size R_E so that $V_{CEQ} = -6.4$ V.

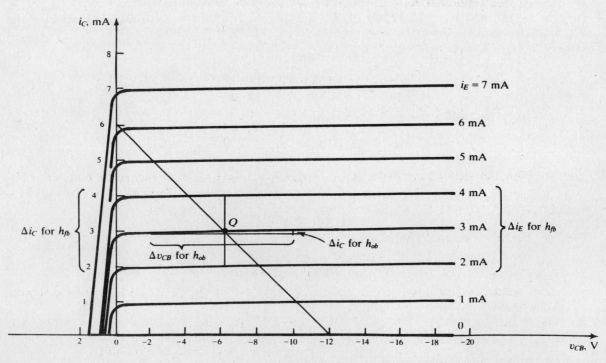

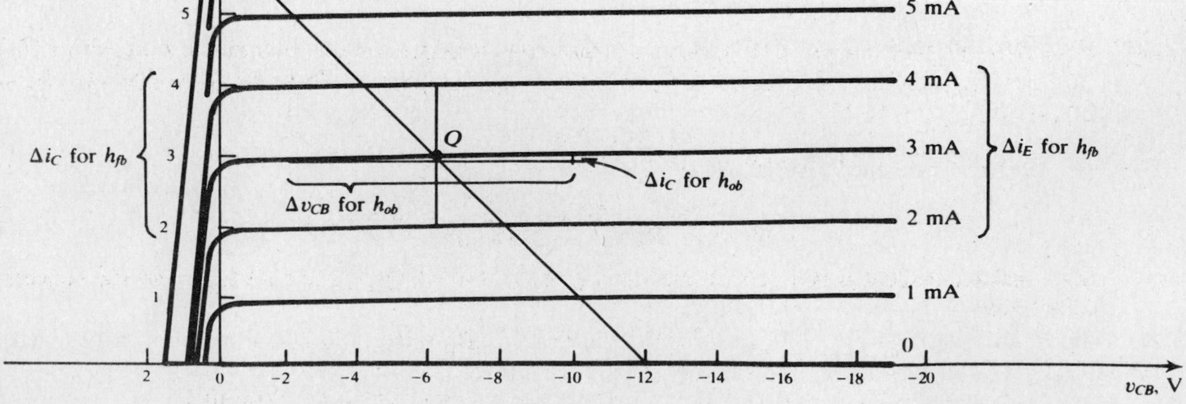

Fig. 3-12

We construct, on Fig. 3-12, a dc load line having v_{CB} intercept $-V_{CC} = -12$ V and i_C intercept $V_{CC}/R_C = 6$ mA. The abscissa of the Q point is given by KVL around the transistor terminals:

$$V_{CBQ} = V_{CEQ} - V_{BEQ} = -6.4 - (-0.3) = -6.1 \text{ V}$$

With the Q point defined, we read $I_{EQ} = 3$ mA from the graph. Now KVL around the emitter-base loop leads to

$$R_E = \frac{V_{EE} + V_{BEQ}}{I_{EQ}} = \frac{2 + (-0.3)}{3 \times 10^{-3}} = 566.7\ \Omega$$

3.13 The circuit of Fig. 3-13 uses current- (or shunt-) feedback bias. The Si transistor has $I_{CEO} \approx 0$, $V_{CE\text{sat}} \approx 0$, and $h_{FE} = 100$. If $R_C = 2$ kΩ and $V_{CC} = 12$ V, size R_F for ideal *maximum symmetrical swing* (that is, location of the quiescent point such that $V_{CEQ} = V_{CC}/2$).

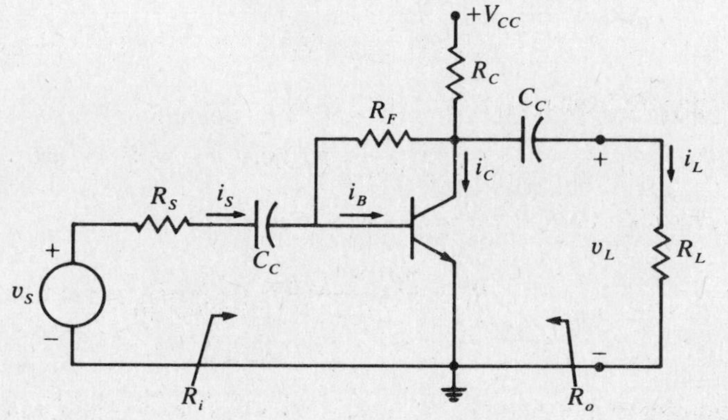

Fig. 3-13

Application of KVL to the collector-emitter bias circuit gives

$$(I_{BQ} + I_{CQ})R_C = V_{CC} - V_{CEQ}$$

With $I_{CQ} = h_{FE}I_{BQ}$, this leads to

$$I_{BQ} = \frac{V_{CC} - V_{CEQ}}{(h_{FE} + 1)R_C} = \frac{12 - 6}{(100 + 1)(2 \times 10^3)} = 29.7\ \mu\text{A}$$

Then, by KVL around the transistor terminals,

$$R_F = \frac{V_{CEQ} - V_{BEQ}}{I_{BQ}} = \frac{6 - 0.7}{29.7 \times 10^{-6}} = 178.5\ \text{k}\Omega$$

3.14 Find the value of the emitter resistor R_E that, when added to the Si transistor circuit of Fig. 3-13, would bias for operation about $V_{CEQ} = 5$ V. Let $I_{CEO} = 0$, $\beta = 80$, $R_F = 220$ kΩ, $R_C = 2$ kΩ, and $V_{CC} = 12$ V.

Application of KVL around the transistor terminals yields

$$I_{BQ} = \frac{V_{CEQ} - V_{BEQ}}{R_F} = \frac{5 - 0.7}{220 \times 10^3} = 19.545\ \mu\text{A}$$

Since leakage current is zero, (*3.1*) and (*3.2*) give $I_{EQ} = (\beta + 1)I_{CQ}$; thus KVL around the collector circuit gives

$$(I_{BQ} + \beta I_{BQ})R_C + (\beta + 1)I_{BQ}R_E = V_{CC} - V_{CEQ}$$

so $$R_E = \frac{V_{CC} - V_{CEQ} - (\beta + 1)I_{BQ}R_C}{(\beta + 1)I_{BQ}} = \frac{12 - 5 - (80 + 1)(19.545 \times 10^{-6})(2 \times 10^3)}{(80 + 1)(19.545 \times 10^{-6})} = 2.42\ \text{k}\Omega$$

3.15 In the circuit of Fig. 3-8, $I_{BQ} = 30\ \mu\text{A}$, $R_E = 1\ \text{k}\Omega$, $V_{CC} = 15$ V, and $\beta = 80$. Find the minimum value of R_C that will maintain the transistor quiescent point at saturation, if $V_{CE\text{sat}} = 0.2$ V, β is constant, and leakage current is negligible.

We first find

$$\alpha = \frac{\beta}{\beta + 1} = \frac{80}{81} = 0.9876$$

Then the use of (*3.2*) and (*3.1*) with negligible leakage current yields

$$I_{CQ} = \beta I_{BQ} = (80)(30 \times 10^{-6}) = 2.4\ \text{mA}$$

and

$$I_{EQ} = \frac{I_{CQ}}{\alpha} = \frac{2.4 \times 10^{-3}}{0.9876} = 2.43\ \text{mA}$$

Now KVL around the collector circuit leads to the minimum value of R_C to ensure saturation:

$$R_C = \frac{V_{CC} - V_{CE\text{sat}} - I_{EQ}R_E}{I_{CQ}} = \frac{15 - 0.2 - (2.43)(1)}{2.4 \times 10^{-3}} = 5.154\ \text{k}\Omega$$

3.16 The Si transistor of Fig. 3-14 has $\beta = 50$ and negligible leakage current. Let $V_{CC} = 18$ V, $V_{EE} = 4$ V, $R_E = 200\ \Omega$, and $R_C = 4\ \text{k}\Omega$. (*a*) Find R_B so that $I_{CQ} = 2$ mA. (*b*) Determine the value of V_{CEQ} for V_B of part (*a*).

(*a*) KVL around the base-emitter-ground loop gives

$$V_{EE} = I_{BQ}R_B + V_{BEQ} + I_{EQ}R_E \tag{1}$$

Also, from (*3.1*) and (*3.2*),

$$I_{EQ} = \frac{\beta + 1}{\beta} I_{CQ} \tag{2}$$

Now, using (*3.2*) and (*2*) in (*1*) and solving for R_B yields

$$R_B = \frac{\beta(V_{EE} - V_{BEQ})}{I_{CQ}} - (\beta + 1)R_E = \frac{50(4 - 0.7)}{2 \times 10^{-3}} - (50 + 1)(200) = 72.3\ \text{k}\Omega$$

(*b*) KVL around the collector-emitter-ground loop gives

$$V_{CEQ} = V_{CC} + V_{EE} - \left(R_C + \frac{\beta + 1}{\beta} R_E\right)I_{CQ}$$

$$= 18 + 4 - \left(4 \times 10^3 + \frac{50 + 1}{50} 200\right)(2 \times 10^{-3}) = 13.59\ \text{V}$$

3.17 The dc current source $I_S = 10\ \mu\text{A}$ of Fig. 3-14 is connected from G to node B. The Si transistor has negligible leakage current and $\beta = 50$. If $R_B = 75\ \text{k}\Omega$, $R_E = 200\ \Omega$, and $R_C = 4\ \text{k}\Omega$, find the dc current-gain ratio I_{CQ}/I_S for (*a*) $V_{CC} = 18$ V and $V_{EE} = 4$ V, and (*b*) $V_{CC} = 22$ V and $V_{EE} = 0$ V.

(*a*) A Thévenin equivalent for the network to the left of terminals B, G has $V_{Th} = R_BI_S$ and $R_{Th} = R_B$. With the Thévenin equivalent circuit in place, KVL around the base-emitter loop yields

$$R_BI_S + V_{EE} = I_{BQ}R_B + V_{BEQ} + I_{EQ}R_E \tag{1}$$

Using (*3.2*) and (*2*) of Problem 3.16 in (*1*), solving for I_{CQ}, and then dividing by I_S result in the desired ratio:

$$\frac{I_{CQ}}{I_S} = \frac{R_BI_S + V_{EE} - V_{BEQ}}{I_S\left(\dfrac{R_B}{\beta} + \dfrac{\beta + 1}{\beta} R_E\right)} = \frac{(75 \times 10^3)(10 \times 10^{-6}) + 4 - 0.7}{(10 \times 10^{-6})\left(\dfrac{75 \times 10^3}{50} + \dfrac{50 + 1}{50} 200\right)} = 237.67 \tag{2}$$

Note that the value of V_{CC} must be large enough so that cutoff does not occur, but otherwise it does not affect the value of I_{CQ}.

(*b*) $V_{EE} = 0$ in (*2*) directly gives

$$\frac{I_{CQ}}{I_S} = \frac{(75 \times 10^3)(10 \times 10^{-6}) - 0.7}{(10 \times 10^{-6})\left(\dfrac{75 \times 10^3}{50} + \dfrac{50+1}{50} 200\right)} = 2.93$$

Obviously, V_{EE} strongly controls the dc current gain of this amplifier.

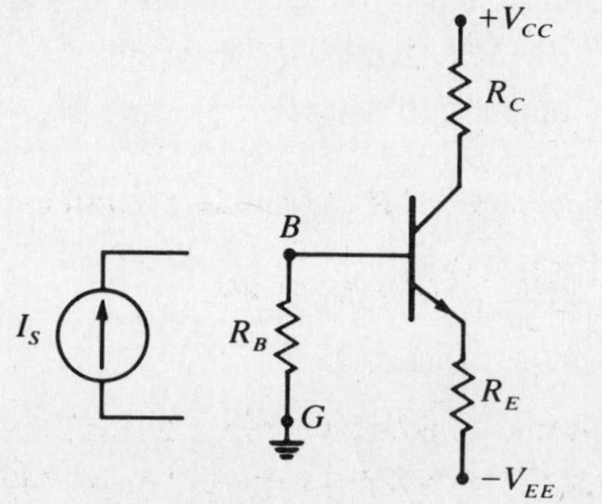

Fig. 3-14

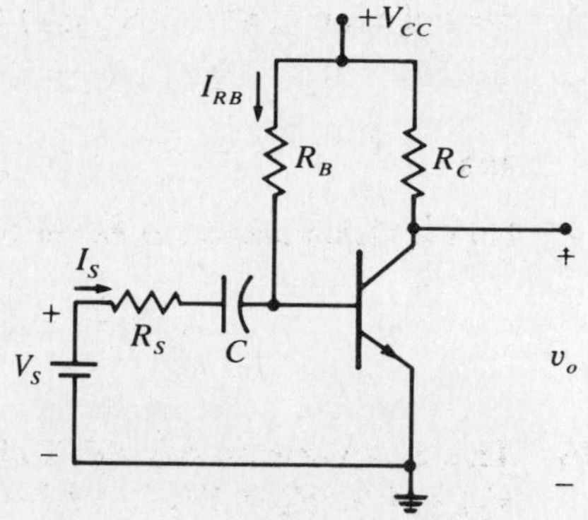

Fig. 3-15

3.18 In the circuit of Fig. 3-15, $V_{CC} = 12$ V, $V_S = 2$ V, $R_C = 4$ kΩ, and $R_S = 100$ kΩ. The Ge transistor is characterized by $\beta = 50$, $I_{CEO} = 0$, and $V_{CE\text{sat}} = 0.2$ V. Find the value of R_B that just results in saturation if (*a*) the capacitor is present, and (*b*) the capacitor is replaced with a short circuit.

(*a*) Application of KVL around the collector loop gives the collector current at the onset of saturation as

$$I_{CQ} = \frac{V_{CC} - V_{CE\text{sat}}}{R_C} = \frac{12 - 0.2}{4 \times 10^3} = 2.95 \text{ mA}$$

With C blocking, $I_S = 0$; hence the use of KVL leads to

$$R_B = \frac{V_{CC} - V_{BEQ}}{I_{BQ}} = \frac{V_{CC} - V_{BEQ}}{I_{CQ}/\beta} = \frac{12 - 0.3}{(2.95 \times 10^{-3})/50} = 198.3 \text{ k}\Omega$$

(*b*) With C shorted, the application of (*3.2*), KCL, and KVL results in

$$I_{BQ} = \frac{I_{CQ}}{\beta} = I_S + I_{RB} = \frac{V_S - V_{BEQ}}{R_S} + \frac{V_{CC} - V_{BEQ}}{R_B}$$

so that
$$R_B = \frac{V_{CC} - V_{BEQ}}{\dfrac{I_{CQ}}{\beta} - \dfrac{V_S - V_{BEQ}}{R_S}} = \frac{12 - 0.3}{\dfrac{2.95 \times 10^{-3}}{50} - \dfrac{2 - 0.3}{100 \times 10^3}} = 278.6 \text{ k}\Omega$$

3.19 The Si *Darlington transistor pair* of Fig. 3-16 has negligible leakage current, and $\beta_1 = \beta_2 = 50$. Let $V_{CC} = 12$ V, $R_E = 1$ kΩ, and $R_2 \to \infty$. (*a*) Find the value of R_1 needed to bias the circuit so that $V_{CEQ2} = 6$ V. (*b*) With R_1 as found in part *a*, find V_{CEQ1}.

(*a*) Since $R_2 \to \infty$, $I_{R2} = 0$ and $I_{BQ1} = I_{R1}$. By KVL,

$$I_{EQ2} = \frac{V_{CC} - V_{CEQ2}}{R_E} = \frac{12 - 6}{1 \times 10^3} = 6 \text{ mA}$$

Now
$$I_{BQ2} = \frac{I_{EQ2}}{\beta_2 + 1} = I_{EQ1}$$

and
$$I_{R1} = I_{BQ1} = \frac{I_{EQ1}}{\beta_1 + 1} = \frac{I_{EQ2}}{(\beta_1 + 1)(\beta_2 + 1)} = \frac{6 \times 10^{-3}}{(50+1)(50+1)} = 2.31 \ \mu\text{A}$$

By KVL (around a path that includes R_1, both transistors, and R_E) and Ohm's law,

$$R_1 = \frac{V_{R1}}{I_{R1}} = \frac{V_{CC} - V_{BEQ1} - V_{BEQ2} - I_{EQ2}R_E}{I_{R1}} = \frac{12 - 0.7 - 0.7 - (6 \times 10^{-3})(1 \times 10^3)}{2.31 \times 10^{-6}} = 1.99\ \text{M}\Omega$$

(*b*) Applying KVL around a path including both transistors and R_E, we have

$$V_{CEQ1} = V_{CC} - V_{BEQ2} - I_{EQ2}R_E = 12 - 0.7 - (6 \times 10^{-3})(1 \times 10^3) = 5.3\ \text{V}$$

3.20 The Si Darlington transistor pair of Fig. 3-16 has negligible leakage current, and $\beta_1 = \beta_2 = 60$. Let $R_1 = R_2 = 1\ \text{M}\Omega$, $R_E = 500\ \Omega$, and $V_{CC} = 12$ V. Find (*a*) I_{EQ2}, (*b*) V_{CEQ2}, and (*c*) I_{CQ1}.

(*a*) A Thévenin equivalent for the circuit to the left of terminals a,b has

$$V_{Th} = \frac{R_2}{R_1 + R_2} V_{CC} = \frac{1 \times 10^6}{1 \times 10^6 + 1 \times 10^6} 12 = 6\ \text{V}$$

and
$$R_{Th} = \frac{R_1 R_2}{R_1 + R_2} = \frac{(1 \times 10^6)(1 \times 10^6)}{1 \times 10^6 + 1 \times 10^6} = 500\ k\Omega$$

With the Thévenin circuit in place, KVL gives

$$V_{Th} = I_{BQ1}R_{Th} + V_{BEQ1} + V_{BEQ2} + I_{EQ2}R_E \qquad (1)$$

Realizing that

$$I_{EQ2} = (\beta_2 + 1)I_{BQ2} = (\beta_2 + 1)(\beta_1 + 1)I_{BQ1}$$

we can substitute for I_{BQ1} in (*1*) and solve for I_{EQ2}, obtaining

$$I_{EQ2} = \frac{(\beta_1 + 1)(\beta_2 + 1)(V_{Th} - V_{BEQ1} - V_{BEQ2})}{R_{Th} + (\beta_1 + 1)(\beta_2 + 1)R_E} = \frac{(60 + 1)(60 + 1)(6 - 0.7 - 0.7)}{500 \times 10^3 + (60 + 1)(60 + 1)(500)} = 7.25\ \text{mA}$$

(*b*) By KVL,

$$V_{CEQ2} = V_{CC} - I_{EQ2}R_E = 12 - (7.25 \times 10^{-3})(500) = 8.375\ \text{V}$$

(*c*) From (*3.1*) and (*3.2*),

$$I_{CQ1} = \frac{\beta_1}{\beta_1 + 1} I_{EQ1} = \frac{\beta_1}{\beta_1 + 1} I_{BQ2} = \frac{\beta_1}{\beta_1 + 1} \frac{I_{EQ2}}{\beta_2 + 1} = \frac{60}{60 + 1} \frac{7.25 \times 10^{-3}}{60 + 1} = 116.9\ \mu\text{A}$$

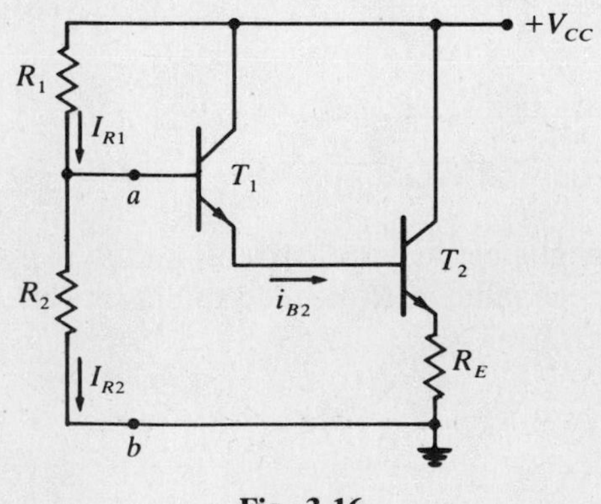

Fig. 3-16

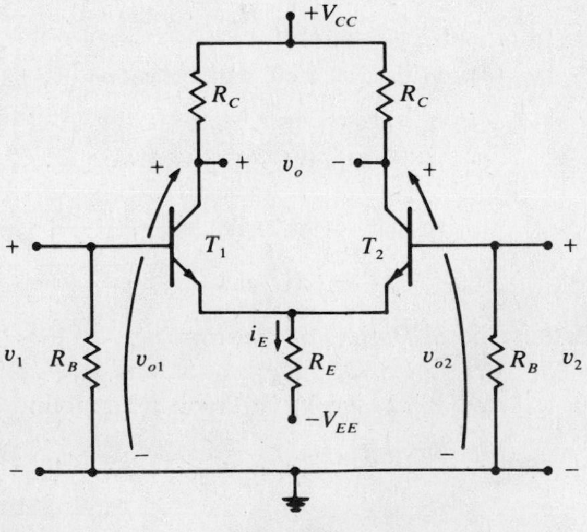

Fig. 3-17

3.21 The Si transistors in the differential amplifier circuit of Fig. 3-17 have negligible leakage current, and $\beta_1 = \beta_2 = 60$. Also, $R_C = 6.8\ \text{k}\Omega$, $R_B = 10\ \text{k}\Omega$, and $V_{CC} = V_{EE} = 15$ V. Find the value of R_E needed to bias the amplifier such that $V_{CEQ1} = V_{CEQ2} = 8$ V.

By symmetry, $I_{EQ1} = I_{EQ2}$. Then, by KCL,

$$i_E = I_{EQ1} + I_{EQ2} = 2I_{EQ1} \qquad (1)$$

Using (*1*) and (*2*) of Problem 3.16 (which apply to the T_1 circuit here), along with KVL around the left collector loop, gives

$$V_{CC} + V_{EE} = \frac{\beta_1}{\beta_1 + 1} I_{EQ1}R_C + V_{CEQ1} + 2I_{EQ1}R_E \qquad (2)$$

Applying KVL around the left base loop gives

$$V_{EE} = I_{BQ1}R_B + V_{BEQ1} + i_E R_E = \frac{I_{EQ1}}{\beta_1 + 1} R_B + V_{BEQ1} + 2I_{EQ1}R_E \qquad (3)$$

Solving (*3*) for $2I_{EQ1}R_E$, substituting the result into (*2*), and solving for I_{EQ1} yield

$$I_{EQ1} = \frac{(\beta_1 + 1)(V_{CC} - V_{CEQ1} + V_{BEQ1})}{\beta_1 R_C - R_B} = \frac{(60 + 1)(15 - 8 + 0.7)}{(60)(6.8 \times 10^3) - 10 \times 10^3} = 1.18 \text{ mA}$$

and, by (*3*),

$$R_E = \frac{V_{EE} - V_{BEQ1} - \dfrac{R_B}{\beta_1 + 1} I_{EQ1}}{2I_{EQ1}} = \frac{15 - 0.7 - \dfrac{10 \times 10^3}{60 + 1} 1.18 \times 10^{-3}}{2(1.18 \times 10^{-3})} = 5.97 \text{ k}\Omega$$

3.22 The Si transistor of Fig. 3-18 has negligible leakage current, and $\beta = 100$. If $V_{CC} = 15$ V, $V_{EE} = 4$ V, $R_E = 3.3$ kΩ, and $R_C = 7.1$ kΩ, find (*a*) I_{BQ} and (*b*) V_{CEQ}.

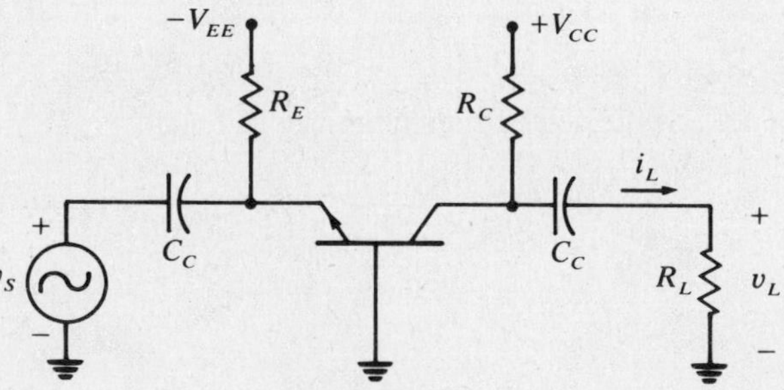

Fig. 3-18

(*a*) By KVL around the base-emitter loop,

$$I_{EQ} = \frac{V_{EE} - V_{BEQ}}{R_E} = \frac{4 - 0.7}{3.3 \times 10^3} = 1 \text{ mA}$$

Then, by (*3.1*) and (*3.2*),

$$I_{BQ} = \frac{I_{EQ}}{\beta + 1} = \frac{1 \times 10^{-3}}{100 + 1} = 9.9 \ \mu\text{A}$$

(*b*) KVL and (*2*) of Problem 3.16 yield

$$V_{CEQ} = V_{CC} + V_{EE} - I_{EQ}R_E - I_{CQ}R_C = V_{CC} + V_{EE} - \left(R_E + \frac{\beta}{\beta + 1} R_C\right) I_{EQ}$$

$$= 15 + 4 - \left(3.3 \times 10^3 + \frac{100}{100 + 1} 7.1 \times 10^3\right)(1 \times 10^{-3}) = 8.67 \text{ V}$$

3.23 Find the proper collector current bias for maximum symmetrical (or undistorted) swing along the ac load line of a transistor amplifier for which $V_{CE\text{sat}} = I_{CEO} = 0$.

For *maximum symmetrical swing*, the Q point must be set at the midpoint of the ac load line. Hence, from (*3.13*), we want

$$I_{CQ} = \frac{1}{2} i_{C\max} = \frac{1}{2}\left(\frac{V_{CEQ}}{R_{ac}} + I_{CQ}\right) \qquad (1)$$

But for a circuit such as that in Fig. 3-5(*a*), KVL gives

$$V_{CEQ} \approx V_{CC} - I_{CQ}R_{dc} \qquad (2)$$

which becomes an equality if no emitter resistor is present. Substituting (*2*) into (*1*), assuming equality, and solving for I_{CQ} yield the desired result:

$$I_{CQ} = \frac{V_{CC}}{R_{ac} + R_{dc}} \qquad (3)$$

3.24 In the circuit of Fig. 3-4(*a*), $R_E = 300\ \Omega$, $R_C = 500\ \Omega$, $V_{CC} = 15$ V, $\beta = 100$, and the Si transistor has β-independent bias. Size R_1 and R_2 for maximum symmetrical swing if $V_{CE\text{sat}} \approx 0$.

For maximum symmetrical swing, the quiescent collector current is

$$I_{CQ} = \frac{1}{2}\frac{V_{CC}}{R_E + R_C} = \frac{15}{2(300 + 500)} = 9.375 \text{ mA}$$

Standard practice is to use a factor of 10 as the margin of inequality for β independence in (*3.8*). Then,

$$R_B = \frac{\beta R_E}{10} = \frac{(100)(300)}{10} = 3 \text{ k}\Omega$$

and, from (*3.7*),

$$V_{BB} \approx V_{BEQ} + I_{CQ}(1.1R_E) = 0.7 + (9.375 \times 10^{-3})(330) = 3.794 \text{ V}$$

Equations (*3.5*) may now be solved simultaneously to obtain

$$R_1 = \frac{R_B}{1 - V_{BB}/V_{CC}} = \frac{3 \times 10^3}{1 - 3.794/15} = 4.02 \text{ k}\Omega$$

and

$$R_2 = R_B\frac{V_{CC}}{V_{BB}} = 3 \times 10^3\frac{15}{3.794} = 11.86 \text{ k}\Omega$$

3.25 In the circuit of Fig. 3-6(*a*), the transistor is a Si device, $R_E = 200\ \Omega$, $R_2 = 10R_1 = 10$ kΩ, $R_L = R_C = 2$ kΩ, $\beta = 100$, and $V_{CC} = 15$ V. Assume that C_C and C_E are very large, that $V_{CE\text{sat}} \approx 0$, and that $i_C = 0$ at cutoff. Find (*a*), I_{CQ}, (*b*)V_{CEQ}, (*c*) the slope of the ac load line, (*d*) the slope of the dc load line, and (*e*) the peak value of undistorted i_L.

(*a*) Equations (*3.5*) and (*3.7*), give

$$R_B = \frac{(1 \times 10^3)(10 \times 10^3)}{11 \times 10^3} = 909\ \Omega \qquad \text{and} \qquad V_{BB} = \frac{1 \times 10^3}{11 \times 10^3}15 = 1.364 \text{ V}$$

so

$$I_{CQ} \approx \frac{V_{BB} - V_{BEQ}}{R_B/(\beta + 1) + R_E} = \frac{1.364 - 0.7}{(909/101) + 200} = 3.177 \text{ mA}$$

(*b*) KVL around the collector-emitter circuit, with $I_{CQ} \approx I_{EQ}$, gives

$$V_{CEQ} = V_{CC} - I_{CQ}(R_E + R_C) = 15 - (3.177 \times 10^{-3})(2.2 \times 10^3) = 8.01 \text{ V}$$

(*c*)

$$\text{Slope} = \frac{1}{R_{ac}} = \frac{1}{R_C} + \frac{1}{R_L} = 2\frac{1}{2 \times 10^3} = 1 \text{ mS}$$

(*d*)

$$\text{Slope} = \frac{1}{R_{dc}} = \frac{1}{R_C + R_E} = \frac{1}{2.2 \times 10^3} = 0.454 \text{ mS}$$

(*e*) From (*3.14*), the ac load line intersects the v_{CE} axis at

$$v_{CE\max} = V_{CEQ} + I_{CQ}R_{\text{ac}} = 8.01 + (3.177 \times 10^{-3})(1 \times 10^{3}) = 11.187 \text{ V}$$

Since $v_{CE\max} < 2V_{CEQ}$, cutoff occurs before saturation and thus sets V_{cem}. With the large capacitors appearing as ac shorts,

$$i_L = \frac{v_L}{R_L} = \frac{v_{ce}}{R_L}$$

or, in terms of peak values,

$$I_{Lm} = \frac{V_{cem}}{R_L} = \frac{v_{CE\max} - V_{CEQ}}{R_L} = \frac{11.187 - 8.01}{2 \times 10^{3}} = 1.588 \text{ mA}$$

3.26 In the circuit of Fig. 3-4(*a*), $R_C = 300\ \Omega$, $R_E = 200\ \Omega$, $R_1 = 2\ \text{k}\Omega$, $R_2 = 15\ \text{k}\Omega$, $V_{CC} = 15$ V, and $\beta = 110$ for the Si transistor. Assume that $I_{CQ} \approx I_{EQ}$ and $V_{CE\text{sat}} \approx 0$. Find the maximum symmetrical swing in collector current (*a*) if an ac base current is injected, and (*b*) if V_{CC} is changed to 10 V but all else remains the same.

(*a*) From (*3.5*) and (*3.7*),

$$R_B = \frac{(2 \times 10^{3})(15 \times 10^{3})}{17 \times 10^{3}} = 1.765\ \text{k}\Omega \qquad \text{and} \qquad V_{BB} = \frac{2 \times 10^{3}}{17 \times 10^{3}}\, 15 = 1.765 \text{ V}$$

so $$I_{CQ} \approx I_{EQ} = \frac{V_{BB} - V_{BEQ}}{R_B/(\beta + 1) + R_E} = \frac{1.765 - 0.7}{1765/111 + 200} = 4.93 \text{ mA}$$

By KVL around the collector-emitter circuit with $I_{CQ} \approx I_{EQ}$,

$$V_{CEQ} = V_{CC} - I_{CQ}(R_C + R_E) = 15 - (4.93 \times 10^{-3})(200 + 300) = 12.535 \text{ V}$$

Since $V_{CEQ} > V_{CC}/2 = 7.5$ V, cutoff occurs before saturation, and i_C can swing ± 4.93 mA about I_{CQ} and remain in the active region.

(*b*) $$V_{BB} = \frac{R_1}{R_1 + R_2} V_{CC} = \frac{2 \times 10^{3}}{17 \times 10^{3}}\, 10 = 1.1765 \text{ V}$$

so that $$I_{CQ} \approx I_{EQ} = \frac{V_{BB} - V_{BEQ}}{R_B/(\beta + 1) + R_E} = \frac{1.1765 - 0.7}{1765/111 + 200} = 2.206 \text{ mA}$$

and $$V_{CEQ} = V_{CC} - I_{CQ}(R_C + R_E) = 10 - (2.206 \times 10^{-3})(0.5) = 8.79 \text{ V}$$

Since $V_{CEQ} > V_{CC}/2 = 5$ V, cutoff again occurs before saturation, and i_C can swing ± 2.206 mA about I_{CQ} and remain in the active region of operation. Here, the 33.3 percent reduction in power supply voltage has resulted in a reduction of over 50 percent in symmetrical collector-current swing.

3.27 If a Si transistor were removed from the circuit of Fig. 3-4(*a*) and a Ge transistor of identical β were substituted, would the *Q* point move in the direction of saturation or of cutoff?

Since R_1, R_2, and V_{CC} are unchanged, R_B and V_{BB} would remain unchanged. However, owing to the different emitter-to-base forward drops for Si (0.7 V) and Ge (0.3 V) transistors,

$$I_{CQ} \approx \frac{V_{BB} - V_{BEQ}}{R_B/(\beta + 1) + R_E}$$

would be higher for the Ge transistor. Thus, the *Q* point would move in the direction of saturation.

3.28 In the circuit of Fig. 3-6(*a*), $V_{CC} = 12$ V, $R_C = R_L = 1\ \text{k}\Omega$, $R_E = 100\ \Omega$, and $C_C = C_E \rightarrow \infty$. The Si transistor has negligible leakage current, and $\beta = 100$. If $V_{CE\text{sat}} = 0$ and the transistor is to have β-independent bias (by having $R_1 \| R_2 = \beta R_E/10$), size R_1 and R_2 for maximum symmetrical swing.

Evaluating R_{ac} and R_{dc}, we find

$$R_{ac} = R_L \| R_C = \frac{(1\times 10^3)(1\times 10^3)}{1\times 10^3 + 1\times 10^3} = 500\ \Omega \qquad R_{dc} = R_C + R_E = 1\times 10^3 + 100 = 1100\ \Omega$$

Thus, according to (*3*) of Problem 3.23, maximum symmetrical swing requires that

$$I_{CQ} = \frac{V_{CC}}{R_{ac} + R_{dc}} = \frac{12}{500 + 1100} = 7.5\ \text{mA}$$

Now,
$$R_B = R_1 \| R_2 = \frac{\beta R_E}{10} = \frac{(100)(100)}{10} = 1\ \text{k}\Omega$$

and, by (*3.6*) and (*2*) of Problem 3.16,

$$V_{BB} = \left(\frac{R_B}{\beta} + \frac{\beta+1}{\beta} R_E\right) I_{CQ} + V_{BEQ} = \left(\frac{1\times 10^3}{100} + \frac{100+1}{100}\,100\right) 7.5\times 10^{-3} + 0.7 = 1.53\ \text{V}$$

Finally, from (*3.5*),

$$R_1 = \frac{R_B}{1 - V_{BB}/V_{CC}} = \frac{1\times 10^3}{1 - 1.53/12} = 1.34\ \text{k}\Omega \qquad \text{and} \qquad R_2 = \frac{R_B V_{CC}}{V_{BB}} = \frac{(1\times 10^3)(12)}{1.53} = 10.53\ \text{k}\Omega$$

3.29 The Si transistor of Fig. 3-6(*a*) has $V_{CE\text{sat}} = I_{CBO} = 0$ and $\beta = 75$. C_E is removed from the circuit, and $C_C \to \infty$. Also, $R_1 = 1\ \text{k}\Omega$, $R_2 = 9\ \text{k}\Omega$, $R_E = R_L = R_C = 1\ \text{k}\Omega$, and $V_{CC} = 15$ V. (*a*) Sketch the dc and ac load lines for this amplifier on a set of i_C-v_{CE} axes. (*b*) Find the maximum undistorted value of i_L, and determine whether cutoff or saturation limits i_L swing.

(*a*)
$$R_{dc} = R_C + R_E = 1\times 10^3 + 1\times 10^3 = 2\ \text{k}\Omega$$

and
$$R_{ac} = R_E + R_C \| R_L = 1\times 10^3 + \frac{(1\times 10^3)(1\times 10^3)}{1\times 10^3 + 1\times 10^3} = 1.5\ \text{k}\Omega$$

By (*3.5*),

$$V_{BB} = \frac{R_1}{R_2} V_{CC} = \frac{1\times 10^3}{9\times 10^3}\,15 = 1.667\ \text{V} \qquad \text{and} \qquad R_B = R_1 \| R_2 = \frac{(1\times 10^3)(9\times 10^3)}{1\times 10^3 + 9\times 10^3} = 900\ \Omega$$

and from (*3.7*),

$$I_{CQ} = \frac{(\beta+1)(V_{BB} - V_{BEQ})}{R_B + (\beta+1)R_E} = \frac{(75+1)(1.667 - 0.7)}{900 + (75+1)(1\times 10^3)} = 0.96\ \text{mA}$$

By KVL around the collector loop and (*2*) of Problem 3.16,

$$V_{CEQ} = V_{CC} - \left(R_C + \frac{\beta+1}{\beta} R_E\right) I_{CQ} = 15 - \left(1\times 10^3 + \frac{75+1}{75}\,1\times 10^3\right) 0.96\times 10^{-3} = 13.07\ \text{V}$$

The ac load-line intercepts now follow directly from (*3.13*) and (*3.14*):

$$i_{C\text{max}} = \frac{V_{CEQ}}{R_{ac}} + I_{CQ} = \frac{13.07}{1.5\times 10^3} + 0.96\times 10^{-3} = 9.67\ \text{mA}$$

$$v_{CE\text{max}} = V_{CEQ} + I_{CQ} R_{ac} = 13.07 + (0.96\times 10^{-3})(1.5\times 10^3) = 14.51\ \text{V}$$

The dc load-line intercepts follow from (*3.9*):

$$i_C\text{-axis intercept} = \frac{V_{CC}}{R_{dc}} = \frac{15}{2\times 10^3} = 7.5\ \text{mA}$$

$$v_{CE}\text{-axis intercept} = V_{CC} = 15\ \text{V}$$

The required load lines are sketched in Fig. 3-19.

(*b*) Since $I_{CQ} < \frac{1}{2} i_{C\text{max}}$, it is apparent that cutoff limits the undistorted swing of i_c to $\pm I_{CQ} = \pm 1.92$ mA. By current division,

$$i_L = \frac{R_E}{R_E + R_L} i_c = \frac{1\times 10^3}{1\times 10^3 + 1\times 10^3} (\pm 0.96\ \text{mA}) = \pm 0.48\ \text{mA}$$

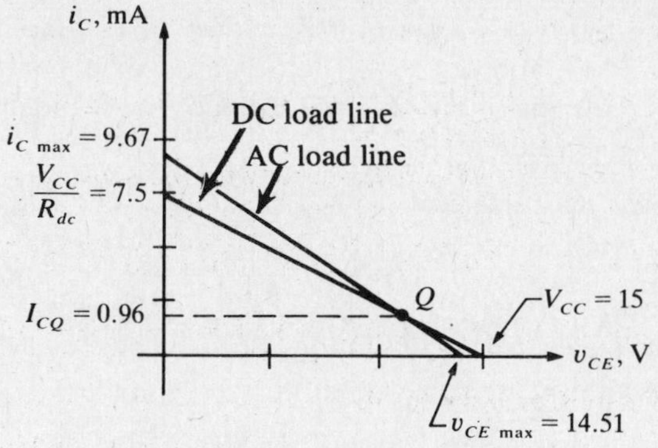

Fig. 3-19

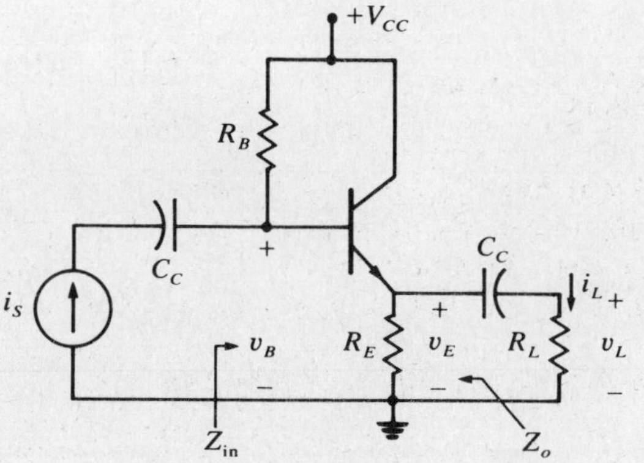

(*a*)

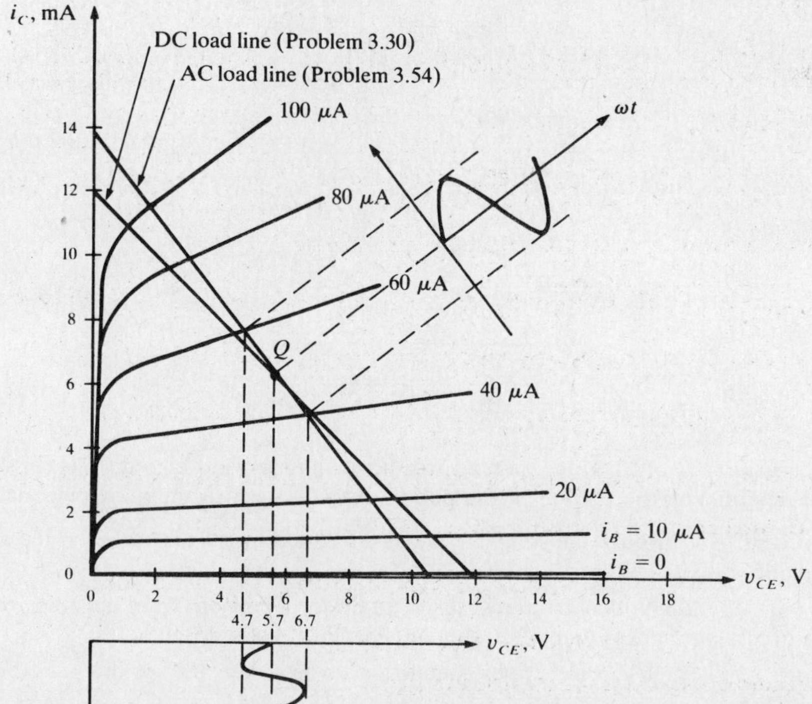

(*b*)

Fig. 3-20

3.30 In the *common-collector* (CC) or *emitter-follower* (EF) amplifier of Fig. 3-20(*a*), $V_{CC} = 12$ V, $R_E = 1$ kΩ, $R_L = 3$ kΩ, and $C_C \to \infty$. The Si transistor is biased so that $V_{CEQ} = 5.7$ V and has the collector characteristic of Fig. 3-20(*b*). (*a*) Construct the dc load line. (*b*) Find the value of β. (*c*) Determine the value of R_B.

(*a*) The dc load line must intercept the v_{CE} axis at $V_{CC} = 12$ V. It intercepts the i_C axis at

$$\frac{V_{CC}}{R_{dc}} = \frac{V_{CC}}{R_E} = \frac{12}{1 \times 10^3} = 12 \text{ mA}$$

The intercepts are connected to form the dc load line shown on Fig. 3-20(*b*).

(*b*) I_{BQ} is determined by entering Fig. 3-20(*b*) at $V_{CEQ} = 5.7$ V and interpolating between i_B curves to find $I_{BQ} \approx 50\ \mu$A. I_{CQ} is then read as ≈ 6.3 mA. Thus,

$$\beta = \frac{I_{CQ}}{I_{BQ}} = \frac{6.3 \times 10^{-3}}{50 \times 10^{-6}} = 126$$

(*c*) By KVL,

$$R_B = \frac{V_{CC} - V_{BEQ} - \dfrac{\beta + 1}{\beta} I_{CQ} R_E}{I_{BQ}} = \frac{12 - 0.7 - \dfrac{126+1}{126}(6 \times 10^{-3})(1 \times 10^3)}{50 \times 10^{-6}} = 105.05 \text{ k}\Omega$$

3.31 The amplifier of Fig. 3-21 uses a Si transistor for which $V_{BEQ} = 0.7$ V. Assuming that the collector-emitter bias does not limit voltage excursion, classify the amplifier according to Table 3-3 if (*a*) $V_B = 1.0$ V and $v_S = 0.25 \cos \omega t$ V, (*b*) $V_B = 1.0$ V and $v_S = 0.5 \cos \omega t$ V, (*c*) $V_B = 0.5$ V and $v_S = 0.6 \cos \omega t$ V, (*d*) $V_B = 0.7$ V and $v_S = 0.5 \cos \omega t$ V.

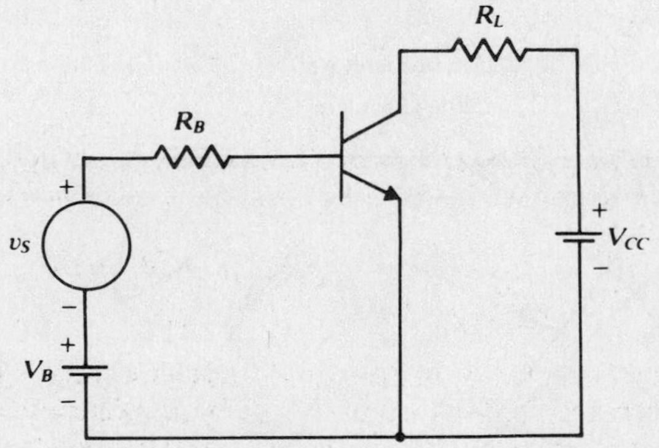

Fig. 3-21

As long as $v_S + V_B > 0.7$ V, the emitter-base junction is forward-biased; thus classification becomes a matter of determining the portion of the period of v_S over which the above inequality holds.

(*a*) $v_S + V_B \geq 0.75$ V through the complete cycle; thus the transistor is always in the active region, and the amplifier is of class A.

(*b*) $0.5 \leq v_S + V_B \leq 1.5$ V; thus the transistor is cut off for a portion of the negative excursion v_S. Since cutoff occurs during less than 180°, the amplifier is of class AB.

(*c*) $-0.1 \leq v_S + V_B \leq 1.1$ V, which gives conduction for less than 180° of the period of v_S, for class C operation.

(*d*) $v_S + V_B \geq 0.7$ V over exactly 180° of the period of v_S, for class B operation.

Supplementary Problems

3.32 The leakage currents of a transistor are $I_{CBO} = 5\ \mu A$ and $I_{CEO} = 0.4$ mA, and $I_B = 30\ \mu A$. Determine the value of I_C. *Ans.* 2.77 mA

3.33 For a BJT, $I_C = 5.2$ mA, $I_B = 50\ \mu A$, and $I_{CBO} = 0.5\ \mu A$. (*a*) Find β and I_{EQ}. (*b*) What is the percentage error in the calculation of β if the leakage current is assumed zero?
Ans. (*a*) 102.96, 5.25 mA; (*b*) 1.01%

3.34 Collector-to-base leakage current can be modeled by a current source as in Fig. 3-22, with the understanding that transistor action relates currents I'_C, I'_B, and I_E ($I'_C = \alpha I_E$, and $I'_C = \beta I'_B$). Prove that

$$(a)\ I_C = \beta I_B + (\beta + 1)I_{CBO} \quad (b)\ I_B = \frac{I_E}{\beta + 1} - I_{CBO} \quad (c)\ I_E = \frac{\beta + 1}{\beta}(I_C - I_{CBO})$$

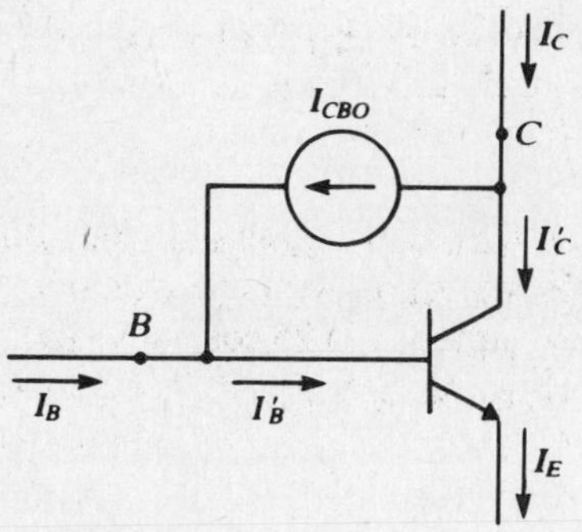

Fig. 3-22

3.35 If the transistor of Problem 3.4 were replaced by a new transistor with 1 percent greater α, what would be the percentage change in emitter current? *Ans.* a 96.07% increase

3.36 In the circuit of Fig. 3-7, $V_{CEsat} = 0.2$ V, $\alpha = 0.99$, $I_{BQ} = 20\ \mu A$, $V_{CC} = 15$ V, and $R_C = 15\ k\Omega$. What is the value of V_{CEQ}? *Ans.* $V_{CEQ} = V_{CEsat} = 0.2$ V

3.37 In many switching applications, the transistor may be utilized without a heat sink, since $P_C \approx 0$ in cutoff and P_C is small in saturation. Support this statement by calculating the collector power dissipated in (*a*) Problem 3.6 (active-region bias) and (*b*) Problem 3.36 (saturation-region bias).
Ans. (*a*) 18 mW; (*b*) 0.39 mW

3.38 The collector characteristics of the transistor of Fig. 3-7 are given in Fig. 3-5(*b*). If $I_{BQ} = 40\ \mu A$, $V_{CC} = 15$ V, and $R_C = 2.2\ k\Omega$, specify the minimum power rating of the transistor to ensure there is no danger of thermal damage. *Ans.* 22.54 mW

3.39 In the circuit of Fig. 3-9, $V_{CC} = 20$ V, $R_C = 5\ k\Omega$, $R_E = 4\ k\Omega$, and $R_B = 500\ k\Omega$. The Si transistor has $I_{CBO} = 0$ and $\beta = 50$. Find I_{CQ} and V_{CEQ}. *Ans.* 1.91 mA, 2.64 V

3.40 The transistor of Problem 3.39 failed and was replaced with a new transistor with $I_{CBO} = 0$ and $\beta = 75$. Is the transistor still biased for active-region operation?
Ans. Since the calculated $V_{CEQ} = -6.0\ V < 0$, the transistor is not in the active region.

3.41 What value of R_B will result in saturation of the Si transistor of Fig. 3-9 if $V_{CC} = 20$ V. $R_C = 5\ k\Omega$, $R_E = 4\ k\Omega$, $\beta = 50$, and $V_{CEsat} = 0.2$ V? *Ans.* $R_B \leq 442.56\ k\Omega$

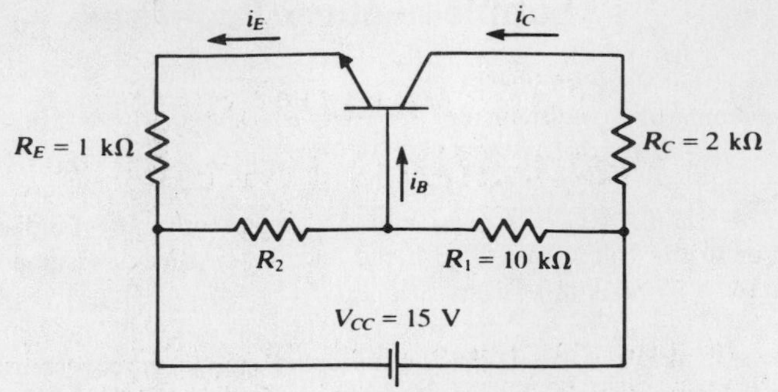

Fig. 3-23

3.42 The circuit of Fig. 3-23 illustrates a method for biasing a CB transistor using a single dc source. The transistor is a Si device ($V_{BEQ} = 0.7$ V), $\beta = 99$, and $I_{BQ} = 30$ μA. Find (*a*) R_2, and (*b*) V_{CEQ}.
Ans. (*a*) 3.36 kΩ; (*b*) 6.06 V

3.43 Rework Problem 3.26(*a*) with $R_2 = 5$ kΩ and all else unchanged.
Ans. ± 13.16 mA about $I_{CQ} = 16.84$ mA

3.44 Because of a poor solder joint, resistor R_1 of Problem 3.26(*a*) becomes open-circuited. Calculate the percentage change in I_{CQ} that will be observed. *Ans.* +508.5%

3.45 The circuit of Problem 3-26(*a*) has β-independent bias ($R_E \geq 10R_B/\beta$). Find the allowable range of β if I_{CQ} can change at most ± 2 percent from its value for $\beta = 110$. *Ans.* $86.4 \leq \beta \leq 149.7$

3.46 For the circuit of Fig. 3-21, $v_S = 0.25 \cos \omega t$ V, $R_B = 30$ kΩ, $V_B = 1$ V, and $V_{CC} = 12$ V. The transistor is a Si device with negligible base-to-emitter resistance. Assume that $V_{CE\text{sat}} \approx 0$ and $I_{CBO} = 0$. Find the range of R_L for class A operation. *Ans.* $R_L \leq 6.545$ kΩ

3.47 If an emitter resistor is added to the circuit of Fig. 3-13, find the value of R_F needed to bias for maximum symmetrical swing. Let $V_{CC} = 15$ V, $R_E = 1.5$ kΩ, and $R_C = 5$ kΩ. Assume the transistor is an Si device with $I_{CEO} = V_{CE\text{sat}} = 0$ and $\beta = 80$. *Ans.* 477.4 kΩ

3.48 In the circuit of Fig. 3-15, the Ge transistor has $I_{CEO} = 0$ and $\beta = 50$. Assume the capacitor is replaced with a short circuit. Let $V_S = 2$ V, $V_{CC} = 12$ V, $R_C = 4$ kΩ, $R_S = 100$ kΩ, and $R_B = 330$ kΩ. Find the ratios (*a*) I_{CQ}/I_s and (*b*) V_{CEQ}/V_S. *Ans.* (*a*) 374.6; (*b*) 0.755

3.49 In the differential amplifier of Fig. 3-17, the Si transistors have negligible leakage current, and $\beta_1 = \beta_2 = 75$. If $R_B = 10$ kΩ, $R_E = R_C = 6.8$ kΩ, and $V_{EE} = V_{CC} = 15$ V, find V_{CEQ1}. *Ans.* 8.85 V

3.50 Obviously, for the balanced transistors of Problem 3.49, $v_o = 0$ since $v_{o1} = v_{o2}$. If T_2 is replaced with a transistor with $\beta_2 = 60$ and all else is unchanged, determine (*a*) V_{CEQ1}, (*b*) V_{CEQ2}, and (*c*) the resulting voltage v_0 (called the *dc offset voltage*). *Ans.* (*a*) 9.66 V; (*b*) 8.11 V; (*c*) 1.55 V

3.51 Find the voltage $v_{o1} = v_{o2}$ for Problem 3.49. *Ans.* 8.01 V

3.52 In the amplifier of Fig. 3-6(*a*), $R_1 = 1$ kΩ, $R_2 = 9$ kΩ, $R_E = 100$ Ω, $R_L = 1$ kΩ, $V_{CC} = 12$ V, $C_C = C_E \to \infty$, and $\beta = 100$. The Si transistor has negligible leakage current, with $V_{CE\text{sat}} = I_{CBO} = 0$. Find R_C so that v_L exhibits maximum symmetrical swing. *Ans.* 1.89 kΩ

3.53 If in Problem 3.29, R_1 is changed to 9 kΩ and all else remains unchanged, determine the maximum undistorted swing of i_c. *Ans.* ± 1.5 mA

3.54 In the CC amplifier of Problem 3.30, let $i_S = 10 \sin \omega t$ μA. Calculate v_L after graphically determining v_{CE}. *Ans.* The ac load line and v_{CE} are sketched on Fig. 3-20: $v_{CE} \approx 5.7 - \sin \omega t$ V; $v_L = \sin \omega t$ V

Chapter 4

Characteristics of Field-Effect Transistors

4.1 INTRODUCTION

The operation of the *field-effect transistor* (FET) can be explained in terms of only majority-carrier (one-polarity) charge flow; the transistor is therefore called *unipolar*. Two kinds of field-effect devices are widely used: the *junction field-effect transistor* (JFET) and the *metal-oxide-semiconductor field-effect transistor* (MOSFET).

4.2 JFET CONSTRUCTION AND SYMBOLS

The physical arrangement of, and symbols for, the two kinds of JFET are shown in Fig. 4-1. Conduction is by the passage of charge carriers from *source* (S) to *drain* (D) through the *channel* between the *gate* (G) elements.

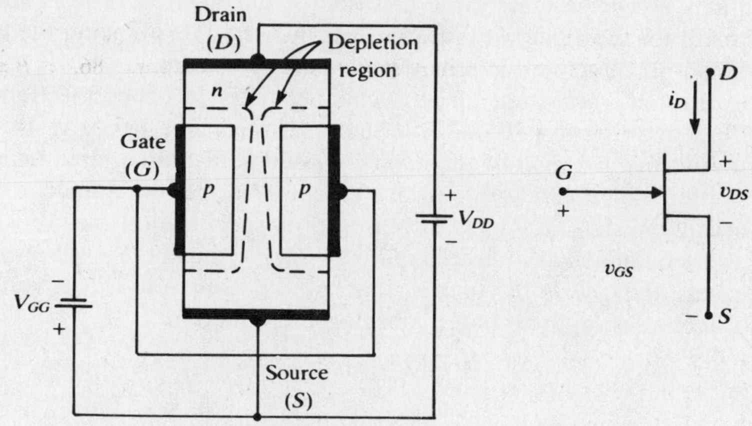

(*a*) *n*-channel JFET

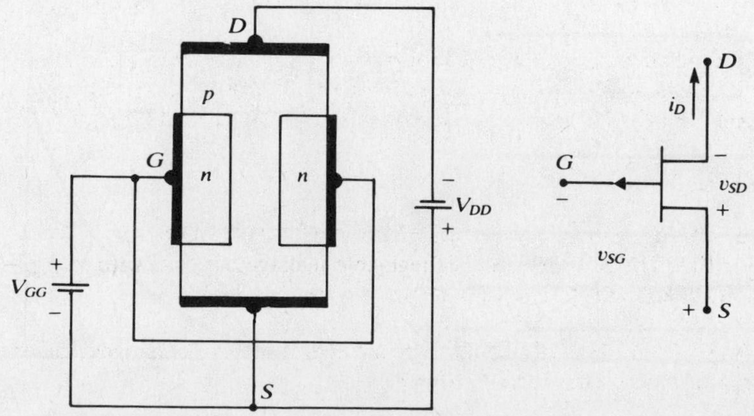

(*b*) *p*-channel JFET

Fig. 4-1

The transistor can be an *n-channel* device (conduction by electrons) or a *p-channel* device (conduction by holes); a discussion of *n*-channel devices applies equally to *p*-channel devices if *complementary* (opposite in sign) voltages and currents are used. Analogies between the JFET and the BJT are shown in Table 4-1. Current and voltage symbology for FETs parallels that given in Table 3-1.

Table 4-1

JFET	BJT
source S	emitter E
drain D	collector C
gate G	base B
drain supply V_{DD}	collector supply V_{CC}
gate supply V_{GG}	base supply V_{BB}
drain current i_D	collector current i_C

4.3 JFET TERMINAL CHARACTERISTICS

The JFET is almost universally applied in the *common-source* (CS) two-port arrangement of Fig. 4-1, where v_{GS} maintains a reverse bias of the gate-source *pn* junction. The resulting gate leakage current is negligibly small for most analyses (usually less than 1 μA), allowing the gate to be treated as an open circuit. Thus, no input characteristic curves are necessary.

Typical *output* or *drain characteristics* for an *n*-channel JFET in CS connection with $v_{GS} \leq 0$ are given in Fig. 4-2(*a*). For a constant value of v_{GS}, the JFET acts as a linear resistive device (in the *ohmic region*) until the *depletion region* of the reverse-biased gate-source junction extends the width of the channel (a condition called *pinchoff*). Above pinchoff but below avalanche breakdown, drain current i_D remains nearly constant as v_{DS} is increased. For specification purposes, the *shorted-gate parameters* I_{DSS} and V_{p0} are defined as indicated in Fig. 4-2(*a*); typically, V_{p0} is between 4 and 5 V. As gate potential decreases, the *pinchoff voltage*, that is, the source-to-drain voltage V_p at which pinchoff occurs, also decreases, approximately obeying the equation

$$V_p = V_{p0} + v_{GS} \tag{4.1}$$

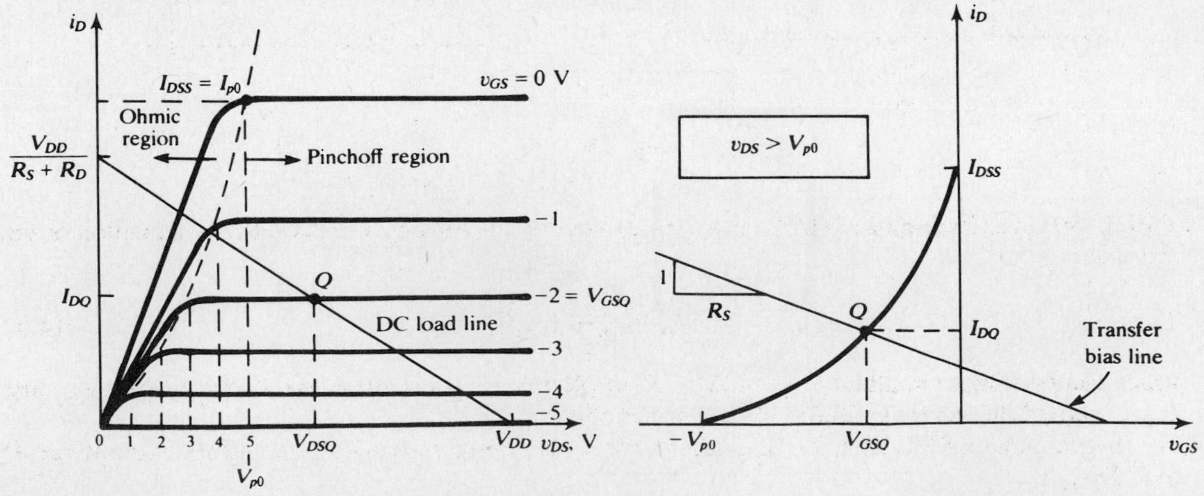

(*a*) Drain characteristics (*b*) Transfer characteristic

Fig. 4-2 CS *n*-channel JFET

The drain current shows an approximate square-law dependence on source-to-gate voltage for constant values of v_{DS} in the pinchoff region:

$$i_D = I_{DSS}\left(1 + \frac{v_{GS}}{V_{p0}}\right)^2 \tag{4.2}$$

This accounts for the unequal vertical spacing of the characteristic curves in Fig. 4-2(*a*). Figure 4-2(*b*) is the graph of (*4.2*), known as the *transfer characteristic* and utilized in bias determination. The transfer characteristic is also determined by the intersections of the drain characteristics with a fixed vertical line, v_{DS} = constant. To the extent that the drain characteristics actually are horizontal in the pinchoff region, one and the same transfer characteristic will be found for all $v_{DS} > V_{p0}$. (See Fig. 4-4 for a slightly nonideal case.)

4.4 JFET BIAS LINE AND LOAD LINE

The commonly used *voltage-divider* bias arrangement of Fig. 4-3(*a*) can be reduced to its equivalent in Fig. 4-3(*b*), where the Thévenin parameters are given by

$$R_G = \frac{R_1R_2}{R_1 + R_2} \quad \text{and} \quad V_{GG} = \frac{R_1}{R_1 + R_2}V_{DD} \tag{4.3}$$

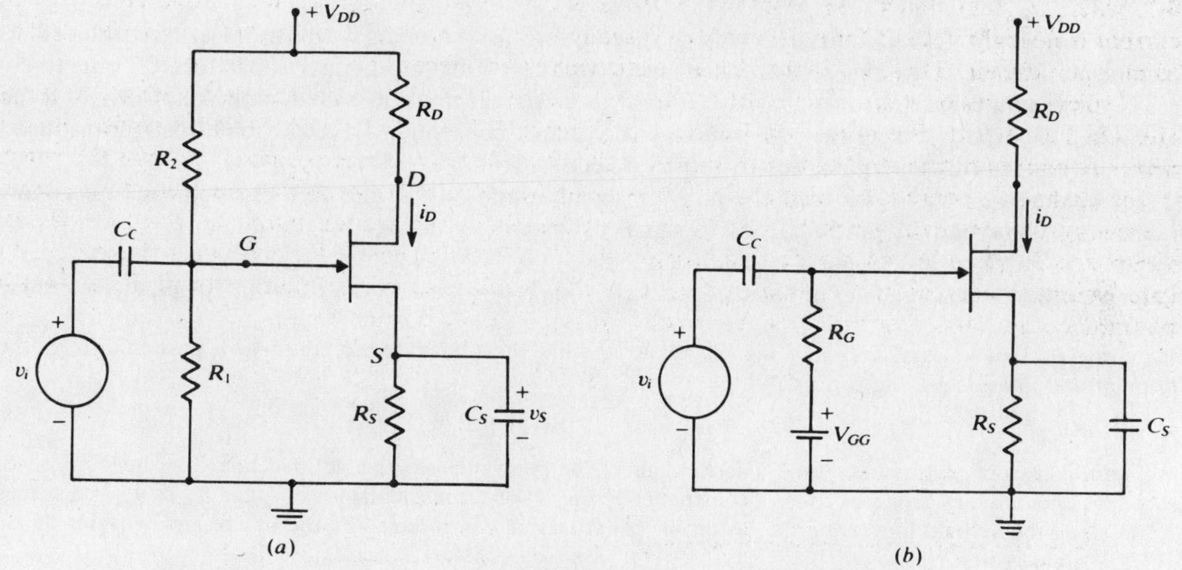

Fig. 4-3

With $i_G = 0$, application of KVL around the gate-source loop of Fig. 4-3(*b*) yields the equation of the *transfer bias line*,

$$i_D = \frac{V_{GG}}{R_S} - \frac{v_{GS}}{R_S} \tag{4.4}$$

which can be solved simultaneously with (*4.2*) or plotted as indicated on Fig. 4-2(*b*) to yield I_{DQ} and V_{GSQ}, two of the necessary three quiescent variables.

Application of KVL around the drain-source loop of Fig. 4-3(*b*) leads to the equation of the *dc load line*,

$$i_D = \frac{V_{DD}}{R_S + R_D} - \frac{v_{DS}}{R_S + R_D} \tag{4.5}$$

which, when plotted on the drain characteristics of Fig. 4-2(a), yields the remaining quiescent value, V_{DSQ}. Alternatively, with I_{DQ} already determined,

$$V_{DSQ} = V_{DD} - (R_S + R_D)I_{DQ}$$

Example 4.1 In the amplifier of Fig. 4-3(a), $V_{DD} = 20$ V, $R_1 = 1$ MΩ, $R_2 = 15.7$ MΩ, $R_D = 3$ kΩ, and $R_S = 2$ kΩ. If the JFET characteristics are given by Fig. 4-4, find (a) I_{DQ}, (b) V_{GSQ}, and (c) V_{DSQ}.

(a) By (*4.3*),

$$V_{GG} = \frac{R_1}{R_1 + R_2} V_{DD} = \frac{1 \times 10^6}{16.7 \times 10^6} 20 = 1.2 \text{ V}$$

On Fig. 4-4(a), we construct the transfer bias line (*4.4*); it intersects the transfer characteristic at the Q point, giving $I_{DQ} = 1.5$ mA.

(b) The Q point of Fig. 4-4(a) also gives $V_{GSQ} = -2$ V.

(c) We construct the dc load line on the drain characteristics, making use of the v_{DS} intercept of $V_{DD} = 20$ V and the i_D intercept of $V_{DD}/(R_S + R_D) = 4$ mA. The Q point was established at $I_{DQ} = 1.5$ mA in part a and at $V_{GSQ} = -2$ V in part b; its abscissa is $V_{DSQ} = 12.5$ V. Analytically,

$$V_{DSQ} = V_{DD} - (R_S + R_D)I_{DQ} = 20 - (5 \times 10^3)(1.5 \times 10^{-3}) = 12.5 \text{ V}$$

4.5 GRAPHICAL ANALYSIS FOR THE JFET

As is done in BJT circuits (Section 3.6), coupling (or blocking) capacitors are introduced to confine dc quantities to the JFET and its bias circuitry. Further, bypass capacitors C_S effectively remove the gain-reducing source resistor insofar as ac signals are concerned, while allowing R_S to be utilized in favorably setting the gate-source bias voltage; consequently, an ac load line is introduced with analysis techniques analogous to those of Section 3.6.

Graphical analysis is favored for large-ac-signal conditions in the JFET, since the square-law relationship between v_{GS} and i_D leads to signal distortion.

Example 4.2 For the amplifier of Example 4.1, let $v_i = \sin t$ ($\omega = 1$ rad/s) and $C_S \rightarrow \infty$. Graphically determine v_{ds} and i_d.

Since C_S appears as a short to ac signals, an ac load line must be added to Fig. 4-4(b), passing through the Q point and intersecting the v_{DS} axis at

$$V_{DSQ} + I_{DQ}R_{ac} = 12.5 + (1.5)(3) = 17 \text{ V}$$

We next construct an auxiliary time axis through Q, perpendicular to the ac load line, for the purpose of showing, on additional auxiliary axes as constructed in Fig. 4-4(b), the excursions of i_d and v_{ds} as $v_{gs} = v_i$ swings ± 1 V along the ac load line. Note the distortion in both signals, introduced by the square-law behavior of the JFET characteristics.

4.6 MOSFET CONSTRUCTION AND SYMBOLS

The n-channel MOSFET (Fig. 4-5) has only a single p region (called the *substrate*), one side of which acts as a conducting channel. A metallic gate is separated from the conducting channel by an insulating metal oxide (usually SiO_2), whence the name *insulated-gate* FET (IGFET) for the device. The p-channel MOSFET, formed by interchanging p and n semiconductor materials, is described by complementary voltages and currents.

4.7 MOSFET TERMINAL CHARACTERISTICS

In an n-channel MOSFET, the gate (positive plate), metal oxide film (dielectric), and substrate (negative plate) form a capacitor, the electric field of which controls channel resistance. When the

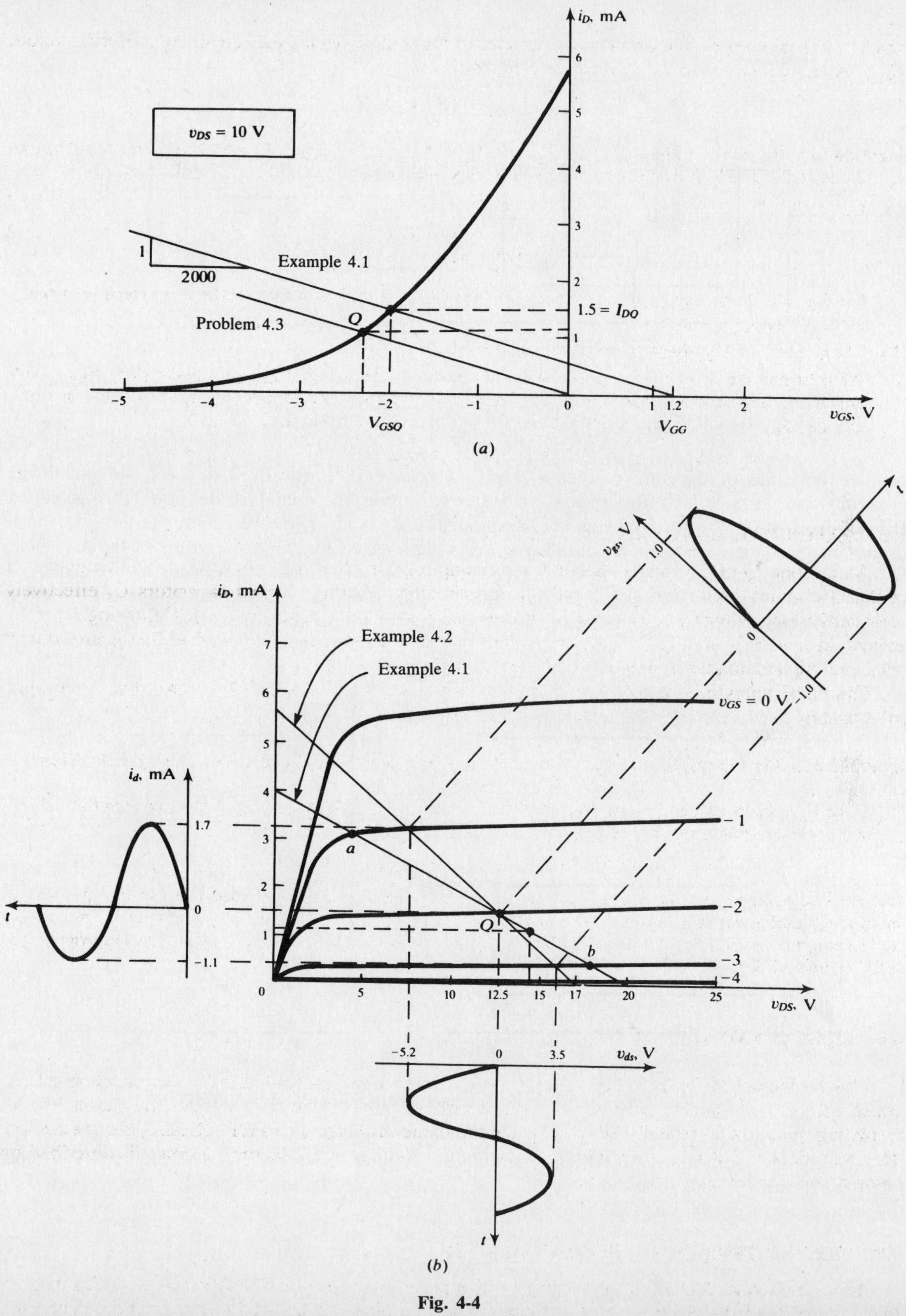

i_D, mA
v_{DS} = 10 V
1
2000
Example 4.1
Problem 4.3
Q
1.5 = I_{DO}
V_{GSO}
V_{GG}
v_{GS}, V
-5
-4
-3
-2
-1
0
1
1.2
2

(a)

i_D, mA
Example 4.2
Example 4.1
v_{GS} = 0 V
−1
−2
−3
−4
a
Q
b
v_{DS}, V
0
5
10
12.5
15
17
20
25
i_d, mA
1.7
0
−1.1
t
v_{gs}, V
1.0
−1.0
−5.2
3.5
v_{ds}, V

(b)

Fig. 4-4

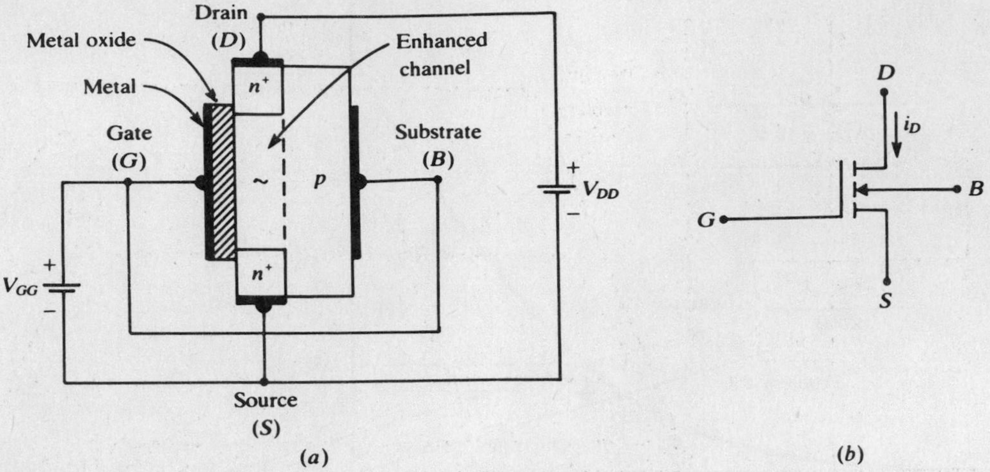

Fig. 4-5

positive potential of the gate reaches a *threshold voltage* V_T (typically 2 to 4 V), sufficient free electrons are attracted to the region immediately beside the metal oxide film (this is called *enhancement-mode* operation) to induce a conducting channel of low resistivity. If the source-to-drain voltage is increased, the enhanced channel is depleted of free charge carriers in the area near the drain, and pinchoff occurs as in the JFET. Typical drain and transfer characteristics are displayed in Fig. 4-6, where $V_T = 2$ V is used for illustration. Commonly, the manufacturer specifies V_T and a value of pinchoff current I_{Don}; the corresponding value of source-to-gate voltage is V_{GSon}.

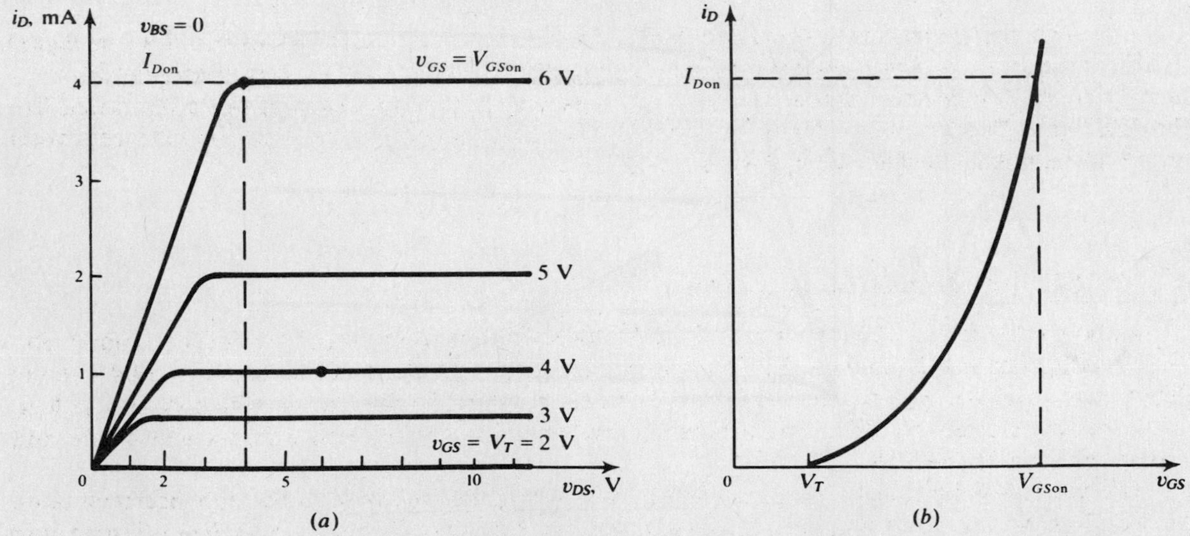

Fig. 4-6 Characteristics of *n*-channel enhancement-mode MOSFET

The enhancement-mode MOSFET, operating in the pinchoff region, is described by (*4.1*) and (*4.2*) if V_{p0} and I_{DSS} are replaced with $-V_T$ and I_{Don}, respectively, and if the substrate is shorted to the source, as in Fig. 4-7(*a*). Then

$$i_D = I_{Don}\left(1 - \frac{v_{GS}}{V_T}\right)^2 \tag{4.6}$$

where $v_{GS} \geq V_T$.

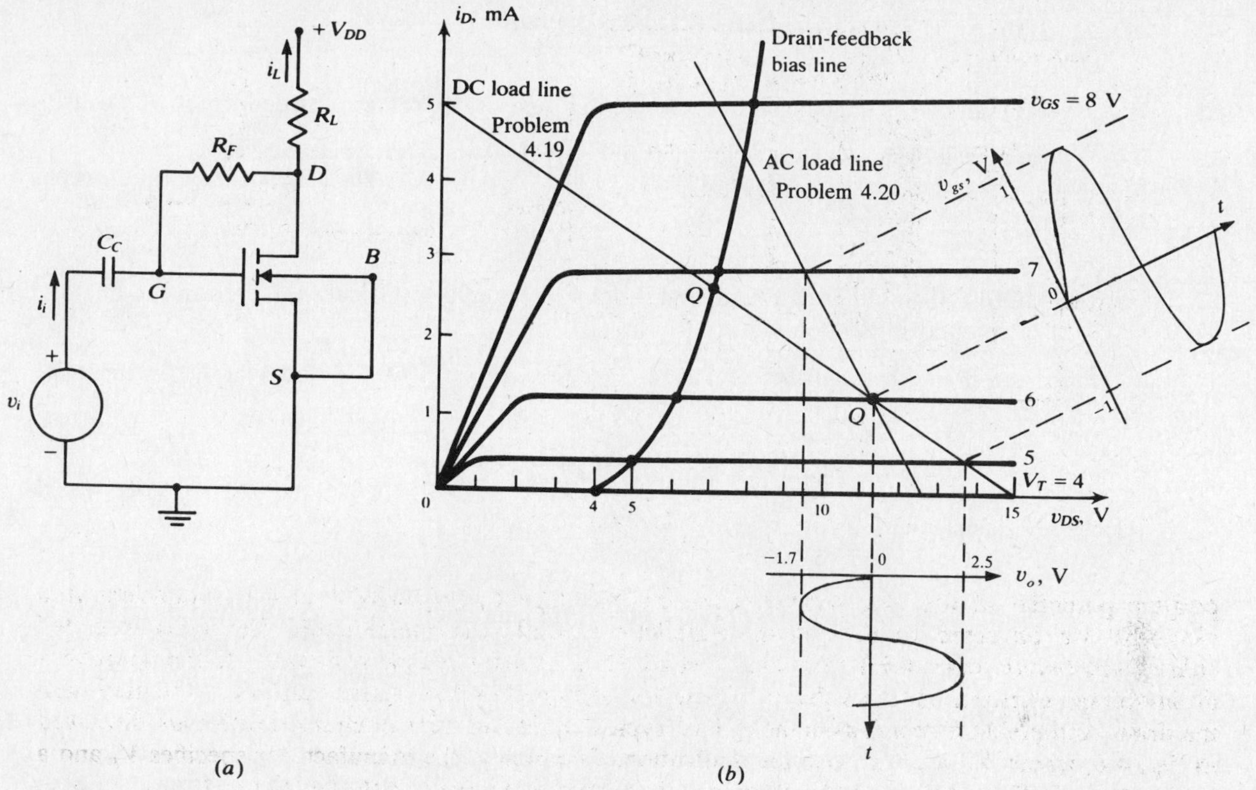

Fig. 4-7

Although the enhancement-mode MOSFET is the more popular (it is widely used in digital switching circuits), a *depletion-mode* MOSFET, characterized by a lightly doped channel between heavily doped source and drain electrode areas, is commercially available that can be operated like the JFET (see Problem 4.21). However, that device displays a gate-source input impedance several orders of magnitude smaller than that of the JFET.

4.8 MOSFET BIAS AND LOAD LINES

Although the transfer characteristic of the MOSFET differs from that of the JFET [compare Fig. 4-2(*b*) with Figs. 4-6(*b*) and 4-20], simultaneous solution with the transfer bias line (*4.4*) allows determination of the gate-source bias V_{GSQ}. Further, graphical procedures in which dc and ac load lines are constructed on drain characteristics can be utilized with both enhancement-mode and depletion-mode MOSFETS.

The voltage-divider bias arrangement (Fig. 4-3) is readily applicable to the enhancement-mode MOSFET; however, since V_{GSQ} and V_{DSQ} are of the same polarity, *drain-feedback bias*, illustrated in Fig. 4-7(*a*), can be utilized to compensate partially for variations in MOSFET characteristics.

Example 4.3 In the amplifier of Fig. 4-7(*a*), $V_{DD} = 15$ V, $R_L = 3$ kΩ, and $R_F = 50$ MΩ. If the MOSFET drain characteristics are given by Fig. 4-7(*b*), determine the values of the quiescent quantities.

The dc load line is constructed on Fig. 4-7(*b*) with v_{DS} intercept of $V_{DD} = 15$ V and i_D intercept of $V_{DD}/R_L = 5$ mA. With gate current negligible (see Section 4.3), no voltage appears across R_F, and so $V_{GS} = V_{DS}$. The *drain-feedback bias line* of Fig. 4-7(*b*) is the locus of all points for which $V_{GS} = V_{DS}$. Since the Q point must lie on both the dc load line and the drain-feedback bias line, their intersection is the Q point. From Fig. 4-7(*b*), $I_{DQ} \approx 2.65$ mA and $V_{DSQ} = V_{GSQ} \approx 6.90$ V.

Solved Problems

4.1 If $C_S = 0$ and all else is unchanged in Example 4.1, find the extremes between which v_S swings.

Voltage v_{gs} will swing along the dc load line of Fig. 4-4(*b*) (which is now identical to the ac load line) from point *a* to point *b*, giving, as extremes of i_D, 3.1 mA and 0.4 mA. The corresponding extremes of $v_S = i_D R_S$ are 6.2 V and 0.8 V.

4.2 For the MOSFET amplifier of Example 4.3, let $V_{GSQ} = 6.90$ V. Calculate I_{DQ} from the analog of (*4.2*) developed in Section 4.7.

From the drain characteristics of Fig. 4-7(*b*), we see that $V_T = 4$ V and that $I_{D\text{on}} = 5$ mA at $V_{GS\text{on}} = 8$ V. Thus,

$$I_{DQ} = I_{D\text{on}}\left(1 - \frac{V_{GSQ}}{V_T}\right)^2 = 5 \times 10^{-3}\left(1 - \frac{6.90}{4}\right)^2 = 2.63 \text{ mA}$$

(Compare Example 4.3.)

4.3 By a method called *self-bias*, the *Q* point of a JFET amplifier may be established using only a single resistor from gate to ground [Fig. 4-3(*b*) with $V_{GG} = 0$]. If $R_D = 3$ kΩ, $R_S = 2$ kΩ, $R_G = 5$ MΩ, and $V_{DD} = 20$ V in Fig. 4-3(*b*), and the JFET characteristics are given by Fig. 4-4, find (*a*) I_{DQ}, (*b*) V_{GSQ}, and (*c*) V_{DSQ}.

(*a*) On Fig. 4-4(*a*) we construct a transfer bias line having a v_{GS} intercept of $V_{GG} = 0$ and a slope of $-1/R_S = -0.5$ mS; the ordinate of its intersection with the transfer characteristic is $I_{DQ} =$ 1.15 mA.

(*b*) The abscissa of the *Q* point of Fig. 4-4(*a*) is $V_{GSQ} = -2.3$ V.

(*c*) The dc load line from Example 4.1, already constructed on Fig. 4-4(*b*), is applicable here. The *Q* point was established at $I_{DQ} = 1.15$ mA in (*a*); the corresponding abscissa is $V_{DSQ} \approx 14.2$ V.

4.4 Replace the JFET of Fig. 4-3 with an *n*-channel enhancement-mode MOSFET characterized by Fig. 4-6. Let $V_{DD} = 8$ V, $V_{GSQ} = 4$ V, $V_{DSQ} = 6$ V, $I_{DQ} = 1$ mA, $R_1 = 5$ MΩ, and $R_2 =$ 3 MΩ. Find (*a*) V_{GG}, (*b*) R_S, and (*c*) R_D.

(*a*) By (*4.3*), $V_{GG} = R_1 V_{DD}/(R_1 + R_2) = 5$ V.

(*b*) Application of KVL around the smaller gate-source loop of Fig. 4-3(*b*) with $i_G = 0$ leads to

$$R_S = \frac{V_{GG} - V_{GSQ}}{I_{DQ}} = \frac{5 - 4}{1 \times 10^{-3}} = 1 \text{ k}\Omega$$

(*c*) Using KVL around the drain-source loop of Fig. 4-3(*b*) and solving for R_D yield

$$R_D = \frac{V_{DD} - V_{DSQ} - I_{DQ}R_S}{I_{DQ}} = \frac{8 - 6 - (1 \times 10^{-3})(1 \times 10^3)}{1 \times 10^{-3}} = 1 \text{ k}\Omega$$

4.5 The JFET amplifier of Fig. 4-8 shows a means of self-bias that allows extremely high input impedance even if low values of gate-source bias voltage are required. Find the Thévenin equivalent voltage and resistance for the network to the left of *a,b*.

With *a,b* open there is no voltage drop across R_3, and the voltage at the open-circuited terminals is determined by the R_1-R_2 voltage divider:

$$v_{Th} = V_{GG} = \frac{R_1}{R_1 + R_2} V_{DD}$$

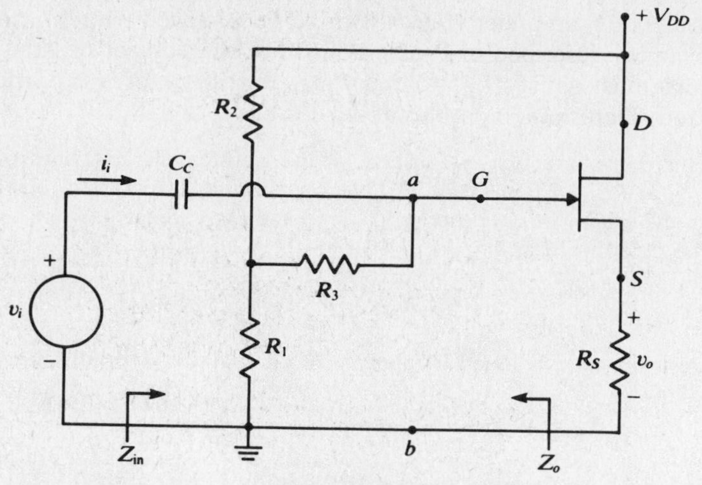

Fig. 4-8

With V_{DD} deactivated (shorted), the resistance to the left of a,b is

$$R_{Th} = R_G = R_3 + \frac{R_1 R_2}{R_1 + R_2}$$

It is apparent that if R_3 is made large, then $R_G = Z_{in}$ is large regardless of the values of R_1 and R_2.

4.6 The manufacturer's specification sheet for a certain kind of n-channel JFET has nominal and worst-case shorted-gate parameters as follows:

Value	I_{DSS}, mA	V_{p0}, V
maximum	7	4.2
nominal	6	3.6
minimum	5	3.0

Sketch the nominal and worst-case transfer characteristics that can be expected from a large sample of the device.

Values can be calculated for the nominal, maximum, and minimum transfer characteristics using (4.2) over the range $-V_{p0} \le v_{GS} \le 0$. The results are plotted in Fig. 4-9.

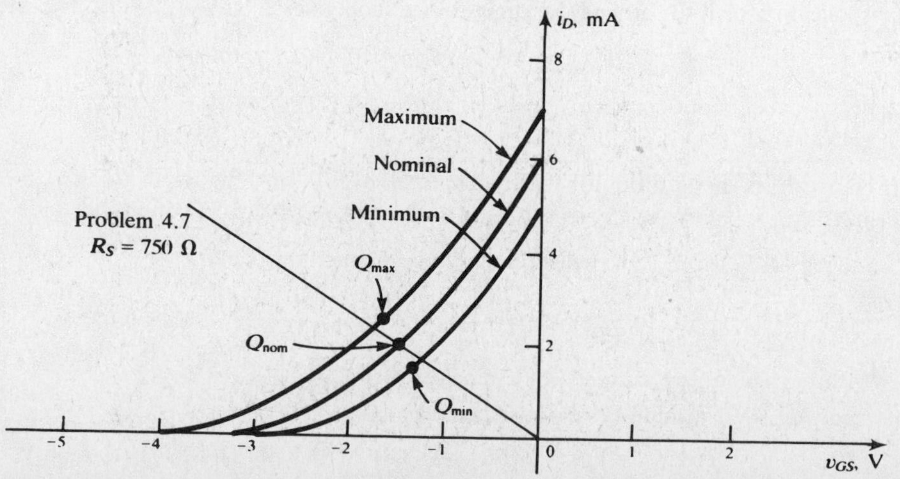

Fig. 4-9

4.7 A self-biased JFET amplifier (Fig. 4-8) is to be designed with $V_{DSQ} = 15$ V and $V_{DD} = 24$ V, using a device as described in Problem 4.6. For the control of gain variation, the quiescent drain current must satisfy $I_{DQ} = 2 \pm 0.4$ mA regardless of the particular parameters of the JFET utilized. Determine appropriate values of R_S and R_D.

Quiescent points are first established on the transfer characteristics of Fig. 4-9: Q_{max} at $I_{DQ} =$ 2.4 mA, Q_{nom} at $I_{DQ} = 2.0$ mA, and Q_{min} at $I_{DQ} = 1.6$ mA. A transfer bias line is then constructed to pass through the origin (i.e., we choose $V_{GG} = 0$) and Q_{nom}. Since its slope is $-1/R_S$, the source resistor value may be determined as

$$R_S = \frac{0-(-3)}{(4-0)\times 10^{-3}} = 750\ \Omega$$

The drain resistor value is found by applying KVL around the drain-source loop and solving for R_D:

$$R_D = \frac{V_{DD} - V_{DSQ} - I_{DQ}R_S}{I_{DQ}} = \frac{24 - 15 - (0.002)(750)}{0.002} = 3.75\ \text{k}\Omega$$

When R_S and R_D have these values, the condition on I_{DQ} is satisfied.

4.8 An n-channel JFET has worst-case shorted-gate parameters given by the manufacturer as follows:

Value	I_{DSS}, mA	V_{p0}, V
maximum	8	6
minimum	4	3

If the JFET is used in the circuit of Fig. 4-3(b), where $R_S = 0$, $R_G = 1$ MΩ, $R_D = 2.2$ kΩ, $V_{GG} = -1$ V, and $V_{DD} = 15$ V, find (a) the maximum and minimum values of I_{DQ} and (b) the maximum and minimum values of V_{DSQ} that could be expected. Assume (4.2) describes the JFET.

(a) By (4.2),

$$I_{DQ\text{max}} = I_{DSS\text{max}}\left(1 + \frac{V_{GSQ}}{V_{p0\text{max}}}\right)^2 = 8\times 10^{-3}\left(1 + \frac{-1}{6}\right)^2 = 5.55\ \text{mA}$$

$$I_{DQ\text{min}} = I_{DSS\text{min}}\left(1 + \frac{V_{GSQ}}{V_{p0\text{min}}}\right)^2 = 4\times 10^{-3}\left(1 + \frac{-1}{3}\right)^2 = 1.77\ \text{mA}$$

(b) Application of KVL around the drain-source loop gives $V_{DSQ} = V_{DD} - I_{DQ}R_D$. Hence,

$$V_{DSQ\text{max}} = 15 - (5.55\times 10^{-3})(2.2\times 10^3) = 2.78\ \text{V}$$

$$V_{DSQ\text{min}} = 15 - (1.77\times 10^{-3})(2.2\times 10^3) = 11.08\ \text{V}$$

4.9 Gate current is negligible for the p-channel JFET of Fig. 4-10. If $V_{DD} = -20$ V, $I_{DSS} = -10$ mA, $I_{DQ} = -8$ mA, $V_{p0} = -4$ V, $R_S = 0$, and $R_D = 1.5$ kΩ, find (a) V_{GG} and (b) V_{DSQ}.

(a) Solving (4.2) for v_{GS} and substituting Q-point conditions yield

$$V_{GSQ} = V_{p0}\left[\left(\frac{I_{DQ}}{I_{DSS}}\right)^{1/2} - 1\right] = -4\left[\left(\frac{-8}{-10}\right)^{1/2} - 1\right] = 0.422\ \text{V}$$

With negligible gate current, KVL requires that $V_{GG} = V_{GSQ} = 0.422$ V.

(b) Applying KVL around the drain-source loop gives

$$V_{DSQ} = V_{DD} - I_{DQ}R_D = (-20) - (-8\times 10^{-3})(1.5\times 10^3) = -8\ \text{V}$$

4.10 The n-channel enhancement-mode MOSFET of Fig. 4-11 is characterized by $V_T = 4$ V and $I_{Don} = 10$ mA. Assume negligible gate current, $R_1 = 50$ kΩ, $R_2 = 0.4$ MΩ, $R_S = 0$, $R_D = 2$ kΩ, and $V_{DD} = 15$ V. Find (*a*) V_{GSQ}, (*b*) I_{DQ}, and (*c*) V_{DSQ}.

(*a*) With negligible gate current, (*4.3*) leads to

$$V_{GSQ} = V_{GG} = \frac{R_2}{R_2 + R_1} V_{DD} = \frac{50 \times 10^3}{50 \times 10^3 + 0.4 \times 10^6} 15 = 1.67 \text{ V}$$

(b) By (*4.6*),

$$I_{DQ} = I_{Don}\left(1 - \frac{V_{GSQ}}{V_T}\right)^2 = 10 \times 10^{-3}\left(1 - \frac{1.67}{4}\right)^2 = 3.39 \text{ mA}$$

(*c*) By KVL around the drain-source loop,

$$V_{DSQ} = V_{DD} - I_{DQ}R_D = 15 - (3.39 \times 10^{-3})(2 \times 10^3) = 8.22 \text{ V}$$

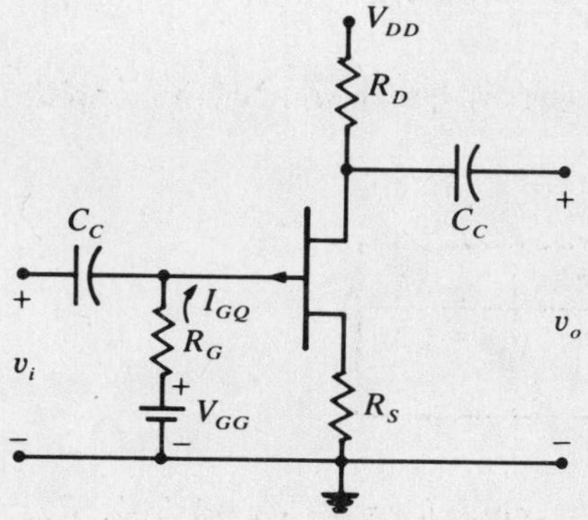

Fig. 4-10

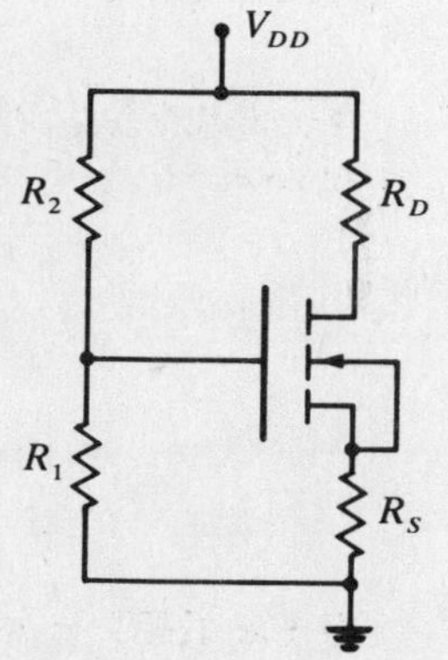

Fig. 4-11

4.11 For the n-channel enhancement-mode MOSFET of Fig. 4-11, gate current is negligible, $I_{Don} = 10$ mA, and $V_T = 4$ V. If $R_S = 0$, $R_1 = 50$ kΩ, $V_{DD} = 15$ V, $V_{GSQ} = 3$ V, and $V_{DSQ} = 9$ V, determine the values of (*a*) R_1 and (*b*) R_D.

(*a*) Since $i_G = 0$, $V_{GSQ} = V_{GG}$ of (*4.3*). Solving for R_2 gives

$$R_2 = R_1\left(\frac{V_{DD}}{V_{GSQ}} - 1\right) = 50 \times 10^3\left(\frac{15}{3} - 1\right) = 200 \text{ k}\Omega$$

(*b*) By (*4.6*),

$$I_{DQ} = I_{Don}\left(1 - \frac{V_{GSQ}}{V_T}\right)^2 = 10 \times 10^{-3}\left(1 - \frac{3}{4}\right)^2 = 0.625 \text{ mA}$$

Then KVL around the drain-source loop requires that

$$R_D = \frac{V_{DD} - V_{DSQ}}{I_{DQ}} = \frac{15 - 9}{0.625 \times 10^{-3}} = 9.6 \text{ k}\Omega$$

4.12 A p-channel MOSFET operating in the enhancement mode is characterized by $V_T = -3$ V and $I_{DQ} = -8$ mA when $V_{GSQ} = -4.5$ V. Find (*a*) V_{GSQ} if $I_{DQ} = -16$ mA and (*b*) I_{DQ} if $V_{GSQ} = -5$ V.

(*a*) Using the given data in (*4.6*) leads to

$$I_{Don} = \frac{I_{DQ}}{(1 - V_{GSQ}/V_T)^2} = \frac{-8 \times 10^{-3}}{(1 - (-4.5/-3)^2} = -32 \text{ mA}$$

Rearrangement of (4.6) now allows solution for V_{GSQ}:

$$V_{GSQ} = V_T\left[1 - \left(\frac{I_{DQ}}{I_{D\text{on}}}\right)^{1/2}\right] = (-3)\left[1 - \left(\frac{-16}{-32}\right)^{1/2}\right] = -0.88 \text{ V}$$

(b) By (4.6),

$$I_{DQ} = I_{D\text{on}}\left(1 - \frac{V_{GSQ}}{V_T}\right)^2 = -32 \times 10^{-3}\left(1 - \frac{-5}{-3}\right)^2 = -14.22 \text{ mA}$$

4.13 The n-channel JFET circuit of Fig. 4-12 employs one of several methods of self-bias. (a) Assume negligible gate leakage current ($i_G \approx 0$), and show that if $V_{DD} > 0$, then $V_{GSQ} < 0$, and hence the device is properly biased. (b) If $R_D = 3$ kΩ, $R_S = 1$ kΩ, $V_{DD} = 15$ V, and $V_{DSQ} =$ 7 V, find I_{DQ} and V_{GSQ}.

(a) By KVL,

$$I_{DQ} = \frac{V_{DD} - V_{DSQ}}{R_S + R_D} \tag{1}$$

Now $V_{DSQ} < V_{DD}$, so it is apparent that $I_{DQ} > 0$. Since $i_G \approx 0$, KVL around the gate-source loop gives

$$V_{GSQ} = -I_{DQ}R_S < 0 \tag{2}$$

(b) By (1),

$$I_{DQ} = \frac{15 - 7}{3 \times 10^3 + 1 \times 10^3} = 2 \text{ mA}$$

and (2),

$$V_{GSQ} = -(2 \times 10^{-3})(1 \times 10^3) = -2 \text{ V}$$

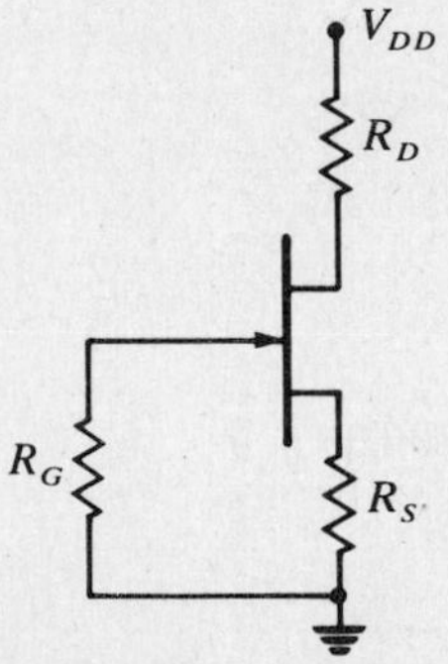

Fig. 4-12

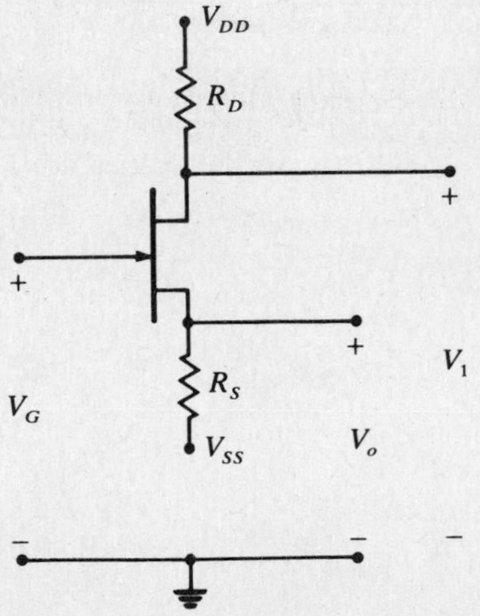

Fig. 4-13

4.14 The n-channel JFET of Fig. 4-13 is characterized by $I_{DSS} = 5$ mA and $V_{p0} = 3$ V. Let $R_D = 3$ kΩ, $R_S = 8$ kΩ, $V_{DD} = 15$ V, and $V_{SS} = -8$ V. Find V_{GSQ} and V_0 (a) if $V_G = 0$ and (b) if $V_G = 10$ V.

(a) Applying KVL around the gate-source loop yields

$$V_G = V_{GSQ} + R_S I_{DQ} + V_{SS} \tag{1}$$

Solving (*1*) for I_{DQ} and equating the result to the right side of (*4.2*) gives

$$\frac{V_G - V_{GSQ} - V_{SS}}{R_S} = I_{DSS}\left(1 + \frac{V_{GSQ}}{V_{p0}}\right)^2 \qquad (2)$$

Rearranging (*2*) leads to the following quadratic in V_{GSQ}:

$$V_{GSQ}^2 + V_{p0}\frac{V_{p0} + 2I_{DSS}R_S}{I_{DSS}R_S}V_{GSQ} + \frac{V_{p0}^2}{I_{DSS}R_S}\left(I_{DSS}R_S - V_G + V_{SS}\right) = 0 \qquad (3)$$

Substituting known values into (*3*) and solving for V_{GSQ} with the quadratic formula lead to

$$V_{GSQ}^2 + 3\frac{3 + (2)(5\times10^{-3})(8\times10^{-3})}{(5\times10^{-3})(8\times10^{3})}V_{GSQ} + \frac{(3)^2}{(5\times10^{-3})(8\times10^{3})}[(5\times10^{-3})(8\times10^{3}) - 0 - 8] = 0$$

so that

$$V_{GSQ}^2 + 6.225V_{GSQ} + 7.2 = 0$$

and $V_{GSQ} = -4.69$ V or -1.53 V. Since $V_{GSQ} = -4.69\text{ V} < -V_{p0}$, this value must be considered extraneous as it will result in $i_D = 0$. Hence, $V_{GSQ} = -1.53$ V. Now, from (*4.2*),

$$I_{DQ} = I_{DSS}\left(1 + \frac{V_{GSQ}}{V_{p0}}\right)^2 = 5\times10^{-3}\left(1 + \frac{-1.53}{3}\right)^2 = 1.2\text{ mA}$$

and, by KVL,

$$V_0 = I_{DQ}R_S + V_{SS} = (1.2\times10^{-3})(8\times10^{3}) + (-8) = 1.6\text{ V}$$

(*b*) Substitution of known values into (*3*) leads to

$$V_{GSQ}^2 + 6.225V_{GSQ} + 4.95 = 0$$

which, after elimination of the extraneous root, results in $V_{GSQ} = -0.936$ V. Then, as in part *a*,

$$I_{DQ} = I_{DSS}\left(1 + \frac{V_{GSQ}}{V_{p0}}\right)^2 = 5\times10^{-3}\left(1 + \frac{-0.936}{4}\right)^2 = 2.37\text{ mA}$$

and

$$V_0 = I_{DQ}R_S + V_{SS} = (2.37\times10^{-3})(8\times10^{3}) + (-8) = 10.96\text{ V}$$

4.15 Find the equivalent of the two identical *n*-channel JFETs connected in parallel in Fig. 4-14.

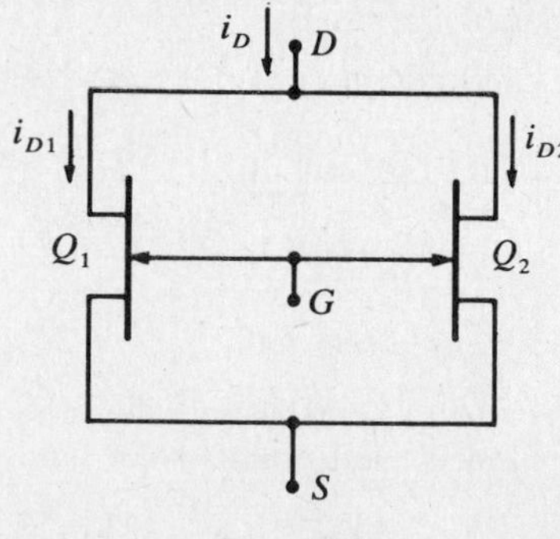

Fig. 4-14

Assume the devices are described by (*4.2*); then

$$i_D = i_{D1} + i_{D2} = I_{DSS}\left(1 + \frac{v_{GS}}{V_{p0}}\right)^2 + I_{DSS}\left(1 + \frac{v_{GS}}{V_{p0}}\right)^2 = 2I_{DSS}\left(1 + \frac{v_{GS}}{V_{p0}}\right)^2$$

Because the two devices are identical and connected in parallel, the equivalent JFET has the same pinchoff voltage as the individual devices. However, it has a value of shorted-gate current I_{DSS} equal to twice that of the individual devices.

4.16 The differential amplifier of Fig. 4-15 includes identical JFETs with $I_{DSS} = 10$ mA and $V_{p0} = 4$ V. Let $V_{DD} = 15$ V, $V_{SS} = 5$ V, and $R_S = 3$ kΩ. If the JFETs are described by (*4.2*), find the value of R_D required to bias the amplifier such that $V_{DSQ1} = V_{DSQ2} = 7$ V.

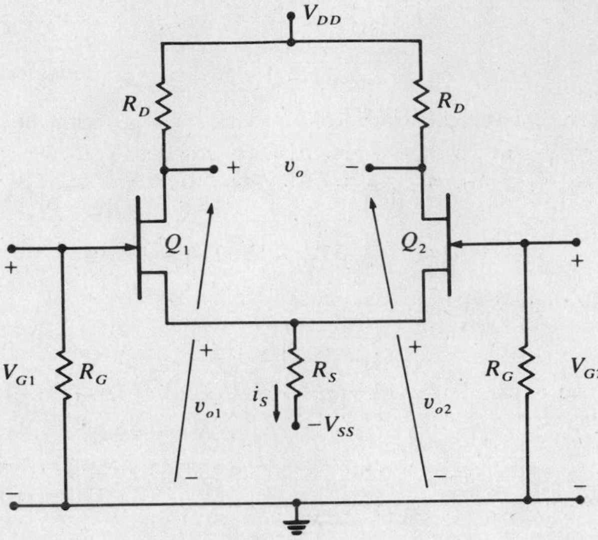

Fig. 4-15

By symmetry, $I_{DQ1} = I_{DQ2}$. KCL at the source node requires that

$$I_{SQ} = I_{DQ1} + I_{DQ2} = 2I_{DQ1} \tag{1}$$

With $i_{G1} = 0$, KVL around the left gate-source loop gives

$$V_{GSQ1} = V_{SS} - I_{SQ}R_S = V_{SS} - 2I_{DQ1}R_S \tag{2}$$

Solving (*4.2*) for V_{GSQ} and equating the result to the right side of (*2*) give

$$V_{p0}\left[\left(\frac{I_{DQ1}}{I_{DSS}}\right)^{1/2} - 1\right] = V_{SS} - 2I_{DQ1}R_S \tag{3}$$

Rearranging (*3*) results in a quadratic in I_{DQ}:

$$I_{DQ1}^2 - \left[\frac{V_{SS} + V_{p0}}{R_S} + \left(\frac{V_{p0}}{2R_S}\right)^2 \frac{1}{I_{DSS}}\right] I_{DQ1} + \left(\frac{V_{SS} + V_{p0}}{2R_S}\right)^2 = 0 \tag{4}$$

Substituting known values into (*4*) yields

$$I_{DQ1}^2 - 3.04 \times 10^{-3} I_{DQ1} + 2.25 \times 10^{-6} = 0 \tag{5}$$

Applying the quadratic formula to (*5*) and disregarding the extraneous root yields $I_{DQ1} = 1.27$ mA. Now the use of KVL around the left drain-source loop gives

$$V_{DD} + V_{SS} - V_{DSQ1} = I_{DQ1}R_D + I_{SQ}R_S \tag{6}$$

Substituting (*1*) into (*6*) and solving the result for R_D leads to the desired result:

$$R_D = \frac{V_{DD} + V_{SS} - V_{DSQ1} - 2I_{DQ1}R_S}{I_{DQ1}} = \frac{15 + 5 - 7 - 2(1.27 \times 10^{-3})(3 \times 10^3)}{1.27 \times 10^{-3}} = 4.20 \text{ k}\Omega$$

4.17 For the series-connected identical JFETs of Fig. 4-16, $I_{DSS} = 8$ mA and $V_{p0} = 4$ V. If $V_{DD} =$ 15 V, $R_D = 5$ kΩ, $R_S = 2$ kΩ, and $R_G = 1$ MΩ, find (*a*) V_{DSQ1}, (*b*) I_{DQ1}, (*c*) V_{GSQ1}, (*d*) V_{GSQ2}, and (*e*) V_{DSQ2}.

(*a*) By KVL,

$$V_{GSQ1} = V_{GSQ2} + V_{DSQ1} \tag{1}$$

But, since $I_{DQ1} \equiv I_{DQ2}$, (*4.2*) leads to

$$I_{DSS}\left(1 + \frac{V_{GSQ1}}{V_{p0}}\right)^2 = I_{DSS}\left(1 + \frac{V_{GSQ2}}{V_{p0}}\right)^2$$

or,

$$V_{GSQ1} = V_{GSQ2} \tag{2}$$

Substitution of (2) into (1) yields $V_{DSQ1} = 0$.

(*b*) With negligible gate current, KVL applied around the lower gate-source loop requires that $V_{GSQ1} = -I_{DQ1}R_S$. Substituting into (*4.2*) and rearranging now give a quadratic in I_{DQ1}:

$$I_{DQ1}^2 - \left(\frac{V_{p0}}{R_S}\right)^2\left(\frac{1}{I_{DSS}} + \frac{2R_S}{V_{p0}}\right)I_{DQ1} + \left(\frac{V_{p0}}{R_S}\right)^2 = 0 \tag{3}$$

Substitution of known values gives

$$I_{DQ1}^2 - 4.5 \times 10^{-3} I_{DQ1} + 4 \times 10^{-6} = 0$$

from which we obtain $I_{DQ1} = 3.28$ mA and 1.22 mA. The value $I_{DQ1} = 3.28$ mA would result in $V_{GSQ1} < -V_{p0}$, so that value is extraneous. Hence, $I_{DQ1} = 1.22$ mA.

(*c*)

$$V_{GSQ1} = -I_{DQ1}R_S = -(1.22 \times 10^{-3})(2 \times 10^3) = -2.44 \text{ V}$$

(*d*) From (*1*) with $V_{DSQ1} = 0$, we have $V_{GSQ2} = V_{GSQ1} = -2.44$ V.

(*e*) By KVL,

$$\begin{aligned} V_{DSQ2} &= V_{DD} - V_{DSQ1} - I_{DQ1}(R_S + R_D) \\ &= 15 - 0 - (1.22 \times 10^{-3})(2 \times 10^3 + 5 \times 10^3) = 6.46 \text{ V} \end{aligned}$$

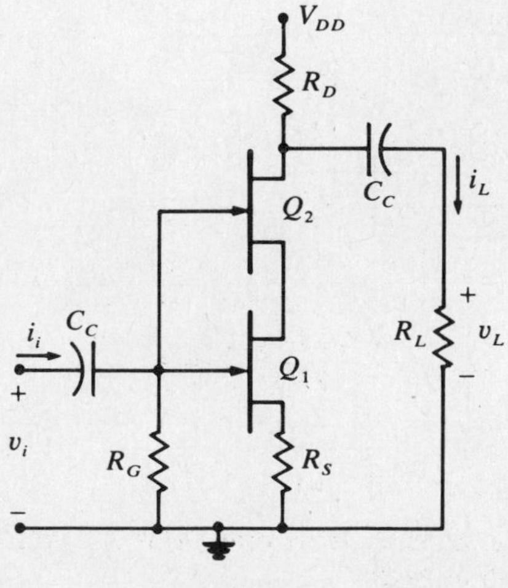

Fig. 4-16

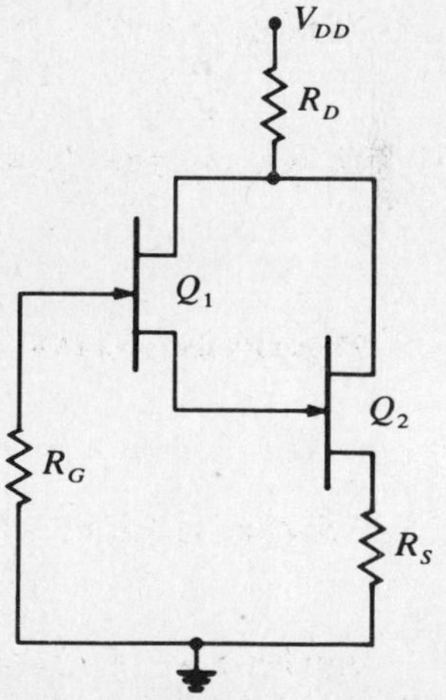

Fig. 4-17

4.18 Identical JFETs characterized by $i_G = 0$, $I_{DSS} = 10$ mA, and $V_{p0} = 4$ V are connected as shown in Fig. 4-17. Let $R_D = 1$ kΩ, $R_S = 2$ kΩ, and $V_{DD} = 15$ V, and find (*a*) V_{GSQ1}, (*b*) I_{DQ2}, (*c*) V_{GSQ2}, (*d*) V_{DSQ1}, and (*e*) V_{DSQ2}.

(*a*) With negligible gate current, (*4.2*) gives

$$i_{G2} = I_{DQ1} = 0 = I_{DSS}\left(1 + \frac{V_{GSQ1}}{V_{p0}}\right)^2$$

so

$$V_{GSQ1} = -V_{p0} = -4 \text{ V}$$

(*b*) With negligible gate current, KVL applied around the lower left-hand loop yields

$$V_{GSQ2} = -V_{GSQ1} - I_{DQ2}R_S \tag{1}$$

Substituting (*1*) into (*4.2*) and rearranging give

$$I_{DQ2}^2 - \left(\frac{V_{p0}}{R_S}\right)^2\left[\frac{1}{I_{DSS}} + 2\left(1 - \frac{V_{GSQ1}}{V_{p0}}\right)\frac{R_S}{V_{p0}}\right]I_{DQ2} + \left(\frac{V_{p0} - V_{GSQ1}}{R_S}\right)^2 = 0$$

which becomes, with known values substituted,

$$I_{DQ2}^2 - 8.4 \times 10^{-3} I_{DQ2} + 1.6 \times 10^{-5} = 0$$

The quadratic formula may be used to find the relevant root $I_{DQ2} = 2.92$ mA.

(*c*) With negligible gate current, KVL leads to

$$V_{GSQ2} = -V_{GSQ1} - I_{DQ2}R_S = -(-4) - (2.92 \times 10^{-3})(2 \times 10^3) = -1.84 \text{ V}$$

(*d*) By KVL,

$$\begin{aligned} V_{DSQ1} &= V_{DD} - (I_{DQ1} + I_{DQ2})R_D - I_{DQ2}R_S - V_{GSQ2} \\ &= 15 - (0 + 2.92 \times 10^{-3})(1 \times 10^3) - (2.92 \times 10^{-3})(2 \times 10^3) - (-1.84) = 8.08 \text{ V} \end{aligned}$$

(*e*) By KVL,

$$\begin{aligned} V_{DSQ2} &= V_{DD} - (I_{DQ1} + I_{DQ2})R_D - I_{DQ2}R_S \\ &= 15 - (0 + 2.92 \times 10^{-3})(1 \times 10^3) - (2.92 \times 10^{-3})(2 \times 10^3) = 6.24 \text{ V} \end{aligned}$$

4.19 Fixed bias can also be utilized for the enhancement-mode MOSFET, as is illustrated by the circuit of Fig. 4-18. The MOSFET is described by the drain characteristic of Fig. 4-7. Let $R_1 = 60$ kΩ, $R_2 = 40$ kΩ, $R_D = 3$ kΩ, $R_L = 1$ kΩ, $V_{DD} = 15$ V, and $C_C \to \infty$. (*a*) Find V_{GSQ}. (*b*) Graphically determine V_{DSQ} and I_{DQ}.

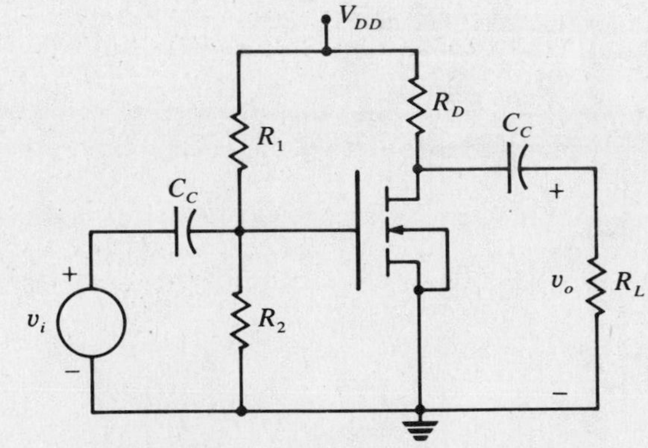

Fig. 4-18

(*a*) Assume $i_G = 0$. Then, by (*4.3*),

$$V_{GSQ} = V_{GG} = \frac{R_2}{R_2 + R_1} V_{DD} = \frac{40 \times 10^3}{40 \times 10^3 + 60 \times 10^3} 15 = 6 \text{ V}$$

(*b*) The dc load line is constructed on Fig. 4-7 with v_{DS} intercept $V_{DD} = 15$ V and i_D intercept $V_{DD}/R_L = 5$ mA. The *Q*-point quantities can be read directly from projections back to the i_D and v_{DS} axes; they are $V_{DSQ} \approx 11.3$ V and $I_{DQ} \approx 1.4$ mA.

4.20 For the enhancement-mode MOSFET amplifier of Problem 4.19, let $v_i = \sin \omega t$ and graphically determine v_o.

We have, first,

$$R_{ac} = R_D \| R_L = \frac{(3 \times 10^3)(1 \times 10^3)}{3 \times 10^3 + 1 \times 10^3} = 0.75 \text{ k}\Omega$$

An ac load line must be added to Fig. 4-7; it passes through the Q point and intersects the v_{DS} axis at

$$V_{DSQ} + I_{DQ}R_{ac} = 11.3 + (1.4 \times 10^{-3})(0.75 \times 10^3) = 12.35 \text{ V}$$

Now we construct an auxiliary time axis through the Q point and perpendicular to the ac load line; on it, we construct the waveform $v_{gs} = v_i$ as it swings ± 1 V along the ac load line about the Q point. An additional auxiliary time axis is constructed perpendicular to the v_{DS} axis, to display the output voltage $v_o = v_{ds}$ as v_{gs} swings along the ac load line.

4.21 If, instead of depending on the enhanced channel (see Fig. 4-5) for conduction, the region between the two heavily doped n^+ regions of the MOSFET is made up of lightly doped n material, a *depletion-enhancement-mode* MOSFET can be formed with drain characteristics as displayed by Fig. 4-19, where v_{GS} may be either positive or negative. Construct a transfer characteristic for the drain characteristics of Fig. 4-19, and clearly label the regions of depletion-mode and enhancement-mode operation.

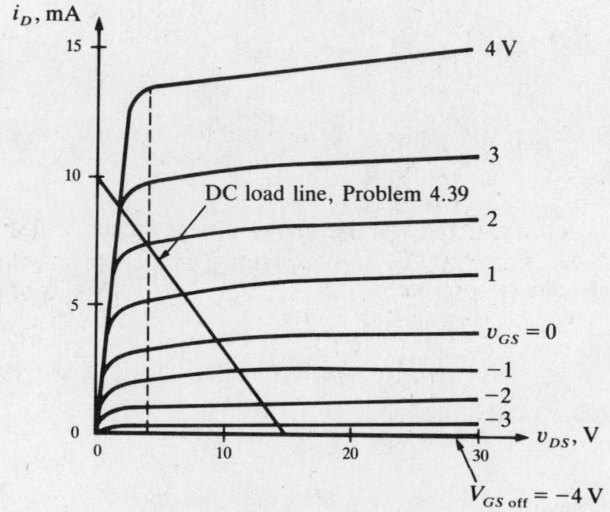

Fig. 4-19

If a constant value of $v_{DS} = V_{GSon} = 4$ V is taken as indicated by the broken line on Fig. 4-19, the transfer characteristic of Fig. 4-20 results. $v_{GS} = 0$ is the dividing line between depletion- and enhancement-mode operation.

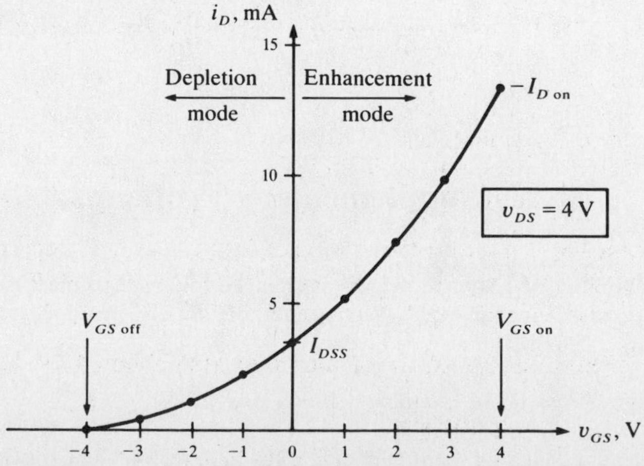

Fig. 4-20

4.22 A common-gate JFET amplifier is shown in Fig. 4-21. The JFET obeys (*4.2*). If $I_{DSS} = 10$ mA, $V_{p0} = 4$ V, $V_{DD} = 15$ V, $R_1 = R_2 = 10$ kΩ, $R_D = 500$ Ω, and $R_S = 2$ kΩ, determine (*a*) V_{GSQ}, (*b*) I_{DQ}, and (*c*) V_{DSQ}. Assume $i_G = 0$.

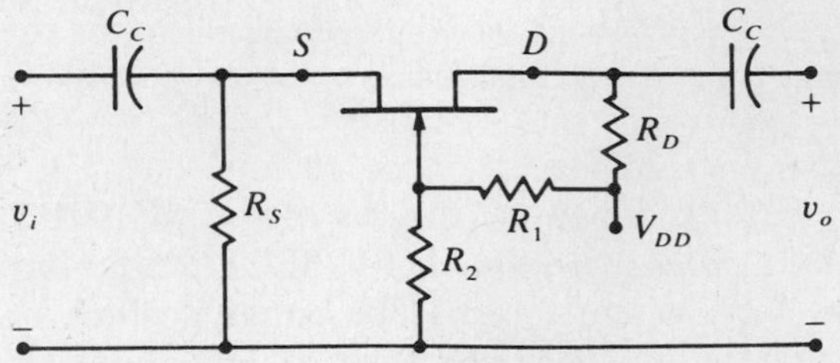

Fig. 4-21

(*a*) By KVL,

$$V_{GSQ} = \frac{R_2}{R_1 + R_2} V_{DD} - I_{DQ} R_S \tag{1}$$

Solving (*1*) for I_{DQ} and equating the result to the right side of (*4.2*) yield

$$\frac{\dfrac{R_2}{R_1 + R_2} V_{DD} - V_{GSQ}}{R_S} = I_{DSS}\left(1 + \frac{V_{GSQ}}{V_{p0}}\right)^2 \tag{2}$$

Rearranging leads to a quadratic in V_{GSQ},

$$V_{GSQ}^2 + \left(2V_{p0} + \frac{V_{p0}^2}{I_{DSS} R_S}\right) V_{GSQ} + V_{p0}^2\left[1 - \frac{R_2 V_{DD}}{(R_1 + R_2) I_{DSS} R_S}\right] = 0 \tag{3}$$

or, with known values substituted,

$$V_{GSQ}^2 + 8.8 V_{GSQ} + 10 = 0 \tag{4}$$

Solving for V_{GSQ} and disregarding the extraneous root $V_{GSQ} = -7.46 < -V_{p0}$, we determine that $V_{GSQ} = -1.34$ V.

(*b*) By (*4.2*),

$$I_{DQ} = I_{DSS}\left(1 + \frac{V_{GSQ}}{V_{p0}}\right)^2 = (10 \times 10^{-3})\left(1 + \frac{-1.34}{4}\right)^2 = 4.42 \text{ mA}$$

(*c*) By KVL,

$$V_{DSQ} = V_{DD} - I_{DQ}(R_S + R_D) = 15 - (4.42 \times 10^{-3})(2 \times 10^3 + 500) = 3.95 \text{ V}$$

Supplementary Problems

4.23 In the JFET amplifier of Example 4.1, R_1 is changed to 2 MΩ to increase the input impedance. If R_D, R_S, and V_{DD} are unchanged, what value of R_2 is needed to maintain the original Q point? *Ans.* 15.67 MΩ

4.24 Find the voltage across R_S in Example 4.1. *Ans.* 3 V

4.25 Find the input impedance as seen by source v_i of Example 4.1 if C_C is large. *Ans.* 940 kΩ

4.26 The method of *source bias*, illustrated in Fig. 4-22, can be employed for both JFETs and MOSFETs. For a JFET with characteristics given by Fig. 4-4 and with $R_D = 1\ \text{k}\Omega$, $R_S = 4\ \text{k}\Omega$, and $R_G = 10\ \text{M}\Omega$, determine V_{DD} and V_{SS} so that the amplifier has the same quiescent conditions as the amplifier of Example 4.1. *Ans.* $V_{SS} = 4$ V, $V_{DD} = 16$ V

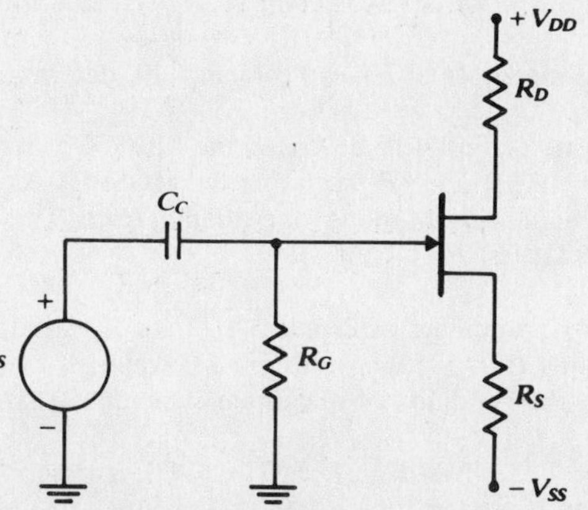

Fig. 4-22

4.27 In the drain-feedback-biased amplifier of Fig. 4-7(*a*), $V_{DD} = 15$ V, $R_F = 5\ \text{M}\Omega$, $I_{DQ} = 0.75$ mA, and $V_{GSQ} = 4.5$ V. Find (*a*) V_{DSQ} and (*b*) R_L. *Ans.* (*a*) 4.5 V; (*b*) 14 kΩ

4.28 A JFET amplifier with the circuit arrangement of Fig. 4-3 is to be manufactured using devices as described in Problem 4.6. For the design, assume a nominal device and use $V_{DD} = 24$ V, $V_{DSQ} = 15$ V, $I_{DQ} = 2$ mA, $R_1 = 2\ \text{M}\Omega$, and $R_2 = 30\ \text{M}\Omega$. (*a*) Determine the values of R_S and R_D for the amplifier. (*b*) Predict the range of I_{DQ} that can be expected.
Ans. (*a*) $R_S = 1.475\ \text{k}\Omega$, $R_D = 3.03\ \text{k}\Omega$; (*b*) 1.8 to 2.2 mA

4.29 To see the effect of a source resistor on *Q*-point conditions, solve Problem 4.9 with $R_S = 500\ \Omega$ and all else unchanged. *Ans.* (*a*) $V_{GG} = -3.58$ V; (*b*) $V_{DSQ} = -4$ V

4.30 Solve Problem 4.11 with a 200-Ω source resistor R_S added to the circuit, and all else unchanged.
Ans. (*a*) $R_2 = 190\ \text{k}\Omega$; (*b*) $R_D = 9.4\ R\Omega$

4.31 For the *n*-channel JFET circuit of Fig. 4-13, $I_{DSS} = 6$ mA, $V_{p0} = 4$ V, $R_D = 5\ \text{k}\Omega$, $R_S = 10\ \text{k}\Omega$, $V_{DD} =$ 15 V, and $V_{SS} = 10$ V. The JFET is described by (*4.2*). (*a*) Find the value of V_G that renders $V_o = 0$, and (*b*) determine V_{DSQ} if $V_o = 0$. *Ans.* (*a*) 17.63 V; (*b*) 10 V

4.32 In the differential amplifier of Fig. 4-15, the identical JFETs are characterized by $I_{DSS} = 10$ mA, $V_{p0} = 4$ V, and $i_G = 0$. If $V_{DD} = 15$ V, $V_{SS} = 5$ V, $R_S = 3\ \text{k}\Omega$, and $R_D = 5\ \text{k}\Omega$, find I_{DQ1} and V_{DSQ1}. *Ans.* 1.27 mA, 6.03 V

4.33 Find the voltage $v_{o1} = v_{o2}$ for the differential amplifier of Problem 4.32. *Ans.* 8.65 V

4.34 A voltage source is connected to the differential amplifier of Fig. 4-15 such that $V_{G1} = 0.5$ V. Let $V_{DD} = 15$ V, $V_{SS} = 2$ V, $I_{DSS} = 10$ mA, $V_{p0} = 4$ V for the identical JFETs, $R_D = 6\ \text{k}\Omega$, and $R_S = 1\ \text{k}\Omega$. Find (*a*) v_{o1} and (*b*) v_{o2}. *Ans.* (*a*) 2.53 V; (*b*) 8.42 V

4.35 For the series-connected, nonidentical JFETs of Fig. 4-16, $i_{G1} = i_{G2} = 0$, $I_{DSS1} = 8$ mA, $I_{DSS2} = 10$ mA, and $V_{p01} = V_{p02} = 4$ V. Let $V_{DD} = 15$ V, $R_G = 1\ \text{M}\Omega$, $R_D = 5\ \text{k}\Omega$, and $R_S = 2\ \text{k}\Omega$. Find (*a*) I_{DQ1}, (*b*) V_{GSQ1}, (*c*) V_{GSQ2}, (*d*) V_{DSQ1}, and (*e*) V_{DSQ2}.
Ans. (*a*) 1.22 mA; (*b*) −2.44 V; (*c*) −2.605 V; (*d*) 0.165 V; (*e*) 6.295 V

4.36 The series-connected, identical JFETs of Fig. 4-16 are characterized by $I_{DSS} = 8$ mA, $V_{p0} = 4$ V, and $i_G = 0.5$ μA. If $V_{DD} = 15$ V, $R_D = 5$ kΩ, $R_S = 2$ kΩ, and $R_G = 1$ MΩ, find (*a*) V_{GSQ1}, (*b*) V_{GSQ2}, (*c*) V_{DSQ1}, and (*d*) V_{DSQ2}. *Ans.* (*a*) −3.44 V; (*b*) −3.44 V; (*c*) 0 V; (*d*) 6.46 V

4.37 In the circuit of Fig. 4-17, the identical JFETs are described by $I_{DSS} = 8$ mA, $V_{p0} = 4$ V, and $i_G = 0.1$ μA. If $R_D = 1$ kΩ, $R_S = 2$ kΩ, $R_G = 1$ MΩ, and $V_{DD} = 15$ V, find (*a*) V_{GSQ1}, (*b*) V_{GSQ2}, (*c*) I_{DQ2}, (*d*) V_{DSQ2}, and (*e*) V_{DSQ1}. *Ans.* (*a*) −3.986 V; (*b*) −1.65 V; (*c*) 2.76 mA; (*d*) 6.72 V; (*e*) 8.37 V

4.38 For the enhancement-mode MOSFET of Problem 4.19, determine the value of I_{Don}. *Ans.* 5.6 mA

4.39 The drain characteristic of Fig. 4-19 describes the MOSFET of Fig. 4-11. Let $V_{DD} = 15$ V, $R_D = 1$ kΩ, $R_S = 500$ Ω, and $R_2 = 10$ kΩ. Find R_1 such that the MOSFET is (*a*) biased for depletion-mode operation and (*b*) biased for enhancement-mode operation. (*Hint*: The dc load line is shown in Fig. 4-19.) *Ans.* (*a*) $R_1 > 71.08$ Ω; (*b*) $R_1 < 71.08$ Ω

4.40 The common-gate JFET amplifier of Problem 4.22 is not biased for maximum symmetrical swing. Shift the bias point by letting $R_1 = 10$ kΩ and $R_2 = 5$ kΩ while all else is unchanged. Does the amplifier bias point move closer to the condition of maximum symmetrical swing? *Ans.* Yes; $V_{DSQ} = 6.59$ V

4.41 In the circuit of Fig. 4-23, $R_G \gg R_{S1}, R_{S2}$. The JFET is described by (*4.2*), $I_{DSS} = 10$ mA, $V_{p0} = 4$ V, $V_{DD} = 15$ V, $V_{DSQ} = 10$ V, and $V_{GSQ} = -2$ V. Find (*a*) R_{S1}, (*b*) R_{S2}, and (*c*) v_S.
Ans. (*a*) 800 Ω; (*b*) 1.2 kΩ; (*c*) 5 V

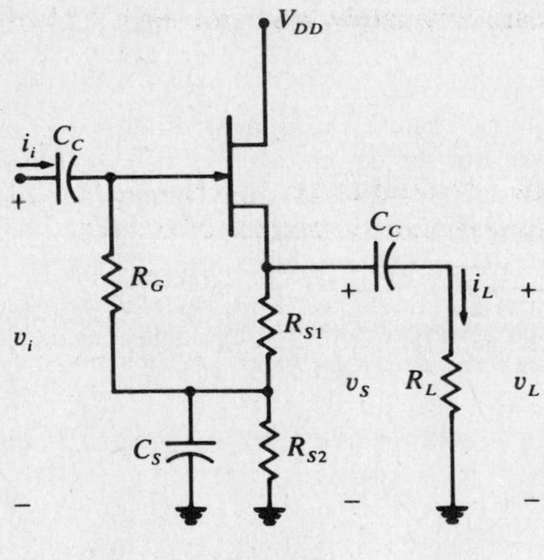

Fig. 4-23

Chapter 5

Transistor Bias Considerations

5.1 INTRODUCTION

In the initial design of transistor circuits, the quiescent operating point is carefully established to ensure that the transistor will operate within specified limits. Completion of the design requires a check of quiescent-point variations due to temperature changes and unit-to-unit parameter differences, to ensure that such variations are within an acceptable range. As the principles of operation of the BJT and FET differ greatly, so do the associated methods of Q-point stabilization.

5.2 β UNCERTAINTY AND TEMPERATURE EFFECTS IN THE BJT

Uncertainty as to the value of β may be due either to unit-to-unit variation (which may reach 200 percent or more) or to temperature variation (approximately 1 percent/°C); however, since unit-to-unit variation has the greater effect, a circuit that has been desensitized to such variation is also insensitive to the effect of temperature on β. The design must, however, directly compensate for the effects of temperature on leakage current I_{CBO} (which doubles for each 10°C rise in temperature) and base-to-emitter voltage V_{BEQ} (which decreases approximately 1.6 mV for each 1°C temperature increase in Ge devices, and approximately 2 mV for each 1°C rise in Si devices).

Constant-Base-Current Bias

The constant-base-current bias arrangement of Fig. 3-10 has the advantage of high current gain; however, the sensitivity of its Q point to changes in β limits its usage.

Example 5.1 The Si transistor of Fig. 3-10 is biased for constant base current. Neglect leakage current I_{CBO}, and let $V_{CC} = 15$ V, $R_B = 500$ kΩ, and $R_C = 5$ kΩ. Find I_{CQ} and V_{CEQ} (*a*) if $\beta = 50$, and (*b*) if $\beta = 100$.

(*a*) By KVL,

$$V_{CC} = V_{BEQ} + I_{BQ}R_B \tag{5.1}$$

Since $I_{BQ} = I_{CQ}/\beta$, we may write, using (*5.1*),

$$I_{CQ} = \beta I_{BQ} = \frac{\beta(V_{CC} - V_{BEQ})}{R_B} = \frac{50(15 - 0.7)}{500 \times 10^3} = 1.43 \text{ mA} \tag{5.2}$$

so that, by KVL,

$$V_{CEQ} = V_{CC} - I_{CQ}R_C = 15 - (1.43)(5) = 7.85 \text{ V} \tag{5.3}$$

(*b*) With β changed to 100, (*5.2*) gives

$$I_{CQ} = \frac{100(15 - 0.7)}{500 \times 10^3} = 2.86 \text{ mA}$$

and, from (*5.3*),

$$V_{CEQ} = 15 - (2.86)(5) = 0.7 \text{ V}$$

Note that, in this example, the collector current I_{CQ} doubled with the doubling of β, and the Q point moved from near the middle of the dc load line to near the saturation region.

Example 5.2 Show that, in the circuit of Fig. 3-10, I_{CQ} varies linearly with β even if leakage current is not neglected, provided $\beta \gg 1$.

Using the result of Problem 3.34(a) and KVL, we have

$$I_{BQ}R_B = \frac{I_{CQ} - (\beta + 1)I_{CBO}}{\beta} R_B = V_{CC} - V_{BEQ}$$

Rearranging and assuming $\beta \gg 1$ lead to the desired result:

$$I_{CQ} = \frac{\beta(V_{CC} - V_{BEQ})}{R_B} + \frac{\beta + 1}{\beta} I_{CBO} \approx \frac{\beta(V_{CC} - V_{BEQ})}{R_B} + I_{CBO}$$

Constant-Emitter-Current Bias

In the CE amplifier circuit of Fig. 5-1, the leakage current is explicitly modeled as a current source.

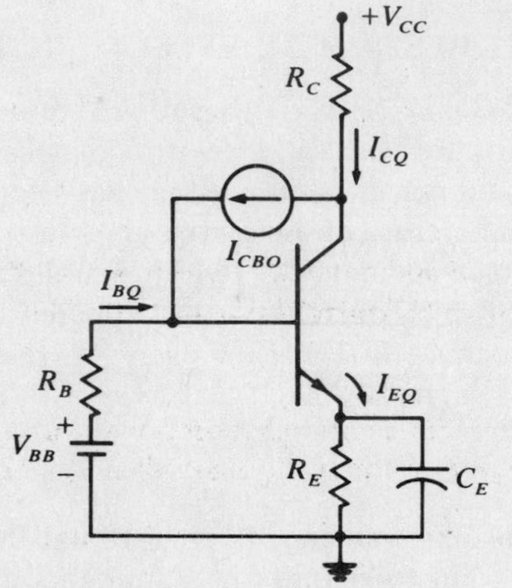

Fig. 5-1

Example 5.3 Use the circuit of Fig. 5-1 to show that (*3.8*) is the condition for β-independent bias even when leakage current is not neglected.

By KVL,

$$V_{BB} = I_{BQ} R_B + V_{BEQ} + I_{EQ} R_E \tag{5.4}$$

Using the results of Problem 3.34 and assuming that $\beta \gg 1$, we may write

$$I_{EQ} = \frac{\beta + 1}{\beta}(I_{CQ} - I_{CBO}) \approx I_{CQ} - I_{CBO} \tag{5.5}$$

and

$$I_{BQ} = \frac{I_{CQ}}{\beta} - \frac{\beta + 1}{\beta} I_{CBO} \approx \frac{I_{CQ}}{\beta} - I_{CBO} \tag{5.6}$$

Substituting (*5.5*) and (*5.6*) into (*5.4*) and rearranging then give

$$I_{CQ} = \frac{V_{BB} - V_{BEQ} + I_{CBO}(R_B + R_E)}{R_B/\beta + R_E} \tag{5.7}$$

From (*5.7*) it is apparent that leakage current I_{CBO} increases I_{CQ}. However, I_{CQ} is relatively independent of β only when $R_B/\beta \ll R_E$.

Shunt-Feedback Bias

A compromise between constant-base-current bias and constant-emitter-current bias is offered by the shunt-feedback-bias circuit of Fig. 3-13, as the following example shows.

Example 5.4 In the shunt-feedback-bias circuit of Fig. 3-13, $V_{CC} = 15$ V, $R_C = 2$ kΩ, $R_F = 150$ kΩ, and $I_{CBO} \approx 0$. The transistor is a Si device. Find I_{CQ} and V_{CEQ} if (*a*) $\beta = 50$ and (*b*) $\beta = 100$.

(*a*) By KVL,

$$V_{CC} = (I_{CQ} + I_{BQ})R_C + I_{BQ}R_F + V_{BEQ} = \left(I_{CQ} + \frac{I_{CQ}}{\beta}\right)R_C + \frac{I_{CQ}}{\beta} R_F + V_{BEQ}$$

so that
$$I_{CQ} = \frac{\beta(V_{CC} - V_{BEQ})}{R_F + (\beta + 1)R_C} = \frac{50(15 - 0.7)}{150 \times 10^3 + (51)(2 \times 10^3)} = 2.84 \text{ mA}$$

Now KVL gives

$$V_{CEQ} = V_{CC} - (I_{BQ} + I_{CQ})R_C = V_{CC} - \left(\frac{1}{\beta} + 1\right)I_{CQ}R_C$$

$$= 15 - \left(\frac{1}{50} + 1\right)(2.84 \times 10^{-3})(2 \times 10^3) = 9.21 \text{ V}$$

(*b*) For $\beta = 100$,

$$I_{CQ} = \frac{100(15 - 0.7)}{150 \times 10^3 + (101)(2 \times 10^3)} = 4.06 \text{ mA}$$

and
$$V_{CEQ} = 15 - \left(\frac{1}{100} + 1\right)(4.06 \times 10^{-3})(2 \times 10^3) = 6.80 \text{ V}$$

With shunt-feedback bias the increase in I_{CQ} is appreciable (here, 43 percent); this case lies between the β-insensitive case of constant-emitter-current bias and the directly sensitive case of constant-base-current bias.

Example 5.5 Neglecting leakage current in the shunt-feedback-bias amplifier of Fig. 3-13, find a set of conditions that will render the collector current I_{CQ} insensitive to small variations in β. Is the condition practical?

From Example 5.4, if $\beta \gg 1$,

$$I_{CQ} = \frac{V_{CC} - V_{BEQ}}{\dfrac{R_F}{\beta} + \dfrac{\beta + 1}{\beta} R_C} \approx \frac{V_{CC} - V_{BEQ}}{\dfrac{R_F}{\beta} + R_C}$$

The circuit would be insensitive to β variations if $R_F/\beta \ll R_C$. However, since $0.3 \leq V_{BEQ} \leq 0.7$, that would lead to $I_{CQ}R_C \rightarrow V_{CC}$; hence, V_{CEQ} would come close to 0 and the transistor would operate near the saturation region.

5.3 STABILITY-FACTOR ANALYSIS

Stability-factor or *sensitivity* analysis is based on the assumption that, for small changes, the variable of interest is a linear function of the other variables, and thus its differential can be replaced by its increment. In a study of BJT Q-point stability, we examine changes in quiescent collector current I_{CQ} due to variations in transistor quantities and/or elements of the surrounding circuit. Specifically, if

$$I_{CQ} = f(\beta, I_{CBO}, V_{BEQ}, \ldots) \tag{5.8}$$

then, by the chain rule, the total differential is

$$dI_{CQ} = \frac{\partial I_{CQ}}{\partial \beta} d\beta + \frac{\partial I_{CQ}}{\partial I_{CBO}} dI_{CBO} + \frac{\partial I_{CQ}}{\partial V_{BEQ}} dV_{BEQ} + \cdots \tag{5.9}$$

We may define a set of *stability factors* or *sensitivity factors* as follows:

$$S_\beta = \left.\frac{\Delta I_{CQ}}{\Delta \beta}\right|_Q \approx \left.\frac{\partial I_{CQ}}{\partial \beta}\right|_Q \tag{5.10}$$

$$S_I = \frac{\Delta I_{CQ}}{\Delta I_{CBO}}\bigg|_Q \approx \frac{\partial I_{CQ}}{\partial I_{CBO}}\bigg|_Q \tag{5.11}$$

$$S_V = \frac{\Delta I_{CQ}}{\Delta V_{BEQ}}\bigg|_Q \approx \frac{\partial I_{CQ}}{V_{BEQ}}\bigg|_Q \tag{5.12}$$

and so on. Then replacing differentials with increments in (*5.9*) yields a first-order approximation to the total change in I_{CQ}:

$$\Delta I_{CQ} \approx S_\beta\, \Delta\beta + S_I\, \Delta I_{CBO} + S_V\, \Delta V_{BEQ} + \cdots \tag{5.13}$$

Example 5.6 For the CE amplifier of Fig. 5-1, use stability-factor analysis to find an expression for the change in I_{CQ} due to variations in β, I_{CBO}, and V_{BEQ}.

The quiescent collector current I_{CQ} is expressed as a function of β, I_{CBO}, and V_{BEQ} in (*5.7*). Thus, by (*5.13*),

$$\Delta I_{CQ} \approx S_\beta\, \Delta\beta + S_I\, \Delta I_{CBO} + S_V\, \Delta V_{BEQ} \tag{5.14}$$

where the stability factors, according to (*5.10*) through (*5.12*), are

$$S_\beta = \frac{\partial I_{CQ}}{\partial\beta} = \frac{\partial}{\partial\beta}\left\{\frac{\beta[V_{BB} - V_{BEQ} + I_{CBO}(R_B + R_E)]}{R_B + \beta R_E}\right\} = \frac{R_B[V_{BB} - V_{BEQ} + I_{CBO}(R_B + R_E)]}{(R_B + \beta R_E)^2} \tag{5.15}$$

$$S_I = \frac{\partial I_{CQ}}{\partial I_{CBO}} = \frac{R_B + R_E}{R_B/\beta + R_E} \tag{5.16}$$

$$S_V = \frac{\partial I_{CQ}}{\partial V_{BEQ}} = -\frac{\beta}{R_B + \beta R_E} \tag{5.17}$$

5.4 NONLINEAR-ELEMENT STABILIZATION OF BJT CIRCUITS

Nonlinear changes in quiescent collector current due to temperature variation can, in certain cases, be eliminated or drastically reduced by judicious insertion of nonlinear devices (such as diodes) into transistor circuits.

Example 5.7 In the CE amplifier circuit of Fig. 5-1, assume that the Si device has negligible leakage current and (*3.8*) holds to the point that R_B/β can be neglected. Also, V_{BEQ} decreases by 2 mV/°C from its value of 0.7 V at 25°C. Find the change in I_{CQ} as the temperature increases from 25°C to 125°C.

Let the subscript 1 denote "at $T = 25$°C," and 2 denote "at $T = 125$°C." Under the given assumptions, (*5.7*) reduces to

$$I_{CQ} = \frac{V_{BB} - V_{BEQ}}{R_E}$$

The change in I_{CQ} is then

$$\Delta I_{CQ} = I_{CQ2} - I_{CQ1} = \frac{0.002(T_2 - T_1)}{R_E} = \frac{0.2}{R_E}$$

Example 5.8 Assume that the amplifier circuit of Fig. 5-2 has been designed so it is totally insensitive to variations of β. Further, $R_B \gg R_D$. As in Example 5.7, V_{BEQ} is equal to 0.7 V at 25°C and decreases by 2 mV/°C. Also assume that V_D varies with temperature exactly as V_{BEQ} does. Find the change in I_{CQ} as the temperature increases from 25°C to 125°C.

A Thévenin equivalent circuit can be found for the network to the left of terminals A,A, under the assumption that the diode can be modeled by a voltage source V_D. The result is

$$R_{TH} = R_D \,\|\, R_B \approx R_D$$

$$V_{Th} = V_D + \frac{V_{BB} - V_D}{R_B + R_D} R_D = \frac{V_{BB}R_D + V_D R_B}{R_D + R_B}$$

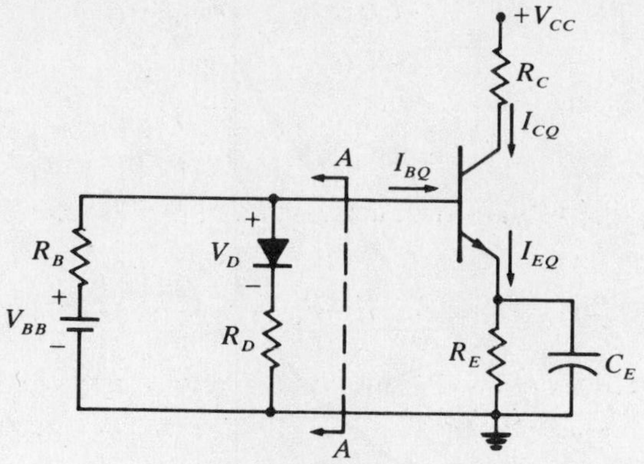

Fig. 5-2

With the Thévenin equivalent in place, KVL and the assumption $I_{BQ} = I_{CQ}/\beta \approx I_{EQ}/\beta$ give

$$I_{CQ} \approx I_{EQ} = \frac{(V_{BB}R_D + V_D R_B)/(R_D + R_B) - V_{BEQ}}{R_D/\beta + R_E}$$

Now if there is total independence of β, then R_D/β must be negligible compared with R_E. Further, since only V_D and V_{BEQ} are dependent on temperature,

$$\frac{\partial I_{CQ}}{\partial T} \approx \frac{\dfrac{R_B}{R_B + R_D}\dfrac{\partial V_D}{\partial T} - \dfrac{\partial V_{BEQ}}{\partial T}}{R_E} = \frac{0.002R_D}{R_E(R_B + R_D)} \approx \frac{0.002}{R_E}\frac{R_D}{R_B}$$

Hence,
$$\Delta I_{CQ} \approx \frac{\partial I_{CQ}}{\partial T}\Delta T = \frac{0.002}{R_E}\frac{R_D}{R_B}100 = \frac{0.2}{R_E}\frac{R_D}{R_B}$$

Because $R_D \ll R_B$ here, the change in I_{CQ} has been reduced appreciably from what it was in the circuit of Example 5.7.

5.5 *Q*-POINT-BOUNDED BIAS FOR THE FET

Just as β may vary in the BJT, the FET shorted-gate parameters I_{DSS} and V_{p0} can vary widely within devices of the same classification. It is, however, possible to set the gate-source bias so that, in spite of this variation, the Q point (and hence the quiescent drain current) is confined within fixed limits.

The extremes of FET parameter variation are usually specified by the manufacturer, and (*4.2*) may be used to establish upper and lower (worst-case) transfer characteristics (Fig. 5-3). The upper and lower quiescent points Q_{max} and Q_{min} are determined by their ordinates I_{DQmax} and I_{DQmin}; we assign I_{DQmax} and I_{DQmin} as the limits of allowable variation of I_{DQ} along a dc load line superimposed on the family of nominal drain characteristics. (These in turn establish V_{DSQmax} and V_{DSQmin}, respectively.) This dc load line is established by choosing $R_D + R_S$ in a circuit like that of Fig. 4-3 so that v_{DS} remains within a desired region of the nominal drain characteristics.

If now a value of R_S is selected such that

$$R_S \geq \frac{|V_{GSQmax} - V_{GSQmin}|}{I_{DQmax} - I_{DQmin}} \tag{5.18}$$

Then the transfer bias line with slope $-1/R_S$ and v_{GS} intercept $V_{GG} \geq 0$ is located as shown in Fig. 5-3, and the nominal Q point is forced to lie beneath Q_{max} and above Q_{min}, so that, as desired,

$$I_{DQmin} \leq I_{DQ} \leq I_{DQmax}$$

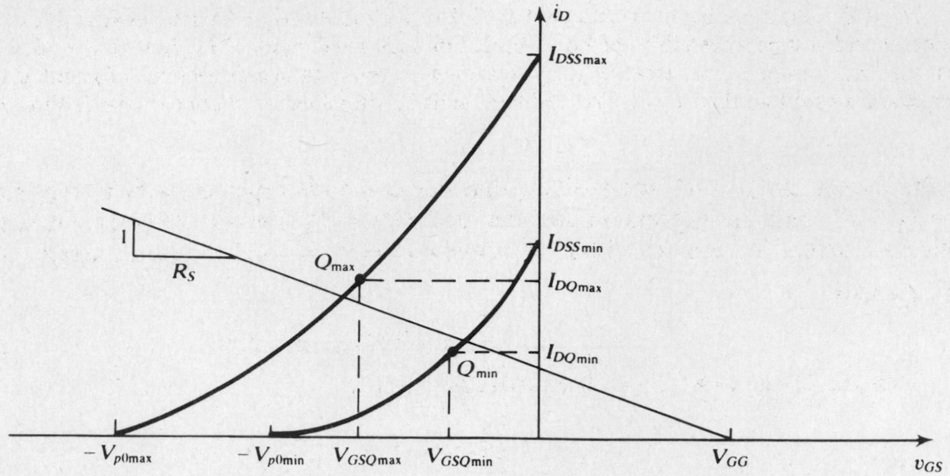

Fig. 5-3

With R_S, R_D, and V_{GG} already assigned, R_G is chosen large enough to give a satisfactory input impedance, and then R_1 and R_2 are determined from (*4.3*). Generally, R_S will be comparable in magnitude to R_D. To obtain desirable ac gains, a bypass capacitor must be used with R_S, and an ac load line introduced; they are analyzed with techniques similar to those of Section 3.6.

Solved Problems

5.1 Leakage current approximately doubles for every 10°C increase in the temperature of a transistor. If a Si transistor has $I_{CBO} = 500$ nA at 25°C, find its leakage current at 90°C.

$$I_{CBO} = (500 \times 10^{-9})2^{(90-25)/10} = (500 \times 10^{-9})(90.51) = 45.25\ \mu\text{A}$$

5.2 Sketch a set of common-emitter output characteristics for each of two different temperatures, indicating which set is for the higher temperature.

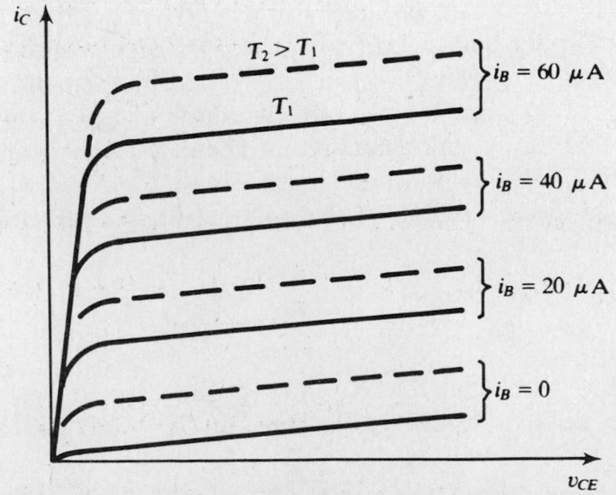

Fig. 5-4

The CE collector characteristics of Fig. 3-3(*c*) are obtained as sets of points (I_C, V_{CE}) from the ammeter and voltmeter readings of Fig. 3-3(*a*). For each fixed value of I_B, $I_C = \beta I_B + (\beta + 1)I_{CBO}$ must increase with temperature, since I_{CBO} increases with temperature (Problem 5.1) and β is much less temperature sensitive than I_{CBO}. The resultant shift in the collector characteristics is shown in Fig. 5-4.

5.3 In the circuit of Fig. 3-9, a transistor that has $\beta = \beta_1$ is replaced with a transistor that has $\beta = \beta_2$. (*a*) Find an expression for the percentage change in collector current. (*b*) Will collector current increase or decrease in magnitude if $\beta_2 > \beta_1$? Neglect leakage current.

(*a*) By KVL,

$$V_{CC} = I_{BQ}R_B + V_{BEQ} + I_{EQ}R_E \tag{1}$$

Using (*3.2*) and (*3.4*) in (*1*) and rearranging lead to

$$V_{CC} - V_{BEQ} = (R_B + R_E)\frac{I_{CQ}}{\beta} + R_E I_{CQ} \tag{2}$$

This equation may be written for the original transistor (with $\beta = \beta_1$ and $I_{CQ} = I_{CQ1}$) and for the replacement transistor (with β_2 and I_{CQ2}). Subtracting the former from the latter then gives

$$0 = (R_B + R_E)\left(\frac{I_{CQ2}}{\beta_2} - \frac{I_{CQ1}}{\beta_1}\right) + R_E(I_{CQ2} - I_{CQ1}) \tag{3}$$

If we define $I_{CQ2} = I_{CQ1} + \Delta I_{CQ}$, then (*3*) can be rewritten as

$$0 = (R_B + R_E)\frac{\beta_1(I_{CQ1} + \Delta I_{CQ}) - \beta_2 I_{CQ1}}{\beta_1\beta_2} + R_E\,\Delta I_{CQ}$$

which, when rearranged, gives the desired ratio:

$$\frac{\Delta I_{CQ}}{I_{CQ1}} = \frac{(\beta_2 - \beta_1)(R_B + R_E)}{\beta_1[R_B + (\beta_2 + 1)R_E]}\,(100\%) \tag{4}$$

(*b*) By inspection of (*4*), it is apparent that ΔI_{CQ} is positive for an increase in β ($\beta_2 > \beta_1$).

5.4 The transistor in the circuit of Fig. 3-14 is a Si device with $I_{CEO} \approx 0$. Let $V_{CC} = 18$ V, $V_{EE} = 4$ V, $R_E = 2$ kΩ, $R_C = 6$ kΩ, and $R_B = 25$ kΩ. Find I_{CQ} and V_{CEQ} (*a*) for $\beta = 50$ and (*b*) for $\beta = 100$.

(*a*) By KVL around the base-emitter loop,

$$V_{EE} - V_{BEQ} = I_{BQ}R_B + I_{EQ}R_E \tag{1}$$

We let $I_{BQ} = I_{CQ}/\beta$ and $I_{EQ} = I_{CQ}(\beta + 1)/\beta$ in (*1*) and rearrange to obtain

$$I_{CQ} = \frac{V_{EE} - V_{BEQ}}{\dfrac{R_B}{\beta} + \dfrac{\beta+1}{\beta}R_E} = \frac{4 - 0.7}{\dfrac{25 \times 10^3}{50} + \dfrac{51}{50}(2 \times 10^3)} = 1.3 \text{ mA}$$

Then KVL around the collector loop with $I_{EQ} = I_{CQ}(\beta + 1)/\beta$ yields

$$V_{CEQ} = V_{CC} + V_{EE} - \left(R_C + \frac{\beta+1}{\beta}R_E\right)I_{CQ} = 18 + 4 - \left(6 + \frac{51}{50}2\right)(1.3) = 11.55 \text{ V}$$

(*b*) For $\beta = 100$,

$$I_{CQ} = \frac{4 - 0.7}{25 \times 10^3/100 + (101/100)(2 \times 10^3)} = 1.45 \text{ mA}$$

$$V_{CEQ} = 18 + 4 - \left(6 + \frac{101}{100}2\right)(1.45) = 10.37 \text{ V}$$

5.5 In the circuit of Fig. 3-14, under what condition will the bias current I_{CQ} be practically independent of β if $I_{CEO} \approx 0$.

With $\beta \gg 1$, the expression for I_{CQ} from Problem 5.4 gives

$$I_{CQ} = \frac{V_{EE} - V_{BEQ}}{\dfrac{R_B}{\beta} + \dfrac{\beta + 1}{\beta} R_E} \approx \frac{V_{EE} - V_{BEQ}}{\dfrac{R_B}{\beta} + R_E}$$

It is apparent that I_{CQ} is practically independent of β if $R_B/\beta \ll R_E$. The inequality is generally considered to be satisfied if $R_B \leq \beta R_E/10$.

5.6 In the circuit of Fig. 3-18, the Si transistor has negligible leakage current, $V_{CC} = 15$ V, $V_{EE} = 5$ V, $R_E = 3$ kΩ, and $R_C = 7$ kΩ. Find I_{CQ}, I_{BQ}, and V_{CEQ} if (*a*) $\beta = 50$ and (*b*) $\beta = 100$.

(*a*) KVL around the base loop yields

$$I_{EQ} = \frac{V_{EE} - V_{BEQ}}{R_E} = \frac{4 - 0.7}{3 \times 10^3} = 1.1 \text{ mA}$$

Now,

$$I_{CQ} = \frac{\beta}{\beta + 1} I_{EQ} = \frac{50}{51} 1.1 = 1.078 \text{ mA}$$

and

$$I_{BQ} = \frac{I_{CQ}}{\beta} = \frac{1.078 \times 10^{-3}}{50} = 21.56\ \mu\text{A}$$

and KVL around the collector loop gives

$$V_{CEQ} = V_{CC} + V_{EE} - I_{EQ}R_E - I_{CQ}R_C = 15 + 5 - (1.1)(3) - (1.078)(7) = 9.154 \text{ V}$$

(*b*) For $\beta = 100$, I_{EQ} is unchanged. However,

$$I_{CQ} = \frac{100}{101} 1.1 = 1.089 \text{ mA}$$

$$I_{BQ} = \frac{1.089 \times 10^{-3}}{100} = 10.89\ \mu\text{A}$$

and

$$V_{CEQ} = 15 + 5 - (1.1)(3) - (1.089)(7) = 9.077 \text{ V}$$

5.7 In the circuit of Fig. 3-10, let $V_{CC} = 15$ V, $R_B = 500$ kΩ, and $R_C = 5$ kΩ. Assume a Si transistor with $I_{CBO} \approx 0$. (*a*) Find the β sensitivity factor S_β and use it to calculate the change in I_{CQ} when β changes from 50 to 100. (*b*) Compare your result with that of Example 5.1.

(*a*) By KVL,

$$V_{CC} = V_{BEQ} + I_{BQ}R_B = V_{BEQ} + \frac{I_{CQ}}{\beta} R_B$$

so that

$$I_{CQ} = \frac{\beta(V_{CC} - V_{BEQ})}{R_B}$$

and by (*5.10*),

$$S_\beta = \frac{\partial I_{CQ}}{\partial \beta} = \frac{V_{CC} - V_{BEQ}}{R_B} = \frac{15 - 0.7}{500 \times 10^3} = 28.6 \times 10^{-6}$$

According to (*5.13*), the change in I_{CQ} due to β alone is

$$\Delta I_{CQ} \approx S_\beta\, \Delta_\beta = (28.6 \times 10^{-6})(100 - 50) = 1.43 \text{ mA}$$

(*b*) From Example 5.1, we have

$$\Delta I_{CQ} = I_{CQ}|_{\beta=100} - I_{CQ}|_{\beta=50} = 2.86 - 1.43 = 1.43 \text{ mA}$$

Because I_{CQ} is of the first degree in β, (*5.13*) produces the exact change.

5.8 For the amplifier of Fig. 3-4, (*a*) find the β sensitivity factor and (*b*) show that the condition under which the β sensitivity factor is reduced to zero is identical to the condition under which the emitter current bias is constant.

(*a*) Since

$$I_{EQ} = \frac{I_{CQ}}{\alpha} = \frac{\beta+1}{\beta} I_{CQ}$$

we have, from (*3.6*),

$$V_{BB} = I_{CQ}\frac{R_B}{\beta} + V_{BEQ} + \frac{\beta+1}{\beta} I_{CQ} R_E$$

Rearranging gives

$$I_{CQ} = \frac{V_{BB} - V_{BEQ}}{\dfrac{R_B}{\beta} + \dfrac{\beta+1}{\beta} R_E} = \frac{\beta(V_{BB} - V_{BEQ})}{R_B + (\beta+1)R_E} \tag{1}$$

and, from (*5.10*),

$$S_\beta = \frac{\partial I_{CQ}}{\partial \beta} = \frac{(R_B + R_E)(V_{BB} - V_{BEQ})}{[R_B + (\beta+1)R_E]^2} \tag{2}$$

(*b*) Note in (*2*) that $\lim_{\beta\to\infty} S_\beta = 0$. Now if $\beta \to \infty$ in (*1*), then $I_{CQ} \approx (V_{BB} - V_{BEQ})/R_E = \text{constant}$.

5.9 Temperature variations can shift the quiescent point by affecting leakage current and base-to-emitter voltage. In the circuit of Fig. 5-1, $V_{BB} = 6$ V, $R_B = 50$ kΩ, $R_E = 1$ kΩ, $R_C = 3$ kΩ, $\beta = 75$, $V_{CC} = 15$ V, and the transistor is a Si device. Initially, $I_{CBO} = 0.5$ μA and $V_{BEQ} = 0.7$ V, but the temperature of the device increases by 20°C. (*a*) Find the exact change in I_{CQ}. (*b*) Predict the new value of I_{CQ} using stability-factor analysis.

(*a*) Let the subscript 1 denote quantities at the original temperature T_1, and 2 denote quantities at $T_1 + 20°\text{C} = T_2$. By (*5.7*),

$$I_{CQ1} = \frac{V_{BB} - V_{BEQ1} + I_{CBO1}(R_B + R_E)}{R_B/\beta + R_E} = \frac{6 - 0.7 + (0.5\times10^{-6})(51\times10^3)}{50\times10^3/75 + 1\times10^3} = 3.1953 \text{ mA}$$

Now, according to Section 5.2,

$$I_{CBO2} = I_{CBO1} 2^{\Delta T/10} = 0.5\times10^{-6} 2^{20/10} = 2\ \mu\text{A}$$

$$\Delta V_{BEQ} = -2\times10^{-3}\,\Delta T = (-2\times10^{-3})(20) = -0.04 \text{ V}$$

so

$$V_{BEQ2} = V_{BEQ1} + \Delta V_{BEQ} = 0.7 - 0.04 = 0.66 \text{ V}$$

Again by (*5.7*),

$$I_{CQ2} = \frac{V_{BB} - V_{BEQ2} + I_{CBO2}(R_B + R_E)}{R_B/\beta + R_E} = \frac{6 - 0.66 + (2\times10^{-6})(51\times10^3)}{50\times10^3/75 + 1\times10^3} = 3.2652 \text{ mA}$$

Thus,

$$\Delta I_{CQ} = I_{CQ2} - I_{CQ1} = 3.2652 - 3.1953 = 0.0699 \text{ mA}$$

(*b*) By (*5.16*) and (*5.17*),

$$S_I = \frac{R_B + R_E}{R_B/\beta + R_E} = \frac{50+1}{50/75 + 1} = 30.6$$

$$S_V = \frac{-\beta}{R_B + \beta R_E} = \frac{-75}{50\times10^3 + (75)(1\times10^3)} = -0.6\times10^{-3}$$

Then, according to (*5.13*),

$$\Delta I_{CQ} \approx S_I\,\Delta I_{CBO} + S_V\,\Delta V_{BEQ} = (30.6)(1.5\times10^{-6}) + (-0.6\times10^{-3})(-0.04) = 0.0699 \text{ mA}$$

and

$$I_{CQ2} = I_{CQ1} + \Delta I_{CQ} = 3.1953 + 0.0699 = 3.2652 \text{ mA}$$

5.10 In Problem 5.9, assume that the given values of I_{CBO} and V_{BEQ} are valid at 25°C (that is, that $T_1 = 25$°C). (*a*) Use stability-factor analysis to find an expression for the change in collector current resulting from a change to any temperature T_2. (*b*) Use that expression to find ΔI_{CQ} when $T_2 = 125$°C. (*c*) What percentage of the change in I_{CQ} is attributable to a change in leakage current?

(*a*) Recalling that leakage current I_{CBO} doubles for each 10°C rise in temperature, we have

$$\Delta I_{CBO} = I_{CBO}|_{T_2} - I_{CBO}|_{T_1} = I_{CBO}|_{25°C}(2^{(T_2-25)/10} - 1)$$

Since V_{BEQ} for a Si device decreases by 2 mV/°C, we have

$$\Delta V_{BEQ} = -0.002(T_2 - 25)$$

Now, substituting S_I and S_V as determined in Problem 5.9 into (*5.13*), we obtain

$$\Delta I_{CQ} = S_I\,\Delta I_{CBO} + S_V\,\Delta V_{BEQ}$$

$$= \frac{\beta(R_B + R_E)}{R_B + \beta R_E} I_{CBO}|_{25°C}(2^{(T_2-25)/10} - 1) + \frac{\beta}{R_B + \beta R_E}(0.002)(T_2 - 25)$$

(*b*) At $T_2 = 125$°C, with the values of Problem 5.9, this expression for ΔI_{CQ} gives

$$\Delta I_{CQ} = (30.6)(0.5 \times 10^{-6})(2^{(125-25)/10} - 1) + (0.0006)(0.002)(125 - 25)$$

$$= 15.65 \text{ mA} + 0.12 \text{ mA} = 15.77 \text{ mA}$$

(*c*) From part *b*, the percentage of ΔI_{CQ} due to I_{CBO} is $(15.65/15.77)(100) = 99.24$ percent.

5.11 In the constant-base-current-bias circuit arrangement of Fig. 5-5, the leakage current is explicitly modeled as a current source I_{CBO}. (*a*) Find I_{CQ} as a function of I_{CBO}, V_{BEQ}, and β. (*b*) Determine the stability factors that should be used in (*5.13*) to express the influence of I_{CBO}, V_{BEQ}, and β on I_{CQ}.

(*a*) By KVL,

$$V_{CC} = I_{BQ}R_b + I_{EQ}R_E \tag{1}$$

Substitution of (*5.5*) and (*5.6*) into (*1*) and rearrangement give

$$I_{CQ} \approx \frac{V_{CC} - V_{BEQ} + I_{CBO}(R_b + R_E)}{R_b/\beta + R_E} \tag{2}$$

(*b*) Based on the symmetry between (*2*) and (*5.7*) we have, from Example 5.6,

$$S_I = \frac{R_b + R_E}{R_b/\beta + R_E} \qquad S_V = \frac{-\beta}{R_b + \beta R_E} \qquad S_\beta = \frac{R_b[V_{CC} - V_{BEQ} + I_{CBO}(R_b + R_E)]}{(R_b + \beta R_E)^2}$$

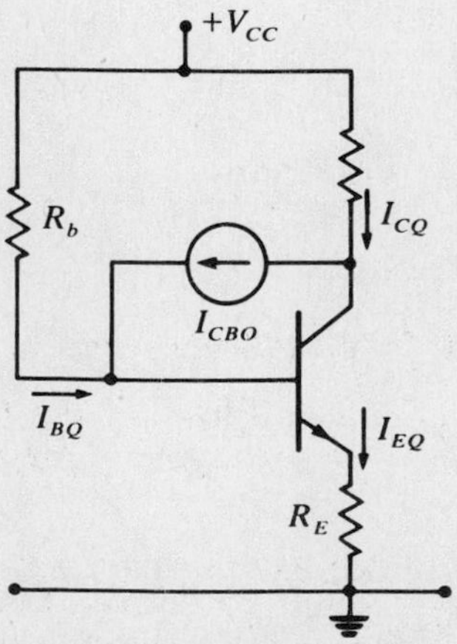

Fig. 5-5

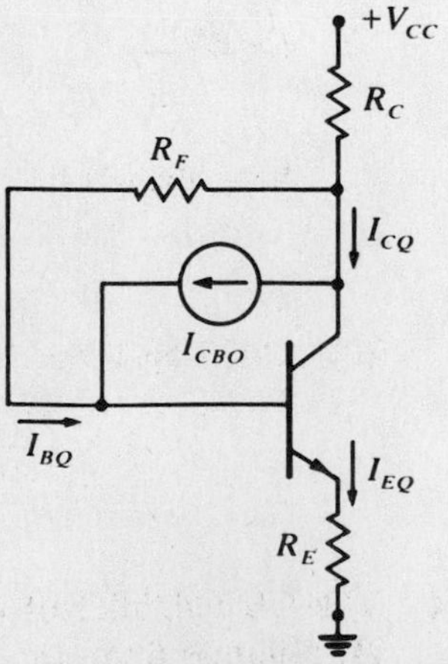

Fig. 5-6

5.12 In the shunt-feedback bias arrangement of Fig. 5-6, the leakage current is explicitly shown as a current source I_{CBO}. (*a*) Find I_{CQ} as a function of I_{CBO}, V_{BEQ}, and β. (*b*) Determine the stability factors that should be used in (*5.13*) to express the influence of I_{CBO}, V_{BEQ}, and β on I_{CQ}.

(*a*) By KVL,

$$V_{CC} = I_{CQ}R_C + I_{BQ}(R_C + R_F) + V_{BEQ} + I_{EQ}R_E \qquad (1)$$

Substituting (*5.5*) and (*5.6*) into (*1*), rearranging, and then assuming $\beta \gg 1$, we obtain

$$I_{CQ} \approx \frac{V_{CC} - V_{BEQ} + I_{CBO}(R_C + R_F + R_E)}{\dfrac{\beta+1}{\beta}R_C + \dfrac{R_F}{\beta} + R_E} \approx \frac{V_{CC} - V_{BEQ} + I_{CBO}(R_C + R_F + R_E)}{R_F/\beta + R_C + R_E} \qquad (2)$$

(*b*) Based on the symmetry between (*2*) and (*5.7*) we have, from Example 5.6,

$$S_I = \frac{R_C + R_F + R_E}{R_F/\beta + R_C + R_E} \qquad S_V = \frac{-\beta}{R_F + \beta(R_C + R_E)}$$

$$S_\beta = \frac{R_F[V_{CC} - V_{BEQ} + I_{CBO}(R_C + R_F + R_E)]}{[R_F + \beta(R_C + R_E)]^2}$$

5.13 In the CB amplifier of Fig. 5-7, the transistor leakage current is shown explicitly as a current source I_{CBO}. (*a*) Find I_{CQ} as a function of I_{CBO}, V_{BEQ}, and β. (*b*) Determine the stability factors that should be used in (*5.13*) to express the influence of I_{CBO}, V_{BEQ}, and β on I_{CQ}.

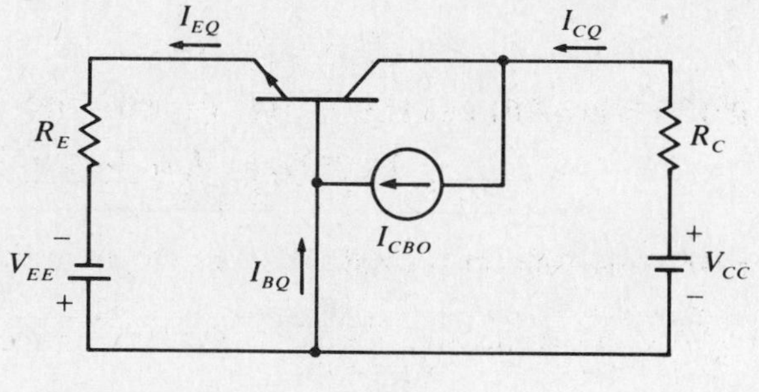

Fig. 5-7

(*a*) By KVL,

$$V_{EE} = V_{BEQ} + I_{EQ}R_E \qquad (1)$$

Substituting (*5.5*) into (*1*) and rearranging yield

$$I_{CQ} = \frac{\beta+1}{\beta}\frac{V_{EE} - V_{BEQ}}{R_E} + I_{CBO} \qquad (2)$$

(*b*) Direct application of (*5.10*) through (*5.12*) to (*2*) gives the desired stability factors as

$$S_\beta = \frac{\partial I_{CQ}}{\partial \beta} = -\frac{1}{\beta^2}\frac{V_{EE} - V_{BEQ}}{R_E} \qquad S_I = \frac{\partial I_{CQ}}{\partial I_{CBO}} = 1 \qquad S_V = \frac{\partial I_{CQ}}{\partial V_{BEQ}} = -\frac{\beta+1}{\beta R_E}$$

5.14 The CB amplifier of Fig. 5-7 has $V_{CC} = 15$ V, $V_{EE} = 5$ V, $R_E = 3$ kΩ, $R_C = 7$ kΩ, and $\beta = 50$. At a temperature of 25°C, the Si transistor has $V_{BEQ} = 0.7$ V and $I_{CBO} = 0.5$ μA. (*a*) Find an expression for I_{CQ} at any temperature. (*b*) Evaluate that expression at $T = 125$°C.

(*a*) Let the subscript 1 denote quantities at $T_1 = 25°C$, and 2 denote them at any other temperature T_2. Then, according to Section 5.2,

$$I_{CBO2} = 2^{(T_2-25)/10} I_{CBO1}$$

$$V_{BEQ2} = V_{BEQ1} + \Delta V_{BEQ} = V_{BEQ1} - 0.002(T_2 - 25)$$

Hence, by (*2*) of Problem 5.13,

$$I_{CQ2} = \frac{\beta+1}{\beta} \frac{V_{EE} - V_{BEQ1} + 0.002(T_2 - 25)}{R_E} + 2^{(T_2-25)/10} I_{CBO1} \qquad (1)$$

(*b*) At $T_2 = 125°C$, (*1*) gives us

$$I_{CQ2} = \frac{51}{50} \frac{5 - 0.7 + (0.002)(125 - 25)}{3 \times 10^3} + (2^{(125-25)/10})(0.5 \times 10^{-6}) = 1.53 + 0.512 = 2.042 \text{ mA}$$

5.15 For the Darlington-pair emitter-follower of Fig. 5-8, find I_{CQ1} as a function of the six temperature-sensitive variables I_{CBO1}, I_{CBO2}, V_{BEQ1}, V_{BEQ2}, β_1, and β_2.

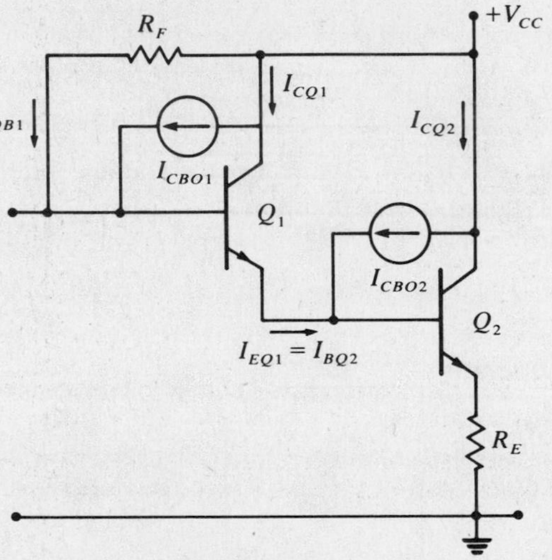

Fig. 5-8

By KVL,

$$V_{CC} = I_{BQ1}R_F + V_{BEQ1} + V_{BEQ2} + I_{EQ2}R_E \qquad (1)$$

By KCL,

$$I_{EQ2} = I_{EQ1} + I_{CQ2} \qquad (2)$$

Using the result of Problem 3.34 in (*2*) and then substituting $I_{BQ2} = I_{EQ1}$, we obtain

$$I_{EQ2} = I_{EQ1} + \beta_2 I_{BQ2} + (\beta_2 + 1)I_{CBO2} = (\beta_2 + 1)I_{EQ1} + (\beta_2 + 1)I_{CBO2}$$

Assuming $\beta_1, \beta_2 \gg 1$ and substituting for I_{EQ1} according to (*5.5*), we obtain

$$I_{EQ2} \approx (\beta_2 + 1)I_{CQ1} + (\beta_2 + 1)(I_{CBO2} - I_{CBO1}) \qquad (3)$$

Also, from (*5.6*),

$$I_{BQ1} \approx \frac{I_{CB1}}{\beta_1} - I_{CBO1} \qquad (4)$$

Now we substitute (*3*) and (*4*) into (*1*) and rearrange to get

$$I_{CQ1} = \frac{V_{CC} - V_{BEQ1} - V_{BEQ2} + I_{CBO1}(R_F + \beta_2 R_E) - I_{CBO2}\beta_2 R_E}{R_F/\beta_1 + \beta_2 R_E} \qquad (5)$$

5.16 (*a*) Determine a first-order approximation for the change in I_{CQ1} in the circuit of Fig. 5-8, in terms of the six variables I_{CBO1}, I_{CBO2}, V_{BEQ1}, V_{BEQ2}, β_1, and β_2. (*b*) Use I_{CQ1} as found in Problem 5.15 to evaluate the sensitivity factors (that is, the coefficients) in the expression determined in part *a*.

(*a*) Since $I_{CQ1} = f(I_{CBO1}, I_{CBO2}, V_{BEQ1}, V_{BEQ2}, \beta_1, \beta_2)$, its total differential is given by

$$dI_{CQ1} = \frac{\partial I_{CQ1}}{\partial I_{CBO1}} dI_{CBO1} + \frac{\partial I_{CQ1}}{\partial I_{CBO2}} dI_{CBO2} + \frac{\partial I_{CQ1}}{\partial V_{BEQ1}} dV_{BEQ1} + \frac{\partial I_{CQ1}}{\partial V_{BEQ2}} dV_{BEQ2} + \frac{\partial I_{CQ1}}{\partial \beta_1} d\beta_1 + \frac{\partial I_{CQ1}}{\partial \beta_2} d\beta_2 \qquad (1)$$

Using the method of Section 5.3, we may write this as

$$\Delta I_{CQ1} \approx S_{I1}\,\Delta I_{CBO1} + S_{I2}\,\Delta I_{CBO2} + S_{V1}\,\Delta V_{BEQ1} + S_{V2}\Delta V_{BEQ2} + S_{\beta 1}\,\Delta\beta_1 + S_{\beta 2}\,\Delta\beta_2 \qquad (2)$$

(*b*) The sensitivity factors in (*1*) may be evaluated with the use of (*5*) of Problem 5.15:

$$S_{I1} = \frac{\partial I_{CQ1}}{\partial I_{CBO1}} = \frac{R_F + \beta_2 R_E}{R_F/\beta_1 + \beta_2 R_E}$$

$$S_{I2} = \frac{\partial I_{CQ1}}{\partial I_{CBO2}} = \frac{-\beta_2 R_E}{R_F/\beta_1 + \beta_2 R_E}$$

$$S_{V1} = \frac{\partial I_{CQ1}}{\partial V_{BEQ1}} = \frac{-1}{R_F/\beta_1 + \beta_2 R_E} = S_{V2} = \frac{\partial I_{CQ1}}{\partial V_{BEQ2}}$$

$$S_{\beta 1} = \frac{\partial I_{CQ1}}{\partial \beta_1} = \frac{R_F[V_{CC} - V_{BEQ1} - V_{BEQ2} + I_{CBO1}(R_F + \beta_2 R_E) - I_{CBO2}\beta_2 R_E]}{(R_F + \beta_1\beta_2 R_E)^2}$$

$$S_{\beta 2} = \frac{\partial I_{CQ2}}{\partial \beta_2} = \frac{\beta_1 R_E[R_F(I_{CBO1} - I_{CBO2}) - \beta_1(V_{CC} - V_{BEQ1} - V_{BEQ2} + I_{CBO1}R_F)]}{(R_F + \beta_1\beta_2 R_E)^2}$$

5.17 It is possible that variations in passive components will have an effect on transistor bias. In the circuit of Fig. 3-4(*a*), let $R_1 = R_C = 500\ \Omega$, $R_2 = 5\ \text{k}\Omega$, $R_E = 100 \pm 10\ \Omega$, $\beta = 75$, $I_{CBO} = 0.2\ \mu\text{A}$, $V_{CC} = 20$ V. (*a*) Find an expression for the change in I_{CQ} due to a change in R_E alone. (*b*) Predict the change that will occur in I_{CQ} as R_E changes from the minimum to the maximum allowable value.

(*a*) We seek a stability factor

$$S_{RE} = \frac{\partial I_{CQ}}{\partial R_E} \qquad \text{such that} \qquad \Delta I_{CQ} \approx S_{RE}\,\Delta R_E$$

Starting with I_{CQ} as given by (*5.7*), we find

$$S_{RE} = \frac{\partial I_{CQ}}{\partial R_E} = \frac{\beta(R_B + \beta R_E)I_{CBO} - \beta^2[V_{BB} - V_{BEQ} + I_{CBO}(R_B + R_E)]}{(R_B + \beta R_E)^2}$$

$$= \frac{\beta R_B I_{CBO} - \beta^2(V_{BB} - V_{BEQ} + I_{CBO}R_B)}{(R_B + \beta R_E)^2}$$

(*b*) We first need to evaluate S_{RE} at $R_E = 100 - 10 = 90\ \Omega$:

$$R_B = R_1 \,\|\, R_2 = 454.5\ \Omega$$

$$V_{BB} = \frac{R_1}{R_1 + R_2} V_{CC} = \frac{500}{5500}\,20 = 1.818\ \text{V}$$

and
$$S_{RE} = \frac{75(454.5)(0.2 \times 10^{-6}) - (75)^2[1.818 - 0.7 + (0.2 \times 10^{-6})(454.5)]}{(454.5 + 75 \times 90)^2}$$

$$= -1.212 \times 10^{-4}\ \text{A}/\Omega$$

Then
$$\Delta I_{CQ} = S_{RE}\,\Delta R_E = (-1.212 \times 10^{-4})(110 - 90) = -2.424\ \text{mA}$$

5.18 The circuit of Fig. 5-9 includes nonlinear diode compensation for variations in V_{BEQ}. (*a*) Neglecting I_{CBO}, find an expression for I_{CQ} that is a function of the temperature-sensitive variables β, V_{BEQ}, and V_D. (*b*) Show that if V_{BEQ} and V_D are equal, then the sensitivity of I_{CQ} to changes in V_{BEQ} is zero. (*c*) Show that it is not necessary that $V_{BEQ} = V_D$, but only (and less restrictively) that $dV_{BEQ}/dT = dV_D/dT$, to ensure the insensitivity of I_{CQ} to temperature T.

(*a*) The usual Thévenin equivalent can be used to replace the R_1-R_2 voltage divider. Then, by KVL,

$$V_{BB} = R_B I_{BQ} + V_{BEQ} + I_{EQ} R_E - V_D \tag{1}$$

Substitution of $I_{BQ} = I_{CQ}/\beta$ and $I_{EQ} = I_{CQ}(\beta + 1)/\beta$ into (*1*) and rearranging yield

$$I_{CQ} = \frac{\beta[V_{BB} - (V_{BEQ} - V_D)]}{R_B + (\beta + 1)R_E} \tag{2}$$

(*b*) From (*2*) it is apparent that if $V_D = V_{BEQ}$, then I_{DQ} is independent of variations in V_{BEQ}.

(*c*) If β is independent of temperature, differentiation of (*2*) with respect to T results in

$$\frac{dI_{CQ}}{dT} = \frac{\beta}{R_B + (\beta + 1)R_E}\left(\frac{dV_D}{dT} - \frac{dV_{BEQ}}{dT}\right)$$

Hence, if $dV_D/dT = dV_{BEQ}/dT$, I_{CQ} is insensitive to temperature.

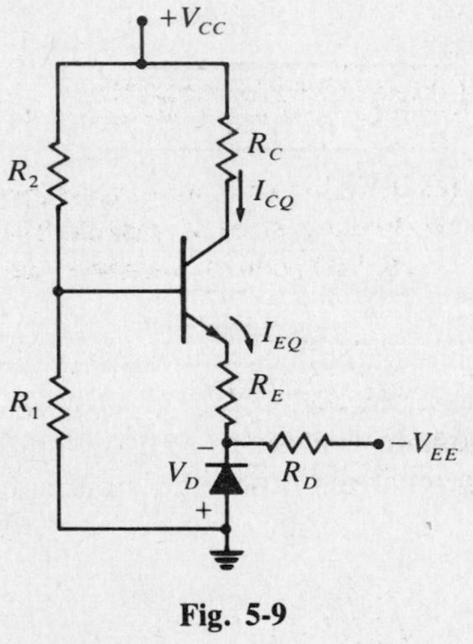

Fig. 5-9

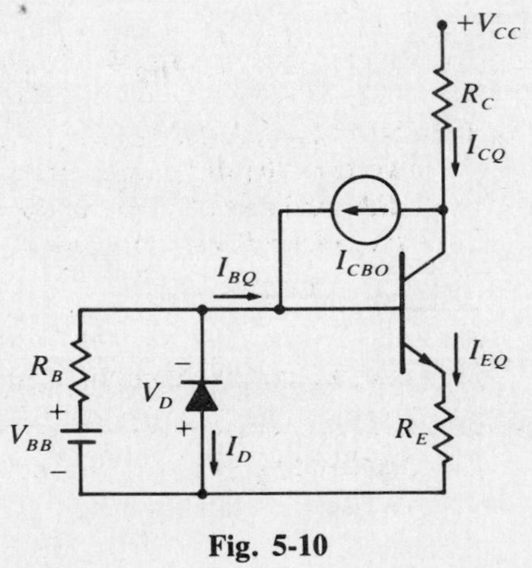

Fig. 5-10

5.19 The circuit of Fig. 5-10 includes nonlinear diode compensation for variations in I_{CBO}. (*a*) Find an expression for I_{CQ} as a function of the temperature-sensitive variables V_{BEQ}, β, I_{CBO}, and V_D. (*b*) What conditions will render I_{CQ} insensitive to changes in I_{CBO}?

(*a*) By KVL,

$$V_{BB} = (I_{BQ} + I_D)R_B + V_{BEQ} + I_{EQ}R_E$$

Substitution for I_{EQ} and I_{BQ} via (*5.5*) and (*5.6*) and rearranging give

$$I_{CQ} = \frac{V_{BB} - V_{BEQ} + I_{CBO}(R_B + R_E) - I_D R_B}{R_B/\beta + R_E} \tag{1}$$

(*b*) According to (*1*), if $R_B \gg R_E$ and $I_D = I_{CBO}$, then I_{CQ} is, in essence, independent of I_{CBO}.

5.20 Show that if a second identical diode is placed in series with the diode of Example 5.8 (see Fig. 5-2), and if R_D is made equal in value to R_B, then the collector current ($I_{CQ} \approx I_{EQ}$) displays zero sensitivity to temperature changes that affect V_{BEQ}. Make the reasonable assumption that $\partial V_D/\partial T = \partial V_{BEQ}/\partial T$.

The equation we found for I_{CQ} in Example 5.8 describes I_{CQ} in this problem if V_D is replaced by $2V_D$; that gives

$$I_{CQ} \approx \frac{(V_{BB}R_D + 2V_DR_B)/(R_D + R_B) - V_{BEQ}}{(2R_D \| R_B)/\beta + R_E} \tag{1}$$

Assuming that only V_{BEQ} and V_D are temperature dependent, we have

$$\frac{\partial I_{CQ}}{\partial T} = \frac{\dfrac{2R_B}{R_D + R_B}\dfrac{\partial V_D}{\partial T} - \dfrac{\partial V_{BEQ}}{\partial T}}{(2R_D \| R_B)/\beta + R_E} \tag{2}$$

With $\partial V_D/\partial T = \partial V_{BEQ}/\partial T$ and $R_B = R_D$, (2) reduces to zero, indicating that I_{CQ} is not a function of temperature.

5.21 A JFET for which (4.2) holds is biased by the voltage-divider arrangement of Fig. 4-3. (a) Find I_{DQ} as a function of I_{DSS}, V_{p0}, and V_{GG}. (b) Find the total differential of I_{DQ}, and make reasonable linearity assumptions that allow you to replace differentials with increments so as to find an expression analogous to (5.13) for the JFET.

(a) We use (4.4) to find an expression for V_{GSQ} and then use (4.2) to obtain

$$I_{DQ} = I_{DSS}\left(1 + \frac{V_{GG} - I_{DQ}R_S}{V_{p0}}\right)^2 \tag{1}$$

which we can solve for I_{DQ}:

$$I_{DQ} = \frac{V_{GG} + V_{p0}}{R_S} + \frac{V_{p0}^2}{2R_S^2 I_{DSS}} \pm \frac{V_{p0}}{2R_S^2}\sqrt{\left(\frac{V_{p0}}{I_{DSS}}\right)^2 + \frac{4(V_{GG} + V_{p0})R_S}{I_{DSS}}} \tag{2}$$

(b) Since V_{GSQ} depends upon the bias network chosen, our result will have more general application if we take the differential of (4.2) and then specialize it to the case at hand, instead of taking the differential of (2). Assuming that I_{DSS}, V_{p0}, and V_{GSQ} are the independent variables, we have, for the total differential of (4.2),

$$dI_{DQ} = \frac{\partial I_{DQ}}{\partial I_{DSS}}\,dI_{DSS} + \frac{\partial I_{DQ}}{\partial V_{p0}}\,dV_{p0} + \frac{\partial I_{DQ}}{\partial V_{GSQ}}\,dV_{GSQ} \tag{3}$$

For the case at hand, V_{GSQ} is given by (4.4), from which

$$dV_{GSQ} = -R_S\,dI_{DQ} \tag{4}$$

Substituting (4) into (3) and rearranging, we find

$$dI_{DQ} = \frac{\partial I_{DQ}/\partial I_{DSS}}{1 + R_S\,\partial I_{DQ}/\partial V_{GSQ}}\,dI_{DSS} + \frac{\partial I_{DQ}/\partial V_{p0}}{1 + R_S\,\partial I_{DQ}/\partial V_{GSQ}}\,dV_{p0} \tag{5}$$

The assumption of linearity allows us to replace the differentials in (5) with increments and define appropriate stability factors:

$$\Delta I_{DQ} \approx S_I\,\Delta I_{DSS} + S_V\,\Delta V_{p0} \tag{6}$$

$$S_I = \frac{\partial I_{DQ}/\partial I_{DSS}}{1 + R_S\,\partial I_{DQ}/\partial V_{GSQ}} = \frac{(1 + V_{GSQ}/V_{p0})^2}{1 + (2R_S I_{DSS}/V_{p0})(1 + V_{GSQ}/V_{p0})} \tag{7}$$

$$S_V = \frac{\partial I_{DQ}/\partial I_{DSS}}{1 + R_S\,\partial I_{DQ}/\partial V_{GSQ}} = \frac{-2I_{DSS}(1 + V_{GSQ}/V_{p0})(V_{GSQ}/V_{p0}^2)}{1 + (2R_S I_{DSS}/V_{p0})(1 + V_{GSQ}/V_{p0})} \tag{8}$$

5.22 The JFET of Fig. 4-3(b) is said to have *fixed bias* if $R_S = 0$. The worst-case shorted-gate parameters are given by the manufacturer of the device as

Value	I_{DSS}, mA	V_{p0}, V
maximum	8	6
minimum	4	3

Let $V_{DD} = 15$ V, $V_{GG} = -1$ V, and $R_D = 2.5$ kΩ. (*a*) Find the range of values of I_{DQ} that could be expected in using this FET. (*b*) Find the corresponding range of V_{DSQ}. (*c*) Comment on the desirability of this bias arrangement.

(*a*) The maximum and minimum transfer characteristics are plotted in Fig. 5-11, based on (*4.2*). Because $V_{GSQ} = V_{GG} = -1$ V is a fixed quantity unaffected by I_{DQ} and V_{DSQ}, the transfer bias line extends vertically at $V_{GS} = -1$, as shown. Its intersections with the two transfer characteristics give $I_{DQ\max} \approx 5.5$ mA and $I_{DQ\min} \approx 1.3$ mA.

(*b*) For $I_{DQ} = I_{DQ\max}$, KVL requires that

$$V_{DSQ\max} = V_{DD} - I_{DQ\max} R_D = 15 - (5.5)(2.5) = 1.25 \text{ V}$$

And, for $I_{DQ\min}$,

$$V_{DSQ\min} = V_{DD} - I_{DQ\min} R_D = 15 - (1.3)(2.5) = 11.75 \text{ V}$$

(*c*) The spread in FET parameters (and thus in transfer characteristics) makes the fixed-bias technique an undesirable one: The value of the Q-point drain current can vary from near the ohmic region to near the cutoff region.

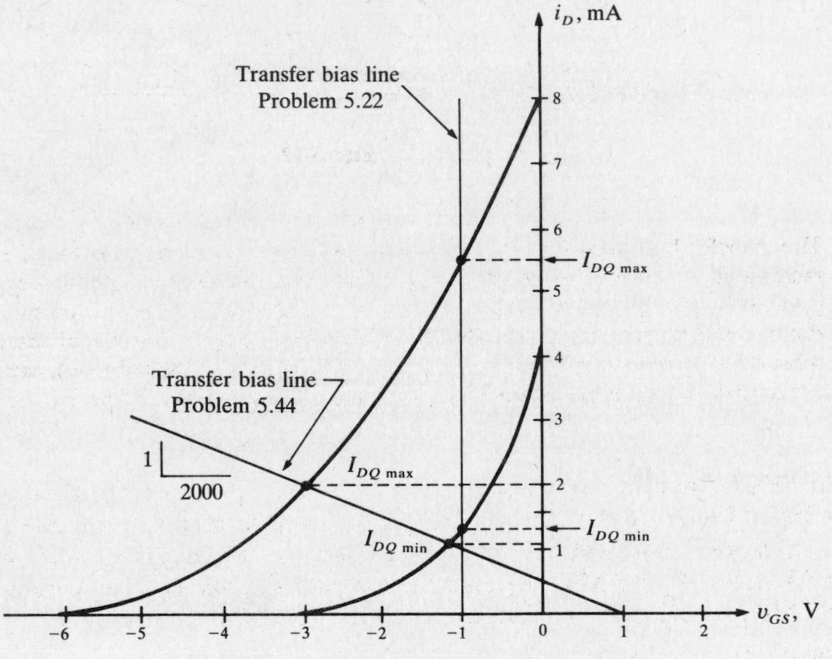

Fig. 5-11

5.23 The self-biased JFET of Fig. 4-12 has a set of worst-case shorted-gate parameters that yield the plots of Fig. 5-12. Let $V_{DD} = 24$ V, $R_D = 3$ kΩ, $R_S = 1$ kΩ, and $R_G = 10$ MΩ. (*a*) Find the range of I_{DQ} that can be expected. (*b*) Find the range of V_{DSQ} that can be expected. (*c*) Discuss the idea of reducing I_{DQ} variation by increasing the value of R_S.

(*a*) Since $V_{GG} = 0$, the transfer bias line must pass through the origin of the transfer characteristics plot, and its slope is $-1/R_S$ (solid line in Fig. 5-12). From the intersections of the transfer bias line and the transfer characteristics, we see that $I_{DQ\max} \approx 2.5$ mA and $I_{DQ\min} \approx 1.2$ mA.

(*b*) For $I_{DQ} = I_{DQ\max}$, KVL requires that

$$V_{DSQ\max} = V_{DD} - I_{DQ\max}(R_S + R_D) = 24 - (2.5)(1 + 3) = 14 \text{ V}$$

And, for $I_{DQ\min}$,

$$V_{DSQ\min} = V_{DD} - I_{DQ\min}(R_S + R_D) = 24 - (1.2)(1 + 3) = 19.2 \text{ V}$$

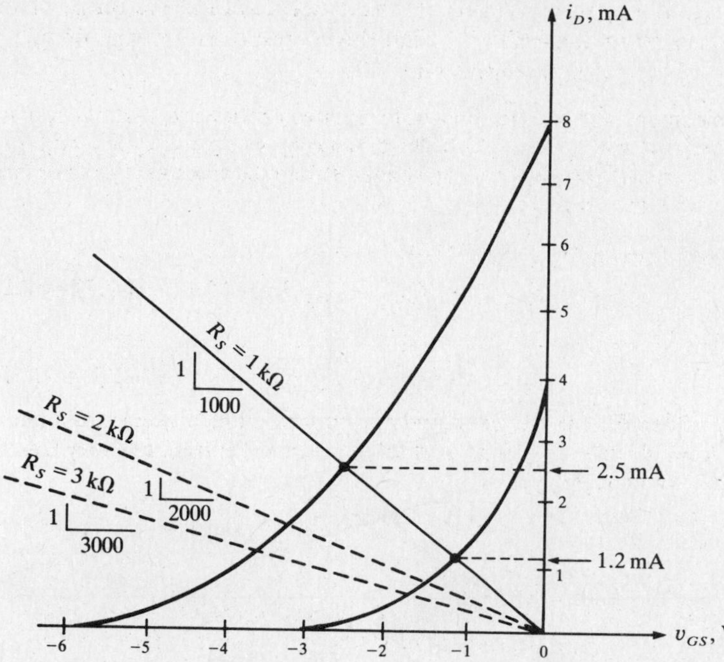

Fig. 5-12

(*c*) The transfer bias lines for $R_S = 2\ \text{k}\Omega$ and $3\ \text{k}\Omega$ are also plotted on Fig. 5-12 (dashed lines). An increase in R_S obviously does decrease the difference between $I_{DQ\max}$ and $I_{DQ\min}$; however, in the process I_{DQ} is reduced to quite low values, so that operation is on the nonlinear portion of the drain characteristics near the ohmic region where appreciable signal distortion results. But if self-bias with an external source is utilized (see Problems 5.24 and 5.44), the transfer bias line can be given a small negative slope without forcing I_{DQ} to approach zero.

5.24 In the JFET circuit of Fig. 4-3(*a*), using self-bias with an external source, $V_{DD} = 24$ V and $R_S = 3\ \text{k}\Omega$. The JFET is characterized by worst-case shorted-gate parameters that result in the transfer characteristics of Fig. 5-13. (*a*) Find the range of I_{DQ} that can be expected if $R_1 = 1\ \text{M}\Omega$ and $R_2 = 3\ \text{M}\Omega$. (*b*) Find the range of I_{DQ} that can be expected if $R_1 = 1\ \text{M}\Omega$ and $R_2 = 7\ \text{M}\Omega$. (*c*) Discuss the significance of the results of parts *a* and *b*.

(*a*) By (*4.3*),

$$V_{GG} = \frac{R_1}{R_1 + R_2} V_{DD} = \frac{1}{1+3} 24 = 6\ \text{V}$$

In this case the transfer bias line, shown on Fig. 5-13, has abscissa intercept $v_{GS} = V_{GG} = 6$ V and slope $-1/R_S$. The range of I_{DQ} is determined by the intersections of the transfer bias line and the transfer characteristics: $I_{DQ\max} \approx 2.8$ mA and $I_{DQ\min} \approx 2.2$ mA.

(*b*) Again by (*4.3*),

$$V_{GG} = \frac{1}{1+7} 24 = 3\ \text{V}$$

The transfer bias line for this case is also drawn on Fig. 5-13; it has abscissa intercept $v_{GS} = V_{GG} =$ 3 V and slope $-1/R_S$. Here $I_{DQ\max} \approx 1.9$ mA and $I_{DQ\min} \approx 1.3$ mA.

(*c*) We changed V_{GG} by altering the R_1-R_2 voltage divider. This allowed us to maintain a small negative slope on the transfer bias line (and, thus, a small difference $I_{DQ\max} - I_{DQ\min}$) while shifting the range of I_{DQ}.

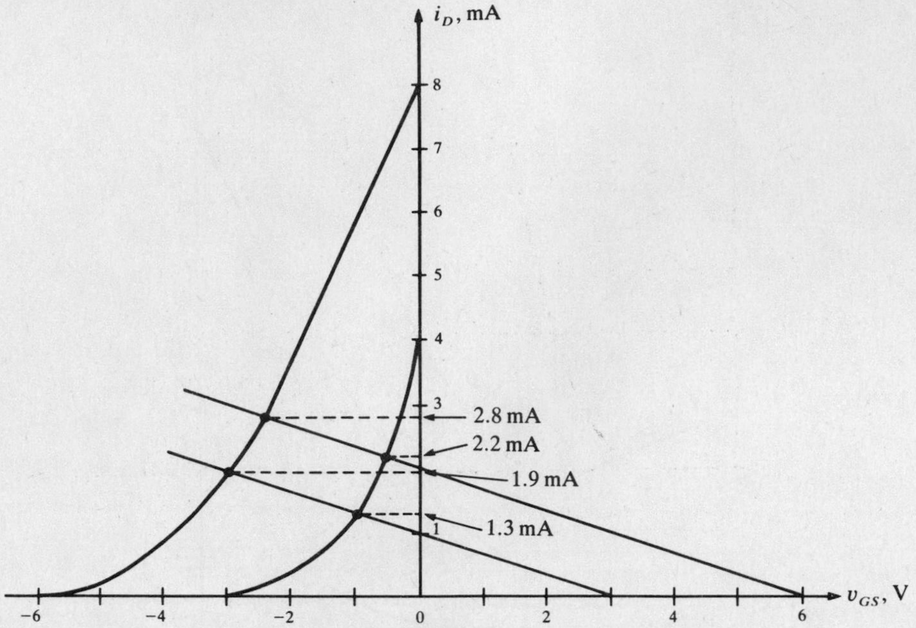

Fig. 5-13

5.25 The MOSFET of Fig. 4-11 is an enhancement-mode device with worst-case shorted-gate parameters as follows:

Value	$I_{D(\text{on})}$, mA	V_T, V
maximum	8	4
minimum	4	2

These parameter values lead to the transfer characteristics of Fig. 5-14 because the device may be assumed to obey (*4.6*). Let $V_{DD} = 24$ V, $R_1 = 2$ MΩ, $R_2 = 2$ MΩ, $R_D = 1$ kΩ, and $R_S = 2$ kΩ. (*a*) Find the range of I_{DQ} that can be expected. (*b*) Find the range of V_{DSQ} to be expected. (*c*) Discuss a technique, suggested by parts *a* and *b*, for minimizing the range of I_{DQ} for this model of MOSFET.

(*a*) By (*4.3*),

$$V_{GG} = \frac{R_1}{R_1 + R_2} V_{DD} = \frac{2}{2+2} 24 = 12 \text{ V}$$

The transfer bias line, with abscissa intercept $v_{GS} = V_{GG} = 12$ V and slope $-1/R_S$, is drawn on Fig. 5-14. From the intersections of the transfer bias line with the transfer characteristics, we see that $I_{DQ\max} \approx 4$ mA and $I_{DQ\min} \approx 2.8$ mA.

(*b*)

$$V_{DSQ\max} = V_{DD} - I_{DQ\max}(R_S + R_D) = 24 - (4)(2+1) = 12 \text{ V}$$

$$V_{DSQ\min} = V_{DD} - I_{DQ\min}(R_S + R_D) = 24 - (2.8)(2+1) = 15.6 \text{ V}$$

(*c*) As in the case of the JFET, the range of I_{DQ} can be decreased by increasing R_S. However, to avoid undesirably small values for I_{DQ}, it is also necessary to increase V_{GG} by altering the R_1-R_2 voltage-divider ratio.

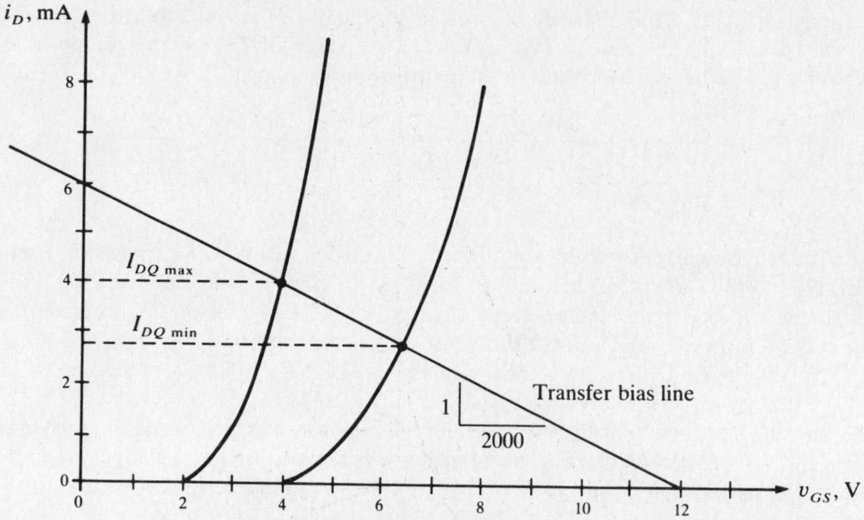

Fig. 5-14

Supplementary Problems

5.26 In the constant-base-current-biased amplifier of Fig. 3-9, $V_{CC} = 15$ V, $R_C = 2.5$ kΩ, $R_E = 500$ Ω, and $R_B = 500$ kΩ. $I_{CBO} \approx 0$ for the Si device. Find I_{CQ} and V_{CEQ} if (*a*) $\beta = 100$ and (*b*) $\beta = 50$.
Ans. (*a*) 2.6 mA, 7.19 V; (*b*) 1.36 mA, 11.09 V

5.27 Under what condition will the bias current I_{CQ} of the amplifier in Fig. 3-10 be practically independent of β? Is this condition practical?
Ans. $R_B/\beta \ll R_E$. It is not practical, as a value of R_B large enough to properly limit I_{BQ} leads, through the condition, to a value of R_E so large that it forces cutoff

5.28 The amplifier of Fig. 3-9 uses a Si transistor for which $I_{CBO} \approx 0$. Let $V_{CC} = 15$ V, $R_C = 2.5$ kΩ, $R_E = 500$ Ω, and $R_B = 500$ kΩ. (*a*) Find the value of the β sensitivity factor $S_\beta = \partial I_{CQ}/\partial\beta$ for $\beta = 50$. (*b*) Use S_β to predict I_{CQ} when $\beta = 100$.
Ans. (*a*) $(R_B + R_E)(V_{CC} - V_{BEQ})/[R_B + (\beta + 1)R_E]^2$; (*b*) 2.65 mA (compare with the result of Problem 5.26)

5.29 (*a*) Solve Problem 3.26(*a*) if $\beta = 75$ and all else is unchanged. (*b*) Use the β sensitivity factor found in Problem 5.8 to predict the change in I_{CQ} when β changes from 110 to 75.
Ans. (*a*) $I_{CQ} = 4.77$ mA; (*b*) $S_\beta = 3.643 \times 10^{-6}$, $\Delta I_{CQ} = -0.127$ mA

5.30 In the shunt-feedback-biased amplifier of Fig. 3-13, $V_{CC} = 15$ V, $R_C = 2$ kΩ, $R_F = 150$ kΩ, $I_{CEO} \approx 0$, and the transistor is a Si device. (*a*) Find an expression for the β sensitivity factor S_β. (*b*) Use S_β to predict the change in quiescent collector current due to a change in β from 50 to 100.
Ans. (*a*) $S_\beta = (R_F + R_C)(V_{CC} - V_{BEQ})/[R_F + (\beta + 1)R_C]^2$; (*b*) $S_\beta = 3.432 \times 10^{-5}$, $\Delta I_{CQ} = 1.71$ mA (compare with Example 5.4)

5.31 In the CB amplifier of Fig. 3-18, $V_{CC} = 15$ V, $V_{EE} = 5$ V, $R_E = 3$ kΩ, $R_C = 7$ kΩ, and $\beta = 50$. (*a*) Find an expression for the β sensitivity factor S_β. (*b*) Evaluate S_β assuming the transistor is a Si device.
Ans. (*a*) $S_\beta = (V_{EE} - V_{BEQ})/(\beta + 1)^2 R_E$; (*b*) $S_\beta = 5.51 \times 10^{-7}$ (very low sensitivity, but see Problem 5.6)

5.32 The circuit of Fig. 5-1 has the values given in Problem 5.9; assume that the initial values of I_{CBO} and V_{BEQ} are for 25°C. (*a*) Find an expression for the value of I_{CQ} at any temperature $T_2 \geq 25°C$ if the transistor is a Si device. (*b*) Evaluate the expression for I_{CQ} at $T_2 = 125°C$.

Ans. (*a*) $$I_{CQ} = \frac{V_{BB} - 0.7 + 0.002(T_2 - 25) + (0.5 \times 10^{-6})(R_B + R_E)2^{(T_2-25)/10}}{(R_B/\beta + R_E)};$$

(*b*) $I_{CQ} = 18.97$ mA

5.33 The constant-base-current-biased amplifier of Fig. 5-5 contains a Si transistor. Let $V_{CC} = 15$ V, $R_C = 2.5$ kΩ, $R_E = 500\ \Omega$, $R_b = 500$ kΩ, and $\beta = 100$. At 25°C, $I_{CBO} = 0.5\ \mu$A and $V_{BEQ} = 0.7$ V. (*a*) Find the exact change in I_{CQ} if the temperature changes to 100°C. (*b*) Use the stability factors developed in Problem 5.11 to predict ΔI_{CQ} for a temperature increase to 100°C.
Ans. (*a*) $\Delta I_{CQ} = I_{CQ2} - I_{CQ1} = 10.864 - 2.645 = 8.219$ mA; (*b*) $\Delta I_{CQ} = 8.22$ mA

5.34 In the constant-base-current-biased amplifier of Fig. 5-5, the Si transistor is characterized by $I_{CBO} = 0.5\ \mu$A and $V_{BEQ} = 0.7$ V at 25°C. (*a*) Find an expression for I_{CQ} at any temperature $T_2 \geq 25°C$. (*b*) Evaluate I_{CQ} at 100°C if $V_{CC} = 15$ V, $R_C = 2.5$ kΩ, $R_E = 500\ \Omega$, $R_b = 500$ kΩ, and $\beta = 100$.

Ans. (*a*) $$I_{CQ} = \frac{V_{CC} - 0.7 + 0.002(T_2 - 25) + (0.5 \times 10^{-6})(R_b + R_E)2^{(T_2-25)/10}}{R_b/\beta + R_E};$$

(*b*) $I_{CQ} = 10.864$ mA

5.35 In the current-feedback-biased amplifier of Fig. 5-6, $V_{CC} = 15$ V, $R_C = 1.5$ kΩ, $R_F = 150$ kΩ, $R_E = 500\ \Omega$, and $\beta = 100$. $I_{CBO} = 0.2\ \mu$A and $V_{BEQ} = 0.7$ V at 25°C for this Si transistor. (*a*) Find the exact change in I_{CQ} when the temperature changes to 125°C. (*b*) Use the stability factors developed in Problem 5.12 to predict ΔI_{CQ} when the temperature is 125°C. *Ans.* (*a*) $\Delta I_{CQ} = 8.943$ mA; (*b*) $\Delta I_{CQ} = 8.943$ mA

5.36 The shunt-feedback-biased amplifier of Fig. 5-6 uses a Si transistor for which $I_{CBO} = 0.2\ \mu$A and $V_{BEQ} = 0.7$ V at 25°C. (*a*) Find an expression for I_{CQ} at any temperature $T_2 \geq 25°C$. (*b*) Evaluate I_{CQ} at $T_2 = 125°C$ if $V_{CC} = 15$ V, $R_C = 1.5$ kΩ, $R_F = 150$ kΩ, $R_E = 500\ \Omega$, and $\beta = 100$.

Ans. (*a*) $$I_{CQ} = \frac{V_{CC} - 0.7 + 0.002(T_2 - 25) + (0.2 \times 10^{-6})(R_C + R_F + R_E)2^{(T_2-25)/10}}{R_F/\beta + R_C + R_E};$$

(*b*) $I_{CQ} = 13.037$ mA

5.37 In the CB amplifier of Fig. 5-7, $V_{CC} = 15$ V, $V_{EE} = 5$ V, $R_E = 3$ kΩ, $R_C = 7$ kΩ, and $\beta = 50$. For the Si transistor, $I_{CBO} = 0.5\ \mu$A and $V_{BEQ} = 0.7$ V at 25°C. (*a*) Find the exact change in I_{CQ} when the temperature changes to 125°C. (*b*) Use the stability factors developed in Problem 5.13 to predict ΔI_{CQ} for the same temperature change.
Ans. (*a*) $\Delta I_{CQ} = 2.042 - 1.4625 = 0.5795$ mA; (*b*) $\Delta I_{CQ} = 0.5769$ mA

5.38 Sensitivity analysis can be extended to handle uncertainties in power-supply voltage. In the circuit of Fig. 3-4(*a*), let $R_1 = R_C = 500\ \Omega$, $R_2 = 5$ kΩ, $R_E = 100\ \Omega$, $\beta = 75$, $V_{BEQ} = 0.7$ V, $I_{CBO} = 0.2\ \mu$A, and $V_{CC} = 20 \pm 2$ V. (*a*) Find an expression for the change in I_{CQ} due to changes in V_{CC} alone. (*b*) Predict the change in I_{CQ} as V_{CC} changes from its minimum to its maximum value.
Ans. (*a*) $\Delta I_{CQ} = S_{VCC}\,\Delta V_{CC}$, where $S_{VCC} = [\beta R_1/(R_1 + R_2)]/[R_B + (\beta + 1)R_E]$; (*b*) $\Delta I_{CQ} = 3.428$ mA

5.39 In the circuit of Fig. 5-9, $R_1 = R_C = 500\ \Omega$, $R_2 = 5$ kΩ, $R_E = 100\ \Omega$, $\beta = 75$, and $V_{CC} = 20$ V. Leakage current is negligible. At 25°C, $V_{BEQ} = 0.7$ V and $V_D = 0.65$ V; however, both change at a rate of -2 mV/°C. (*a*) Find the exact change in I_{CQ} due to an increase in temperature to 125°C. (*b*) Use sensitivity-analysis to predict the change in I_{CQ} when the temperature increases to 125°C.
Ans. (*a*) $\Delta I_{CQ} = 0$; (*b*) $\Delta I_{CQ} = 0$

5.40 In Problem 5.21, it was assumed that V_{GG}, and hence V_{DD}, was constant. Suppose now that the power-supply voltage does vary, and find an expression for ΔI_{DQ} using stability factors.

Ans. $\Delta I_{DQ} \approx S_I \, \Delta I_{DSS} + S_V \, \Delta V_{p0} + S_{VGG} \, \Delta V_{GG}$, where

$$S_{VGG} = \frac{\partial I_{DQ}/\partial V_{GSQ}}{1 + R_S \, \partial I_{DQ}/\partial V_{GSQ}} = \frac{(2I_{DSS}/V_{p0})(1 + V_{GSQ}/V_{p0})}{1 + (2R_S I_{DSS}/V_{p0})(1 + V_{GSQ}/V_{p0})}$$

and S_I and S_V are given by (7) and (8) of Problem 5.21.

5.41 The MOSFET of Fig. 4-11 is characterized by $V_T = 4$ V and $I_{D(\mathrm{on})} = 10$ mA. The device obeys (4.6). Let $i_G \approx 0$, $R_1 = 0.4$ MΩ, $R_2 = 5$ kΩ, $R_S = 0$, $R_D = 2$ kΩ, and $V_{DD} = 20$ V. (a) Find the exact change in I_{DQ} when the MOSFET is replaced with a new device characterized by $V_T = 3.8$ V and $I_{D(\mathrm{on})} = 9$ mA. (b) Find the change in I_{DQ} predicted by sensitivity analysis when the original device is replaced as in part a. *Ans.* (a) $\Delta I_{DQ} = 2.836 - 3.402 = -0.566$ mA; (b) $\Delta I_{DQ} = -0.548$ mA

5.42 In the JFET amplifier of Fig. 4-3, $V_{DD} = 20$ V, $R_1 = 1$ MΩ, $R_2 = 15.7$ MΩ, $R_D = 3$ kΩ, $R_S = 2$ kΩ, and $i_G \approx 0$. The JFET obeys (4.2) and is characterized by $I_{DSS} = 5$ mA and $V_{p0} = 5$ V. Due to aging, the resistance of R_1 increases by 20 percent. (a) Find the exact change in I_{DQ} due to the increase in resistance. (b) Predict the change in I_{DQ} due to the increase in resistance, using sensitivity analysis. *Ans.* (a) $\Delta I_{DQ} = 1.735 - 1.658 = 0.077$ mA; (b) $\Delta I_{DQ} = S_{VGG} \, \Delta V_{GG} = 0.0776$ mA

5.43 For a FET, the temperature dependence of V_{GSQ} is very small when I_{DQ} is held constant. Moreover, for constant V_{DSQ}, the temperature dependency of V_{GSQ} is primarily due to changes in the shorted-gate current; those changes are given by

$$I_{DSS} = I_{DSSO}(k \, \Delta T + 1.1) \tag{1}$$

where I_{DSSO} = value of I_{DSS} at 0°C
ΔT = change in temperature from 0°C
k = constant (typically 0.003 °C^{-1})

For the JFET of Fig. 4-3, $V_{DD} = 20$ V, $R_1 = 1$ MΩ, $R_2 = 15.7$ MΩ, $R_D = 3$ kΩ, $R_S = 2$ kΩ, $i_G \approx 0$, $I_{DSSO} = 5$ mA, and $V_{p0} = 5$ V (and is temperature-independent). (a) Find the exact value of I_{DQ} at 100°C. (b) Use sensitivity analysis to predict I_{DQ} at 100°C. *Ans.* (a) $I_{DQ} = 1.82$ mA; (b) $I_{DQ} = 1.84$ mA

5.44 Solve parts a and b of Problem 5.22 if $R_S = 2$ kΩ, $V_{GG} = 1$ V, and all else remains unchanged. *Ans.* (a) The transfer bias line is drawn on Fig. 5-11: $I_{DQ\max} \approx 2$ mA, $I_{DQ\min} \approx 1.1$ mA; (b) $V_{DSQ\max} \approx 6$ V, $V_{DSQ\min} \approx 10.05$ V

Chapter 6

Small-Signal Midfrequency BJT Amplifiers

6.1 INTRODUCTION

For sufficiently small emitter-collector voltage and current excursions about the quiescent point (*small signals*), the BJT is considered linear; it may then be replaced with any of several two-port networks of impedances and controlled sources (called *small-signal equivalent-circuit models*), to which standard network analysis methods are applicable. Moreover, there is a range of signal frequencies which are large enough so that coupling or bypass capacitors (see Section 3.6) can be considered short circuits, yet low enough so that inherent capacitive reactances associated with BJTs can be considered open circuits. In this chapter, all BJT voltage and current signals are assumed to be in this *midfrequency range*.

In practice, the design of small-signal amplifiers is divided into two parts: (1) setting the dc bias or Q point (Chapters 3 and 5), and (2) determining voltage- or current-gain ratios and impedance values at signal frequencies.

6.2 HYBRID-PARAMETER MODELS

General hybrid-parameter analysis of two-port networks was introduced in Section 1.6. Actually, different sets of h parameters are defined, depending on which element of the transistor (E, B, or C) shares a common point with the amplifier input and output terminals.

Common-Emitter Transistor Connection

From Fig. 3-3(b) and (c), we see that if i_C and v_{BE} are taken as dependent variables in the CE transistor configuration, then

$$v_{BE} = f_1(i_B, v_{CE}) \tag{6.1}$$

$$i_C = f_2(i_B, v_{CE}) \tag{6.2}$$

If the total emitter-to-base voltage v_{BE} goes through only *small* excursions (ac signals) about the Q point, then $\Delta v_{BE} = v_{be}$, $\Delta i_C = i_c$, and so on. Therefore, after applying the chain rule to (*6.1*) and (*6.2*), we have, respectively,

$$v_{be} = \Delta v_{BE} \approx dv_{BE} = \left.\frac{\partial v_{BE}}{\partial i_B}\right|_Q i_b + \left.\frac{\partial v_{BE}}{\partial v_{CE}}\right|_Q v_{ce} \tag{6.3}$$

$$i_c = \Delta i_C \approx di_C = \left.\frac{\partial i_C}{\partial i_B}\right|_Q i_b + \left.\frac{\partial i_C}{\partial v_{CE}}\right|_Q v_{ce} \tag{6.4}$$

The four partial derivatives, evaluated at the Q point, that occur in (*6.3*) and (*6.4*) are called *CE hybrid parameters* and are denoted as follows:

$$\textit{Input resistance} \qquad h_{ie} \equiv \left.\frac{\partial v_{BE}}{\partial i_B}\right|_Q \approx \left.\frac{\Delta v_{BE}}{\Delta i_B}\right|_Q \tag{6.5}$$

$$\textit{Reverse voltage ratio} \qquad h_{re} \equiv \left.\frac{\partial v_{BE}}{\partial v_{CE}}\right|_Q \approx \left.\frac{\Delta v_{BE}}{\Delta v_{CE}}\right|_Q \tag{6.6}$$

$$\textit{Forward current gain} \qquad h_{fe} \equiv \left.\frac{\partial i_C}{\partial i_B}\right|_Q \approx \left.\frac{\Delta i_C}{\Delta i_B}\right|_Q \tag{6.7}$$

$$\textit{Output admittance} \qquad h_{oe} \equiv \left.\frac{\partial i_C}{\partial v_{CE}}\right|_Q \approx \left.\frac{\Delta i_C}{\Delta v_{CE}}\right|_Q \tag{6.8}$$

The equivalent circuit for (*6.3*) and (*6.4*) is shown in Fig. 6-1(*a*). The circuit is valid for use with signals whose excursion about the Q point is sufficiently small so that the h parameters may be treated as constants.

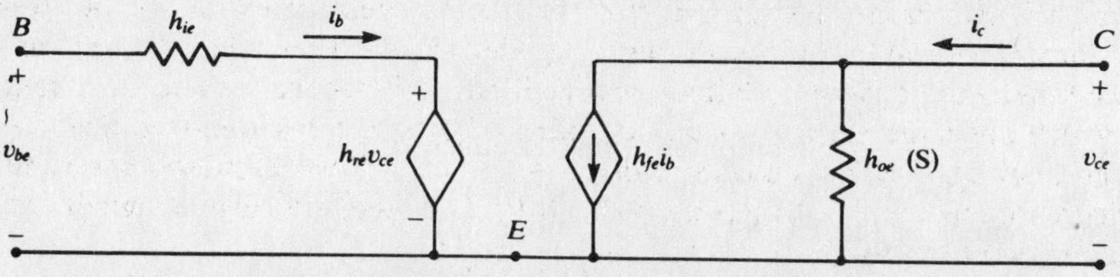

(*a*) CE small-signal equivalent circuit

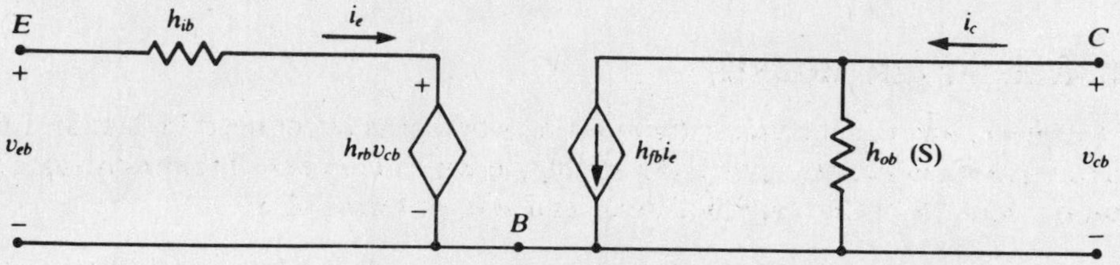

(*b*) CB small-signal equivalent circuit

Fig. 6-1

Common-Base Transistor Connection

If v_{EB} and i_C are taken as the dependent variables for the CB transistor characteristics of Fig. 3-2(*b*) and (*c*), then, as in the CE case, equations can be found specifically for small excursions about the Q point. The results are

$$v_{eb} = h_{ib} i_e + h_{rb} v_{cb} \tag{6.9}$$

$$i_c = h_{fb} i_e + h_{ob} v_{cb} \tag{6.10}$$

The partial-derivative definitions of the CB h-parameters are:

$$\textit{Input resistance} \qquad h_{ib} \equiv \left.\frac{\partial v_{EB}}{\partial i_E}\right|_Q \approx \left.\frac{\Delta v_{EB}}{\Delta i_E}\right|_Q \tag{6.11}$$

$$\textit{Reverse voltage ratio} \qquad h_{rb} \equiv \left.\frac{\partial v_{EB}}{\partial v_{CB}}\right|_Q \approx \left.\frac{\Delta v_{EB}}{\Delta v_{CB}}\right|_Q \tag{6.12}$$

$$\textit{Forward current gain} \qquad h_{fb} \equiv \left.\frac{\partial i_C}{\partial i_E}\right|_Q \approx \left.\frac{\Delta i_C}{\Delta i_E}\right|_Q \tag{6.13}$$

$$\textit{Output admittance} \qquad h_{ob} \equiv \left.\frac{\partial i_C}{\partial v_{CB}}\right|_Q \approx \left.\frac{\Delta i_C}{\Delta v_{CB}}\right|_Q \tag{6.14}$$

A small-signal, h-parameter equivalent circuit satisfying (*6.9*) and (*6.10*) is shown in Fig. 6-1(*b*).

Common-Collector Amplifier

The *common-collector* (CC) or *emitter-follower* (EF) amplifier, with the universal-bias circuitry of Fig. 6-2(*a*), can be modeled for small-signal ac analysis by replacing the CE-connected transistor with its *h*-parameter model, Fig. 6-1(*a*). Assuming, for simplicity, that $h_{re} = h_{oe} = 0$, we obtain the equivalent circuit of Fig. 6-2(*b*).

An even simpler model can be obtained by finding a Thévenin equivalent for the circuit to the right of *a*,*a* in Fig. 6-2(*b*). Application of KVL around the outer loop gives

$$v = i_b h_{ie} + i_e R_E = i_b h_{ie} + (h_{fe} + 1) i_b R_E \tag{6.15}$$

The Thévenin impedance is the driving-point impedance:

$$R_{Th} = \frac{v}{i_b} = h_{ie} + (h_{fe} + 1) R_E \tag{6.16}$$

The Thévenin voltage is zero (computed with terminals *a*,*a* open); thus, the equivalent circuit consists only of R_{Th}. This is shown, in a base-current frame of reference, in Fig. 6-2(*c*). (See Problem 6.12 for a development of the CC *h*-parameter model.)

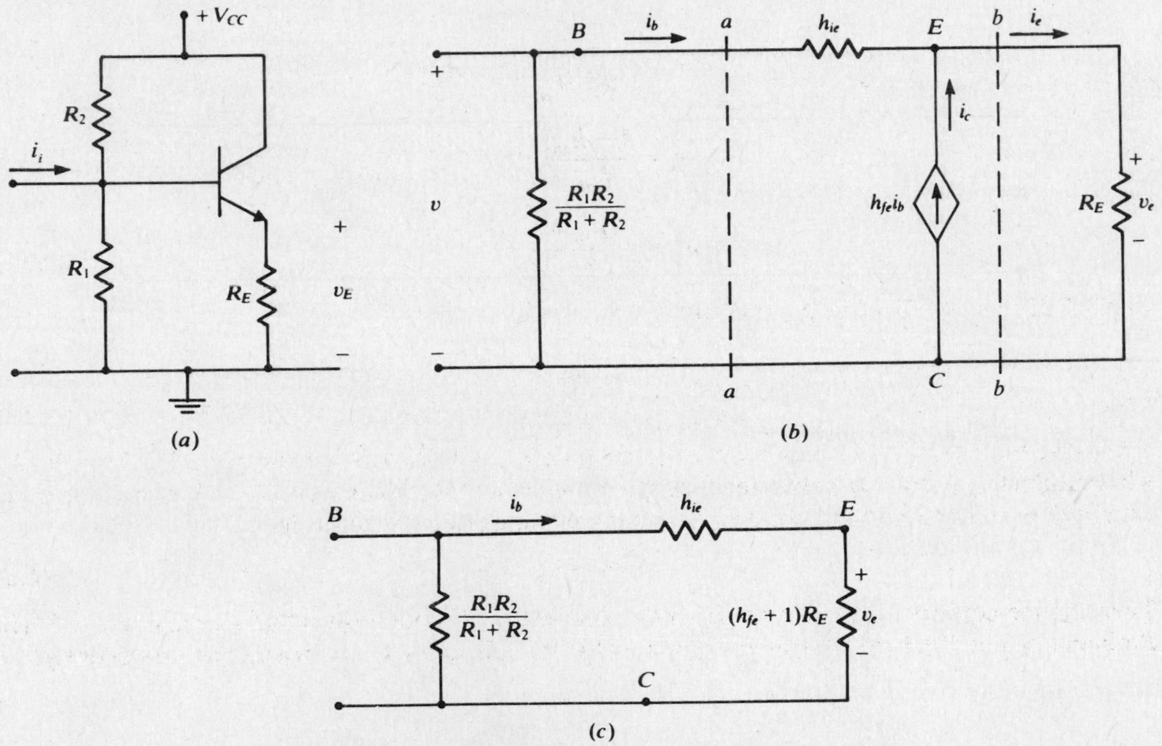

Fig. 6-2 CC amplifier

6.3 TEE-EQUIVALENT CIRCUIT

The *tee-equivalent circuit* or *r-parameter model* is a circuit realization based on the *z* parameters of Chapter 1. Applying the *z*-parameter definitions of (*1.10*) to (*1.13*) to the CB small-signal equivalent circuit of Fig. 6-1(*b*) leads to

$$z_{11} = h_{ib} - \frac{h_{rb} h_{fb}}{h_{ob}} \tag{6.17}$$

$$z_{12} = \frac{h_{rb}}{h_{ob}} \tag{6.18}$$

$$z_{21} = -\frac{h_{fb}}{h_{ob}} \tag{6.19}$$

$$z_{22} = \frac{1}{h_{ob}} \tag{6.20}$$

(See Problem 6.16.) Substitution of these z parameters into (*1.8*) and (*1.9*) yields

$$v_{eb} = \left(h_{ib} - \frac{h_{rb}h_{fb}}{h_{ob}}\right)i_e + \frac{h_{rb}}{h_{ob}}(-i_c) \tag{6.21}$$

$$v_{cb} = -\frac{h_{fb}}{h_{ob}}i_e + \frac{1}{h_{ob}}(-i_c) \tag{6.22}$$

If we now define

$$r_b = \frac{h_{rb}}{h_{ob}} \tag{6.23}$$

$$r_e = h_{ib} - \frac{h_{rb}}{h_{ob}}(1 + h_{fb}) \tag{6.24}$$

$$r_c = \frac{1 - h_{rb}}{h_{ob}} \tag{6.25}$$

$$\alpha' = -\frac{h_{fb} + h_{rb}}{1 - h_{rb}} \tag{6.26}$$

then (*6.21*) and (*6.22*) can be written

$$v_{eb} = (r_e + r_b)i_e - r_b i_c \tag{6.27}$$

and

$$v_{cb} = (\alpha' r_c + r_b)i_e - (r_b + r_c)i_c \tag{6.28}$$

Typically, $-0.9 > h_{fb} > -1$ and $0 \le h_{rb} \ll 1$. Letting $h_{rb} \approx 0$ in (*6.26*), comparing (*6.13*) with (*3.1*) while neglecting thermally generated leakage currents, and assuming that $h_{FB} = h_{fb}$ (which is a valid assumption *except* near the boundary of active-region operation) result in

$$\alpha' \approx -h_{fb} = \alpha \tag{6.29}$$

Then the tee-equivalent circuit or r-parameter model for CB operation is that shown in Fig. 6-3. (See Problems 6.3 and 6.5 for r-parameter models for the CE and CC configurations, respectively.)

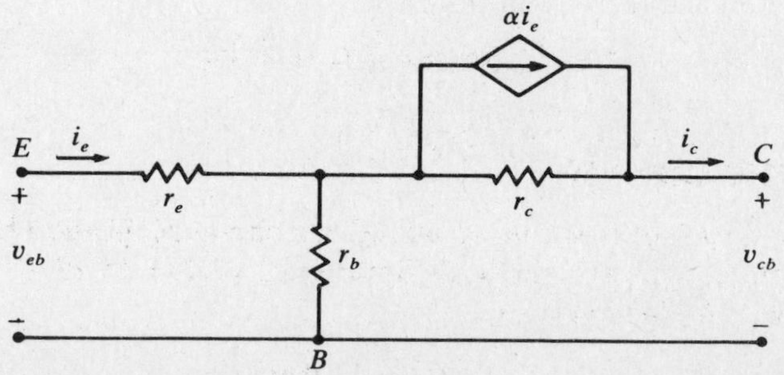

Fig. 6-3

6.4 CONVERSION OF PARAMETERS

Transistor manufacturers typically specify h_{FE} ($\approx h_{fe}$) and a set of input characteristics and collector characteristics for either CE or CB connection. Thus the necessity arises for conversion of h parameters among the CE, CB, and CC configurations or for calculation of r parameters from h parameters. Formulas can be developed to allow ready conversion from a known parameter set to a desired parameter set.

Example 6.1 Apply KVL and KCL to Fig. 6-1(a) to obtain $v_{eb} = g_1(i_e, v_{cb})$ and $i_c = g_2(i_e, v_{eb})$. Compare these equations with (6.9) and (6.10) to find the CB h parameters in terms of the CE h parameters. Use the typically reasonable approximations $h_{re} \ll 1$ and $h_{fe} + 1 \gg h_{ie}h_{oe}$ to simplify the computations and results.

KVL around the E,B loop of Fig. 6-1(a) (with assumed current directions reversed) yields

$$v_{eb} = -h_{ie}i_b - h_{re}v_{ce} \tag{6.30}$$

But KCL at node E requires that

$$i_b = -i_e - i_c = -i_e - h_{fe}i_b - h_{oe}v_{ce}$$

or

$$-i_b = \frac{1}{h_{fe}+1}\, i_e + \frac{h_{oe}}{h_{fe}+1}\, v_{ce} \tag{6.31}$$

In addition, KVL requires that

$$v_{ce} = v_{cb} - v_{eb} \tag{6.32}$$

Substituting (6.31) and (6.32) into (6.30) and rearranging give

$$\frac{(1-h_{re})(h_{fe}+1) + h_{ie}h_{oe}}{h_{fe}+1}\, v_{eb} = \frac{h_{ie}}{h_{fe}+1}\, i_e + \left(\frac{h_{ie}h_{oe}}{h_{fe}+1} - h_{re}\right) v_{cb} \tag{6.33}$$

Use of the given approximations reduces the coefficient of v_{eb} in (6.33) to unity, so that

$$v_{eb} \approx \frac{h_{ie}}{h_{fe}+1}\, i_e + \left(\frac{h_{ie}h_{oe}}{h_{fe}+1} - h_{re}\right) v_{cb} \tag{6.34}$$

Now KCL at node C of Fig. 6-1(a) (again with assumed current directions reversed) yields

$$i_c = h_{fe}i_b + h_{oe}v_{ce} \tag{6.35}$$

Substituting (6.31), (6.32), and (6.34) into (6.35) and solving for i_c give

$$i_c = -\left[\frac{h_{fe}}{h_{fe}+1} + \frac{h_{oe}h_{ie}}{(h_{fe}+1)^2}\right] i_e - h_{oe}\left[\frac{h_{ie}h_{oe}}{(h_{fe}+1)^2} - \frac{h_{re}+1}{h_{fe}+1}\right] v_{cb} \tag{6.36}$$

Use of the given approximations then leads to

$$i_c \approx -\frac{h_{fe}}{h_{fe}+1}\, i_e + \frac{h_{oe}}{h_{fe}+1}\, v_{cb} \tag{6.37}$$

Comparing (6.34) with (6.9) and (6.37) with (6.10), we see that

$$h_{ib} = \frac{h_{ie}}{h_{fe}+1} \tag{6.38}$$

$$h_{rb} = \frac{h_{ie}h_{oe}}{h_{fe}+1} - h_{re} \tag{6.39}$$

$$h_{fb} = -\frac{h_{fe}}{h_{fe}+1} \tag{6.40}$$

$$h_{ob} = \frac{h_{oe}}{h_{fe}+1} \tag{6.41}$$

6.5 MEASURES OF AMPLIFIER GOODNESS

Amplifiers are usually designed to emphasize one or more of the following interrelated performance characteristics, whose quantitative measures of goodness are defined in terms of the quantities of Fig. 6-4:

1. *Current amplification*, measured by the current-gain ratio $A_i = i_o/i_{in}$.
2. *Voltage amplification*, measured by the voltage-gain ratio $A_v = v_o/v_{in}$.
3. *Power amplification*, measured by the ratio $A_p = A_v A_i = v_o i_o / i_o i_{in}$.
4. *Phase shift of signals*, measured by the phase angle of the frequency-domain ratio $A_v(j\omega)$ or $A_i(j\omega)$.
5. *Impedance match or change*, measured by the input impedance Z_{in} (the driving-point impedance looking into the input port).
6. *Power transfer ability*, measured by the output impedance Z_o (the driving-point impedance looking into the output port with the load removed). If $Z_o = Z_L$, the maximum power transfer occurs.

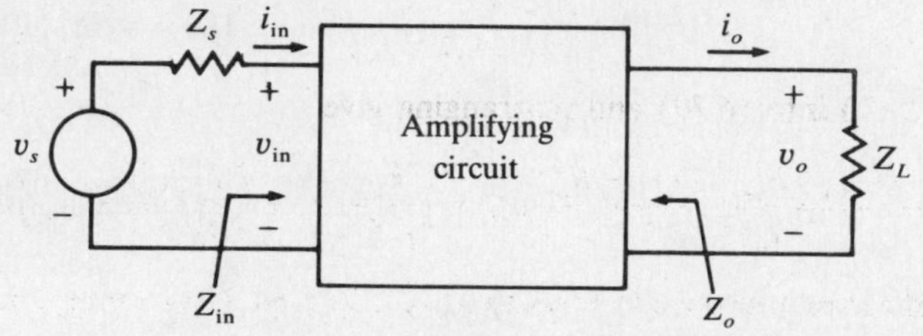

Fig. 6-4

6.6 CE AMPLIFIER ANALYSIS

A simplified (bias network omitted) CE amplifier is shown in Fig. 6-5(a), and the associated small-signal equivalent circuit in Fig. 6-5(b).

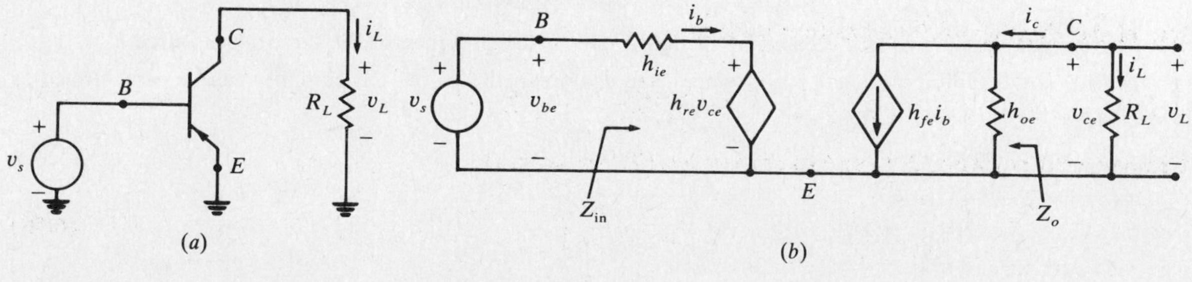

Fig. 6-5

Example 6.2 In the CE amplifier of Fig. 6-5(b), let $h_{ie} = 1\ \text{k}\Omega$, $h_{re} = 10^{-4}$, $h_{fe} = 100$, $h_{oe} = 12\ \mu\text{S}$, and $R_L = 2\ \text{k}\Omega$. (These are typical CE amplifier values.) Find expressions for the (a) current-gain ratio A_i, (b) voltage-gain ratio A_v, (c) input impedance Z_{in}, and (d) output impedance Z_o. (e) Evaluate this typical CE amplifier.

(*a*) By current division at node C,

$$i_L = \frac{1/h_{oe}}{1/h_{oe} + R_L}(-h_{fe} i_b) \tag{6.42}$$

and

$$A_i = \frac{i_L}{i_b} = -\frac{h_{fe}}{1 + h_{oe} R_L} = -\frac{100}{1 + (12 \times 10^{-6})(2 \times 10^3)} = -97.7 \tag{6.43}$$

Note that $A_i \approx -h_{fe}$, where the minus sign indicates a 180° phase shift between input and output currents.

(*b*) By KVL around B,E mesh,

$$v_s = v_{be} = h_{ie} i_b + h_{re} v_{ce} \tag{6.44}$$

Ohm's law applied to the output network requires that

$$v_{ce} = -h_{fe} i_b \left(\frac{1}{h_{oe}} \parallel R_L\right) = \frac{-h_{fe} R_L i_b}{1 + h_{oe} R_L} \tag{6.45}$$

Solving (*6.45*) for i_b, substituting the result into (*6.44*), and rearranging yield

$$A_v = \frac{v_s}{v_{ce}} = -\frac{h_{fe} R_L}{h_{ie} + R_L(h_{ie} h_{oe} - h_{fe} h_{re})}$$

$$= -\frac{(100)(2 \times 10^3)}{1 \times 10^3 + (2 \times 10^3)[(1 \times 10^3)(12 \times 10^{-6}) - (100)(1 \times 10^{-4})]} = -199.2 \tag{6.46}$$

Observe that $A_v \approx -h_{fe} R_L / h_{ie}$, where the minus sign indicates a 180° phase shift between input and output voltages.

(*c*) Substituting (*6.45*) into (*6.44*) and rearranging yield

$$Z_{\text{in}} = \frac{v_s}{i_b} = h_{ie} - \frac{h_{re} h_{fe} R_L}{1 + h_{oe} R_L} = 1 \times 10^3 - \frac{(1 \times 10^{-4})(100)(2 \times 10^3)}{1 + (12 \times 10^{-6})(2 \times 10^3)} = 980.5\ \Omega \tag{6.47}$$

Note that for typical CE amplifier values, $Z_{\text{in}} \approx h_{ie}$.

(*d*) We deactivate (short) v_s and replace R_L with a driving-point source so that $v_{dp} = v_{ce}$. Then, for the input mesh, Ohm's law requires that

$$i_b = -\frac{h_{re}}{h_{ie}} v_{dp} \tag{6.48}$$

However, at node C (with, now, $i_c = i_{dp}$), KCL yields

$$i_c = i_{dp} = h_{fe} i_b + h_{oe} v_{dp} \tag{6.49}$$

Using (*6.48*) in (*6.49*) and rearranging then yield

$$Z_o = \frac{v_{dp}}{i_{dp}} = \frac{1}{h_{oe} - h_{fe} h_{re} / h_{ie}} = \frac{1}{12 \times 10^{-6} - (100)(1 \times 10^{-4})/(1 \times 10^3)} = 500\ \text{k}\Omega \tag{6.50}$$

The output impedance is increased by feedback due to the presence of the controlled source $h_{re} v_{ce}$.

(*e*) Based on the typical values of this example, the characteristics of the CE amplifier can be summarized as follows:

1. Large current gain
2. Large voltage gain
3. Large power gain ($A_i A_v$)
4. Current and voltage phase shifts of 180°
5. Moderate input impedance
6. Moderate output impedance

6.7 CB AMPLIFIER ANALYSIS

A simplified (bias network omitted) CB amplifier is shown in Fig. 6-6(*a*), and the associated small-signal equivalent circuit in Fig. 6-6(*b*).

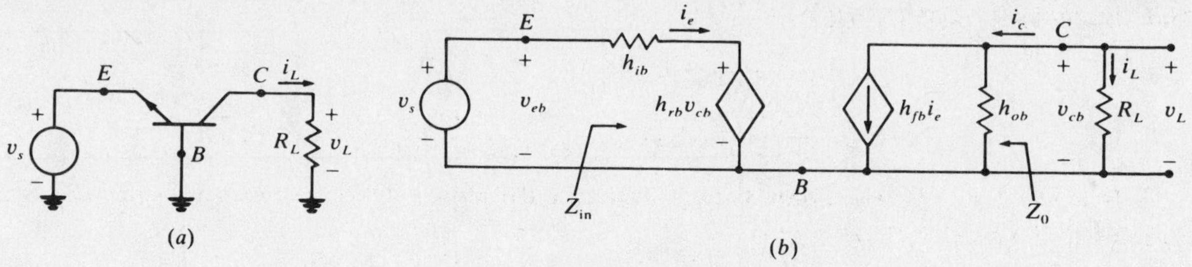

Fig. 6-6

Example 6.3 In the CB amplifier of Fig. 6-6(*b*), let $h_{ib} = 30\ \Omega$, $h_{rb} = 4 \times 10^{-6}$, $h_{fb} = -0.99$, $h_{ob} = 8 \times 10^{-7}$ S, and $R_L = 20$ kΩ. (These are typical CB amplifier values.) Find expressions for the (*a*) current-gain ratio A_i, (*b*) voltage-gain ratio A_v, (*c*) input impedance Z_{in}, and (*d*) output impedance Z_o. (*e*) Evaluate this typical CE amplifier.

(*a*) By direct analogy with Fig. 6-5(*b*) and (*6.43*)

$$A_i = -\frac{h_{fb}}{1 + h_{ob}R_L} = -\frac{-0.99}{1 + (8 \times 10^{-7})(20 \times 10^3)} = 0.974 \qquad (6.51)$$

Note that $A_i \approx -h_{fb} < 1$, and that the input and output currents are in phase because $h_{fb} < 0$.

(*b*) By direct analogy with Fig. 6-5(*b*) and (*6.46*),

$$A_v = -\frac{h_{fb}R_L}{h_{ib} + R_L(h_{ib}h_{oc} - h_{fb}h_{rb})} = -\frac{(-0.99)(20 \times 10^3)}{30 + (20 \times 10^3)[(30)(8 \times 10^{-7}) - (-0.99)(4 \times 10^{-6})]} = 647.9 \qquad (6.52)$$

Observe that $A_v \approx -h_{fb}R_L/h_{ib}$, and the output and input voltages are in phase because $h_{fb} < 0$.

(*c*) By direct analogy with Fig. 6-5(*b*) and (*6.47*)

$$Z_{in} = h_{ib} - \frac{h_{rb}h_{fb}R_L}{1 + h_{ob}R_L} = 30 - \frac{(4 \times 10^{-6})(-0.99)(20 \times 10^3)}{1 + (8 \times 10^{-7})(20 \times 10^3)} = 30.08\ \Omega \qquad (6.53)$$

It is apparent that $Z_{in} \approx h_{ib}$.

(*d*) By analogy with Fig. 6-5(*b*) and (*6.50*),

$$Z_o = \frac{1}{h_{ob} - h_{fb}h_{rb}/h_{ib}} = \frac{1}{8 \times 10^{-7} - (-0.99)(4 \times 10^{-6})/30} = 1.07\ \text{M}\Omega \qquad (6.54)$$

Note that Z_o is decreased because of the feedback from the output mesh to the input mesh through $h_{rb}v_{cb}$.

(*e*) Based on the typical values of this example, the characteristics of the CB amplifier can be summarized as follows:

1. Current gain of less than 1
2. High voltage gain
3. Power gain approximately equal to voltage gain
4. No phase shift for current or voltage
5. Small input impedance
6. Large output impedance

6.8 CC AMPLIFIER ANALYSIS

Figure 6-7(*a*) shows a CC amplifier with the bias network omitted. The small-signal equivalent circuit is drawn in Fig. 6-7(*b*).

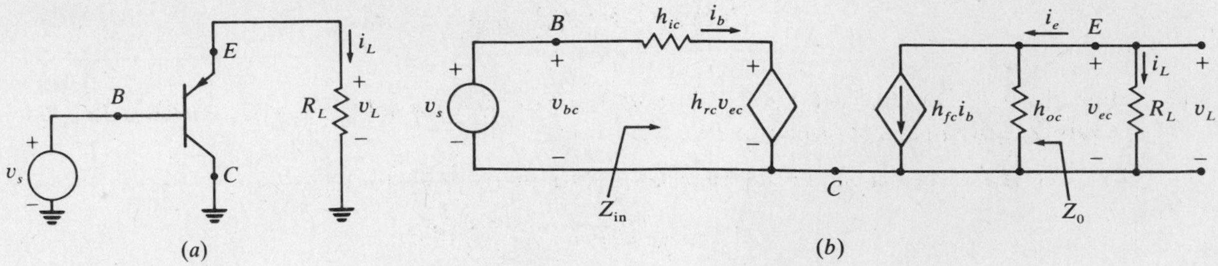

Fig. 6-7

Example 6.4 In the CC amplifier of Fig. 6-7(*b*), let $h_{ic} = 1\ \text{k}\Omega$, $h_{rc} = 1$, $h_{fc} = -101$, $h_{oc} = 12\ \mu\text{S}$, and $R_L = 2\ \text{k}\Omega$. Drawing direct analogies with the CE amplifier of Example 6.2, find expressions for the (*a*) current-gain ratio A_i, (*b*) voltage-gain ratio A_v, (*c*) input impedance Z_{in}, and (*d*) output impedance Z_o. (*e*) Evaluate this typical CC amplifier.

(*a*) In parallel with (*6.43*),

$$A_i = \frac{h_{fc}}{1 + h_{oc}R_L} = -\frac{-101}{1 + (12 \times 10^{-6})(2 \times 10^3)} = 98.6 \tag{6.55}$$

Note that $A_i \approx -h_{fc}$, and that the input and output currents are in phase because $h_{fc} < 0$.

(*b*) In parallel with (*6.46*),

$$A_v = -\frac{h_{fc}R_L}{h_{ic} + R_L(h_{ic}h_{oc} - h_{fc}h_{rc})} = -\frac{(-101)(2 \times 10^3)}{1 \times 10^3 + (2 \times 10^3)[(1 \times 10^3)(12 \times 10^{-6}) - (-101)(1)]} = 0.995 \tag{6.56}$$

Observe that $A_v \approx 1/(1 - h_{ic}h_{oc}/h_{fc}) \approx 1$. Since the gain is approximately 1 and the output voltage is in phase with the input voltage, this amplifier is commonly called a *unity follower*.

(*c*) In parallel with (*6.47*),

$$Z_{\text{in}} = h_{ic} - \frac{h_{rc}h_{fc}R_L}{1 + h_{oc}R_L} = 1 \times 10^3 - \frac{(1)(-101)(2 \times 10^3)}{1 + (12 \times 10^{-6})(2 \times 10^3)} = 8.41\ \text{M}\Omega \tag{6.57}$$

Note that $Z_{\text{in}} \approx -h_{fc}/h_{oc}$.

(*d*) In parallel with (*6.50*),

$$Z_o = \frac{1}{h_{oc} - h_{fc}h_{rc}/h_{ic}} = \frac{1}{12 \times 10^{-6} - (-101)(1)/(1 \times 10^3)} = 9.9\ \Omega$$

Note that $Z_o \approx -h_{ic}/h_{fc}$.

(*e*) Based on the typical values of this example, the characteristics of the CB amplifier can be summarized as follows:

1. High current gain
2. Voltage gain of approximately unity
3. Power gain approximately equal to current gain
4. No current or voltage phase shift
5. Large input impedance
6. Small output impedance

Solved Problems

6.1 For the CB amplifier of Fig. 3-18, find the voltage-gain ratio $A_v = v_L/v_S$ using the tee-equivalent small-signal circuit of Fig. 6-3.

The small-signal circuit for the amplifier is given by Fig. 6-8. By Ohm's law,

$$i_c = \frac{v_{cb}}{R_C \| R_L} = \frac{(R_C + R_L)v_L}{R_C R_L} \tag{1}$$

Substituting (*1*) into (*6.27*) and (*6.28*) gives, respectively,

$$v_S = v_{eb} = (r_e + r_b)i_e - r_b \frac{(R_C + R_L)v_L}{R_C R_L} \tag{2}$$

$$v_L = v_{cb} = (\alpha r_c + r_b)i_e - (r_b + r_c)\frac{(R_C + R_L)v_L}{R_C R_L} \tag{3}$$

where we also made use of (*6.29*). Solving (*2*) for i_e and substituting the result into (*3*) yield

$$v_L = (\alpha r_c + r_b)\frac{v_S + \dfrac{r_b(R_C + R_L)}{R_C + R_L} v_L}{r_e + r_b} - (r_b + r_c)\frac{R_C + R_L}{R_C R_L} v_L \tag{4}$$

The voltage-gain ratio follows directly from (*4*) as

$$A_v = \frac{v_L}{v_S} = \frac{(\alpha r_c + r_b)R_C R_L}{R_C R_L(r_e + r_b) + (R_C + R_L)[(1-\alpha)r_c r_b + r_e(r_b + r_c)]}$$

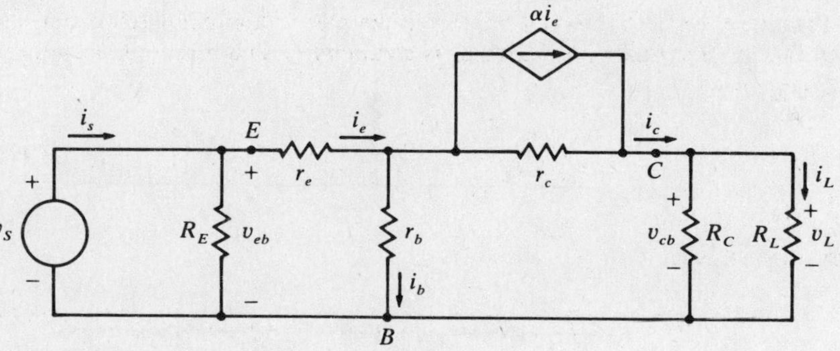

Fig. 6-8

6.2 Assume that r_c is large enough so that $i_c \approx \alpha i_e$ for the CB amplifier of Fig. 3-18, whose small-signal circuit is given by Fig. 6-8. Find an expression for the current-gain ratio $A_i = i_L/i_s$ and evaluate it if $r_e = 30\ \Omega$, $r_b = 300\ \Omega$, $r_c = 1\ \text{M}\Omega$, $R_E = 5\ \text{k}\Omega$, $R_C = R_L = 4\ \text{k}\Omega$, and $\alpha = 0.99$.

Letting $i_c \approx \alpha i_e$ in (*6.27*) allows us to determine the input resistance R_{in}:

$$v_{eb} = (r_e + r_b)i_e - r_b(\alpha i_e)$$

from which

$$R_{in} = \frac{v_{eb}}{i_e} = r_e + (1-\alpha)r_b$$

By current division at node E,

$$i_e = \frac{R_E}{R_E + R_{in}} i_s$$

Solving for i_s gives

$$i_s = \frac{R_E + R_{\text{in}}}{R_E} i_e = \frac{R_E + r_e + (1-\alpha)r_b}{R_E} i_e \qquad (1)$$

Current division at node C, again with $i_c \approx \alpha i_e$, yields

$$i_L = \frac{R_C}{R_C + R_L} i_c = \frac{R_C \alpha i_e}{R_C + R_L} \qquad (2)$$

The current gain is now the ratio of (2) to (1):

$$A_i = \frac{i_L}{i_s} = \frac{\alpha R_C/(R_C + R_L)}{[R_E + r_e + (1-\alpha)r_b]/R_E} = \frac{\alpha R_C R_E}{(R_C + R_L)[R_E + r_e + (1-\alpha)r_b]}$$

Substituting the given values results in

$$A_i = \frac{(0.99)(4\times 10^3)(5\times 10^3)}{(4\times 10^3 + 4\times 10^3)[5\times 10^3 + 30 + (1-0.99)(300)]} = 0.492$$

6.3 The transistor of a CE amplifier can be modeled with the tee-equivalent circuit of Fig. 6-3 if the base and emitter terminals are interchanged, as shown by Fig. 6-9(*a*); however, the controlled source is no longer given in terms of a port current—an analytical disadvantage. Show that the circuits of Fig. 6-9(*b*) and (*c*), where the controlled variable of the dependent source is the input current i_b, can be obtained by application of Thévenin's and Norton's theorems to the circuit of Fig. 6-9(*a*).

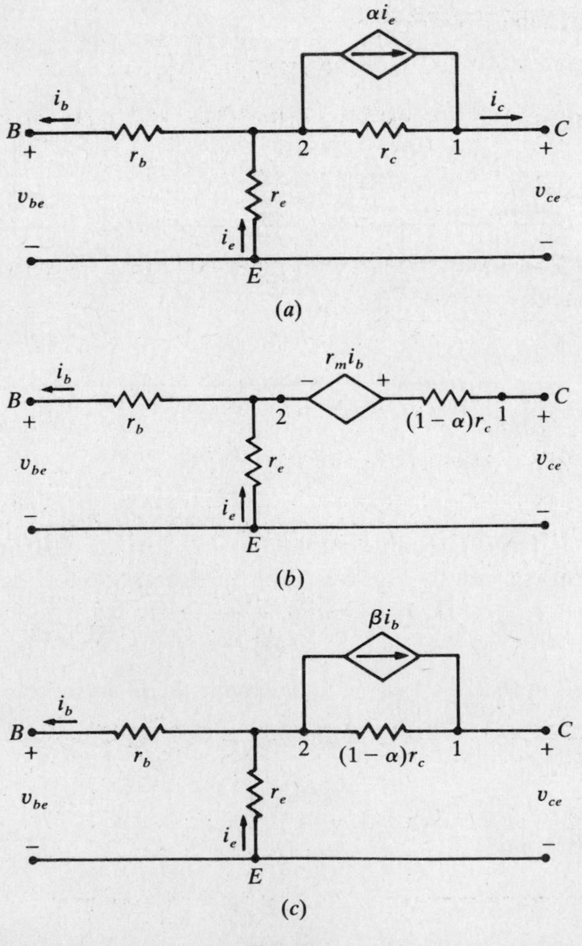

Fig. 6-9

The Thévenin equivalent for the circuit above terminals 1,2 of Fig. 6-9(a) has

$$v_{th} = \alpha r_c ie \qquad Z_{th} = r_c$$

By KCL, $i_e = i_c + i_b$, so that

$$v_{th} = \alpha r_c i_c + \alpha r_c i_b \tag{1}$$

We recognize that if the Thévenin elements are placed in the network, the first term on the right side of (1) must be modeled by using a "negative resistance." The second term represents a controlled voltage source. Thus, a modified Thévenin equivalent can be introduced, in which the "negative resistance" is combined with Z_{th} to give

$$v'_{th} = \alpha r_c i_b = r_m i_b \qquad Z'_{th} = (1-\alpha) r_c \tag{2}$$

With the modified Thévenin elements of (2) in position, we obtain Fig. 6-9(b).

The elements of the Norton equivalent circuit can be determined directly from (2) as

$$Z_N = \frac{1}{Y_N} = Z'_{th} = (1-\alpha) r_c \qquad I_N = \frac{v'_{th}}{Z'_{th}} = \frac{\alpha r_c i_b}{(1-\alpha) r_c} = \beta i_b \tag{3}$$

The elements of (3) give the circuit of Fig. 6-9(c).

6.4 Utilize the r-parameter equivalent circuit of Fig. 6-9(b) to find the voltage-gain ratio $A_v = v_L/v_i$ for the CE amplifier circuit of Fig. 3-6.

The small-signal equivalent circuit for the amplifier is drawn in Fig. 6-10. After finding the Thévenin equivalent for the network to the left of terminals B,E, we may write

$$v_{be} = \frac{R_B}{R_B + R_i} v_i + \frac{R_B R_i}{R_B + R_i} i_b \tag{1}$$

Ohm's law at the output requires that

$$v_{ce} = v_L = \frac{R_C R_L}{R_C + R_L} i_c \tag{2}$$

Applying KVL around the B,E mesh and around the C,E mesh while noting that $i_e = i_c + i_b$ yields, respectively,

$$v_{be} = -r_b i_b - r_e i_e = -(r_b + r_e) i_b - r_e i_c \tag{3}$$

and

$$v_{ce} = -r_e i_e + r_m i_b - (1-\alpha) r_c i_c = -(r_e - r_m) i_b - [(1-\alpha) r_c + r_e] i_c \tag{4}$$

Equating (1) to (3) and (2) to (4) allows formulation of the system of linear equations

$$\begin{bmatrix} -\left(r_b + r_e + \dfrac{R_B R_i}{R_B + R_i}\right)\dfrac{R_B + R_i}{R_B} & -\dfrac{r_e (R_B + R_i)}{R_B} \\ -(r_e - r_m) & -\left[(1-\alpha) r_c + r_e + \dfrac{R_C R_L}{R_C + R_L}\right] \end{bmatrix} \begin{bmatrix} i_b \\ i_c \end{bmatrix} = \begin{bmatrix} v_i \\ 0 \end{bmatrix}$$

from which, by Cramer's rule, $i_c = \Delta_2/\Delta$, where

$$\Delta = \frac{R_B + R_i}{R_B}\left\{\left(r_b + r_e + \frac{R_B R_i}{R_B + R_i}\right)\left[(1-\alpha) r_c + r_e + \frac{R_C R_L}{R_C + R_L}\right] - r_e (r_e - r_m)\right\}$$

$$\Delta_2 = (r_e - r_m) v_i$$

Then

$$A_v = \frac{v_L}{v_i} = \frac{(R_L \| R_C) i_c}{v_i} = \frac{R_L R_C}{R_L + R_C} \frac{r_e - r_m}{\Delta}$$

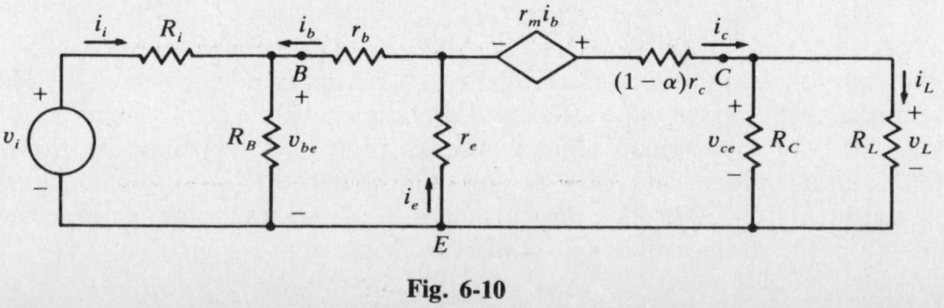

Fig. 6-10

6.5 The CE tee-equivalent circuit of Fig. 6-9(*b*) is suitable for use in the analysis of an EF amplifier if the collector and emitter branches are interchanged. Use this technique to calculate (*a*) the voltage-gain ratio $A_v = v_L/v_B$ and (*b*) the input impedance for the amplifier of Fig. 3-20(*a*).

(*a*) The appropriate small-signal equivalent circuit is given in Fig. 6-11. By KVL around the B,C loop, with $r_m = \alpha r_c$ (from Problem 6.3),

$$v_B = r_b i_b + r_m i_b + (1-\alpha) r_c (i_b - i_e) = (r_b + r_c) i_b - (1-\alpha) r_c i_e \qquad (1)$$

Application of KVL around the C,E loop, again with $r_m = \alpha r_c$, gives

$$0 = r_e i_e - r_m i_b - (1-\alpha) r_c (i_b - i_e) + \frac{R_E R_L}{R_E + R_L} i_e = -r_c i_b + \left[r_e + (1-\alpha) r_c + \frac{R_E R_L}{R_E + R_L} \right] i_e \qquad (2)$$

By Cramer's rule applied to the system consisting of (*1*) and (*2*), $i_e = \Delta_2/\Delta$, where

$$\Delta = r_b \left[r_e + (1-\alpha) r_c + \frac{R_E R_L}{R_E + R_L} \right] + r_c \left(r_e + \frac{R_E R_L}{R_E + R_L} \right)$$

$$\Delta_2 = r_c v_B$$

Now, by Ohm's law,

$$v_L = (R_E \| R_L) i_e = \frac{R_E R_L}{R_E + R_L} \frac{\Delta_2}{\Delta}$$

Then $$A_v = \frac{v_L}{v_B} = \frac{R_E R_L r_c/(R_E + R_L)}{r_b[r_e + (1-\alpha) r_c + R_E R_L/(R_E + R_L)] + r_c[r_e + R_E R_L/(R_E + R_L)]}$$

(*b*) The input impedance can be found as $Z_{in} = R_B \| (v_B/i_b)$. Now, in the system consisting of (*1*) and (*2*), by Cramer's rule, $i_b = \Delta_1/\Delta$, where

$$\Delta_1 = \left[r_e + (1-\alpha) r_c + \frac{R_E R_L}{R_E + R_L} \right] v_B$$

Hence, $$Z_{in} = R_B \| \left(\frac{\Delta}{\Delta_1} v_B \right) = \frac{R_B r_b \left[r_e + (1-\alpha) r_c + \dfrac{R_E R_L}{R_E + R_L} \right] + R_B r_c \left(r_e + \dfrac{R_E R_L}{R_E + R_L} \right)}{(R_B + r_b) \left[r_e + (1-\alpha) r_c + \dfrac{R_E R_L}{R_E + R_L} \right] + r_c \left(r_e + \dfrac{R_E R_L}{R_E + R_L} \right)}$$

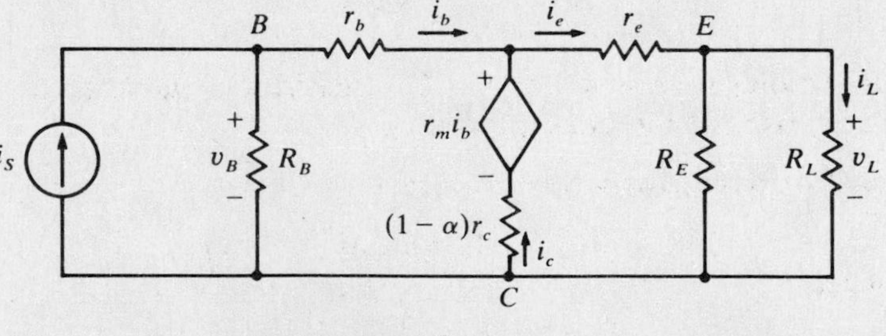

Fig. 6-11

6.6 Answer the following questions relating to a CE-connected transistor: (*a*) How are the input characteristics (i_B versus v_{BE}) affected if there is negligible feedback of v_{CE}? (*b*) What might be the effect of a too-small emitter-base junction bias? (*c*) Suppose the transistor has an infinite output impedance; how would that affect the output characteristics? (*d*) With reference to Fig. 3-5(*b*), does the current gain of the transistor increase or decrease as the mode of operation approaches saturation from the active region?

(*a*) The family of input characteristics degenerates to a single curve—one that is frequently used to approximate the family.

(*b*) If I_{BQ} were so small that operation occurred near the knee of an input characteristic curve, distortion would result.

(*c*) The slope of the output characteristic curves would be zero in the active region.

(*d*) Δi_C decreases for constant Δi_B; hence, the current gain decreases.

6.7 Use a small-signal *h*-parameter equivalent circuit to analyze the amplifier of Fig. 3-6(*a*), given $R_C = R_L = 800\ \Omega$, $R_i = 0$, $R_1 = 1.2\ \text{k}\Omega$, $R_2 = 2.7\ \text{k}\Omega$, $h_{re} \approx 0$, $h_{oe} = 100\ \mu\text{S}$, $h_{fe} = 90$, and $h_{ie} = 200\ \Omega$. Calculate (*a*) the voltage gain A_v and (*b*) the current gain A_i.

(*a*) The small-signal circuit is shown in Fig. 6-12, where $R_B = R_1 R_2/(R_1 + R_2) = 831\ \Omega$. By current division in the collector circuit,

$$-i_L = \frac{R_C(1/h_{oe})}{R_C(1/h_{oe}) + R_L(1/h_{oe}) + R_L R_C} h_{fe} i_b$$

The voltage gain is then

$$A_v \equiv \frac{v_L}{v_i} = \frac{R_L i_L}{h_{ie} i_b} = -\frac{h_{fe} R_L R_C}{h_{ie}(R_C + R_L + h_{oe} R_L R_C)} = -\frac{(90)(800)^2}{200[1600 + (100 \times 10^{-6})(800)^2]} = -173.08 \tag{1}$$

(*b*) By current division,

$$i_b = \frac{R_B}{R_B + h_{ie}} i_i$$

so
$$A_i \equiv \frac{i_L}{i_i} = \frac{R_B}{R_B + h_{ie}} \frac{i_L}{i_b} = \frac{R_B h_{ie}}{R_L(R_B + h_{ie})} A_v = \frac{(831)(200)(-173.08)}{(800)(1031)} = -34.87$$

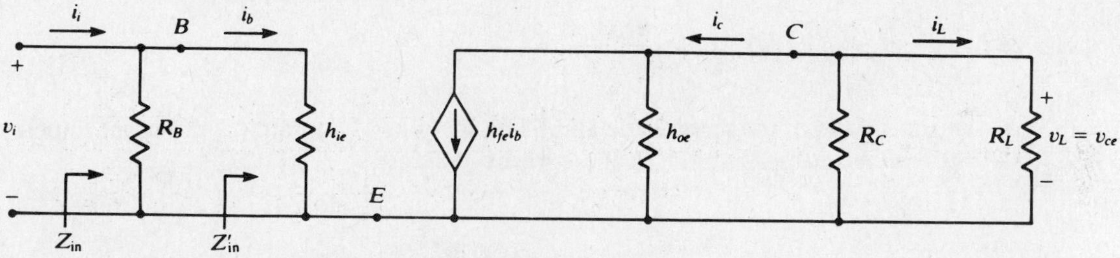

Fig. 6-12

6.8 Suppose the emitter-base junction of a Ge transistor is modeled as a forward-biased diode. Express h_{ie} in terms of the emitter current.

The use of transistor notation in (*2.1*) gives

$$i_B = I_{CBO}(e^{v_{BE}/v_T} - 1) \tag{1}$$

Then, by (*6.5*),

$$\frac{1}{h_{ie}} = \frac{\partial i_B}{\partial v_{BE}}\bigg|_Q = \frac{1}{V_T} I_{CBO}\, e^{v_{BEQ}/v_T} \tag{2}$$

But, by (*1*) and Problem 2.1,

$$I_{BQ} = I_{CBO}(e^{v_{BEQ}/v_T} - 1) \approx I_{CBO}\, e^{v_{BEQ}/v_T} \tag{3}$$

and

$$I_{BQ} = \frac{I_{EQ}}{\beta + 1} \tag{4}$$

Equations (*2*), (*3*), and (*4*) imply

$$h_{ie} = \frac{V_T(\beta + 1)}{I_{EQ}}$$

6.9 For the CB amplifier of Problem 3.12, determine graphically (*a*) h_{fb} and (*b*) h_{ob}.

(*a*) The Q point was established in Problem 3.12 and is indicated in Fig. 3-12. By (*6.13*),

$$h_{fb} \approx \left.\frac{\Delta i_C}{\Delta i_E}\right|_{v_{CBQ}=-6.1\text{ V}} = \frac{(3.97-2.0)\times 10^{-3}}{(4-2)\times 10^{-3}} = 0.985$$

(*b*) By (*6.14*),

$$h_{ob} \approx \left.\frac{\Delta i_C}{\Delta v_{CB}}\right|_{I_{EQ}=3\text{ mA}} = \frac{(3.05-2.95)\times 10^{-3}}{-10-(-2)} = 12.5\ \mu\text{S}$$

6.10 Find the input impedance Z_{in} of the circuit of Fig. 3-6(*a*) in terms of the h parameters, all of which are nonzero.

The small-signal circuit of Fig. 6-12, with $R_B = R_1R_2/(R_1+R_2)$, is applicable if a dependent source $h_{re}v_{ce}$ is added in series with h_{ie}, as in Fig. 6-1(*a*). The admittance of the collector circuit is given by

$$G = h_{oe} + \frac{1}{R_L} + \frac{1}{R_C}$$

and, by Ohm's law,

$$v_{ce} = \frac{-h_{fe}i_b}{G} \tag{1}$$

By KVL applied to the input circuit,

$$i_b = \frac{v_i - h_{re}v_{ce}}{h_{ie}} \tag{2}$$

Now (*1*) may be substituted in (*2*) to eliminate v_{ce}, and the result rearranged into

$$Z'_{\text{in}} = \frac{v_i}{i_b} = h_{ie} - \frac{h_{re}h_{fe}}{G} \tag{3}$$

Then

$$Z_{\text{in}} = \frac{R_BZ'_{\text{in}}}{R_B + Z'_{\text{in}}} = \frac{R_B(h_{ie} - h_{re}h_{fe}/G)}{R_B + h_{ie} - h_{re}h_{fe}/G} \tag{4}$$

6.11 In terms of the CB h parameters for the amplifier of Fig. 6-13(*a*), find (*a*) the input impedance Z_{in}, (*b*) the voltage gain A_v, and (*c*) the current gain A_i.

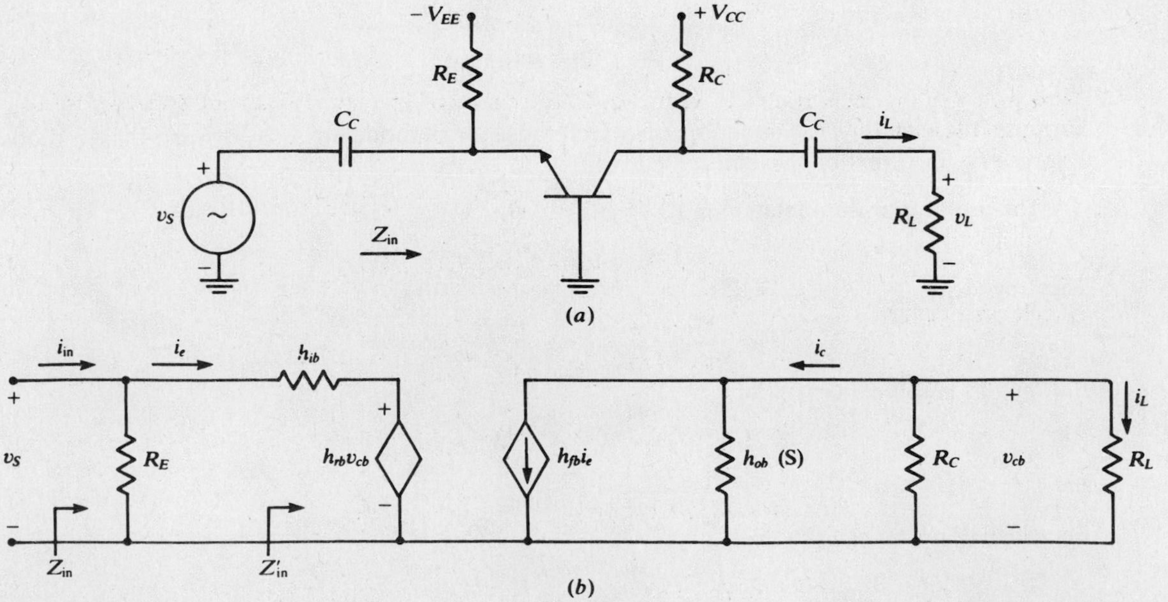

Fig. 6-13

(*a*) The h-parameter equivalent circuit is given in Fig. 6-13(*b*). By Ohm's law,

$$v_{cb} = -\frac{h_{fb}i_e}{h_{ob} + 1/R_C + 1/R_L} \equiv -\frac{h_{fb}i_e}{G} \tag{1}$$

Application of KVL at the input gives

$$v_S = h_{rb}v_{cb} + h_{ib}i_e \tag{2}$$

Now (*1*) may be substituted into (*2*) and the result solved for $Z'_{\text{in}} \equiv v_S/i_e$. Finally, Z_{in} may be found as the parallel combination of Z'_{in} and R_E:

$$Z_{\text{in}} = \frac{R_E(h_{ib}G - h_{rb}h_{fb})}{R_EG + h_{ib}G - h_{rb}h_{fb}} \tag{3}$$

(*b*) By elimination of i_e between (*1*) and (*2*) followed by rearrangement,

$$A_v = \frac{v_{cb}}{v_S} = -\frac{h_{fb}}{h_{ib}G - h_{rb}h_{fb}}$$

(*c*) From (*1*),

$$i_L = \frac{v_{cb}}{R_L} = -\frac{h_{fb}i_e}{R_LG} \tag{4}$$

By KCL at the emitter node,

$$i_e = i_{\text{in}} - \frac{v_S}{R_E} = i_{\text{in}} - \frac{i_{\text{in}}Z_{\text{in}}}{R_E} = i_{\text{in}}\left(1 - \frac{Z_{\text{in}}}{R_E}\right) \tag{5}$$

Now elimination of i_e between (*4*) and (*5*) and rearrangement give

$$A_i = \frac{i_L}{i_{\text{in}}} = -\frac{h_{fb}}{R_LG}\left(1 - \frac{Z_{\text{in}}}{R_E}\right)$$

6.12 The CE h-parameter transistor model (with $h_{re} = h_{oe} = 0$) was applied to the CC amplifier in Section 6.2. Taking i_B and v_{EC} as independent variables, develop a CC h-parameter model which allows for more accurate representation of the transistor than the circuit of Fig. 6-2(*c*).

CC characteristics are not commonly given by transistor manufacturers, but they would be plots of i_B vs. v_{BC} with v_{EC} as parameter (input characteristics) and plots of i_E vs. v_{EC} with i_B as parameter (output or emitter characteristics). With i_B and v_{EC} as independent variables, we have

$$v_{BC} = f_1(i_B, v_{EC}) \tag{1}$$

$$i_E = f_2(i_B, v_{EC}) \tag{2}$$

Next we apply the chain rule to form the total differentials of (*1*) and (*2*), assuming that $v_{bc} = \Delta v_{BC} \approx dv_{BC}$, and similarly for i_e:

$$v_{bc} = \Delta v_{BC} \approx dv_{BC} = \left.\frac{\partial v_{BC}}{\partial i_B}\right|_Q i_b + \left.\frac{\partial v_{BC}}{\partial v_{EC}}\right|_Q v_{ec} \tag{3}$$

$$i_e = \Delta i_E \approx di_E = \left.\frac{\partial i_E}{\partial i_B}\right|_Q i_b + \left.\frac{\partial i_E}{\partial v_{EC}}\right|_Q v_{ec} \tag{4}$$

Finally, we define:

Input resistance $$h_{ic} \equiv \left.\frac{\partial v_{BC}}{\partial i_B}\right|_Q \approx \left.\frac{\Delta v_{BC}}{\Delta i_B}\right|_Q \tag{5}$$

Reverse voltage ratio $$h_{rc} \equiv \left.\frac{\partial v_{BC}}{\partial v_{EC}}\right|_Q \approx \left.\frac{\Delta v_{BC}}{\Delta v_{EC}}\right|_Q \tag{6}$$

Forward current gain $$h_{fc} \equiv \left.\frac{\partial i_E}{\partial i_B}\right|_Q \approx \left.\frac{\Delta i_E}{\Delta i_B}\right|_Q \tag{7}$$

Output admittance $$h_{oc} \equiv \left.\frac{\partial i_E}{\partial v_{EC}}\right|_Q \approx \left.\frac{\Delta i_E}{\Delta v_{EC}}\right|_Q \tag{8}$$

A circuit that satisfies (*3*) and (*4*) with definitions (*5*) to (*8*) is displayed by Fig. 6-14.

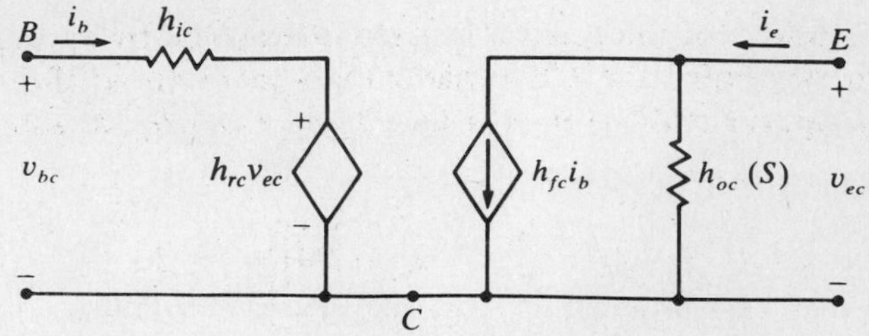

Fig. 6-14 CC small-signal equivalent circuit

6.13 Redraw the CE small-signal equivalent circuit of Fig. 6-1(*a*) so that the collector C is common to the input and output ports. Then apply KVL at the input port and KCL at the output port to find a set of equations that can be compared with (*3*) and (*4*) of Problem 6.12 to determine the CC h parameters in terms of the CE h parameters.

Figure 6-1(*a*) is rearranged, to make the collector common, in Fig. 6-15. Applying KVL around the B,C loop, with $v_{ce} = -v_{ec}$, results in

$$v_{bc} = h_{ie}i_b + h_{re}v_{ce} + v_{ec} = h_{ie}i_b + (1 - h_{re})v_{ec} \tag{1}$$

Applying KCL at node E gives

$$i_e = -i_b - h_{fe}i_b + h_{oe}v_{ec} = -(h_{fe} + 1)i_b + h_{oe}v_{ec} \tag{2}$$

Comparison of (*1*) and (*2*) above with (*3*) and (*4*) of Problem 6.12 yields, by direct analogy,

$$h_{ic} = h_{ie} \qquad h_{rc} = 1 - h_{re} \qquad h_{fc} = -(h_{fe} + 1) \qquad h_{oc} = h_{oe} \tag{3}$$

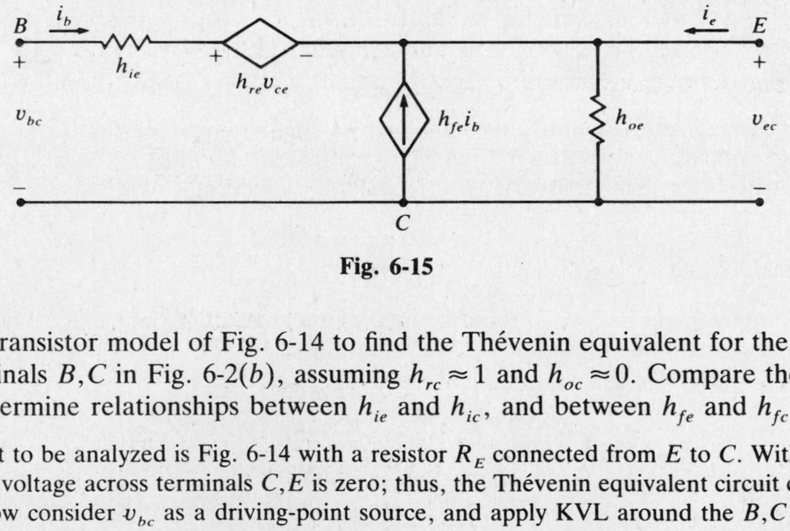

Fig. 6-15

6.14 Use the CC transistor model of Fig. 6-14 to find the Thévenin equivalent for the circuit to the right of terminals B,C in Fig. 6-2(*b*), assuming $h_{rc} \approx 1$ and $h_{oc} \approx 0$. Compare the results with (*6.16*) to determine relationships between h_{ie} and h_{ic}, and between h_{fe} and h_{fc}.

The circuit to be analyzed is Fig. 6-14 with a resistor R_E connected from E to C. With terminal pair B,C open, the voltage across terminals C,E is zero; thus, the Thévenin equivalent circuit consists only of $Z_{Th} = R_{Th}$. Now consider v_{bc} as a driving-point source, and apply KVL around the B,C loop to obtain

$$v_{dp} = v_{bc} = h_{ic}i_b + h_{rc}v_{ec} \approx h_{ic}i_b + v_{ec} \tag{1}$$

Use KCL at node E to obtain

$$v_{ec} = -i_eR_E = -(h_{fc}i_b + h_{oc}v_{ce})R_E \approx -h_{fc}R_Ei_b \tag{2}$$

Substitute (*2*) into (*1*), and solve for the driving-point impedance:

$$R_{Th} = \frac{v_{bc}}{i_b} = h_{ic} - h_{fc}R_E \tag{3}$$

Now (*3*) is compared with (*6.16*), it becomes apparent that $h_{ic} = h_{ie}$ and $h_{fc} = -(h_{fe} + 1)$, as given in (*3*) of Problem 6.13.

6.15 Apply the definitions of the general h parameters given by (*1.16*) to (*1.19*) to the circuit of Fig. 6-1(*b*) to determine the CE h parameters in terms of the CB h parameters. Use the typically good approximations $h_{rb} \ll 1$ and $h_{ob}h_{ib} \ll 1 + h_{fb}$ to simplify the results.

By (*1.16*),

$$h_{ie} = \left.\frac{v_{be}}{i_b}\right|_{v_{ce}=0} \tag{1}$$

If $v_{ce} = 0$ (short-circuited) in the network of Fig. 6-1(*b*), then $v_{cb} = -v_{be}$, so that, by KVL around the E,B loop,

$$v_{be} = -h_{ib}i_e - h_{rb}v_{cb} = -h_{ib}i_e + h_{rb}v_{be}$$

which gives

$$i_e = \frac{h_{rb} - 1}{h_{ib}}\, v_{be} \tag{2}$$

KCL at node B then gives

$$i_b = -(1 + h_{fb})i_e - h_{ob}v_{cb} = \left[\frac{(1 + h_{fb})(1 - h_{rb})}{h_{ib}} + h_{ob}\right]v_{be}$$

Now, (*1*) and the given approximations,

$$h_{ie} = \frac{h_{ib}}{h_{ib}h_{ob} + (1 + h_{fb})(1 - h_{rb})} \approx \frac{h_{ib}}{1 + h_{fb}}$$

By (*1.17*),

$$h_{re} = \left.\frac{v_{be}}{v_{ce}}\right|_{i_b=0} \tag{3}$$

If $i_b = 0$, then $i_c = -i_e$ in Fig. 6-1(*b*). By KVL,

$$v_{ce} = v_{cb} - h_{rb}v_{cb} - h_{ib}i_e = (1 - h_{rb})v_{cb} - h_{ib}i_e \tag{4}$$

KCL at node C then gives

$$i_c = -i_e = h_{fb}i_e + h_{ob}v_{cb}$$

so that

$$i_e = -\frac{h_{ob}}{1 + h_{fb}} \tag{5}$$

Substituting (*5*) into (*4*) with $v_{cb} = v_{ce} - v_{be}$ gives

$$v_{ce} = (1 - h_{rb})(v_{ce} - v_{be}) + \frac{h_{ib}h_{ob}}{1 + h_{fb}}(v_{ce} - v_{be})$$

After rearranging, (*3*) and the given approximations lead to

$$h_{re} = \frac{h_{rb}(1 + h_{fb}) - h_{ib}h_{ob}}{-h_{ib}h_{ob} + (h_{rb} - 1)(1 + h_{fb})} \approx \frac{h_{ib}h_{ob}}{1 + h_{fb}} - h_{rb}$$

By (*1.18*),

$$h_{fe} = \left.\frac{i_c}{i_b}\right|_{v_{ce}=0} \tag{6}$$

By KCL at node B of Fig. 6-1(*b*), with $v_{ce} = 0$ (and thus $v_{cb} = v_{eb} = -v_{be}$),

$$i_b = -(1 + h_{fb})i_e - h_{ob}v_{cb} = -(1 + h_{fb})i_e + h_{ob}v_{be}$$

Solving (*2*) for v_{be} with $i_e = -i_b - i_c$ and substituting now give

$$i_b = (1 + h_{fb})(i_b + i_c) + \frac{h_{ib}h_{ob}}{1 - h_{rb}}(i_b + i_c)$$

After rearranging, (6) and the given approximations lead to

$$h_{fe}=\frac{-h_{fb}(1-h_{rb})-h_{ib}h_{ob}}{(1+h_{fb})(1-h_{rb})+h_{ib}h_{ob}}\approx\frac{-h_{fb}}{1+h_{fb}}$$

By (1.19),

$$h_{oe}=\frac{i_c}{v_{ce}}\bigg|_{i_b=0} \tag{7}$$

If $i_b=0$, then $-i_c=i_e$. Replacing i_e with $-i_c$ in (4) and (5), solving (4) for v_{cb}, and substituting into (5) give

$$i_c=\frac{h_{ob}}{1+h_{fb}}\left(\frac{v_{ce}}{1-h_{rb}}-\frac{h_{ib}}{1-h_{rb}}i_c\right)$$

After rearranging, (7) and the given approximations lead to

$$h_{oe}=\frac{h_{ob}}{(1-h_{fb})(1+h_{rb})+h_{ib}h_{ob}}\approx\frac{h_{ob}}{1+h_{fb}}$$

6.16 Apply the definitions of the z parameters given by (1.10) through (1.13) to the CB h-parameter circuit of Fig. 6-1(b) to find values for the z parameters in terms of the CB h parameters.

The circuit of Fig. 6-1(b) is described by the linear system of equations

$$\begin{bmatrix} h_{ib} & h_{rb} \\ h_{fb} & h_{ob}\end{bmatrix}\begin{bmatrix} i_e \\ v_{cb}\end{bmatrix}=\begin{bmatrix} v_{eb} \\ i_c\end{bmatrix} \tag{1}$$

By (1.10) and Fig. 1-6,

$$z_{11}=\frac{v_{eb}}{i_e}\bigg|_{i_c=0} \tag{2}$$

Setting $i_c=0$ in (1) yields

$$v_{cb}=-\frac{h_{fb}}{h_{ob}}i_e \tag{3}$$

Substituting (3) into the first equation of (1) and applying (2) yield

$$z_{11}=h_{ib}-\frac{h_{rb}h_{fb}}{h_{ob}}$$

By (1.12) and (3),

$$z_{21}=\frac{v_{cb}}{i_e}\bigg|_{i_c=0}=-\frac{h_{fb}}{h_{ob}}$$

By (1.11),

$$z_{12}=\frac{v_{eb}}{i_c}\bigg|_{i_e=0} \tag{4}$$

Setting $i_e=0$ in (1), solving the two equations for v_{cb}, and equating the results give

$$\frac{v_{eb}}{h_{rb}}=\frac{i_c}{h_{ob}} \qquad \text{from which} \qquad z_{12}=\frac{h_{rb}}{h_{ob}}$$

Finally, by (1.13),

$$z_{22}=\frac{v_{cb}}{i_c}\bigg|_{i_e=0} \tag{5}$$

Letting $i_e=0$ in the second equation of (1) and applying (5) yield $z_{22}=1/h_{ob}$ directly.

6.17 For the CE amplifier of Fig. 3-13, assume that $h_{re} = h_{oe} \approx 0$, $h_{ie} = 1.1\ \text{k}\Omega$, $h_{fe} = 50$, $C_c \to \infty$, $R_F = 100\ \text{k}\Omega$, $R_S = 5\ \text{k}\Omega$, and $R_C = R_L + 20\ \text{k}\Omega$. Using CE h parameters, find and evaluate expressions for (*a*) $A_i = i_L/i_S$, (*b*) $A_i' = i_L/i_b$, (*c*) $A_v = v_L/v_S$, and (*d*) $A_v' = v_L/v_{be}$.

(*a*) The small-signal equivalent circuit for the amplifier is given in Fig. 6-16. By the method of node voltages,

$$\frac{v_s - v_{be}}{R_S} + \frac{v_{ce} - v_{be}}{R_F} - \frac{v_{be}}{h_{ie}} = 0 \tag{1}$$

$$\frac{v_{be} - v_{ce}}{R_F} - h_{fe} i_b - \frac{R_C + R_L}{R_C R_L} v_{ce} = 0 \tag{2}$$

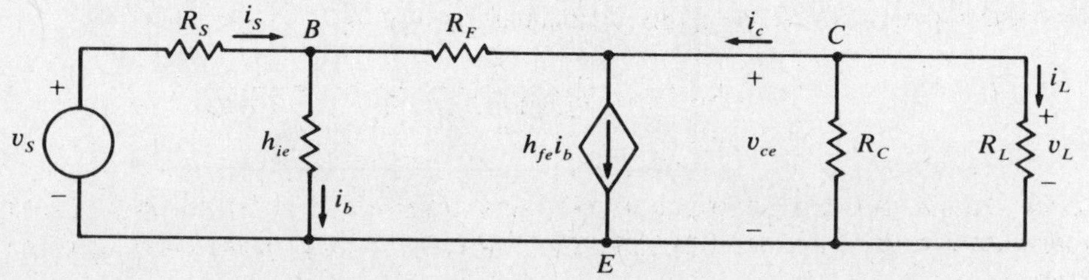

Fig. 6-16

Rearranging (*1*) and (*2*) and substituting $i_b = v_{be}/h_{ie}$ lead to

$$\begin{bmatrix} \dfrac{1}{R_S} + \dfrac{1}{R_F} + \dfrac{1}{h_{ie}} & -\dfrac{1}{R_F} \\ \dfrac{h_{fe}}{h_{ie}} - \dfrac{1}{R_F} & \dfrac{1}{R_F} + \dfrac{R_C + R_L}{R_C R_L} \end{bmatrix} \begin{bmatrix} v_{be} \\ v_{ce} \end{bmatrix} = \begin{bmatrix} \dfrac{v_S}{R_S} \\ 0 \end{bmatrix}$$

The determinant of coefficients is then

$$\Delta = \left(\frac{1}{R_S} + \frac{1}{R_F} + \frac{1}{h_{ie}}\right)\left(\frac{1}{R_F} + \frac{R_C + R_L}{R_C R_L}\right) + \frac{1}{R_F}\left(\frac{h_{fe}}{h_{ie}} - \frac{1}{R_F}\right)$$

$$= \left(\frac{1}{5} + \frac{1}{100} + \frac{1}{1.1}\right)\left(\frac{1}{100} + \frac{10 + 10}{10 \times 10}\right)(10^{-6}) + \frac{1}{100}\left(\frac{50}{1.1} - \frac{1}{100}\right)10^{-6} = 4.557 \times 10^{-6}$$

By Cramer's rule,

$$v_{be} = \frac{\Delta_1}{\Delta} = \frac{\left(\dfrac{1}{R_F} + \dfrac{R_C + R_L}{R_C R_L}\right) v_S}{R_S\,\Delta} = \frac{\left(\dfrac{1}{100} + \dfrac{1}{10}\right)(10^{-3}) v_S}{(5 \times 10^3)(4.557 \times 10^{-6})} = 4.828 \times 10^{-3} v_S \tag{3}$$

and

$$v_L = v_{ce} = \frac{\Delta_2}{\Delta} = \frac{\left(\dfrac{1}{R_F} - \dfrac{h_{fe}}{h_{ie}}\right) v_S}{R_S\,\Delta} = \frac{\left(\dfrac{1}{100} - \dfrac{50}{1.1}\right)(10^{-3}) v_S}{(5 \times 10^3)(4.557 \times 10^{-6})} = -1.995 v_S \tag{4}$$

So

$$A_i = \frac{i_L}{i_S} = \frac{v_L/R_L}{(v_S - v_{be})/R_S} = \frac{R_S v_L}{R_L(v_S - v_{be})} = \frac{(5 \times 10^3)(-1.995 v_S)}{(20 \times 10^3)(v_S - 4.828 \times 10^{-3} v_S} = -0.501$$

(*b*)

$$A_i' = \frac{i_L}{i_b} = \frac{v_L/R_L}{v_{be}/h_{ie}} = \frac{h_{ie} v_L}{R_L v_{be}} = \frac{(1.1 \times 10^3)(-1.995 v_s)}{(20 \times 10^3)(4.828 \times 10^{-3} v_S)} = -22.73$$

(*c*)

$$A_v = \frac{v_L}{v_S} = \frac{-1.995 v_S}{v_S} = -1.995$$

(*d*)

$$A_v' = \frac{v_L}{v_{be}} = \frac{-1.995 v_S}{4.828 \times 10^{-3} v_S} = -413.2$$

6.18 In the CB amplifier of Fig. 6-17(a), let $R_1 = R_2 = 50\ \text{k}\Omega$, $R_C = 2.2\ \text{k}\Omega$, $R_E = 3.3\ \text{k}\Omega$, $R_L = 1.1\ \text{k}\Omega$, $C_C = C_B \rightarrow \infty$, $h_{rb} \approx 0$, $h_{ib} = 25\ \Omega$, $h_{ob} = 10^{-6}$ S, and $h_{fb} = -0.99$. Find and evaluate expressions for (a) the voltage-gain ratio $A_v = v_L/v_s$ and (b) the current-gain ratio $A_i = i_L/i_s$.

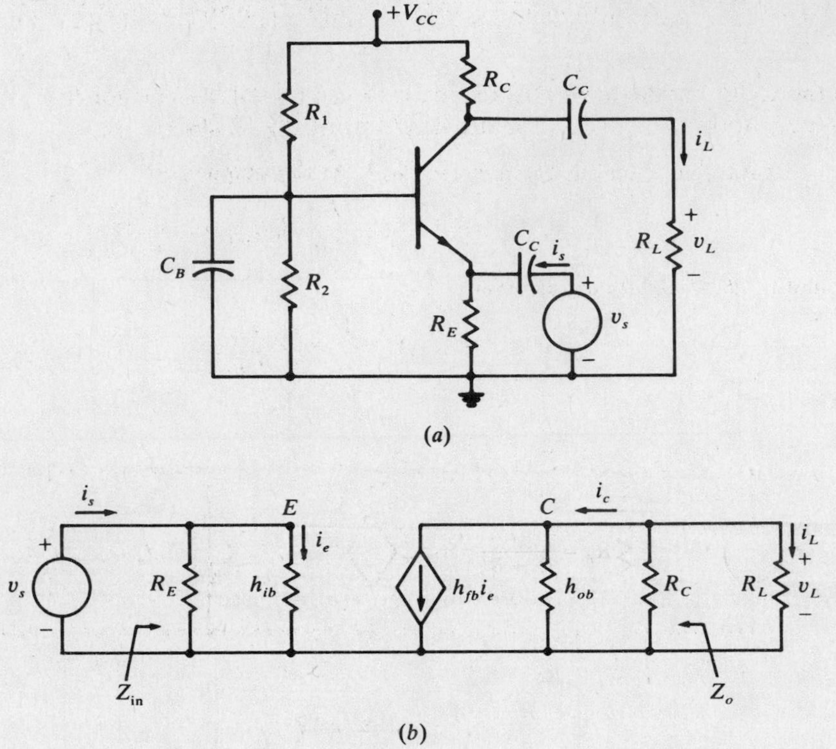

Fig. 6-17

(a) With $h_{rb} = 0$, the CB h-parameter model of Fig. 6-1(b) can be used to draw the small-signal circuit of Fig. 6-17(b). By Ohm's law at the input mesh,

$$i_e = \frac{v_s}{h_{ib}} \tag{1}$$

Ohm's law at the output mesh requires that

$$v_L = \left(\frac{1}{h_{ob}} \| R_C \| R_L\right)(-h_{fb} i_e) = -\frac{R_C R_L h_{fb} i_e}{R_C + R_L + h_{ob} R_C R_L} \tag{2}$$

Substitution of (*1*) into (*2*) allows the formation of A_v:

$$A_v = \frac{v_L}{v_s} = -\frac{R_C R_L h_{fb}}{h_{ib}(R_C + R_L + h_{ob} R_C R_L)}$$

$$= -\frac{(2.2 \times 10^3)(1.1 \times 10^3)(-0.99)}{(25)[2.2 \times 10^3 + 1.1 \times 10^3 + (10^{-6})(2.2 \times 10^3)(1.1 \times 10^3)]} = 29.02$$

(b) By current division at node E,

$$i_e = \frac{R_E}{R_E + h_{ib}} i_s \tag{3}$$

Current division at node C gives

$$i_L = \frac{(1/h_{ob}) \| R_C}{(1/h_{ob}) \| R_C + R_L}(-h_{fb} i_e) = -\frac{R_C h_{fb} i_e}{R_C + R_L + h_{ob} R_L R_C} \tag{4}$$

Now substitution of (3) into (4) allows direct calculation of A_i:

$$A_i = \frac{i_L}{i_s} = -\frac{R_E R_C h_{fb}}{(R_E + h_{ib})(R_C + R_L + h_{ob} R_L R_C)}$$

$$= -\frac{(3.3\times 10^3)(2.2\times 10^3)(-0.99)}{(3.3\times 10^3 + 25)[2.2\times 10^3 + 1.1\times 10^3 + (10^{-6})(1.1\times 10^3)(2.2\times 10^3)]} = 0.655$$

6.19 Use the CC h-parameter model of Fig. 6-14 to find expressions for the current-gain ratios (*a*) $A_i' = i_e/i_b$ and (*b*) $A_i = i_e/i_i$ for the amplifier of Fig. 6-2(*a*).

(*a*) The equivalent circuit is given in Fig. 6-18. At the output port,

$$-i_e R_E = v_{ec} = -h_{fc} i_b \left(\frac{1}{h_{oc}} \parallel R_E\right) = -\frac{h_{fc} R_E}{h_{oc} R_E + 1}\, i_b \qquad (1)$$

and A_i' is obtained directly from (1) as

$$A_i' = \frac{i_e}{i_b} = \frac{h_{fc}}{h_{oc} R_E + 1}$$

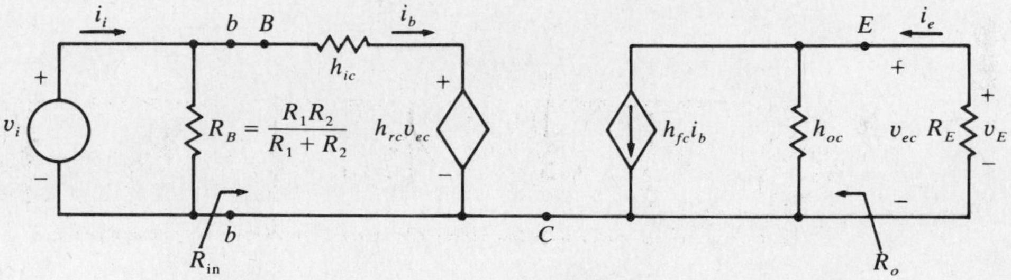

Fig. 6-18

(*b*) With $R_{Th} = R_{in} = h_{ic} - h_{rc} h_{fc} R_E/(h_{oc} R_E + 1)$, current division at node B gives

$$\frac{i_b}{i_e} = \frac{1}{A_i'} = \frac{R_B}{R_B + R_{in}} \frac{i_i}{i_e} = \frac{R_B}{R_B + R_{in}} \frac{1}{A_i}$$

so $$A_i = \frac{R_B}{R_B + R_{in}} A_i' = \frac{R_B}{R_B + h_{ic} + h_{rc} h_{fc} R_E/(h_{oc} R_E + 1)} \frac{h_{fc}}{h_{oc} R_E + 1}$$

$$= \frac{h_{fc} R_B}{(R_B + h_{ic})(h_{oc} R_E + 1) + h_{rc} h_{fc} R_E}$$

6.20 In the two-stage amplifier of Fig. 6-19, the transistors are identical, having $h_{ie} = 1500\ \Omega$, $h_{fe} = 40$, $h_{re} \approx 0$, and $h_{oe} = 30\ \mu\text{S}$. Also, $R_i = 1\ \text{k}\Omega$, $R_{C2} = 20\ \text{k}\Omega$, $R_{C1} = 10\ \text{k}\Omega$,

$$R_{B1} \equiv \frac{R_{11} R_{12}}{R_{11} + R_{12}} = 5\ \text{k}\Omega \qquad \text{and} \qquad R_{B2} = \frac{R_{21} R_{22}}{R_{21} + R_{22}} = 5\ \text{k}\Omega$$

Find (*a*) the final-stage voltage gain $A_{v2} \equiv v_o/v_{o1}$; (*b*) the final-stage input impedance Z_{in2}; (*c*) the initial-stage voltage gain $A_{v1} \equiv v_{o1}/v_{in}$; (*d*) the amplifier input impedance Z_{in1}; and (*e*) the amplifier voltage gain $A_v \equiv v_o/v_i$.

(*a*) The final-stage voltage gain is given by the result of Problem 6.7(*a*) if the parallel combination of R_L and R_C is replaced with R_{C2}:

$$A_{v2} = -\frac{h_{fe} R_{C2}}{h_{ie}(1 + h_{oe} R_{C2})} = -\frac{(40)(20\times 10^3)}{(1500)[(1 + (30\times 10^{-6})(20\times 10^3)]} = -333.3$$

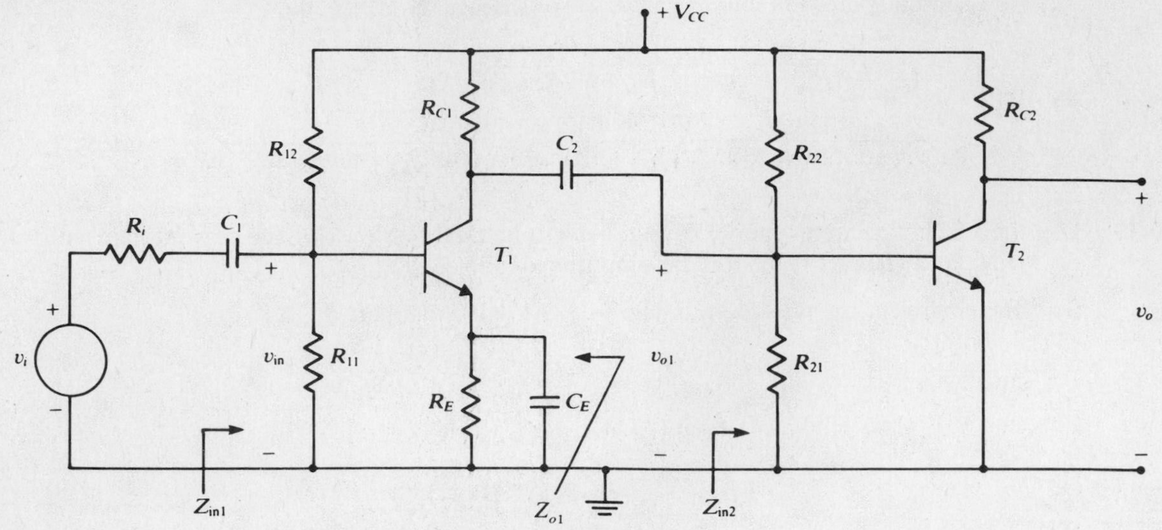

Fig. 6-19

(*b*) From (*4*) of Problem 6.10 with $h_{re} \approx 0$,

$$Z_{\text{in}2} = \frac{R_{B2}h_{ie}}{R_{B2} + h_{ie}} = \frac{(5 \times 10^3)(1500)}{5 \times 10^3 + 1500} = 1.154 \text{ k}\Omega$$

(*c*) The initial-stage voltage gain is given by the result of Problem 6.7(*a*) if R_C and R_L are replaced with R_{C1} and $Z_{\text{in}2}$, respectively:

$$A_{v1} = -\frac{h_{fe}Z_{\text{in}2}R_{C1}}{h_{ie}(R_{C1} + Z_{\text{in}2} + h_{oe}Z_{\text{in}2}R_{C1})} = -\frac{(40)(1154)(10^4)}{(1500)(10^4 + 1154 + 346.2)} = -26.8$$

(*d*) As in part *b*,

$$Z_{\text{in}1} = \frac{R_{B1}h_{ie}}{R_{B1} + h_{ie}} = 1.154 \text{ k}\Omega$$

(*e*) By voltage division,

$$\frac{v_{\text{in}}}{v_i} = \frac{Z_{\text{in}1}}{Z_{\text{in}1} + R_i} = \frac{1154}{1154 + 1000} = 0.5357$$

and $$A_v \equiv \frac{v_o}{v_i} = \frac{v_{\text{in}}}{v_i} A_{v1}A_{v2} = (0.5357)(-26.8)(-333.3) = 4786$$

6.21 In the amplifier of Fig. 6-20(*a*), the transistors are identical and have $h_{re} = h_{oe} \approx 0$. Use the CE *h*-parameter model to draw an equivalent circuit and find expressions for (*a*) the current-gain ratio $A_i = i_E/i_i$, (*b*) the input resistance R_{in}, (*c*) the voltage-gain ratio $A_v = v_o/v_i$, and (*d*) the output resistance R_o.

(*a*) With $h_{re} = h_{oe} \approx 0$, the small-signal equivalent circuit is given by Fig. 6-20(*b*). KCL at node *E* gives

$$i_E = h_{fe}i_{b1} + h_{fe}i_{b2} + i_{b1} + i_{b2} = (h_{fe} + 1)(i_{b1} + i_{b2}) \qquad (1)$$

Since $i_i = i_{b1} + i_{b2}$, the current-gain ratio follows directly from (*1*) and is $A_i = h_{fe} + 1$.

(*b*) KVL applied around the outer loop gives

$$v_i = (h_{ie} \| h_{ie})i_i + R_E(h_{fe} + 1)i_i$$

so that $$R_{\text{in}} = \frac{v_i}{i_i} = \frac{1}{2}h_{ie} + (h_{fe} + 1)R_E \qquad (2)$$

(*c*) By KVL,

$$v_o = v_i - (h_{ie} \| h_{ie})i_i = v_i - \tfrac{1}{2}h_{ie}i_i \qquad (3)$$

But

$$i_i = \frac{v_i}{R_{\text{in}}} \qquad (4)$$

Substitution of (*4*) and then (*2*) into (*3*) allows solution for the voltage-gain ratio as

$$A_v = \frac{v_o}{v_i} = 1 - \frac{1}{2}\frac{h_{ie}}{R_{\text{in}}} = 1 - \frac{\frac{1}{2}h_{ie}}{\frac{1}{2}h_{ie} + (h_{fe}+1)R_E} = \frac{(h_{fe}+1)R_E}{\frac{1}{2}h_{ie} + (h_{fe}+1)R_E}$$

(*d*) If R_E is replaced by a driving-point source with v_i shorted, KCL requires that

$$i_{dp} = -h_{fe}(i_{b1} + i_{b2}) + \frac{v_{dp}}{h_{ie} \| h_{ie}} \qquad (5)$$

But

$$i_{b1} + i_{b2} = i_i = -\frac{v_{dp}}{h_{ie} \| h_{ie}} = -\frac{v_{dp}}{\frac{1}{2}h_{ie}} \qquad (6)$$

Substituting (*6*) into (*5*) leads to

$$R_o = \frac{v_{dp}}{i_{dp}} = \frac{1}{h_{fe}/\frac{1}{2}h_{ie} + 1/\frac{1}{2}h_{ie}} = \frac{h_{ie}}{2(h_{fe}+1)}$$

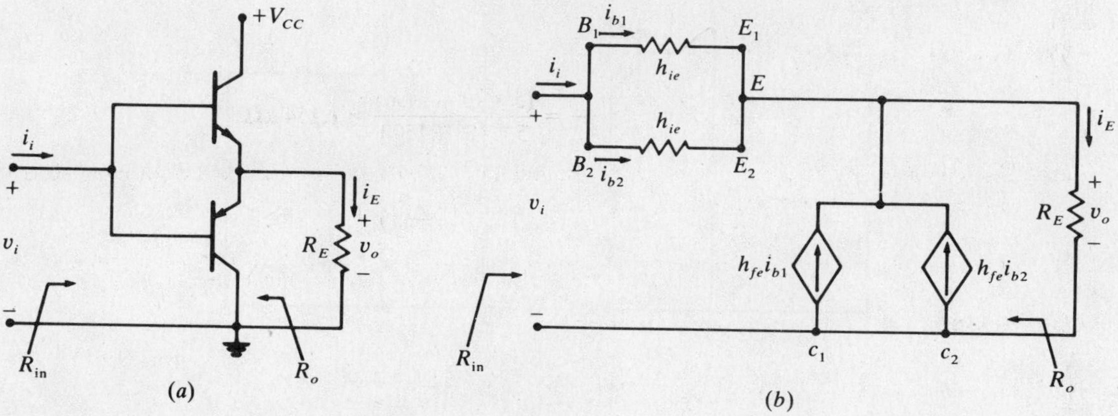

Fig. 6-20

6.22 The cascaded amplifier of Fig. 6-21(*a*) uses a CC first stage followed by a CE second stage. Let $R_S = 0$, $R_{11} = 100\ \text{k}\Omega$, $R_{12} = 90\ \text{k}\Omega$, $R_{21} = 10\ \text{k}\Omega$, $R_{22} = 90\ \text{k}\Omega$, $R_L = R_C = 5\ \text{k}\Omega$, and $R_E = 9\ \text{k}\Omega$. For transistor Q_1, $h_{oc} \approx 0$, $h_{ic} = 1\ \text{k}\Omega$, $h_{rc} \approx 1$, and $h_{fc} = -100$. For Q_2, $h_{re} = h_{oe} \approx 0$, $h_{fe} = 100$, and $h_{ie} = 1\ \text{k}\Omega$. Find (*a*) the overall voltage-gain ratio $A_v = v_L/v_s$ and (*b*) the overall current-gain ratio $A_i = i_L/i_s$.

(*a*) The small-signal equivalent circuit is drawn in Fig. 6-21(*b*), where

$$R_{B1} = R_{11} \| R_{12} = \frac{(90\times10^3)(100\times10^3)}{90\times10^3 + 100\times10^3} = 47.37\ \text{k}\Omega$$

and

$$R_{B2} = R_{22} \| R_{21} = \frac{(90\times10^3)(10\times10^3)}{90\times10^3 + 10\times10^3} = 4.5\ \text{k}\Omega$$

From the results of Problem 6.41,

$$A_{v1} = -\frac{h_{fc}(R_E \| R_{B2} \| h_{ie})}{h_{ic} - h_{rc}h_{fc}(R_E \| R_{B2} \| h_{ie})} = -\frac{(-100)(818.2)}{1\times10^3 - (1)(-100)(818.2)} = 0.9879$$

and from the results of Problem 6.7,

$$A_{v2} = -\frac{h_{fe}R_LR_C}{h_{ie}(R_L + R_C)} = -\frac{(100)(5\times10^3)(5\times10^3)}{(1\times10^3)(5\times10^3 + 5\times10^3)} = -100$$

Then

$$A_v = A_{v1}A_{v2} = (0.9879)(-100) = -98.79$$

(*b*) From the results of Problem 6.19,

$$A_{i1} = \frac{-i_{e1}}{i_s} = -\frac{h_{fc}R_{B1}}{R_{B1} + h_{ic} + h_{rc}h_{fc}(R_E \| R_{B2} \| h_{ie})} = -\frac{(-100)(47.37 \times 10^3)}{47.37 \times 10^3 + 1 \times 10^3 + (1)(-100)(818.2)}$$

$$= 36.38$$

and again from Problem 6.7,

$$A_{i2} = \frac{(R_E \| R_{B2})h_{ie}}{R_L(R_E \| R_{B2} + h_{ie})} A_{v2} = \frac{(4.5 \times 10^3)(91 \times 10^3)}{(5 \times 10^3)(4.5 \times 10^3 + 1 \times 10^3)}(-100) = -16.36$$

Then $$A_i = A_{i1}A_{i2} = (36.38)(-16.36) = -595.2$$

Note that, in this problem, we made use of the labor-saving technique of applying results determined for single-stage amplifiers to the individual stages of a cascaded (multistage) amplifier.

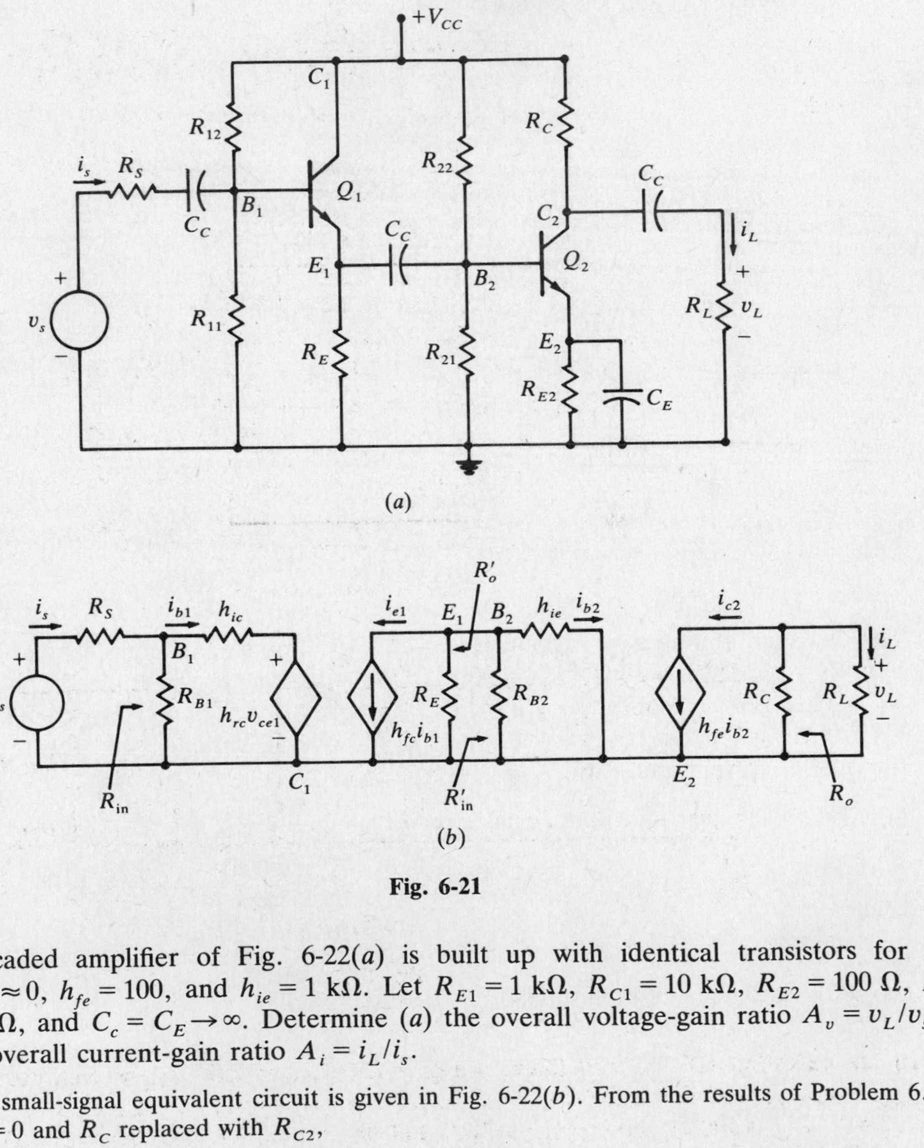

Fig. 6-21

6.23 The cascaded amplifier of Fig. 6-22(*a*) is built up with identical transistors for which $h_{re} = h_{oe} \approx 0$, $h_{fe} = 100$, and $h_{ie} = 1$ kΩ. Let $R_{E1} = 1$ kΩ, $R_{C1} = 10$ kΩ, $R_{E2} = 100$ Ω, $R_{C2} = R_L = 3$ kΩ, and $C_c = C_E \to \infty$. Determine (*a*) the overall voltage-gain ratio $A_v = v_L/v_s$, and (*b*) the overall current-gain ratio $A_i = i_L/i_s$.

(*a*) The small-signal equivalent circuit is given in Fig. 6-22(*b*). From the results of Problem 6.7 with $h_{oe} = 0$ and R_C replaced with R_{C2},

$$A_{v2} = -\frac{h_{fe}R_LR_{C2}}{h_{ie}(R_L + R_{C2})} = -\frac{(100)(3 \times 10^3)(3 \times 10^3)}{(1 \times 10^3)(3 \times 10^3 + 3 \times 10^3)} = -150$$

From the results of Problem 6.45, in which R_C, R_L, and R_E are replaced with R_{C1}, h_{ie}, and R_{E1}, respectively,

$$A_{v1} = -\frac{h_{fe}R_{C1}h_{ie}}{(R_{C1}+h_{ie})[(h_{fe}+1)R_{E1}+h_{ie}]} = -\frac{(100)(10\times10^3)(1\times10^3)}{(11\times10^3)[(100+1)(1\times10^3)+1\times10^3]} = -0.891$$

Thus,

$$A_v = A_{v1}A_{v2} = (-0.891)(-150) = 133.6$$

(*b*) From the results of Problem 6.45 with R_C and R_L replaced with R_{C1} and h_{ie}, respectively,

$$A_{i1} = -\frac{h_{fe}R_{C1}}{R_{C1}+h_{ie}} = -\frac{(100)(10\times10^3)}{10\times10^3+1\times10^3} = -90.91$$

Now, by current division at the output network,

$$i_L = -h_{fe}i_{b2}\frac{R_{C2}}{R_{C2}+R_L}$$

Hence,

$$A_{i2} = \frac{i_L}{i_{b2}} = -\frac{h_{fe}R_{C2}}{R_{C2}+R_L} = -\frac{(100)(3\times10^3)}{3\times10^3+3\times10^3} = -50$$

and

$$A_i = A_{i1}A_{i2} = (-90.91)(-50) = 4545.4$$

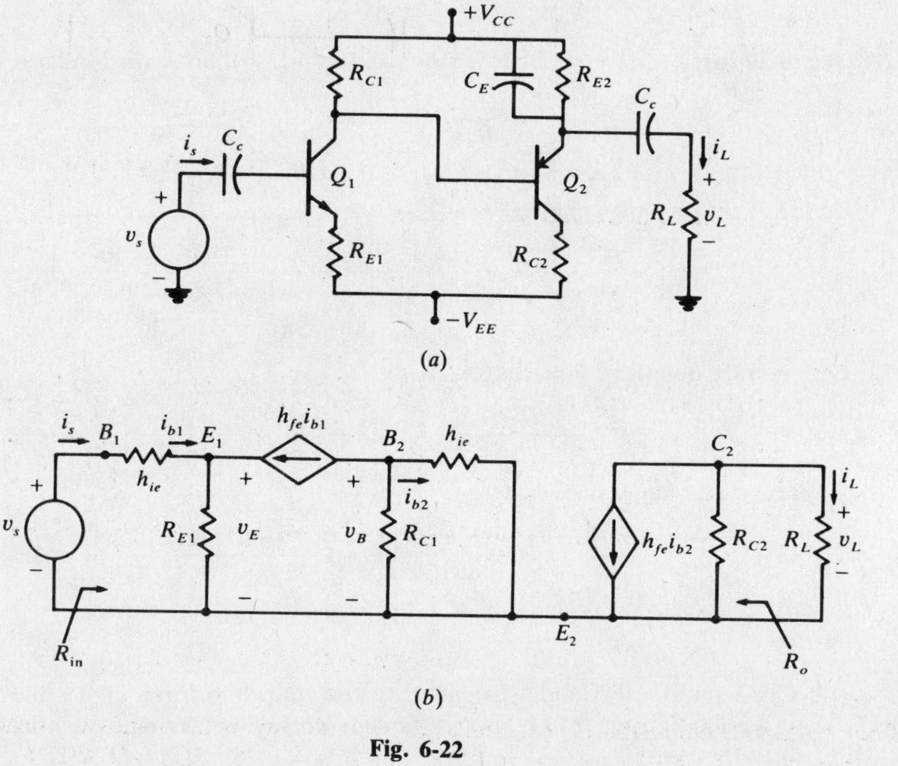

Fig. 6-22

6.24 In the cascaded CB-CC amplifier of Fig. 6-23(*a*), transistor Q_1 is characterized by $h_{rb1} = h_{ob1} \approx 0$, $h_{ib1} = 50\ \Omega$, and $h_{fb1} = -0.99$. The h parameters of transistor Q_2 are $h_{oc2} \approx 0$, $h_{rc2} = 1$, $h_{ic2} = 500\ \Omega$, and $h_{fc2} = -100$. Let $R_L = R_{E2} = 2\ \text{k}\Omega$, $R_{B1} = 30\ \text{k}\Omega$, $R_{B2} = 60\ \text{k}\Omega$, $R_1 = 50\ \text{k}\Omega$, $R_2 = 100\ \text{k}\Omega$, $R_{E1} = 5\ \text{k}\Omega$, and $C_B = C_c \to \infty$. Find (*a*) the overall voltage-gain ratio $A_v = v_L/v_s$ and (*b*) the overall current-gain ratio $A_i = i_L/i_s$.

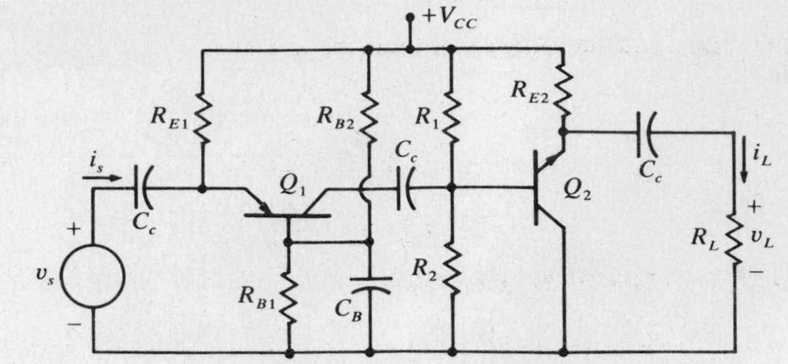

(a)

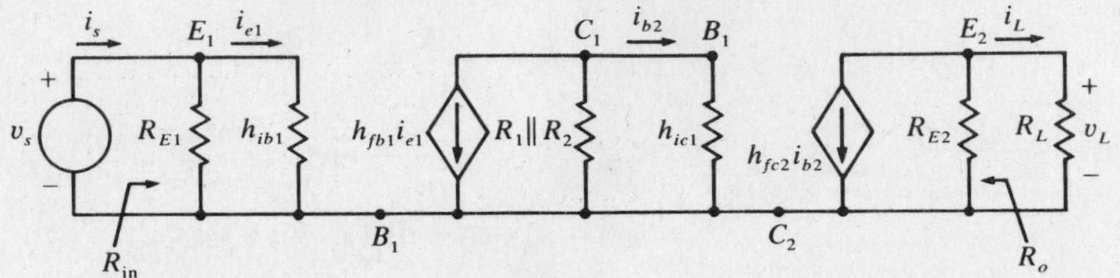

(b)

Fig. 6-23

(a) The small-signal equivalent circuit is shown in Fig. 6-23(b). From the results of Problem 6.18, with $R_B = R_1 \| R_2$,

$$A_{v1} = -\frac{h_{fb1}R_B h_{ic2}}{h_{ib1}(R_B + h_{ic2})} = -\frac{(-0.99)(33.3 \times 10^3)(500)}{(50)(33.3 \times 10^3 + 500)} = 9.75$$

By the results of Problem 6.41,

$$A_{v2} = -\frac{h_{fc2}(R_{E2} \| R_L)}{h_{ic2} - h_{rc2}h_{fc2}(R_{E2} \| R_L)} = -\frac{(-100)(1 \times 10^3)}{500 - (1)(-100)(1 \times 10^3)} = 0.995$$

Thus, $$A_v = A_{v1}A_{v2} = (9.75)(0.995) = 9.70$$

(b) Based on the results of Problem 6.18,

$$A_{i1} = -\frac{h_{fb2}R_{E1}R_B}{(R_{E1} + h_{ib1})(R_B + h_{ic2})} = -\frac{(-0.99)(5 \times 10^3)(33.3 \times 10^3)}{(5 \times 10^3 + 50)(33.3 \times 10^3 + 500)} = 0.966$$

By current division at node E_2,

$$\frac{i_L}{i_{b2}} = A_{i2} = -\frac{h_{fc2}R_{E2}}{R_{E2} + R_L} = -\frac{(-100)(2 \times 10^3)}{(2 \times 10^3) + (2 \times 10^3)} = 50$$

Then, $$A_i = A_{i1}A_{i2} = (0.966)(50) = 48.3$$

6.25 Use the CE h-parameter model to calculate the output voltage v_o for the amplifier of Fig. 3-17, thus demonstrating that it is a *difference amplifier*. Assume identical transistors with $h_{re} = h_{oe} \approx 0$.

The small-signal circuit is given in Fig. 6-24. Let $a = h_{ie} + (h_{fe} + 1)R_E$ and $b = (h_{fe} + 1)R_E$; then, by KVL,

$$v_1 = ai_{b1} + bi_{b2} \tag{1}$$

$$v_2 = b_1 i_{b1} + ai_{b2} \tag{2}$$

$$v_o = h_{fe}R_C(i_{b1} - i_{b2}) \tag{3}$$

Solving (*1*) and (*2*) simultaneously using Cramer's rule gives

$$\Delta = a^2 - b^2$$

and

$$i_{b1} = \frac{\Delta_1}{\Delta} = \frac{av_1 - bv_2}{a^2 - b^2} \tag{4}$$

$$i_{b2} = \frac{\Delta_2}{\Delta} = \frac{av_2 - bv_1}{a^2 - b^2} \tag{5}$$

Substituting (*4*) and (*5*) into (*3*) gives, finally,

$$v_o = \frac{h_{fe}R_C}{a^2 - b^2}(av_1 - bv_2 - av_2 + bv_1) = \frac{h_{fe}R_C}{a - b}(v_1 - v_2) = \frac{h_{fe}}{h_{ie}}R_C(v_1 - v_2)$$

which clearly shows that the circuit amplifies the *difference* between signals v_1 and v_2.

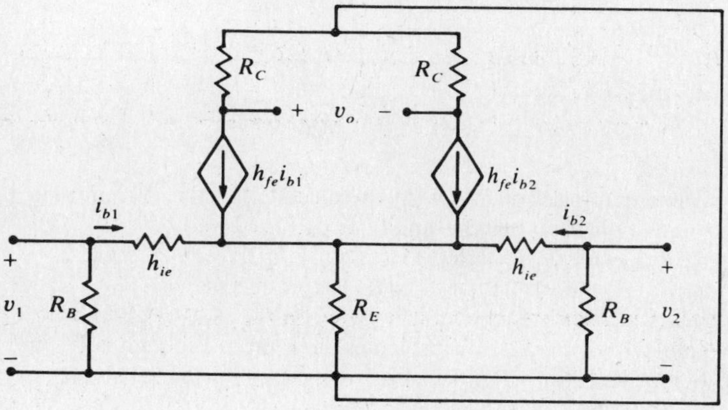

Fig. 6-24

Supplementary Problems

6.26 For the CB amplifier of Fig. 3-18, find the voltage-gain ratio $A_v = v_L/v_{eb}$ using the tee-equivalent circuit of Fig. 6-3 if r_c is large enough that $i_c \approx \alpha i_e$. *Ans.* $A_v = (\alpha R_C R_L)/\{(R_C + R_L)[r_e + (1-\alpha)r_b]\}$

6.27 For the CB amplifier of Fig. 3-18 and Problem 6.26, $R_C = R_L = 4\ \text{k}\Omega$, $r_e = 30\ \Omega$, $r_b = 300\ \Omega$, $r_c = 1\ \text{M}\Omega$, and $\alpha = 0.99$. Determine the percentage error in the approximate voltage gain of Problem 6.26 (in which we assumed $i_c \approx \alpha i_e$), relative to the exact gain as determined in Problem 6.1.
Ans. Approximate gain is 1.99 percent greater

6.28 Use the *r*-parameter equivalent circuit of Fig. 6-9(*b*) to find the current-gain ratio $A_i = i_L/i_b$ for the CE amplifier of Fig. 3-6.

Ans. $$A_i = \frac{R_C(r_e - r_m)}{(R_C + R_L)[(1-\alpha)r_c + r_e] + R_C R_L}$$

6.29 For the EF amplifier of Fig. 3-20(*a*), use an appropriate *r*-parameter model of the transistor to calculate the current-gain ratio $A_i = i_L/i_s$.

Ans. $$A_i = \frac{R_E R_B r_c}{(R_E + R_L)(R_B + r_b)[r_e + (1-\alpha)r_c + R_E \| R_L] + (R_E + R_L)r_c(r_e + R_E \| R_L)}$$

6.30 Apply the definitions of the *h* parameters, given by (*1.16*) through (*1.19*), to the *r*-parameter circuit of Fig. 6-3 to find the CB *h* parameters in terms of the *r* parameters.
Ans. $h_{ib} = r_e + (1-\alpha)r_b r_c/(r_b + r_c)$, $h_{rb} = r_b/(r_b + r_c)$, $h_{fb} = -(r_b + \alpha r_c)/(r_b + r_c)$, $h_{ob} = 1/(r_b + r_c)$

6.31 Apply the definitions of the z parameters, given by (*1.10*) through (*1.13*), to the circuit of Fig. 6-3 to find values for the z parameters in the equivalent circuit of Fig. 6-25, which contains two dependent voltage sources.
Ans. $z_{11} = r_e + r_b$, $z_{12} = r_b$, $z_{21} = r_b + \alpha r_c$, $z_{22} = r_b + r_c$

6.32 Apply the definitions of the z parameters, given by (*1.10*) through (*1.13*), to the CE h-parameter circuit of Fig. 6-1(*a*) to find values for the z parameters in the equivalent circuit of Fig. 6-25 in terms of the CE h parameters.
Ans. $z_{11} = h_{ie} - h_{fe}h_{re}/h_{oe}$, $z_{12} = h_{re}/h_{oe}$, $z_{21} = -h_{fe}/h_{oe}$, $z_{22} = 1/h_{oe}$

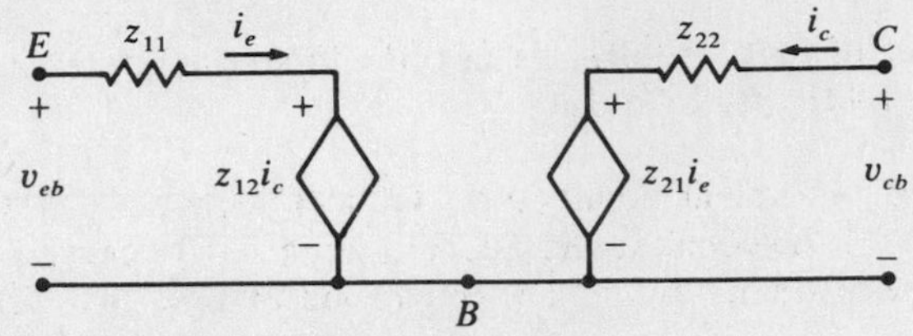

Fig. 6-25

6.33 Use the z-parameter model of Fig. 6-25 to calculate (*a*) the current-gain ratio $A_i = i_L/i_i$ and (*b*) the voltage-gain ratio $A_v = v_L/v_i$ for the amplifier of Fig. 3-6(*a*).
Ans. $A_i = R_C R_B z_{21}/(R_C + R_L)[(R_B + z_{11})(z_{22} + R_C \| R_L) + z_{12}z_{21}]$,
$A_v = z_{21}R_B(R_L \| R_C)/\{(z_{22} + R_C \| R_L)[R_B R_i + z_{11}(R_B + R_i)]\}$

6.34 For the CE amplifier of Fig. 3-13 with values as given in Problem 6.17, find (*a*) the input resistance R_i and (*b*) the output resistance R_o. *Ans.* (*a*) 24.26 Ω; (*b*) 2.154 kΩ

6.35 A CE transistor amplifier is operating in the active region, with $V_{CC} = 12$ V and $R_{dc} = 2$ kΩ. If the collector characteristics are given by Fig. 3-5(*b*) and the quiescent base current is 30 μA, determine (*a*) h_{fe} and (*b*) h_{oe}. *Ans.* (*a*) 190; (*b*) 83.33 μS

6.36 In the circuit of Fig. 6-26, $h_{re} = 10^{-4}$, $h_{ie} = 200$ Ω, $h_{fe} = 100$, and $h_{oe} = 100$ μS. (*a*) Find the power gain as $A_p = |A_i A_v|$, the product of the current and voltage gains. (*b*) Determine the numerical value of R_L that maximizes the power gain.
Ans. (*a*) $h_{fe}^2/|(h_{oe}R_L + 1)(h_{oe}h_{ie} - h_{re}h_{fe}h_{ie}R_L^{-1})|$; (*b*) 14.14 kΩ

6.37 The EF amplifier of Fig. 3-20(*a*) utilizes a Si transistor with negligible leakage current and $\beta = 59$. Also, $V_{CC} = 15$ V, $V_L = 3$ V (V_L is the dc component of v_L), and $R_E = 1.5$ kΩ. Calculate (*a*) R_B, (*b*) the output impedance Z_o, and (*c*) the input impedance Z_{in}. *Ans.* (*a*) 339 kΩ; (*b*) 1.185 kΩ; (*c*) 50.98 kΩ

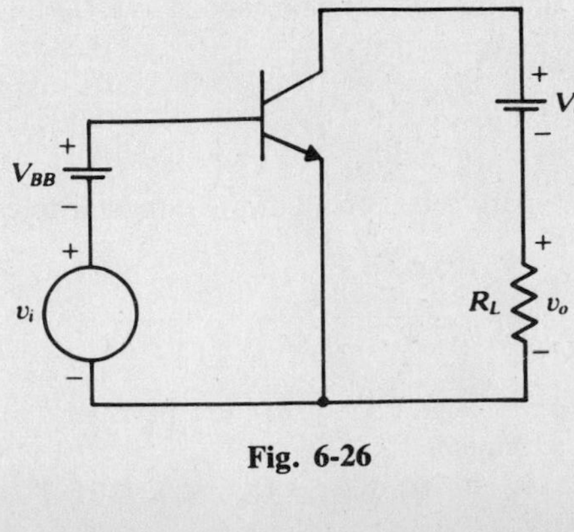

Fig. 6-26

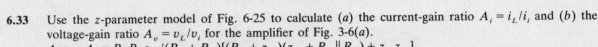

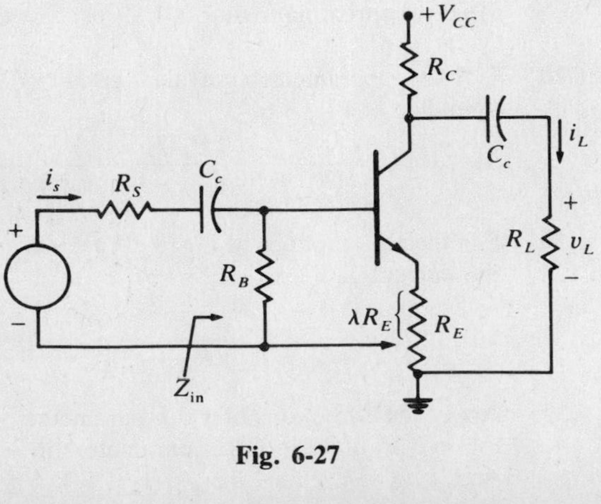

Fig. 6-27

6.38 The amplifier of Fig. 6-27 has an adjustable emitter resistor R_E, as indicated, with $0 \le \lambda \le 1$. Assume that $h_{re} = h_{oe} \approx 0$ and $C_c \to \infty$, and find expressions for (*a*) the current-gain ratio $A_i = i_L/i_s$, (*b*) the voltage-gain ratio $A_v = v_L/v_s$, and (*c*) the input impedance Z_{in}.

Ans. (*a*) $A_i = -\dfrac{h_{fe}R_BR_L}{(R_C + R_L)[R_B + h_{ie} + (h_{fe}+1)\lambda R_E]}$;

(*b*) $A_v = -\dfrac{h_{fe}R_CR_BR_L}{(R_C + R_L)\{R_SR_B + (R_S + R_B)[h_{ie} + (h_{fe}+1)\lambda R_E]\}}$;

(*c*) $R_i = \dfrac{R_B[h_{ie} + (h_{fe}+1)\lambda R_E]}{R_B + h_{ie} + (h_{fe}+1)\lambda R_E}$

6.39 For the CB amplifier of Problem 6.18, find (*a*) the input impedance Z_{in} and (*b*) the output impedance Z_o. *Ans.* (*a*) 24.8 Ω; (*b*) 2.195 kΩ

6.40 The exact small-signal equivalent circuit for the CC amplifier of Fig. 6-2(*a*) is given by Fig. 6-18. Find the Thévenin equivalent for the circuit to the right of terminals b,b, assume that $h_{re} = h_{oe} \approx 0$, and show that the circuit of Fig. 6-2(*c*) results. (*Hint:* The conversion from CE to CC h parameters was worked out in Problem 6.13.)

6.41 Apply the CC h-parameter model of Fig. 6-14 to the amplifier of Fig. 6-2(*a*) to find an expression for the voltage-gain ratio $A_v = v_E/v_i$. Evaluate A_v if $h_{ic} = 100\ \Omega$, $h_{rc} = 1$, $h_{fc} = -100$, $h_{oc} = 10^{-5}$ S, and $R_E = 1$ kΩ. *Ans.* $A_v = -h_{fc}R_E/[h_{ic}(h_{oc}R_E + 1) - h_{rc}h_{fc}R_E] \approx 0.999$

6.42 Find an expression for R_o in the CC amplifier of Fig. 6-18; use the common approximations $h_{rc} \approx 1$ and $h_{oc} \approx 0$ to simplify the expression; and then evaluate it if $R_1 = 1$ kΩ, $R_2 = 10$ kΩ, $h_{fc} = -100$, and $h_{ic} = 100\ \Omega$. *Ans.* $R_o = h_{ic}/(h_{oc}h_{ic} - h_{fc}h_{rc}) \approx -h_{ic}/h_{fc} = 1\ \Omega$

6.43 The cascaded amplifier circuit of Fig. 6-21(*a*) matches a high-input-impedance CC first stage with a high-output-impedance CE second stage to produce an amplifier with high input and output impedances. To illustrate this claim, refer to Fig. 6-21(*b*) and determine values for (*a*) $Z_{in} = R_{in}$, (*b*) Z'_{in}, (*c*) Z_o, and (*d*) Z'_o if $R_S = 5$ kΩ and all other circuit values are as given in Problem 6.22.
Ans. (*a*) 29.18 kΩ; (*b*) 818.2 Ω; (*c*) 5 kΩ; (*d*) 9.99 Ω

6.44 To illustrate the effect of signal-source internal impedance, calculate the voltage-gain ratio $A_v = v_L/v_s$ for the cascaded amplifier of Fig. 6-21(*a*) if $R_S = 20$ kΩ and all other values are as given in Problem 6.22; then compare your result with the value of A_v found in Problem 6.22.
Ans. $A_v = -58.61$, which represents a reduction of approximately 40 percent.

6.45 For the amplifier of Fig. 6-28, find expressions for (*a*) the voltage-gain ratio $A_v = v_L/v_s$ and (*b*) the current-gain ratio $A_i = i_L/i_s$. Assume that $h_{re} = h_{oe} \approx 0$.
Ans. (*a*) $A_v = -h_{fe}R_CR_L/\{(R_C + R_L)[(h_{fe}+1)R_E + h_{ie}]\}$; (*b*) $A_i = -h_{fe}R_C/(R_C + R_L)$

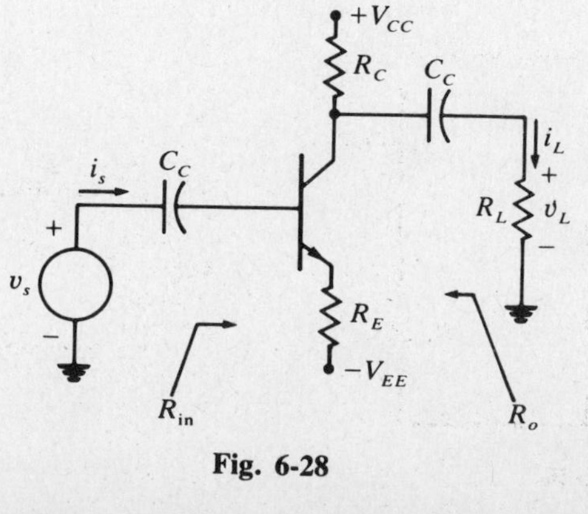

Fig. 6-28

6.46 Find expressions for (*a*) R_{in} and (*b*) R_o for the amplifier of Fig. 6-28 if $h_{re} = h_{oe} \approx 0$.
Ans. (*a*) $R_{\text{in}} = h_{ie} + (h_{fe} + 1)R_E$; (*b*) $R_o = R_C$

6.47 Suppose v_2 is replaced with a short circuit in the differential amplifier of Fig. 3-17. Find the input impedance $R_{\text{in}1}$ looking into the terminal across which v_1 appears if $R_B = 20\ \text{k}\Omega$, $R_E = 1\ \text{k}\Omega$, $h_{ie} = 25\ \Omega$, $h_{fe} = 100$, and $h_{re} = h_{oe} \approx 0$. *Ans.* 9.11 kΩ

6.48 For the Darlington-pair emitter-follower of Fig. 6-29, $h_{re1} = h_{re2} = h_{oe1} = h_{oe2} = 0$. In terms of the (nonzero) h parameters, find expressions for (*a*) Z'_{in}; (*b*) the voltage gain $A_v \equiv v_E/v_s$; (*c*) the current gain $A_i \equiv i_{e2}/i_{\text{in}}$; (*d*) Z_{in}; and (*e*) Z_o (if the signal source has internal resistance R_S).

Ans. (*a*) $Z'_{\text{in}} = h_{ie1} + (h_{fe1} + 1)[h_{ie2} + (h_{fe2} + 1)R_E]$; (*b*) $A_v = \dfrac{(h_{fe1} + 1)(h_{fe2} + 1)R_E}{Z'_{\text{in}}}$;

(*c*) $A_i = \dfrac{(h_{fe1} + 1)(h_{fe2} + 1)R_F}{R_F + Z'_{\text{in}}}$; (*d*) $Z_{\text{in}} = \dfrac{R_F Z'_{\text{in}}}{R_F + Z'_{\text{in}}}$;

(*e*) $Z_o = \dfrac{h_{ie2}}{h_{fe2} + 1} \dfrac{[R_S R_F/(R_S + R_F) + h_{ie1}]}{(h_{fe1} + 1)(h_{fe2} + 1)}$

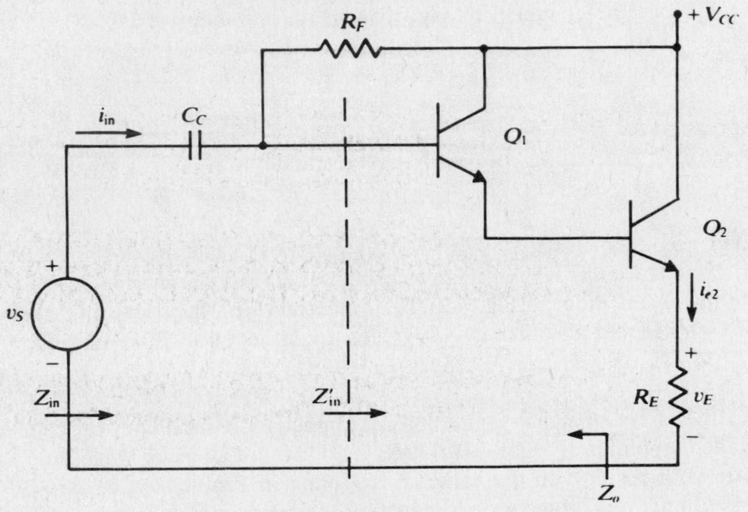

Fig. 6-29

Chapter 7

Small-Signal Midfrequency FET Amplifiers

7.1 INTRODUCTION

Several two-port linear network models are available that allow accurate analysis of thc FET for small drain-source voltage and small current excursions about a quiescent point (small-signal operation). In this chapter, all voltage and current signals are considered to be in the midfrequency range, where all capacitors appear as short circuits (see Section 4.5).

There are three basic FET amplifier configurations: the *common-source* (CS), *common-drain* (CD) or *source-follower* (SF), and *common-gate* (CG) configurations. The CS amplifier, which provides good voltage amplification, is most frequently used. The CD and CG amplifiers are applied as buffer amplifiers (with high input impedance and near-unity voltage gain) and high-frequency amplifiers, respectively.

7.2 SMALL-SIGNAL EQUIVALENT CIRCUITS FOR THE FET

From the FET drain characteristics of Fig. 4-2(*a*), it is seen that if i_D is taken as the dependent variable, then

$$i_D = f(v_{GS}, v_{DS}) \tag{7.1}$$

For small excursions (ac signals) about the Q point, $\Delta i_D = i_d$; thus, application of the chain rule to (*7.1*) leads to

$$i_d = \Delta i_D \approx di_D = g_m v_{gs} + \frac{1}{r_{ds}} v_{ds} \tag{7.2}$$

where g_m and r_{ds} are defined as follows:

$$\textit{Transconductance} \qquad g_m \equiv \left.\frac{\partial i_D}{\partial v_{GS}}\right|_Q \approx \left.\frac{\Delta i_D}{\Delta v_{GS}}\right|_Q \tag{7.3}$$

$$\textit{Source-drain resistance} \qquad r_{ds} \equiv \left.\frac{\partial v_{DS}}{\partial i_D}\right|_Q \approx \left.\frac{\Delta v_{DS}}{\Delta i_D}\right|_Q \tag{7.4}$$

As long as the JFET is operated in the pinchoff region, $i_G = i_g = 0$, so that the gate acts as an open circuit. This, along with (*7.2*), leads to the current-source equivalent circuit of Fig. 7-1(*a*). The voltage-source model of Fig. 7-2(*b*) is derived in Problem 7.2. Either of these models may be used in analyzing an amplifier, but one may be more efficient than the other in a particular circuit.

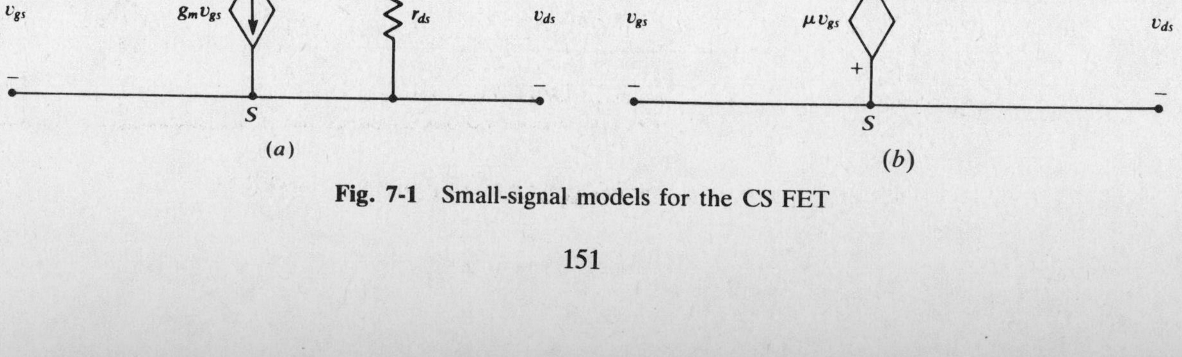

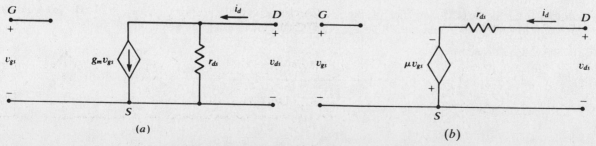

Fig. 7-1 Small-signal models for the CS FET

7.3 CS AMPLIFIER ANALYSIS

A simple common-source amplifier is shown in Fig. 7-2(*a*); its associated small-signal equivalent circuit, incorporating the voltage-source model of Fig. 7-1(*b*), is displayed in Fig. 7-2(*b*). Source resistor R_s is used to set the Q point but is bypassed by C_s for midfrequency operation.

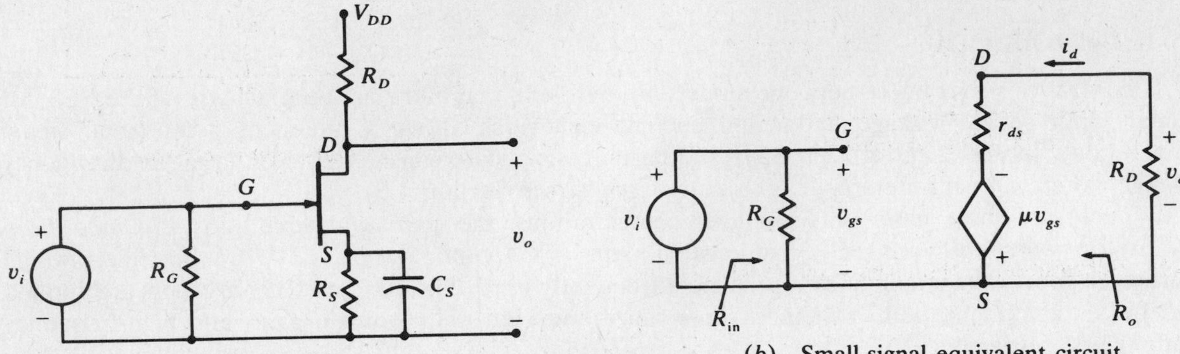

(*a*) CS amplifier

(*b*) Small-signal equivalent circuit

Fig. 7-2

Example 7.1 In the CS amplifier of Fig. 7-2(*b*), let $R_D = 3\ \text{k}\Omega$, $\mu = 60$, and $r_{ds} = 30\ \text{k}\Omega$. (*a*) Find an expression for the voltage-gain ratio $A_v = v_o/v_i$. (*b*) Evaluate A_v using the given typical values.

(*a*) By voltage division,

$$v_o = -\frac{R_D}{R_D + r_{ds}}\,\mu v_{gs}$$

Substitution of $v_{gs} = v_i$ and rearrangement give

$$A_v = \frac{v_o}{v_i} = -\frac{\mu R_D}{R_D + r_{ds}} \tag{7.5}$$

(*b*) The given values lead to

$$A_v = -\frac{(60)(3\times 10^3)}{3\times 10^3 + 30\times 10^3} = -5.45$$

where the minus sign indicates a 180° phase shift between v_i and v_o.

7.4 CD AMPLIFIER ANALYSIS

A simple common-drain (or source-follower) amplifier is shown in Fig. 7-3(*a*); its associated small-signal equivalent circuit is given in Fig. 7-3(*b*), where the voltage-source equivalent of Fig. 7-1(*b*) is used to model the FET.

Example 7.2 In the CD amplifier of Fig. 7-3(*b*), let $R_S = 5\ \text{k}\Omega$, $\mu = 60$, and $r_{ds} = 30\ \text{k}\Omega$. (*a*) Find an expression for the voltage-gain ratio $A_v = v_o/v_i$. (*b*) Evaluate A_v using the given typical values.

(*a*) By voltage division,

$$v_o = \frac{R_S}{R_S + r_{ds}/(\mu+1)}\,\frac{\mu}{\mu+1}\,v_{gd} = \frac{\mu R_S v_{gd}}{(\mu+1)R_S + r_{ds}}$$

Replacement of v_{gd} by v_i and rearrangement give

$$A_v = \frac{v_o}{v_i} = \frac{\mu R_S}{(\mu+1)R_S + r_{ds}} \tag{7.6}$$

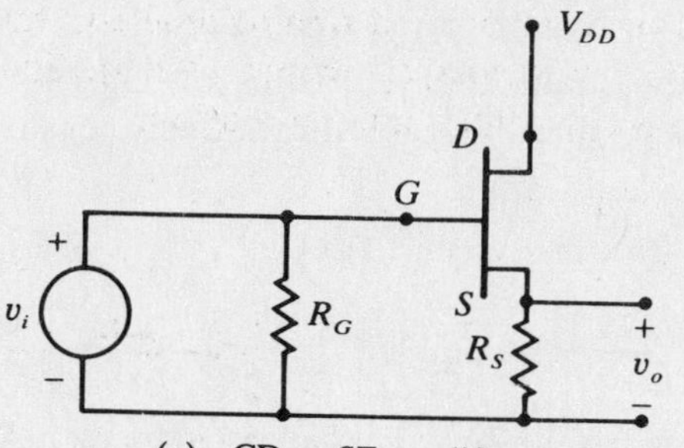

(a) CD or SF amplifier

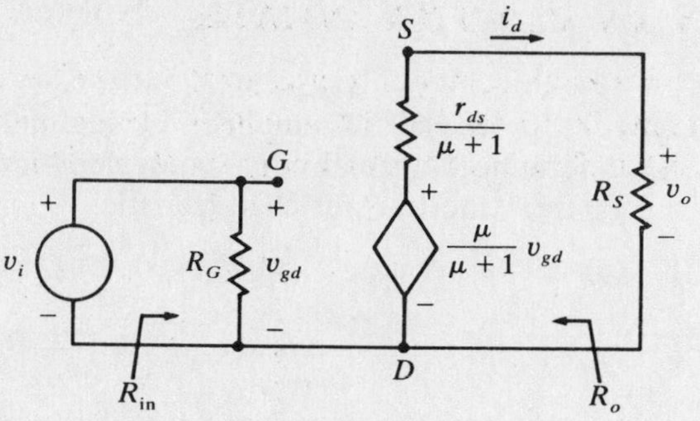

(b) Small-signal equivalent circuit

Fig. 7-3

(b) Substitution of the given values leads to

$$A_v = \frac{(60)(5 \times 10^3)}{(61)(5 \times 10^3) + (30 \times 10^3)} = 0.895$$

Note that the gain is less than unity; its positive value indicates that v_o and v_i are in phase.

7.5 CG AMPLIFIER ANALYSIS

Figure 4-21 is a simple common-gate amplifier circuit. Its small-signal equivalent circuit, incorporating the current-source model of Fig. 7-1(a), is given in Fig. 7-4.

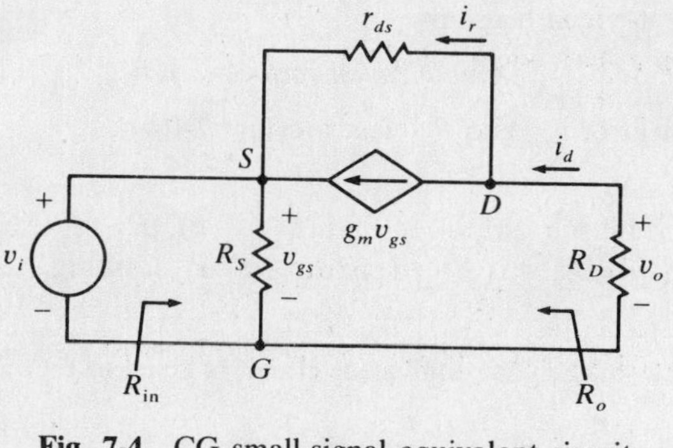

Fig. 7-4 CG small-signal equivalent circuit

Example 7.3 In the CG amplifier of Fig. 7-4, let $R_D = 1\ \text{k}\Omega$, $g_m = 2 \times 10^{-3}$ S, and $r_{ds} = 30\ \text{k}\Omega$. (a) Find an expression for the voltage-gain ratio $A_v = v_o/v_i$. (b) Evaluate A_v using the given typical values.

(a) By KCL, $i_r = i_d - g_m v_{gs}$. Applying KVL around the outer loop gives

$$v_o = (i_d - g_m v_{gs}) r_{ds} - v_{gs}$$

But $v_{gs} = -v_i$ and $i_d = -v_o/R_D$; thus,

$$v_o = \left(-\frac{v_o}{R_D} + g_m v_i\right) r_{ds} + v_i$$

and

$$A_v = \frac{v_o}{v_i} = \frac{(g_m r_{ds} + 1) R_D}{R_D + r_{ds}} \tag{7.7}$$

(b) Substitution of the given values yields

$$A_v = \frac{(61)(1 \times 10^3)}{1 \times 10^3 + 30 \times 10^3} = 1.97$$

Solved Problems

7.1 (*a*) For the JFET amplifier of Example 4.1, use the drain characteristics of Fig. 4-4 to determine the small-signal equivalent-circuit constants g_m and r_{ds}. (*b*) Alternatively, evaluate g_m from the transfer characteristic.

(*a*) Let v_{gs} change by ±1 V about the *Q* point of Fig. 4-4(*b*); then, by (*7.3*),

$$g_m \approx \left.\frac{\Delta i_D}{\Delta v_{GS}}\right|_Q = \frac{(3.3 - 0.3) \times 10^{-3}}{2} = 1.5 \text{ mS}$$

At the *Q* point of Fig. 4-4(*b*), while v_{DS} changes from 5 V to 20 V, i_D changes from 1.4 mA to 1.6 mA; thus, by (*7.4*),

$$r_{ds} \approx \left.\frac{\Delta v_{DS}}{\Delta i_D}\right|_Q = \frac{20 - 5}{(1.6 - 1.4) \times 10^{-3}} = 75 \text{ k}\Omega$$

(*b*) At the *Q* point of Fig. 4-4(*a*), while i_D changes from 1 mA to 2 mA, v_{GS} changes from −2.4 V to −1.75 V; by (*7.3*),

$$g_m \approx \left.\frac{\Delta i_D}{\Delta v_{GS}}\right|_Q = \frac{(2 - 1) \times 10^{-3}}{-1.75 - (-2.4)} = 1.54 \text{ mS}$$

7.2 Derive the small-signal voltage-source model of Fig. 7-1(*b*) from the current-source model of Fig. 7-1(*a*).

We find the Thévenin equivalent for the network to the left of the output terminals of Fig. 7-1(*a*). If all independent sources are deactivated, $v_{gs} = 0$; thus, $g_m v_{gs} = 0$, so that the dependent source also is deactivated (open circuit for a current source), and the Thévenin resistance is $R_{Th} = r_{ds}$. The open-circuit voltage appearing at the output terminals is $v_{Th} = v_{ds} = -g_m v_{gs} r_{ds} = -\mu v_{gs}$, where we have defined a new equivalent-circuit constant,

$$\textit{Amplification factor} \qquad \mu \equiv g_m r_{ds}$$

Proper series arrangement of v_{Th} and R_{Th} leads to Fig. 7-1(*b*).

7.3 In the drain-feedback-biased amplifier of Fig. 4-7(*a*), $R_F = 5$ MΩ, $R_L = 14$ kΩ, $r_{ds} = 40$ kΩ, and $g_m = 1$ mS. Find (*a*) $A_v = v_{ds}/v_i$, (*b*) Z_{in}, (*c*) Z_o looking back through the drain-source terminals, and (*d*) $A_i = i_i/i_L$.

(*a*) The voltage-source small-signal equivalent circuit is given in Fig. 7-5. With v_{ds} as a node voltage,

$$\frac{v_i - v_{ds}}{R_F} = \frac{v_{ds}}{R_L} + \frac{v_{ds} + \mu v_i}{r_{ds}}$$

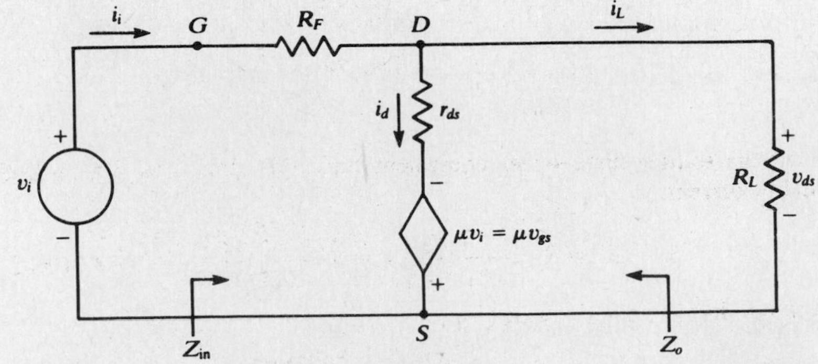

Fig. 7-5

Substituting for $\mu = g_m r_{ds}$ and rearranging yield

$$A_v = \frac{v_{ds}}{v_i} = \frac{R_L r_{ds}(1 - R_F g_m)}{R_F r_{ds} + R_L r_{ds} + R_L R_F}$$

$$= \frac{(14 \times 10^3)(40 \times 10^3)[1 - (5 \times 10^6)(1 \times 10^{-3})]}{(5 \times 10^6)(40 \times 10^3) + (14 \times 10^3)(40 \times 10^3) + (14 \times 10^3)(5 \times 10^6)} = -10.35$$

(*b*) KVL around the outer loop of Fig. 7-5 gives $v_i = i_i R_F + v_{ds} = i_i R_F + A_v v_i$, from which

$$Z_{in} = \frac{v_i}{i_i} = \frac{R_F}{1 - A_v} = \frac{5 \times 10^6}{1 - (-10.35)} = 440\ \text{k}\Omega$$

(*c*) The driving-point impedance Z_o is found after deactivating the independent source v_i. With $v_i = 0$, $\mu v_{gs} = \mu v_i = 0$ and

$$Z_o = \frac{r_{ds} R_F}{r_{ds} + R_F} = \frac{(40 \times 10^3)(5000 \times 10^3)}{5040 \times 10^3} = 39.68\ \text{k}\Omega$$

(*d*)
$$A_i = \frac{i_L}{i_i} = \frac{v_{ds}/R_L}{v_i/Z_{in}} = \frac{A_v Z_{in}}{R_L} = \frac{(-10.35)(440 \times 10^3)}{14 \times 10^3} = -325.3$$

7.4 For the JFET amplifier of Fig. 7-6, $g_m = 2$ mS, $r_{ds} = 30$ kΩ, $R_S = 3$ kΩ, $R_D = R_L = 2$ kΩ, $R_1 = 200$ kΩ, $R_2 = 800$ kΩ, and $r_i = 5$ kΩ. If C_C and C_S are large and the amplifier is biased in the pinchoff region, find (*a*) Z_{in}, (*b*) $A_v = v_L/v_i$, and (*c*) $A_i = i_L/i_i$.

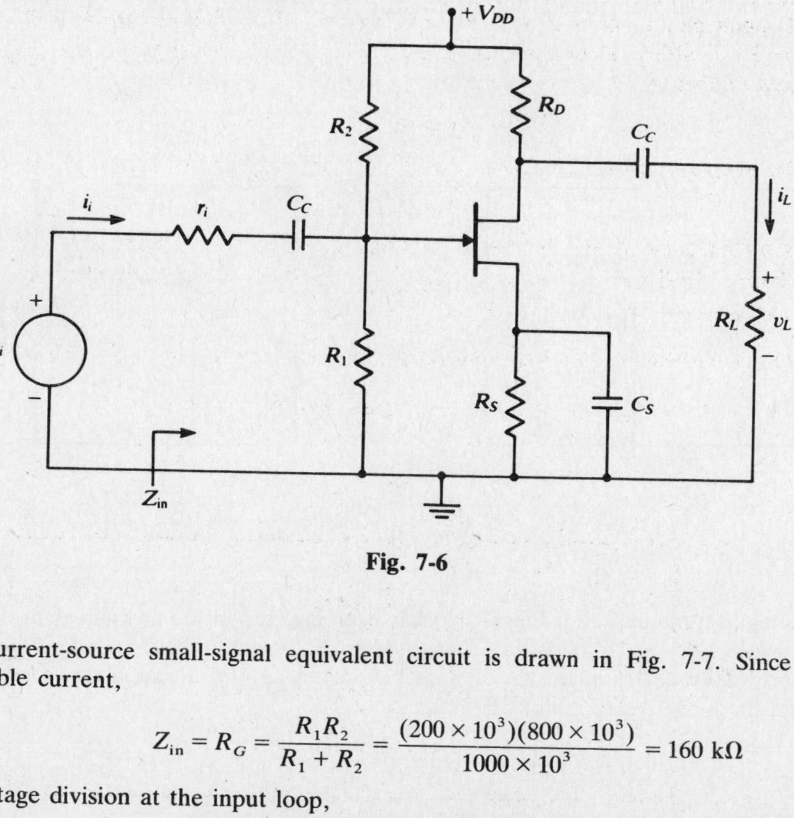

Fig. 7-6

(*a*) The current-source small-signal equivalent circuit is drawn in Fig. 7-7. Since the gate draws negligible current,

$$Z_{in} = R_G = \frac{R_1 R_2}{R_1 + R_2} = \frac{(200 \times 10^3)(800 \times 10^3)}{1000 \times 10^3} = 160\ \text{k}\Omega$$

(*b*) By voltage division at the input loop,

$$v_{gs} = \frac{R_G}{R_G + r_i} v_i = \frac{160 \times 10^3}{165 \times 10^3} v_i = 0.97 v_i \qquad (1)$$

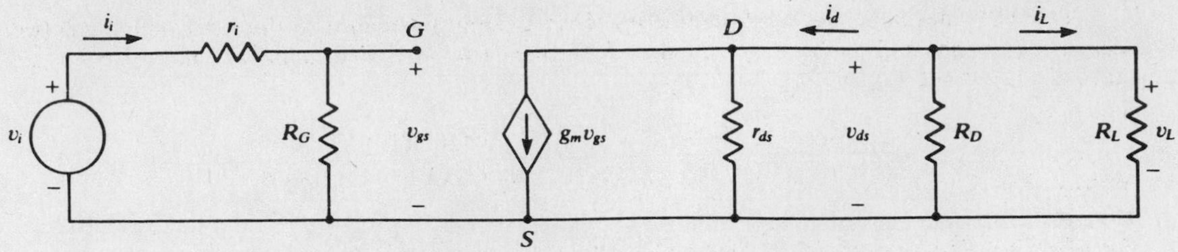

Fig. 7-7

The dependent current source drives into R_{ep}, where

$$\frac{1}{R_{ep}} = \frac{1}{r_{ds}} + \frac{1}{R_D} + \frac{1}{R_L} = \frac{1}{30 \times 10^3} + \frac{1}{2 \times 10^3} + \frac{1}{2 \times 10^3} = \frac{1}{967.74}\ S$$

and so $$v_L = -g_m v_{gs} R_{ep} \qquad (2)$$

Eliminating v_{gs} between (1) and (2) yields

$$A_v = \frac{v_L}{v_i} = 0.97(-g_m R_{ep}) = -(0.97)(2 \times 10^{-3})(967.74) = -1.88$$

(c) $$A_i = \frac{i_L}{i_i} = \frac{v_L/R_L}{v_i(R_G + r_i)} = \frac{A_v(R_G + r_i)}{R_L} = \frac{(-1.88)(165 \times 10^3)}{2 \times 10^3} = -155.1$$

7.5 Show that a small-signal equivalent circuit for the common-drain FET amplifier of Fig. 4-8 is given by Fig. 7-8(*b*).

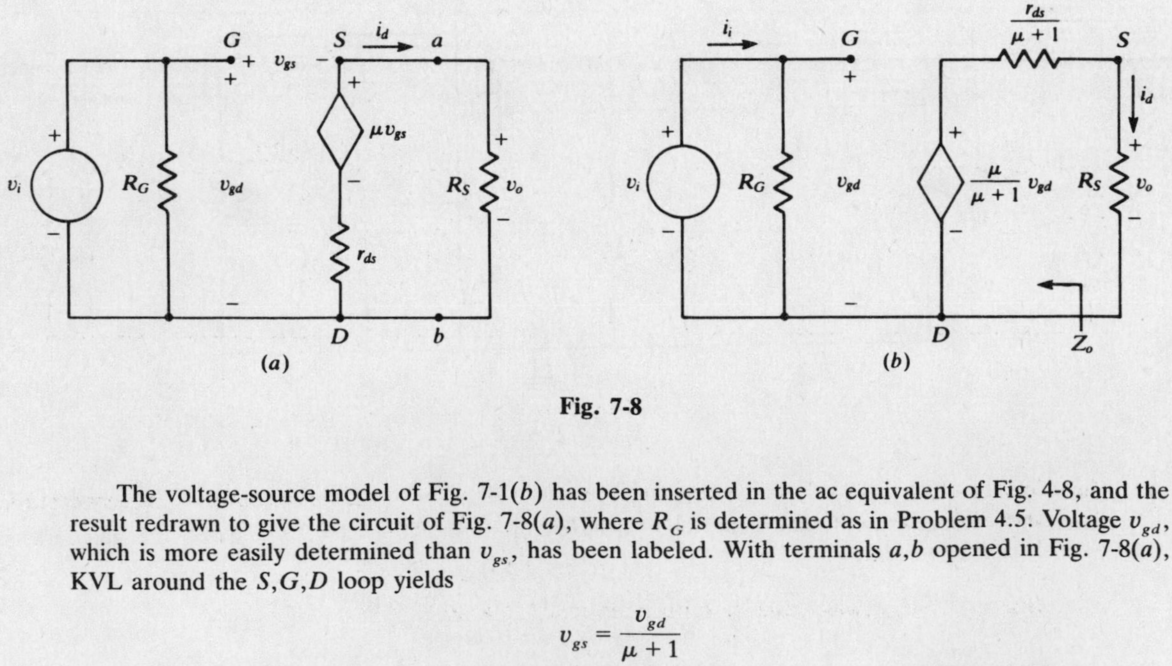

Fig. 7-8

The voltage-source model of Fig. 7-1(*b*) has been inserted in the ac equivalent of Fig. 4-8, and the result redrawn to give the circuit of Fig. 7-8(*a*), where R_G is determined as in Problem 4.5. Voltage v_{gd}, which is more easily determined than v_{gs}, has been labeled. With terminals *a*,*b* opened in Fig. 7-8(*a*), KVL around the *S*,*G*,*D* loop yields

$$v_{gs} = \frac{v_{gd}}{\mu + 1}$$

Then the Thévenin voltage at the open-circuited terminals *a*,*b* is

$$v_{Th} = \mu v_{gs} = \frac{\mu}{\mu + 1} v_{gd} \qquad (1)$$

The Thévenin impedance is found as the driving-point impedance to the left through a,b (with v_i deactivated or shorted), as seen by a source v_{ab} driving current i_a into terminal a. Since $v_{gs} = -v_{ab}$, KVL around the output loop of Fig. 7-8(a) gives

$$v_{ab} = \mu v_{gs} + i_a r_{ds} = -\mu v_{ab} + i_a r_{ds}$$

from which

$$R_{Th} = \frac{v_{ab}}{i_a} = \frac{r_{ds}}{\mu + 1} \tag{2}$$

Expressions (1) and (2) lead directly to the circuit of Fig. 7-8(b).

7.6 Figure 7-9(a) is a small-signal equivalent circuit (voltage-source model) of a common-gate JFET amplifier. Use the circuit to verify two *rules of impedance and voltage reflection* for FET amplifiers:

(a) Voltages and impedances in the drain circuit are reflected to the source circuit divided by $\mu + 1$. [Verify this rule by finding the Thévenin equivalent for the circuit to the right of a,a' in Fig. 7-9(a) and showing that Fig. 7-9(b) results.]

(b) Voltages and impedances in the source circuit are reflected to the drain circuit multiplied by $\mu + 1$. [Verify this rule by finding the Thévenin equivalent for the circuit to the left of b,b' in Fig. 7-9(a) and showing that Fig. 7-9(c) results.]

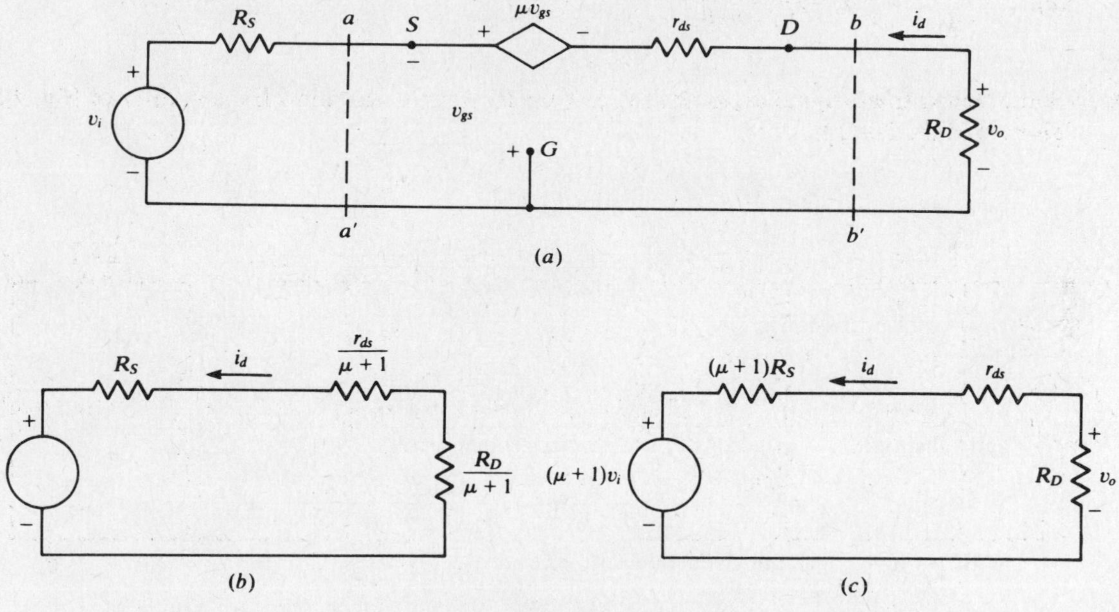

Fig. 7-9

(a) With a,a' open, $i_d = 0$; hence, $v_{gs} = 0$ and $v_{Th} = 0$. After a driving-point source $v_{aa'}$ is connected to terminals a,a' to drive current i_a into terminal a, KVL gives

$$v_{aa'} = \mu v_{gs} + i_a(r_{ds} + R_D) \tag{1}$$

But $v_{gs} = -v_{aa'}$, which can be substituted into (1) to give

$$R_{Th} = \frac{v_{aa'}}{i_a} = \frac{r_{ds}}{\mu + 1} + \frac{R_D}{\mu + 1} \tag{2}$$

With $v_{Th} = 0$, insertion of R_{Th} in place of the network to the right of a,a' in Fig. 7-9(a) leads directly to Fig. 7-9(b).

(*b*) Applying KVL to the left of b,b' in Fig. 7-9(*a*) with b,b' open, while noting that $v_i = -v_{gs}$, yields

$$v_{Th} = v_i - \mu v_{gs} = (\mu + 1)v_i \tag{3}$$

Deactivating (shorting) v_i, connecting a driving-point source $v_{bb'}$ to terminals b,b' to drive current i_b into terminal b, noting that $v_{gs} = -i_b R_S$, and applying KVL around the outer loop of Fig. 7-9(*a*) yield

$$v_{bb'} = i_b(r_{ds} + R_S) - \mu v_{gs} = i_b[r_{ds} + (\mu + 1)R_S] \tag{4}$$

The Thévenin impedance follows from (*4*) as

$$R_{Th} = \frac{v_{bb'}}{i_b} = r_{ds} + (\mu + 1)R_S \tag{5}$$

When the Thévenin source of (*3*) and impedance of (*5*) are used to replace the network to the left of b,b', the circuit of Fig. 7-9(*c*) results.

7.7 Suppose capacitor C_S is removed from the circuit of Problem 7.4 (Fig. 7-6), and all else remains unchanged. Find (*a*) the voltage-gain ratio $A_v = v_L/v_i$, (*b*) the current-gain ratio $A_i = i_L/i_i$, and (*c*) the output impedance R_o looking to the left through the output port with R_L removed.

(*a*) The voltage-source small-signal equivalent circuit is given in Fig. 7-10 (the current-source model was utilized in Problem 7.4). Voltage division and KVL give

$$v_{gs} = \frac{R_G}{R_G + r_i} v_i - i_d R_S \tag{1}$$

But by Ohm's law,

$$i_d = \frac{\mu v_{gs}}{r_{ds} + R_S + R_D \| R_L} \tag{2}$$

Substituting (*2*) into (*1*) and solving for v_{gs} yield

$$v_{gs} = \frac{R_G(r_{ds} + R_S + R_D \| R_L)v_i}{(R_G + r_i)[r_{ds} + (\mu + 1)R_S + R_D \| R_L]} \tag{3}$$

Now voltage division gives

$$v_L = -\frac{R_D \| R_L}{r_{ds} + R_S + R_D \| R_L} \mu v_{gs} \tag{4}$$

and substitution of (*3*) into (*4*) and rearrangement give

$$A_v = \frac{v_L}{v_i} = \frac{-\mu R_G R_D R_L}{(R_G + r_i)\{(R_D + R_L)[r_{ds} + (\mu + 1)R_S] + R_D R_L\}} \tag{5}$$

With $\mu = g_m r_{ds}$ and the given values, (*5*) becomes

$$A_v = \frac{-(2 \times 10^{-3})(30 \times 10^3)(160)(2)(2)}{(160 + 5)\{(2 + 2)[30 + (60 + 1)3] + (2)(2)\}} = -0.272$$

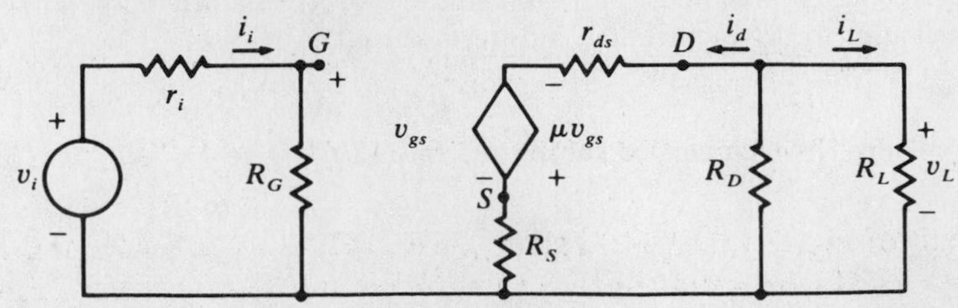

Fig. 7-10

(*b*) The current gain is found as

$$A_i = \frac{i_L}{i_i} = \frac{v_L/R_L}{v_i/(R_G + r_i)} = \frac{A_v(R_G + r_i)}{R_L} = \frac{(-0.272)(160+5)}{2} = -22.4$$

(*c*) R_L is disconnected, and a driving-point source is added such that $v_{dp} = v_L$. With v_i deactivated (short-circuited), $v_{gs} = 0$ and

$$R_o = R_D \| (r_{ds} + R_S) = \frac{R_D(r_{ds} + R_S)}{R_D + r_{ds} + R_S} = \frac{(2\times10^3)(30\times10^3 + 3\times10^3)}{2\times10^3 + 30\times10^3 + 3\times10^3} = 1.89\ \text{k}\Omega$$

Note that when R_S is not bypassed, the voltage- and current-gain ratios are significantly reduced.

7.8 Find a small-signal equivalent circuit for the two parallel-connected JFETs of Fig. 7-11 if the devices are not identical.

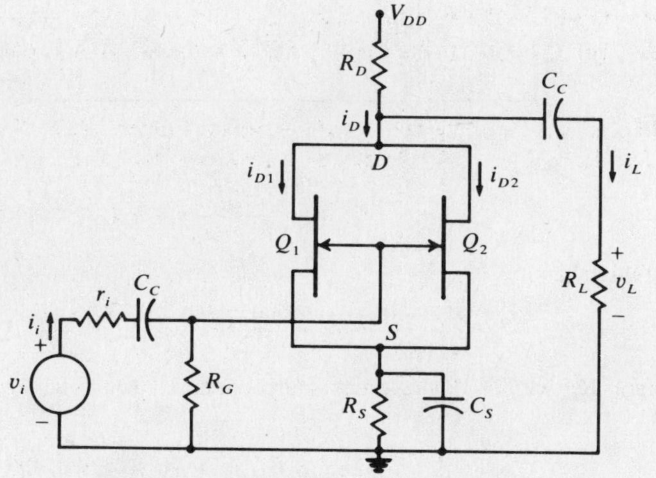

Fig. 7-11

By KCL,

$$i_D = i_{D1} + i_{D2} \tag{1}$$

Since the parallel connection assures that the gate-source and drain-source voltages are the same for both devices, (*1*) can be written as

$$i_D = f_1(v_{GS}, v_{DS}) + f_2(v_{GS}, v_{DS}) \tag{2}$$

Application of the chain rule to (*2*) yields

$$i_d = \Delta i_D \approx di_D = (g_{m1} + g_{m2})v_{gs} + \left(\frac{1}{r_{ds1}} + \frac{1}{r_{ds2}}\right)v_{ds} \tag{3}$$

where $$g_{m1} = \left.\frac{\partial i_{D1}}{\partial v_{GS}}\right|_Q \quad g_{m2} = \left.\frac{\partial i_{D2}}{\partial v_{GS}}\right|_Q \quad r_{ds1} = \left.\frac{\partial v_{DS}}{\partial i_{D1}}\right|_Q \quad r_{ds2} = \left.\frac{\partial v_{DS}}{\partial i_{D2}}\right|_Q$$

Equation (*3*) is satisfied by the current-source circuit of Fig. 7-1(*a*) if $g_m = g_{m1} + g_{m2}$ and $r_{ds} = r_{ds1} \| r_{ds2}$.

7.9 In the circuit of Fig. 7-11, $R_s = 3\ \text{k}\Omega$, $R_D = R_L = 2\ \text{k}\Omega$, $r_i = 5\ \text{k}\Omega$, and $R_G = 100\ \text{k}\Omega$. Assume that the two JFETs are identical with $r_{ds} = 25\ \text{k}\Omega$ and $g_m = 0.0025$ S. Find (*a*) the voltage-gain ratio $A_v = v_L/v_i$, (*b*) the current-gain ratio $A_i = i_L/i_i$, and (*c*) the output impedance R_o.

(*a*) The small-signal equivalent circuit is given in Fig. 7-12, which includes the model for two parallel JFETs as determined in Problem 7.8. By voltage division,

$$v_{gs} = \frac{R_G}{R_G + r_i} v_i = \frac{100}{100 + 5} v_i = 0.952 v_i \tag{1}$$

Now let

$$R_{eq} = \tfrac{1}{2} r_{ds} \| R_D \| R_L = \frac{r_{ds} R_D R_L}{2R_L R_D + r_{ds}(R_L + R_D)} = \frac{(25)(2)(2) \times 10^3}{(2)(2)(2) + (25)(2 + 2)} = 962\ \Omega \tag{2}$$

Then, by Ohm's law, $v_L = -2g_m v_{gs} R_{eq}$; with (*1*) and (*2*), this gives

$$A_v = \frac{v_L}{v_i} = -2g_m \frac{R_G}{R_G + r_i} R_{eq} = -2(0.0025)(0.952)(962) = -4.58$$

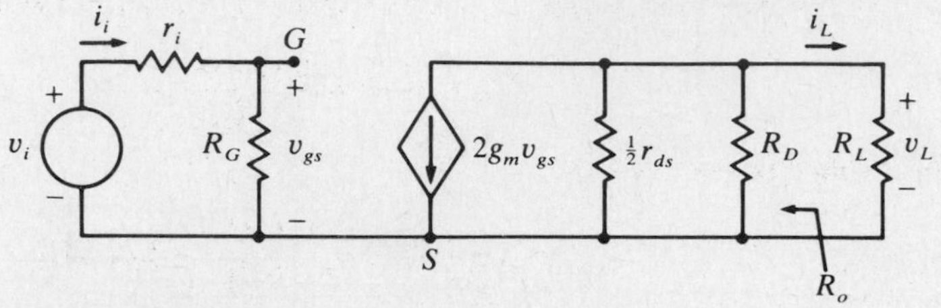

Fig. 7-12

(*b*) The current-gain ratio is

$$A_i = \frac{i_L}{i_i} = \frac{v_L / R_L}{v_i/(R_G + r_i)} = \frac{A_v(R_G + r_i)}{R_L} = \frac{(-4.58)(100 + 5)}{2} = -240.4$$

(*c*) We replace R_L with a driving-point source oriented such that $v_{dp} = v_L$. With v_i deactivated (short-circuited), $v_{gs} = 0$; thus,

$$R_o = R_D \| (\tfrac{1}{2} r_{ds}) = \frac{R_D r_{ds}}{2R_D + r_{ds}} = \frac{(2)(25) \times 10^3}{(2)(2) + 25} = 1.72\ \text{k}\Omega$$

7.10 Move capacitor C_S from its parallel connection across R_{S2} to a position across R_{S1} in Fig. 4-23. Let $R_G = 1\ \text{M}\Omega$, $R_{S1} = 800\ \Omega$, $R_{S2} = 1.2\ \text{k}\Omega$, and $R_L = 1\ \text{k}\Omega$. The JFET is characterized by $g_m = 0.002$ S and $r_{ds} = 30\ \text{k}\Omega$. Find (*a*) the voltage-gain ratio $A_v = v_L/v_i$, (*b*) the current-gain ratio $A_i = i_L/i_i$, (*c*) the input impedance R_{in}, and (*d*) the output impedance R_o.

(*a*) The equivalent circuit (with current-source JFET model) is given in Fig. 7-13. By KVL,

$$v_{gs} = v_i - v_L \tag{1}$$

Using v_i and v_L as node voltages, we have

$$i_i = \frac{v_i - v_L}{R_G} \tag{2}$$

Now let

$$\frac{1}{R_{eq}} = \frac{1}{r_{ds}} + \frac{1}{R_{S2}} + \frac{1}{R_L} = \frac{1}{30 \times 10^3} + \frac{1}{1.2 \times 10^3} + \frac{1}{1 \times 10^3} = \frac{1}{536}$$

By KCL and Ohm's law,

$$v_L = (i_i + g_m v_{gs}) R_{eq} \tag{3}$$

Substitution of (*1*) and (*2*) into (*3*) and rearrangement lead to

$$A_v = \frac{v_L}{v_i} = \frac{(g_m R_G + 1) R_{eq}}{R_G + (g_m R_G + 1) R_{eq}} = \frac{[(0.002)(1 \times 10^6) + 1](536)}{1 \times 10^6 + [(0.002)(1 \times 10^6) + 1](536)} = 0.517$$

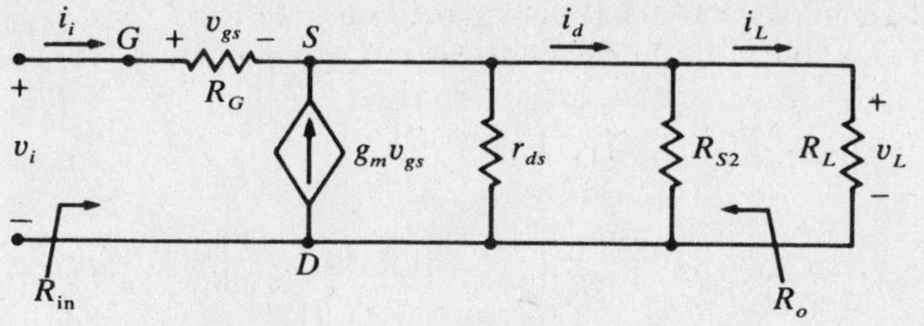

Fig. 7-13

(*b*) The current-gain ratio follows from part *a* as

$$A_i = \frac{i_L}{i_i} = \frac{v_L/R_L}{(v_i - v_L)/R_G} = \frac{A_v R_G}{(1-A_v)R_L} = \frac{(0.517)(1\times 10^6)}{(1-0.517)(1\times 10^3)} = 1070.4$$

(*c*) From (*2*),

$$i_i = \frac{v_i - v_L}{R_G} = \frac{v_i(1-A_v)}{R_G} \tag{4}$$

R_{in} is found directly from (*4*) as

$$R_{in} = \frac{v_i}{i_i} = \frac{R_G}{1-A_v} = \frac{1\times 10^6}{1-0.517} = 2.07\ \text{M}\Omega$$

(*d*) We remove R_L and connect a driving-point source oriented such that $v_{dp} = v_L$. With v_i deactivated (shorted), $v_{gs} = -v_{dp}$. Then, by KCL,

$$i_{dp} = v_{dp}\left(\frac{1}{R_{S2}} + \frac{1}{r_{ds}} + \frac{1}{R_G}\right) - g_m v_{gs} = v_{dp}\left(\frac{1}{R_{S2}} + \frac{1}{r_{ds}} + \frac{1}{R_G} + g_m\right)$$

and $$R_o = \frac{v_{dp}}{i_{dp}} = \frac{1}{\dfrac{1}{R_{S2}} + \dfrac{1}{r_{ds}} + \dfrac{1}{R_G} + g_m} = \frac{1}{\dfrac{1}{1.2\times 10^3} + \dfrac{1}{30\times 10^3} + \dfrac{1}{1\times 10^6} + 0.002} = 348.7\ \Omega$$

7.11 Use the small-signal equivalent circuit to predict the peak values of i_d and v_{ds} in Example 4.2. Compare your result with that of Example 4.2, and comment on any differences.

The values of g_m and r_{ds} for operation near the *Q* point of Fig. 4-4 were determined in Problem 7.1. We may use the current-source model of Fig. 7-1(*a*) to form the equivalent circuit for Fig. 4-3. In that circuit, with $v_{gs} = \sin t$ V, Ohm's law requires that

$$v_{ds} = -g_m v_{gs}(r_{ds} \parallel R_D) = \frac{-g_m r_{ds} R_D v_{gs}}{r_{ds} + R_D} = \frac{-(1.5\times 10^{-3})(75\times 10^3)(3\times 10^3)v_{gs}}{75\times 10^3 + 3\times 10^3} = -4.33 v_{gs}$$

Thus, $$V_{dsm} = 4.33 V_{gsm} = 4.33(1) = 4.33\ \text{V}$$

Also, from Fig. 7-1(*a*),

$$i_d = g_m v_{gs} + \frac{v_{ds}}{r_{ds}}$$

so $$I_{dm} = g_m V_{gsm} + \frac{V_{dsm}}{r_{ds}} = (1.5\times 10^{-3})(1) + \frac{1}{75\times 10^3} = 1.513\ \text{mA}$$

The ±1-V excursion of v_{gs} leads to operation over a large portion of the nonlinear drain characteristics. Consequently, the small-signal equivalent circuit predicts greater positive peaks and smaller negative peaks of i_d and v_{ds} than the graphical solution of Example 4.2, which inherently accounts for the nonlinearities.

7.12 For the JFET drain characteristics of Fig. 4-2(*a*), take v_{DS} as the dependent variable [so that $v_{DS} = f(v_{GS}, i_D)$] and derive the voltage-source small-signal model.

For small variations about a Q point, the chain rule gives

$$v_{ds} = \Delta v_{DS} \approx dv_{DS} = \left.\frac{\partial v_{DS}}{\partial v_{GS}}\right|_Q v_{gs} + \left.\frac{\partial v_{DS}}{\partial i_D}\right|_Q i_d \tag{1}$$

Now we may define

$$\left.\frac{\partial v_{DS}}{\partial v_{GS}}\right|_Q = \mu \quad \text{and} \quad \left.\frac{\partial v_{DS}}{\partial i_D}\right|_Q = r_{ds}$$

If the JFET operates in the pinchoff region, then gate current is negligible and (*1*) is satisfied by the equivalent circuit of Fig. 7-1(*b*).

7.13 Find a current-source small-signal equivalent circuit for the CD FET amplifier.

Norton's theorem can be applied to the voltage-source model of Fig. 7-8(*b*). The open-circuit voltage at terminals S,D (with R_S removed) is

$$v_{oc} = \frac{\mu}{\mu + 1} v_{gd} \tag{1}$$

The short-circuit current at terminals S,D is

$$i_{SC} = \frac{\dfrac{\mu}{\mu+1} v_{gd}}{r_{ds}/(\mu + 1)} = \frac{\mu}{r_{ds}} v_{gd} = g_m v_{gd} \tag{2}$$

The Norton impedance is found as the ratio of (*1*) to (*2*):

$$R_N = \frac{v_{oc}}{i_{SC}} = \frac{\dfrac{\mu}{\mu+1} v_{gd}}{g_m v_{gd}} = \frac{\mu}{(\mu + 1) g_m}$$

The equivalent circuit is given in Fig. 7-14. Usually, $\mu \gg 1$ and, thus, $R_N \approx 1/g_m$.

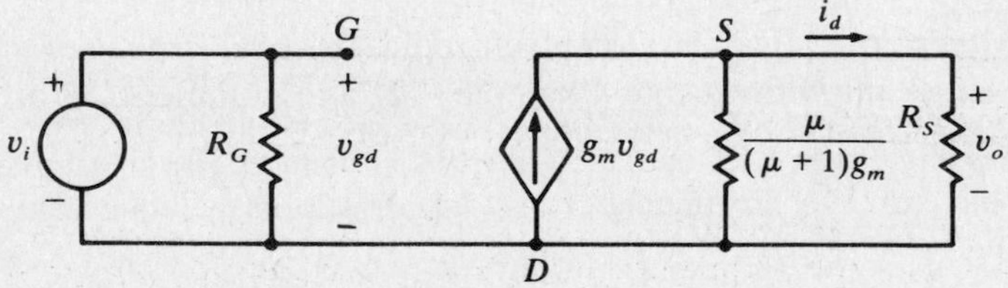

Fig. 7-14

7.14 In the cascaded MOSFET amplifier of Fig. 7-15, $C_C \to \infty$. Find (*a*) the voltage-gain ratio $A_v = v_L/v_i$ and (*b*) the current-gain ratio $A_i = i_L/i_i$.

(*a*) The small-signal equivalent circuit is given in Fig. 7-16. Using the result of Example 7.1, but replacing R_D with $R_{D1} \| R_{G2}$ where $R_{G2} = R_{21} \| R_{22}$, we have

$$A_{v1} = \frac{-g_{m1} r_{ds1}(R_{D1} \| R_{G2})}{r_{ds1} + (R_{D1} \| R_{G2})} \tag{1}$$

Similarly,

$$A_{v2} = \frac{-g_{m2} r_{ds2}(R_{D2} \| R_L)}{r_{ds2} + (R_{D2} \| R_L)} \tag{2}$$

Then

$$A_v = A_{v1} A_{v2} = \frac{g_{m1} g_{m2} r_{ds1} r_{ds2}(R_{D1} \| R_{G2})(R_{D2} \| R_L)}{[r_{ds1} + (R_{D1} \| R_{G2})][r_{ds2} + (R_{D2} \| R_L)]} \tag{3}$$

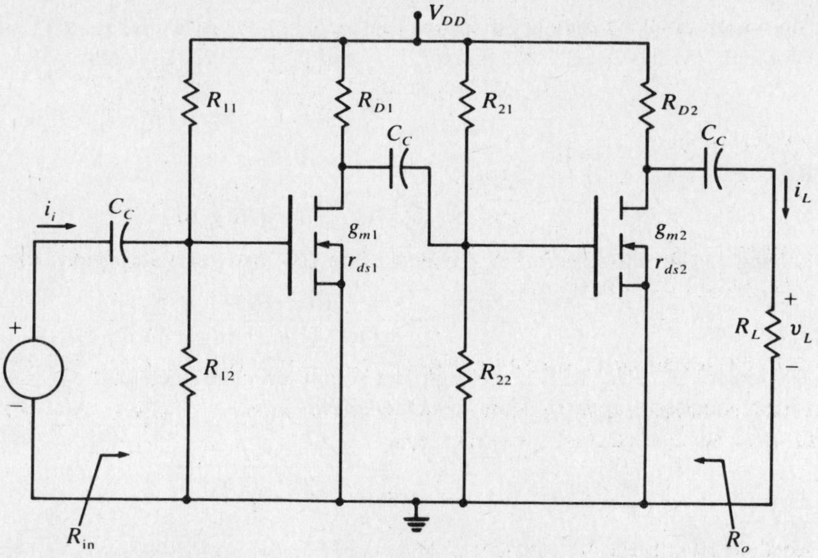

Fig. 7-15

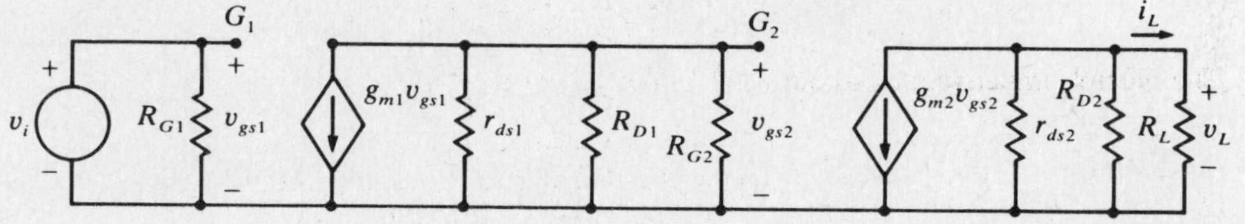

Fig. 7-16

(*b*) Realizing that $R_{G1} = R_{11} \| R_{12}$, we have

$$A_i = \frac{i_L}{i_i} = \frac{v_o/R_L}{v_i/R_{G1}} = A_v \frac{R_{G1}}{R_L}$$

where A_v is given by (*3*).

7.15 For the JFET-BJT Darlington amplifier of Fig. 7-17(*a*), find (*a*) the voltage-gain ratio $A_v = v_e/v_i$ and (*b*) the output impedance R_o. Assume $h_{re} = h_{oe} = 0$ and that $R_G \gg R_1, R_2$.

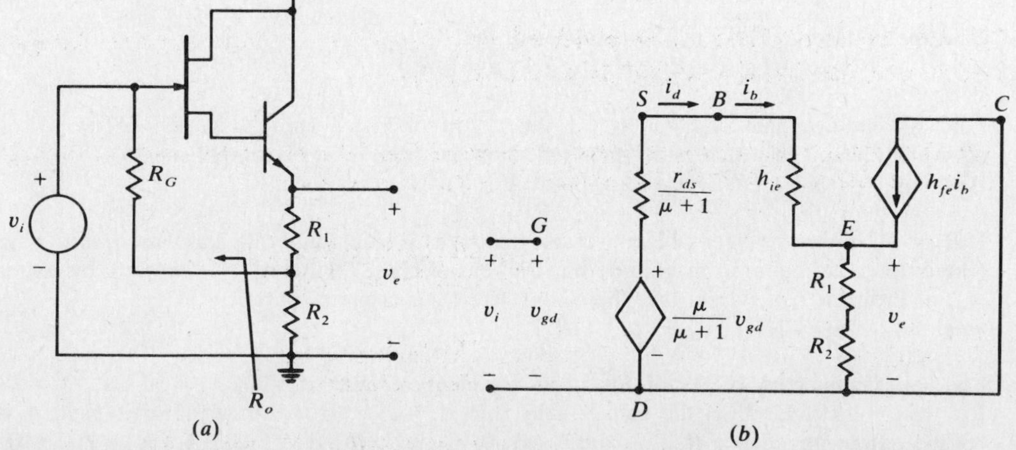

Fig. 7-17

(*a*) The small-signal equivalent circuit is given in Fig. 7-17(*b*), where the CD model of the JFET (see Problem 7.5) has been used. Since $i_b = i_d$ and $v_{gd} = v_i$, KVL yields

$$\frac{\mu}{\mu+1} v_i = i_d\left(\frac{r_{ds}}{\mu+1} + h_{ie}\right) + (h_{fe}+1)i_d(R_1+R_2) \tag{1}$$

By Ohm's law,

$$v_e = (h_{fe}+1)i_d(R_1+R_2) \tag{2}$$

Solving (*1*) for i_d, substituting the result into (*2*), and rearranging give

$$A_v = \frac{v_e}{v_i} = \frac{\mu(h_{fe}+1)(R_1+R_2)}{r_{ds} + (\mu+1)[h_{ie} + (h_{fe}+1)(R_1+R_2)]}$$

(*b*) We replace $R_1 + R_2$ with a driving-point source oriented such that $v_{dp} = v_e$. With v_i deactivated (short circuited), $v_{gd} = 0$. Then, by Ohm's law,

$$i_b = -\frac{v_{dp}}{h_{ie} + r_{ds}/(\mu+1)} \tag{3}$$

and by KCL,

$$i_{dp} = -(h_{fe}+1)i_b \tag{4}$$

Substituting (*3*) into (*4*) and rearranging give

$$R_o = \frac{v_{dp}}{i_{dp}} = \frac{r_{ds} + (\mu+1)h_{ie}}{(\mu+1)(h_{fe}+1)}$$

Supplementary Problems

7.16 Find the input impedance as seen by the source v_i of Example 4.1 if C_C is large. *Ans.* 940 kΩ

7.17 Show that the transconductance of a JFET varies as the square root of the drain current. *Ans.* $g_m = (2\sqrt{I_{DSS}}/V_{p0})\sqrt{i_D}$

7.18 In the amplifier of Fig. 4-8, $R_1 = 20$ kΩ, $R_2 = 100$ kΩ, $R_3 = 1$ MΩ, $r_{ds} = 30$ kΩ, $\mu = 150$ (see Problem 7.2), and $R_S = 1$ kΩ. Find (*a*) $A_v = v_o/v_i$, (*b*) $A_i = i_d / i_i$, and (*c*) Z_o.
Ans. (*a*) 0.829; (*b*) 843; (*c*) 198.7 Ω

7.19 Find the voltage gain of the CG amplifier of Fig. 7-9(*a*).
Ans. $A_v = v_o/v_i = (\mu+1)R_D/[R_D + r_{ds} + (\mu+1)R_S]$

7.20 Find the voltage gain $A_{v2} = v_2/v_i$ for the circuit of Fig. 7-18(*a*). Figure 7-18(*b*) is a small-signal equivalent circuit in which impedance reflection has been used for simplification.
Ans. $A_{v2} = -\mu R_D/[R_D + r_{ds} + (\mu+1)R_S]$

7.21 If $R_D = R_S$ in the amplifier of Fig. 7-18(*a*), the circuit is commonly called a *phase splitter*, since $v_2 = -v_1$ (the outputs are equal in magnitude but 180° out of phase). Find $A_{v1} = v_1/v_i$ and, by comparison with A_{v2} of Problem 7.20, verify that the circuit actually is a phase splitter.
Ans. $A_{v1} = \mu R_S/[R_D + r_{ds} + (\mu+1)R_S]$

7.22 The series-connected JFETs of Fig. 4-16 are identical, with $\mu = 70$, $r_{ds} = 30$ kΩ, $R_G = 100$ kΩ, and $R_D = R_L = 4$ kΩ. Find (*a*) the voltage-gain ratio $A_v = v_L/v_i$, (*b*) the current-gain ratio $A_i = i_L/i_i$, and (*c*) the output impedance R_o. *Ans.* (*a*) $A_v = -9.32$; (*b*) $A_i = -233$; (*c*) $R_o = 2.16$ MΩ

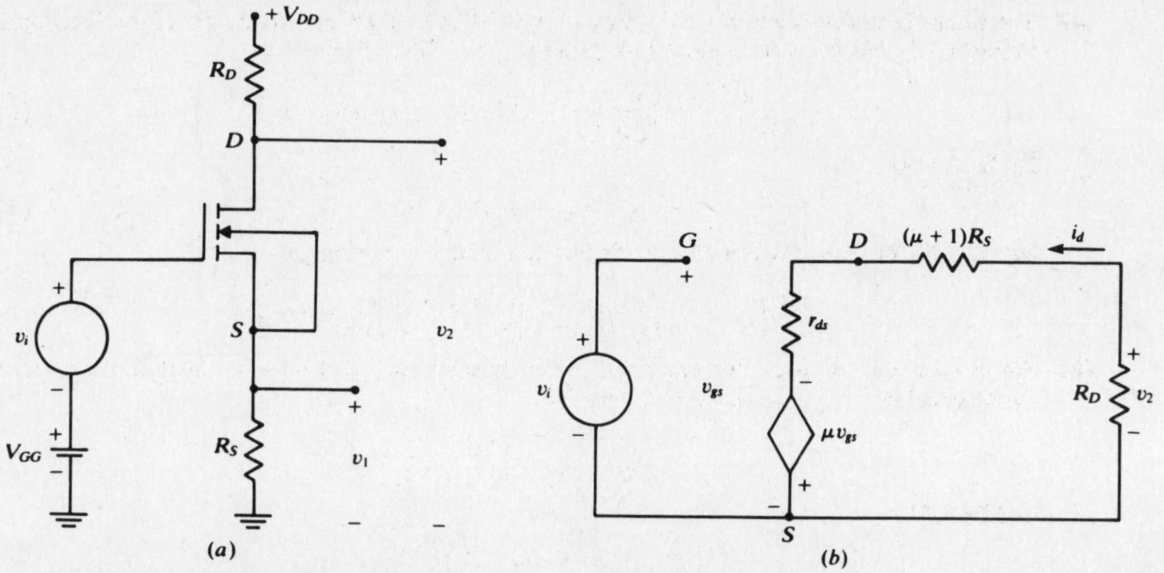

Fig. 7-18

7.23 The JFET amplifier of Fig. 4-23 has $R_G = 1\ \text{M}\Omega$, $R_{S1} = 800\ \Omega$, $R_{S2} = 1.2\ \text{k}\Omega$, and $R_L = 1\ \text{k}\Omega$. The JFET obeys (*4.2*) and is characterized by $I_{DSS} = 10$ mA, $V_{p0} = 4$ V, $V_{GSQ} = -2$ V, and $\mu = 60$. Determine (*a*) g_m by use of (*7.3*), (*b*) r_{ds}, and (*c*) the voltage-gain ratio $A_v = v_L/v_i$.
Ans. (*a*) 2.5 mS; (*b*) 24 kΩ; (*c*) 0.52

7.24 For the JFET amplifier of Fig. 4-13, find expressions for (*a*) the voltage-gain ratio $A_{v1} = V_o/V_G$ and (*b*) the voltage-gain ratio $A_{v2} = V_1/V_G$.
Ans. (*a*) $A_{v1} = \mu R_S/[(\mu+1)R_S + R_D + r_{ds}]$; (*b*) $A_{v2} = -\mu R_D/[(\mu+1)R_S + R_D + r_{ds}]$

7.25 Frequently, in integrated circuits, the gate of a FET is connected to the drain; then the drain-to-source terminals are considered the terminals of a resistor. Starting with (*7.2*), show that if $\mu \gg 1$, then the small-signal equivalent circuit is no more than a resistor of value $1/g_m$.

7.26 For the CS amplifier of Fig. 7-2(*b*), find (*a*) the input impedance R_{in} and (*b*) the output impedance R_o. *Ans.* (*a*) $R_{\text{in}} = R_G$; (*b*) $R_o = r_{ds}$.

7.27 For the CD amplifier of Fig. 7-3(*b*), find (*a*) the input impedance R_{in} and (*b*) the output impedance R_o. *Ans.* (*a*) $R_{\text{in}} = R_G$; (*b*) $R_o = r_{ds}/(\mu+1)$

7.28 For the CG amplifier of Fig. 7-4, find (*a*) the input impedance R_{in} and (*b*) the output impedance R_o. *Ans.* (*a*) $R_{\text{in}} = R_S(R_D + r_{ds})/[(\mu+1)R_S + R_D + r_{ds}]$; (*b*) $R_o = r_{ds}$

7.29 In the circuit of Fig. 7-19, the two FETs are identical. Find (*a*) the voltage-gain ratio $A_v = v_o/v_i$ and (*b*) the output impedance R_o.
Ans. (*a*) $A_v = -\mu R_L | \{2R_L + 2[(\mu+1)R + r_{ds}]\}$; (*b*) $R_o = \frac{1}{2}[(\mu+1)R + r_{ds}]$

7.30 For the cascaded MOSFET amplifier of Fig. 7-15 with equivalent circuit in Fig. 7-16, find (*a*) the input impedance R_{in} and (*b*) the output impedance R_o.
Ans. (*a*) $R_{\text{in}} = R_{11}R_{12}/(R_{11} + R_{12})$; (*b*) $R_o = r_{ds2}R_{D2}/(r_{ds2} + R_{D2})$

7.31 In the cascaded FET-BJT circuit of Fig. 7-20, assume $h_{re} = h_{oe} = 0$ and $h_{ie} \ll R_D$. Find expressions for (*a*) $A_{v1} = v_{o1}/v_i$ and (*b*) $A_{v2} = v_{o2}/v_i$.
Ans. (*a*) $A_{v1} = \mu(h_{fe}+1)R_S/[(\mu+1)(h_{fe}+1)R_S + h_{ie} + r_{ds}]$;
(*b*) $A_{v2} = [\mu h_{fe}R_C + \mu(h_{fe}+1)R_S]/[(\mu+1)(h_{fe}+1)R_S + h_{ie} + r_{ds}]$

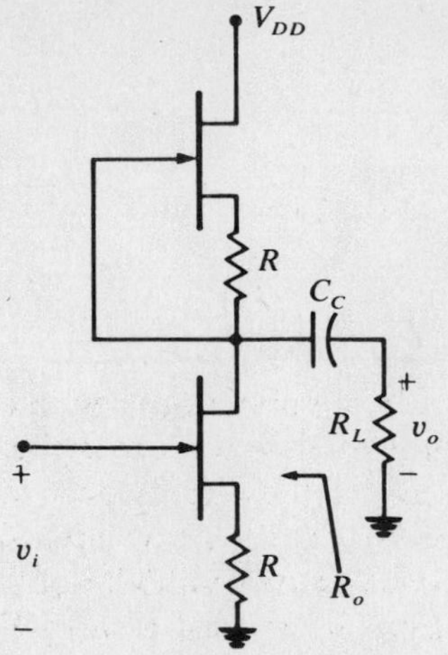

Fig. 7-19

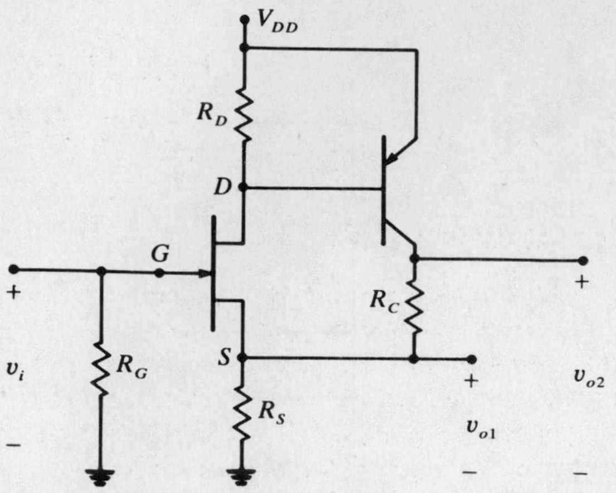

Fig. 7-20

Chapter 8

Frequency Effects in Amplifiers

8.1 INTRODUCTION

In the analyses of the two preceding chapters, we assumed *operation* in the midfrequency range, in which the reactances of all bypass and coupling capacitors can be considered to be zero while all inherent capacitive reactances associated with transistors are infinitely large. However, over a wide range of signal frequencies, the response of an amplifier is that of a *band-pass filter*: Low and high frequencies are attenuated, but signals over a band (or range) of frequencies between high and low are not attenuated. The typical frequency behavior of an *RC*-coupled amplifier is illustrated by Fig. 8-1(*a*). In practical amplifiers the midfrequency range spans several orders of magnitude, so that terms in the gain ratio expression which alter low-frequency gain are essentially constant over the high-frequency range. Conversely, terms that alter high-frequency gain are practically constant over the low-frequency range. Thus the high- and low-frequency analyses of amplifiers are treated as two independent problems.

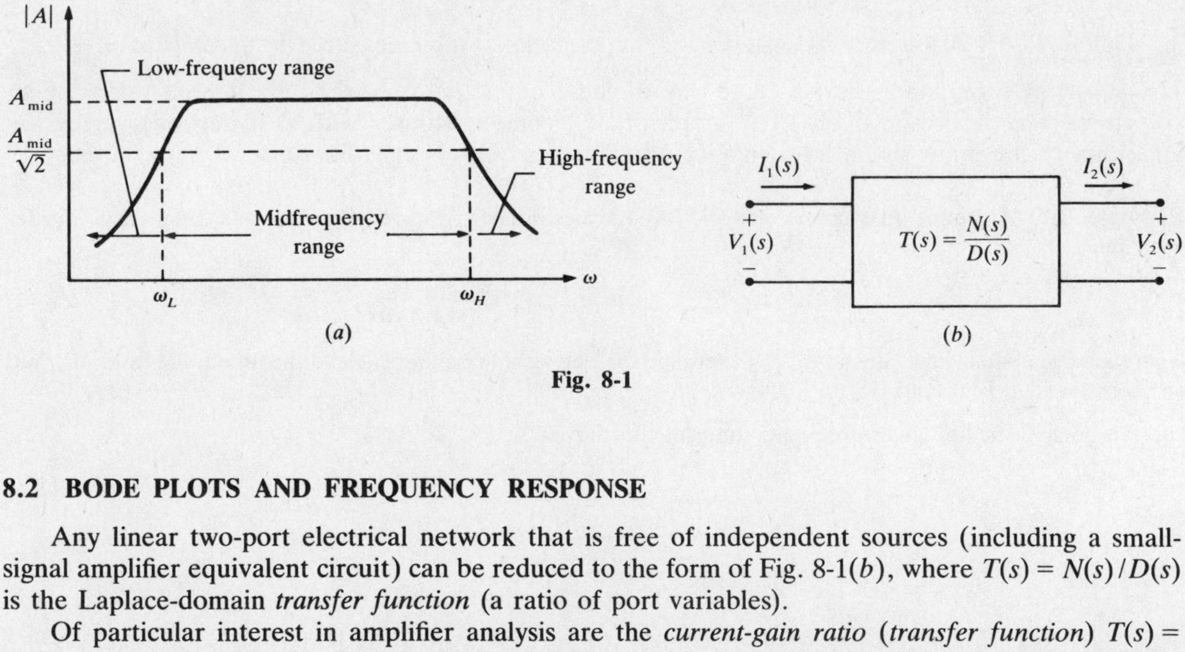

Fig. 8-1

8.2 BODE PLOTS AND FREQUENCY RESPONSE

Any linear two-port electrical network that is free of independent sources (including a small-signal amplifier equivalent circuit) can be reduced to the form of Fig. 8-1(*b*), where $T(s) = N(s)/D(s)$ is the Laplace-domain *transfer function* (a ratio of port variables).

Of particular interest in amplifier analysis are the *current-gain ratio* (*transfer function*) $T(s) = A_i(s)$ and *voltage-gain ratio* (*transfer function*) $T(s) = A_v(s)$. For a sinusoidal input voltage signal, the Laplace transform pair

$$v_1(t) = V_{1m} \sin \omega t \qquad \leftrightarrow \qquad V_1(s) = \frac{V_{1m}\omega}{s^2 + \omega^2}$$

is applicable, and the network response is given by

$$V_2(s) = A_v(s)V_1(s) = \frac{A_v(s)V_{1m}\omega}{s^2 + \omega^2} \tag{8.1}$$

Without loss of generality, we may assume that the polynomial $D(s) = 0$ has n distinct roots. Then the partial-fraction expansion of (*8.1*) yields

$$V_2(s) = \frac{k_1}{s - j\omega} + \frac{k_2}{s + j\omega} + \frac{k_3}{s + p_1} + \frac{k_4}{s + p_2} + \cdots + \frac{k_{n+2}}{s + p_n} \tag{8.2}$$

where the first two terms on the right-hand side are forced-response terms (called the *frequency response*), and the balance of the terms constitute the transient response. The transient response diminishes to zero with time, provided the roots of $D(s) = 0$ are located in the left half plane of complex numbers (the condition for a *stable* system).

The coefficients k_1 and k_2 are evaluated by the method of residues, and the results are used in an inverse transformation to the time-domain steady-state sinusoidal response given by

$$v_2(t) = V_{1m}|A_v(j\omega)| \sin(\omega t + \phi) = V_{2m} \sin(\omega t + \phi) \tag{8.3}$$

(see Problem 8.21). The *network phase angle* ϕ is defined as

$$\phi = \tan^{-1} \frac{\text{Im}\{A_v(j\omega)\}}{\text{Re}\{A_v(j\omega)\}} \tag{8.4}$$

From (*8.4*), it is apparent that a sinusoidal input to a stable, linear, two-port network results in a steady-state output that is also sinusoidal; the input and output waveforms differ only in amplitude and phase angle.

For convenience, we make the following definitions:

1. Call $A(j\omega)$ the *frequency transfer function*.
2. Define $M \equiv |A(j\omega)|$, the *gain ratio*.
3. Define $M_{db} \equiv 20 \log M = 20 \log |A(j\omega)|$, the *amplitude ratio*, measured in *decibels* (db).

The subscript v or i may be added to any of these quantities to specifically denote reference to voltage or current, respectively. The graph of M_{db} (simultaneously with ϕ if desired) versus the logarithm of the input signal frequency (positive values only) is called a *Bode plot*.

Example 8.1 A simple first-order network has Laplace-domain transfer function and frequency transfer function

$$A(s) = \frac{1}{\tau s + 1} \qquad \text{and} \qquad A(j\omega) = \frac{1}{1 + j\omega\tau}$$

where τ is the system time constant. (*a*) Determine the network phase angle ϕ and the amplitude ratio M_{db} and (*b*) construct the Bode plot for the network.

(*a*) In polar form, the given frequency transfer function is

$$A(j\omega) = \frac{1}{\sqrt{1 + (\omega\tau)^2}\ \underline{|\tan^{-1}(\omega\tau/1)}} = \frac{1}{\sqrt{1 + (\omega\tau)^2}}\ \underline{|-\tan^{-1}\omega\tau}$$

Hence,
$$\phi = -\tan^{-1} \omega\tau \tag{8.5}$$

and
$$M_{db} = 20 \log |A(j\omega)| = 20 \log \frac{1}{\sqrt{1 + (\omega\tau)^2}} = -10 \log [1 + (\omega\tau)^2] \tag{8.6}$$

(*b*) If values of (*8.5*) and (*8.6*) are calculated and plotted for various values of ω, then a Bode plot is generated. This is done in Fig. 8-2, where ω is given in terms of time constants τ rather than, say, hertz. This particular system is called a *lag network* because its phase angle ϕ is negative for all ω.

Example 8.2 A simple first-order network has Laplace-domain transfer function and frequency transfer function

$$A(s) = \tau s + 1 \qquad \text{and} \qquad A(j\omega) = 1 + j\omega\tau$$

Determine the network phase angle ϕ and the amplitude ratio M_{db}, and discuss the nature of the Bode plot.

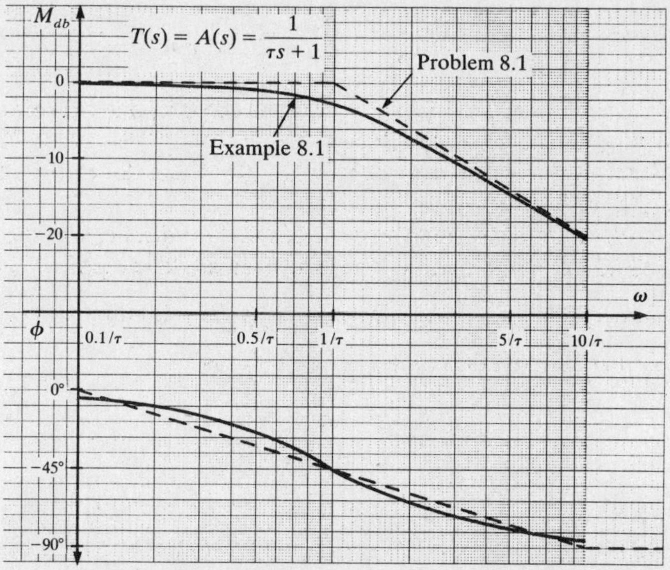

Fig. 8-2

After $A(j\omega)$ is converted to polar form, it becomes apparent that

$$\phi = \tan^{-1} \omega\tau \tag{8.7}$$

and

$$M_{db} = 20 \log |A(j\omega)| = 20 \log \sqrt{1 + (\omega\tau)^2} = 10 \log [1 + (\omega\tau)^2] \tag{8.8}$$

Comparison of (*8.5*) and (*8.7*) reveals that the network phase angle is the mirror image of the phase angle for the network of Example 8.1. (As ω increases, ϕ ranges from 0° to 90°.) Further, (*8.8*) shows that the amplitude ratio is the mirror image of the amplitude ratio of Example 8.1. (As ω increases, M_{db} ranges from 0 to positive values.) Thus, the complete Bode plot consists of the mirror images about zero of M_{db} and ϕ of Fig. 8-2. Since here the phase angle ϕ is everywhere positive, this network is called a *lead network*.

A *break frequency* or *corner frequency* is the frequency $1/\tau$. For a simple lag or lead network, it is the frequency at which $M^2 = |A(j\omega)|^2$ has changed by 50 percent from its value at $\omega = 0$; at that frequency, M_{db} has changed by 3 db from its value at $\omega = 0$. Corner frequencies serve as key points in the construction of Bode plots.

Example 8.3 Describe the Bode plot of a network whose output is the time derivative of its input.

The network has Laplace-domain transfer function $A(s) = s$ and frequency transfer function $A(j\omega) = j\omega$. Converting $A(j\omega)$ to polar form shows that

$$\phi = \tan^{-1} \frac{\omega}{0} = 90° \tag{8.9}$$

and

$$M_{db} = 20 \log \omega \tag{8.10}$$

Obviously, the network phase angle is a constant 90°. By (*8.10*), $M_{db} = 0$ when $\omega = 1$; further, M_{db} increases by 20 db for each order-of-magnitude (*decade*) change in ω. A graph of M_{db} versus the logarithm of ω would thus have a slope of 20 db per decade of frequency. A complete Bode plot is shown in Fig. 8-3.

The exact Bode plot of a network frequency transfer function is tedious to construct. Frequently, sufficiently accurate information can be obtained from an *asymptotic* Bode plot (see Problem 8.1).

Example 8.4 The exact Bode plot for the first-order system of Example 8.1 is given in Fig. 8-2. (*a*) Add the asymptotic Bode plot to that figure. (*b*) Describe the asymptotic Bode plot for the system of Example 8.2.

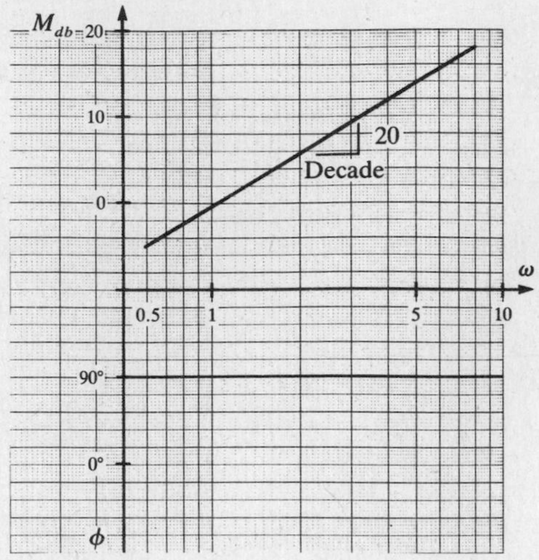

Fig. 8-3

(a) Asymptotic Bode plots are piecewise-linear approximations. The asymptotic plot of M_{db} for a simple lag network has value zero out to the single break frequency $\omega = 1/\tau$ and then *decreases* at 20 db per decade. The asymptotic plot of ϕ has the value zero out to $\omega = 0.1/\tau$, decreases linearly to $-90°$ at $\omega = 10/\tau$, and then is constant at $-90°$. Both asymptotic plots are shown dashed in Fig. 8-2.

(b) The asymptotic Bode plot for a simple lead network is the mirror image of that for a simple lag network. Thus, the asymptotic plot of M_{db} in Example 8.2 is zero out to $\omega = 1/\tau$ and then *increases* at 20 db per decade; the plot of ϕ is zero out to $\omega = 0.1/\tau$, increases to 90° at $\omega = 10/\tau$, and then remains constant.

8.3 LOW-FREQUENCY EFFECT OF BYPASS AND COUPLING CAPACITORS

As the frequency of the input signal to an amplifier decreases below the midfrequency range, the voltage (or current) gain ratio decreases in magnitude. The *low-frequency cutoff point* ω_L is the frequency at which the gain ratio equals $1/\sqrt{2}\,(= 0.707)$ times its midfrequency value [Fig. 8-1(*a*)], or at which M_{db} has decreased by exactly 3 db from its midfrequency value. The range of frequencies below ω_L is called the *low-frequency region*. Low-frequency amplifier performance (attenuation, really) is a consequence of the use of bypass and coupling capacitors to fashion the dc bias characteristics. When viewed from the low-frequency region, such amplifier response is analogous to that of a *high-pass filter* (signals for which $\omega < \omega_L$ are appreciably attenuated, whereas higher-frequency signals with $\omega \geq \omega_L$ are unattenuated).

Example 8.5 For the amplifier of Fig. 3-6, assume that $C_C \to \infty$ but that the bypass capacitor C_E cannot be neglected. Also, let $h_{re} = h_{oe} \approx 0$ and $R_i = 0$. Find an expression that is valid for small signals and that gives (*a*) the voltage-gain ratio $A_v(s)$ at any frequency; then find (*b*) the voltage-gain ratio at low frequencies, (*c*) the voltage-gain ratio at higher frequencies, and (*d*) the low-frequency cutoff point. (*e*) Sketch the asymptotic Bode plot for the amplifier (amplitude ratio only).

(a) The small-signal low-frequency equivalent circuit (with the approximation implemented) is displayed in Fig. 8-4. In the Laplace domain, we have

$$Z_E = R_E \parallel \frac{1}{sC_E} = \frac{(R_E)(1/sC_E)}{R_E + 1/sC_E} = \frac{R_E}{sR_EC_E + 1} \tag{8.11}$$

We next note that

$$I_E = I_b + h_{fe}I_b = (h_{fe} + 1)I_b \tag{8.12}$$

Then KVL and (*8.12*) yield

$$V_i = h_{ie}I_b + Z_EI_E = [h_{ie} + (h_{fe} + 1)Z_E]I_b \tag{8.13}$$

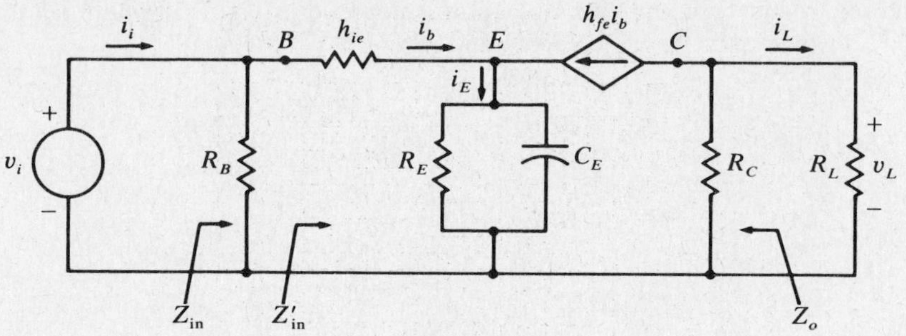

Fig. 8-4

But, by Ohm's law,

$$V_L = -(h_{fe}I_b)(R_C \| R_L) = -\frac{h_{fe}R_CR_L}{R_C+R_L}I_b \tag{8.14}$$

Solving (*8.13*) for I_b, substituting the result into (*8.14*), using (*8.11*), and rearranging give the desired voltage-gain ratio:

$$A_v(s) = \frac{V_L}{V_i} = -\frac{h_{fe}R_CR_L}{R_C+R_L}\,\frac{sR_EC_E+1}{sR_EC_Eh_{ie}+h_{ie}+(h_{fe}+1)R_E} \tag{8.15}$$

(*b*) The low-frequency voltage-gain ratio is obtained by letting $s \to 0$ in (*8.15*):

$$A_v(0) = \lim_{s\to 0}\frac{V_L}{V_i} = \frac{-h_{fe}R_CR_L}{(R_C+R_L)[h_{ie}+(h_{fe}+1)R_E]} \tag{8.16}$$

Comparison of (*8.16*) with (*1*) of Problem 6.7 (but with $h_{oe}=0$) shows that inclusion of the bypass capacitor in the analysis can significantly change the expression one obtains for the voltage-gain ratio.

(*c*) The higher-frequency (midfrequency) voltage-gain ratio is obtained by letting $s \to \infty$ in (*8.15*):

$$A_v(\infty) = \lim_{s\to\infty}\frac{V_L}{V_i} = \lim_{s\to\infty}\left\{-\frac{h_{fe}R_CR_L}{R_C+R_L}\,\frac{R_EC_E+1/s}{R_EC_Eh_{ie}+[h_{ie}+(h_{fe}+1)R_E]/s}\right\} = \frac{-h_{fe}R_CR_L}{h_{ie}(R_C+R_L)} \tag{8.17}$$

(*d*) Equation (*8.15*) can be rearranged to give

$$A_v(s) = \frac{-h_{fe}R_CR_L}{(R_C+R_L)[h_{ie}+(h_{fe}+1)R_E]}\;\frac{sR_EC_E+1}{s\dfrac{R_EC_Eh_{ie}}{h_{ie}+(h_{fe}+1)R_E}+1} \tag{8.18}$$

which clearly is of the form

$$A_v(s) = k_v\frac{\tau_1 s+1}{\tau_2 s+1}$$

Thus, we may use (*8.18*) to write

$$\omega_1 = \frac{1}{\tau_1} = \frac{1}{C_ER_E} \tag{8.19}$$

and

$$\omega_2 = \frac{1}{\tau_2} = \frac{h_{ie}+(h_{fe}+1)R_E}{R_EC_Eh_{ie}} \tag{8.20}$$

Typically, $h_{fe} \gg 1$ and $h_{fe}R_E \gg h_{ie}$, so a reasonable approximation of ω_2 is

$$\omega_2 \approx \frac{1}{C_Eh_{ie}/h_{fe}} \tag{8.21}$$

Since h_{ie}/h_{fe} is typically an order of magnitude smaller than R_E, ω_2 is an order of magnitude greater than ω_1, and $\omega_L = \omega_2$.

(*e*) The low- and midfrequency asymptotic Bode plot is depicted in Fig. 8-5, where ω_1 and ω_2 are given by (*8.19*) and (*8.21*), respectively. From (*8.16*) and (*8.17*),

$$M_{dbL} = 20 \log \frac{h_{fe} R_C R_L}{(R_C + R_L)[h_{ie} + (h_{fe} + 1)R_E]} \tag{8.22}$$

and

$$M_{dbM} = 20 \log \frac{h_{fe} R_C R_L}{h_{ie}(R_C + R_L)} \tag{8.23}$$

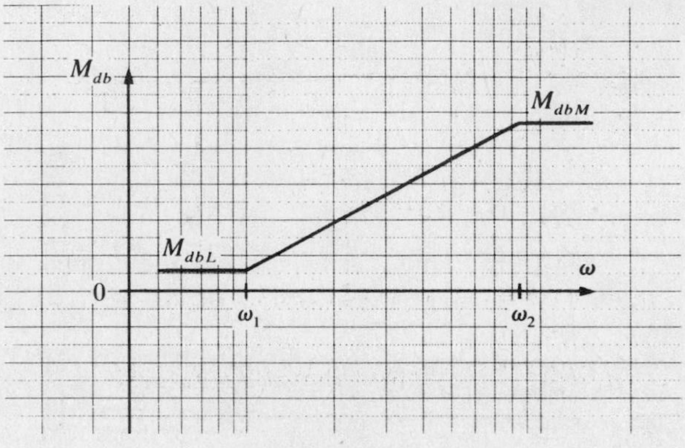

Fig. 8-5

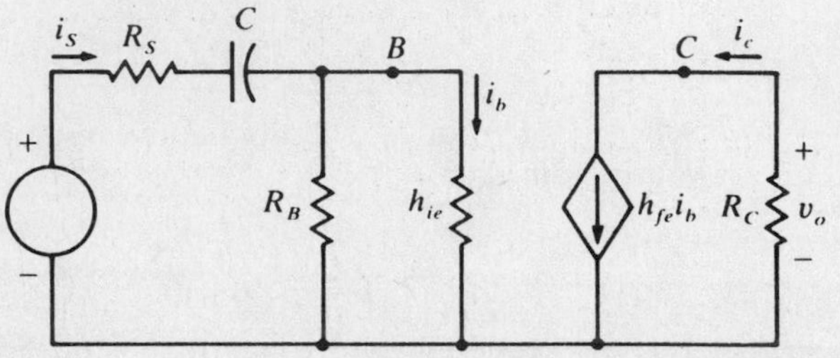

Fig. 8-6

Example 8.6 In the circuit of Fig. 3-15, battery V_S is replaced with a sinusoidal source v_S. The impedance of the coupling capacitor is not negligibly small. (*a*) Find an expression for the voltage-gain ratio $M = |A_v(j\omega)| = |v_o/v_S|$. (*b*) Determine the midfrequency gain of this amplifier. (*c*) Determine the low-frequency cutoff point ω_L, and sketch an asymptotic Bode plot.

(*a*) The small-signal low-frequency equivalent circuit is shown in Fig. 8-6. By Ohm's law,

$$I_S = \frac{V_S}{R_S + h_{ie} \| R_B + 1/sC} \tag{8.24}$$

Then current division gives

$$I_b = \frac{R_B}{R_B + h_{ie}} I_S = \frac{R_B V_S}{(R_B + h_{ie})(R_S + h_{ie} \| R_B + 1/sC)} \tag{8.25}$$

But Ohm's law requires that

$$V_o = -h_{fe} R_C I_b \tag{8.26}$$

Substituting (*8.25*) into (*8.26*) and rearranging give

$$A(s) = \frac{V_o}{V_S} = \frac{-h_{fe} R_C R_B Cs}{(R_B + h_{ie})[1 + sC(R_S + h_{ie} \| R_B)]} \tag{8.27}$$

Now, with $s = j\omega$ in (*8.27*), its magnitude is

$$M = |A(j\omega)| = \frac{h_{fe} R_C R_B C\omega}{(R_B + h_{ie})\sqrt{1 + (\omega C)^2 (R_S + h_{ie} \| R_B)^2}} \tag{8.28}$$

(*b*) The midfrequency gain follows from letting $s = j\omega \to \infty$ in (*8.27*). We may do so because reactances associated with inherent capacitances have been assumed infinitely large (neglected) in the equivalent circuit. We have, then,

$$A_{\text{mid}} = \frac{-h_{fe} R_C R_B}{(R_B + h_{ie})(R_S + h_{ie} \| R_B)} \tag{8.29}$$

(*c*) From (*8.27*),

$$\omega_L = 1/\tau = \frac{1}{C(R_S + h_{ie} \| R_B)} = \frac{R_B + h_{ie}}{C[R_S(h_{ie} + R_B) + h_{ie}R_B]} \tag{8.30}$$

The asymptotic Bode plot is sketched in Fig. 8-7.

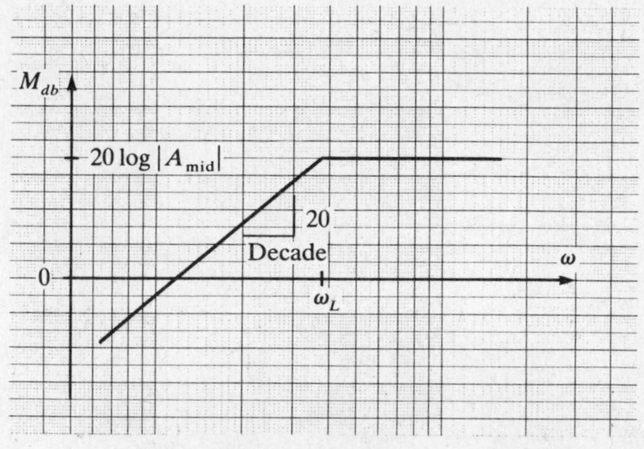

Fig. 8-7

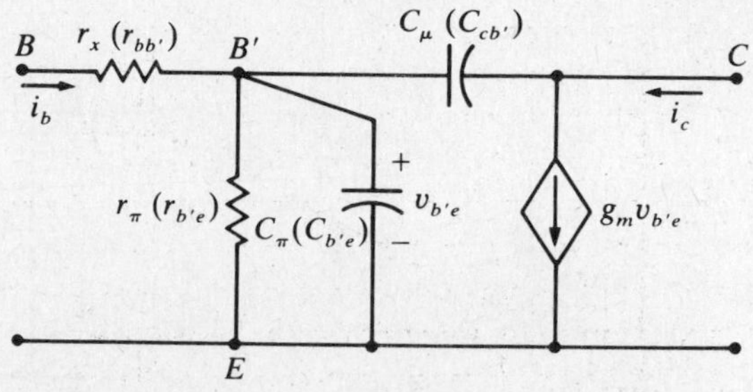

Fig. 8-8 Hybrid-π model for the BJT

8.4 HIGH-FREQUENCY HYBRID-π BJT MODEL

Because of capacitance that is inherent within the transistor, amplifier current- and voltage-gain ratios decrease in magnitude as the frequency of the input signal increases beyond the midfrequency range. The *high-frequency cutoff point* ω_H is the frequency at which the gain ratio equals $1/\sqrt{2}$ times its midfrequency value [see Fig. 8-1(*a*)], or at which M_{db} has decreased by 3 db from its midfrequency value. The range of frequencies above ω_H is called the *high-frequency region*. Like ω_L, ω_H is a break frequency.

The most useful high-frequency model for the BJT is called the *hybrid-π* equivalent circuit (see Fig. 8-8). In this model, the reverse voltage ratio h_{re} and output admittance h_{oe} are assumed negligible. The *base ohmic resistance* $r_{bb'}$, assumed to be located between the base terminal B and the base junction B', has a constant value (typically 10 to 50 Ω) that depends directly on the base width. The *base-emitter-junction resistance* $r_{b'e}$ is usually much larger than $r_{bb'}$ and can be calculated as

$$r_{b'e} = \frac{V_T(\beta + 1)}{I_{EQ}} = \frac{V_T\beta}{I_{CQ}} \tag{8.31}$$

(see Problem 6.8). Capacitance C_μ is the depletion capacitance (see Section 2.3) associated with the reverse-biased collector-base junction; its value is a function of V_{BCQ}. Capacitance C_π ($\gg C_\mu$) is the diffusion capacitance associated with the forward-biased base-emitter junction; its value is a function of I_{EQ}.

Example 8.7 Apply the hybrid-π model of Fig. 8-8 to the amplifier of Fig. 3-6 to find an expression for its voltage-gain ratio $A_v(s)$ valid at high frequencies. Assume $R_i = 0$.

The high-frequency hybrid-π, small-signal equivalent circuit is drawn in Fig. 8-9(*a*). To simplify the analysis, a Thévenin equivalent circuit may be found for the network to the left of terminal pair B',E, with

$$V_{Th} = \frac{r_\pi}{r_\pi + r_x} V_S \tag{8.32}$$

and

$$R_{Th} = r_\pi \| r_x = \frac{r_\pi r_x}{r_\pi + r_x} \tag{8.33}$$

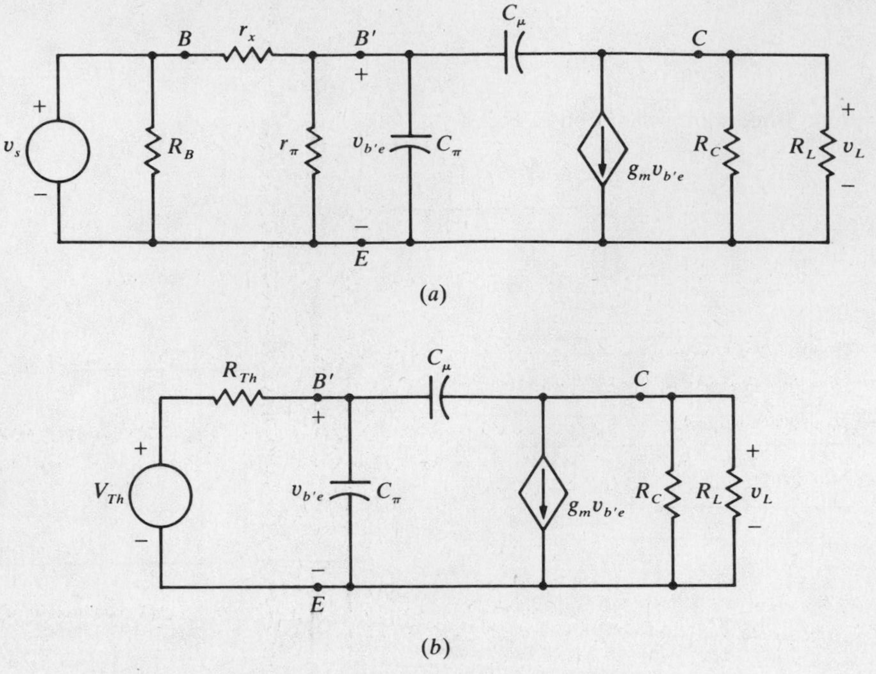

Fig. 8-9

Figure 8-9(*b*) shows the circuit with the Thévenin equivalent in position. Using $v_{b'e}$ and v_L as node voltages and working in the Laplace domain, we may write the following two equations:

$$\frac{V_{b'e} - V_{Th}}{R_{Th}} + \frac{V_{b'e}}{1/sC_\pi} + \frac{V_{b'e} - V_L}{1/sC_\mu} = 0 \tag{8.34}$$

$$\frac{V_L}{R_C \parallel R_L} + g_m V_{b'e} + \frac{V_L - V_{b'e}}{1/sC_\mu} = 0 \tag{8.35}$$

The latter equation can be solved for $V_{b'e}$, then substituted into (*8.34*), and the result rearranged to give the voltage ratio V_{Th}/V_L:

$$\frac{V_{Th}}{V_L} = \frac{s^2 C_\mu C_\pi R_{Th}(R_C \parallel R_L) + s[(1 - g_m)C_\mu(R_C \parallel R_L)] + 1}{(R_C \parallel R_L)(sC_\mu - g_m)} \tag{8.36}$$

For typical values, the coefficient of s^2 on the right side of (*8.36*) is several orders of magnitude smaller than the other terms; by approximating this coefficient as zero (i.e., neglecting the s^2 term), we neglect a breakpoint at a frequency much greater than ω_H. Doing so and using (*8.32*), we obtain the desired high-frequency voltage-gain ratio:

$$A_v(s) = \frac{V_L}{V_S} = \frac{r_\pi}{r_\pi + r_x} \frac{R_C \parallel R_L(sC_\mu - g_m)}{s(1 - g_m)C_\mu(R_C \parallel R_L) + 1} \tag{8.37}$$

8.5 HIGH-FREQUENCY FET MODELS

The small-signal high-frequency model for the FET is an extension of the midfrequency model of Fig. 7-1. Three capacitors are added: C_{gs} between gate and source, C_{gd} between gate and drain, and C_{ds} between drain and source. They are all of the same order of magnitude—typically 1 to 10 pF. Figure 8-10 shows the small-signal high-frequency model based on the current-source model of Fig. 7-1(*a*). Another model, based on the voltage-source model of Fig. 7-1(*b*), can also be drawn.

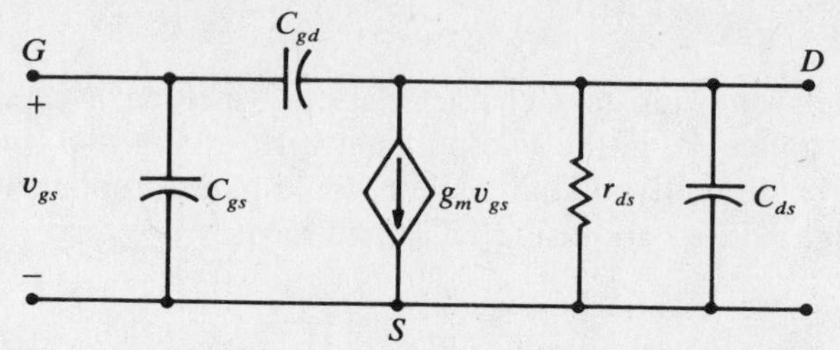

Fig. 8-10 High-frequency small-signal current-source FET model

Example 8.8 For the JFET amplifier of Fig. 4-3(b), (a) find an expression for the high-frequency voltage-gain ratio $A_v(s)$ and (b) determine the high-frequency cutoff point.

(a) The high-frequency small-signal equivalent circuit is displayed in Fig. 8-11, which incorporates Fig. 8-10. We first find a Thévenin equivalent for the network to the left of terminal pair a,a'. Noting that $v_{gs} = v_i$, we see that the open-circuit voltage is given by

$$V_{Th} = V_i - \frac{g_m}{sC_{gd}} V_i = \frac{sC_{gd} - g_m}{sC_{gd}} V_i \tag{8.38}$$

If V_i is deactivated, $V_i = V_{gs} = 0$ and the dependent current source is zero (open-circuited). A driving-point source connected to a,a' sees only

$$Z_{Th} = \frac{V_{dp}}{I_{dp}} = \frac{1}{sC_{gd}} \tag{8.39}$$

Now, with the Thévenin equivalent in place, voltage division leads to

$$V_L = \frac{Z_{eq}}{Z_{eq} + Z_{Th}} V_{Th} = \frac{1}{1 + Z_{Th}/Z_{eq}} \frac{sC_{gd} - g_m}{sC_{gd}} V_i \tag{8.40}$$

where
$$\frac{1}{Z_{eq}} = Y_{eq} = sC_{ds} + \frac{1}{r_{ds}} + \frac{1}{R_D} + \frac{1}{R_L} = sC_{ds} + g_{ds} + G_D + G_L \tag{8.41}$$

Rearranging (8.40) and using (8.41), we get

$$A_v(s) = \frac{V_L}{V_i} = \frac{sC_{gd} - g_m}{s(C_{ds} + C_{gd}) + g_{ds} + G_D + G_L} \tag{8.42}$$

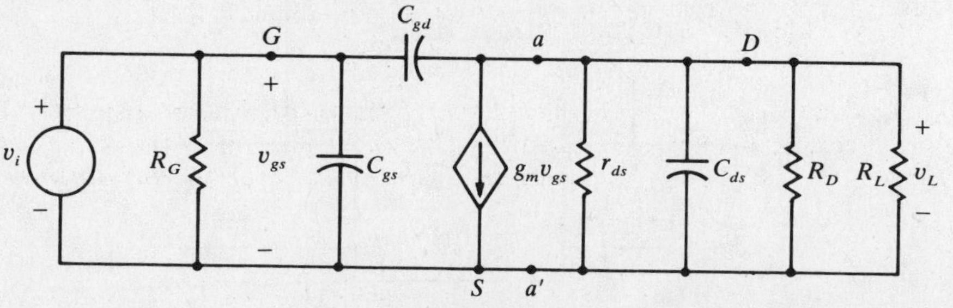

Fig. 8-11

(b) From (8.42), the high-frequency cutoff point is obviously

$$\omega_H = \frac{g_{ds} + G_D + G_L}{C_{ds} + C_{gd}} \tag{8.43}$$

Note that the high-frequency cutoff point is independent of C_{gs} as long as the source internal impedance is negligible. (See Problem 8.38.)

8.6 MILLER CAPACITANCE

High-frequency models of transistors characteristically include a capacitor path from input to output, modeled as admittance Y_F in the two-port network of Fig. 8-12(*a*). This added conduction path generally increases the difficulty of analysis; we would like to replace it with an equivalent shunt element. Referring to Fig. 8-12(*a*) and using KCL, we have

$$Y_{in} = \frac{I_i}{V_1} = \frac{I_1 + I_F}{V_1} \tag{8.44}$$

But

$$I_F = (V_1 - V_2)Y_F \tag{8.45}$$

Substitution of (*8.45*) into (*8.44*) gives

$$Y_{in} = \frac{I_1}{V_1} + \frac{(V_1 - V_2)Y_F}{V_1} = Y_1 + (1 - K_F)\, Y_F \tag{8.46}$$

where $K_F = V_2 / V_I$ is obviously the forward voltage-gain ratio of the amplifier.

In a similar manner,

$$Y_o = \frac{-I_o}{V_2} = \frac{-(I_2 + I_F)}{V_2} \tag{8.47}$$

and the use of (*8.45*) in (*8.47*) gives us

$$Y_o = -\left(\frac{I_2}{V_2} + \frac{V_1 - V_2}{V_2}\, Y_F\right) = -\left[-Y_2 + \left(K_R - 1\right)Y_F\right] = Y_2 + \left(1 - K_R\right)Y_F \tag{8.48}$$

where $K_R = V_1 / V_2$ is the reverse voltage-gain ratio of the amplifier.

Equations (*8.46*) and (*8.48*) suggest that the feedback admittance Y_F can be replaced with two shunt-connected admittances as shown in Fig. 8-12(*b*). When this two-port network is used to model an amplifier, the voltage gain K_F usually turns out to have a large negative value, so that $(1 - K_F)\, Y_F \approx |K_F|\, Y_F$. Hence, a small feedback capacitance appears as a large shunt capacitance (called the *Miller capacitance*). On the other hand, K_R is typically small so that $(1 - K_R)\, Y_F \approx Y_F$.

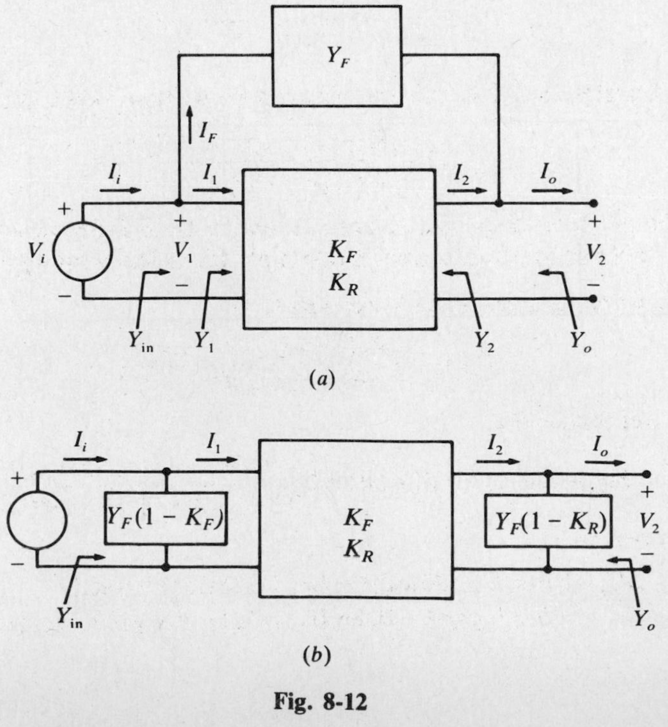

Fig. 8-12

Solved Problems

8.1 Calculate and tabulate the difference between the asymptotic and exact plots of Fig. 8-2, for use in correcting asymptotic plots to exact plots.

The difference ε may be found by subtraction. For the M_{db} plot,

For $0 \le \omega \le \dfrac{1}{\tau}$:
$$\varepsilon_{Mdb} = 0 - \{-10 \log [1 + (\omega\tau)^2]\} = 10 \log [1 + (\omega\tau)^2] \tag{1}$$

For $\omega > \dfrac{1}{\tau}$:
$$\varepsilon_{Mdb} = -10 \log (\omega\tau)^2 - (-10 \log [1 + (\omega\tau)^2]) = 10 \log [1 + 1/(\omega\tau)^2] \tag{2}$$

and for the ϕ plot,

For $0 \le \omega \le \dfrac{0.1}{\tau}$:
$$\varepsilon_\phi = 0 - (-\tan^{-1} \omega\tau) = \tan^{-1} \omega\tau \tag{3}$$

For $\dfrac{0.1}{\tau} < \omega < \dfrac{10}{\tau}$:
$$\varepsilon_\phi = -45° \log 10\omega\tau + \tan^{-1} \omega\tau \tag{4}$$

For $\omega \ge \dfrac{10}{\tau}$:
$$\varepsilon_\phi = -90° - (-\tan^{-1} \omega\tau) = \tan^{-1} \omega\tau - 90° \tag{5}$$

Application of (*1*) to (*5*) yields Table 8-1.

Table 8-1
Bode-Plot Corrections

ω	ε_{Mdb}	ε_ϕ
$0.1/\tau$	0.04	5.7°
$0.5/\tau$	1	−4.9°
$0.76/\tau$	2	−2.4°
$1/\tau$	3	0°
$1.32/\tau$	2	2.4°
$2/\tau$	1	4.9°
$10/\tau$	0.04	−5.7°

8.2 The s-domain transfer function for a system can be written in the form

$$T(s) = \frac{K_b(\tau_{z1}s + 1)(\tau_{z2}s + 1)\cdots}{s^n(\tau_{p1}s + 1)(\tau_{p2}s + 1)\cdots} \tag{1}$$

where n may be positive, negative, or zero. Show that the Bode plot (for M_{db} only) may be generated as a composite of individual Bode plots for three basic types of terms.

The frequency transfer function corresponding to (*1*) is

$$T(j\omega) = \frac{K_b(1 + j\omega\tau_{z1})(1 + j\omega\tau_{z2})\cdots}{(j\omega)^n(1 + j\omega\tau_{p1})(1 + j\omega\tau_{p2})\cdots} \tag{2}$$

From definition 3 of Section 8.2,

$$M_{db} = 20 \log |T(j\omega)| = 20 \log \left[\frac{K_b|1 + j\omega\tau_{z1}|\,|1 + j\omega\tau_{z2}|\cdots}{|(j\omega)^n|\,|1 + j\omega\tau_{p1}|\,|1 + j\omega\tau_{p2}|\cdots}\right] \tag{3}$$

which may be written as

$$M_{db} = 20 \log K_b + 20 \log |1 + j\omega\tau_{z1}| + 20 \log |1 + j\omega\tau_{z2}| + \cdots$$
$$- 20n \log |j\omega| - 20 \log |1 + j\omega\tau_{p1}| - 20 \log |1 + j\omega\tau_{p2}| - \cdots \tag{4}$$

It is apparent from (4) that the Bode plot of $T(j\omega)$ can be formed by point-by-point addition of the plots of three types of terms:

1. A frequency-invariant or gain-constant term K_b whose Bode plot is a horizontal line at $M_{db} = 20 \log K_b$.
2. Poles or zeroes of multiplicity n, $(j\omega)^{\pm n}$, whose amplitude ratio is $M_{db} = \pm 20n \log \omega$, where the plus sign corresponds to zeroes and the minus sign to poles of the transfer function. (See Example 8.3.)
3. First-order lead and lag factors, $(1 + j\omega\tau)^{\pm n}$, as discussed in Examples 8.1 and 8.2. They are usually approximated with asymptotic Bode plots; if greater accuracy is needed, the asymptotic plots are corrected using Table 8-1.

8.3 The circuit of Fig. 8-13(*a*) is driven by a sinusoidal source v_S. (*a*) Sketch the asymptotic Bode plot (M_{db} only) associated with the Laplace-domain transfer function $T(s) = V_o/V_S$. (*b*) Use Table 8-1 to correct asymptotic plot, so as to show the exact Bode plot.

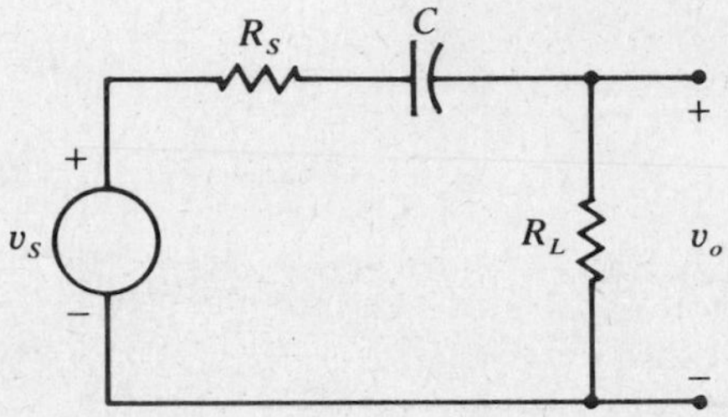

(*a*)

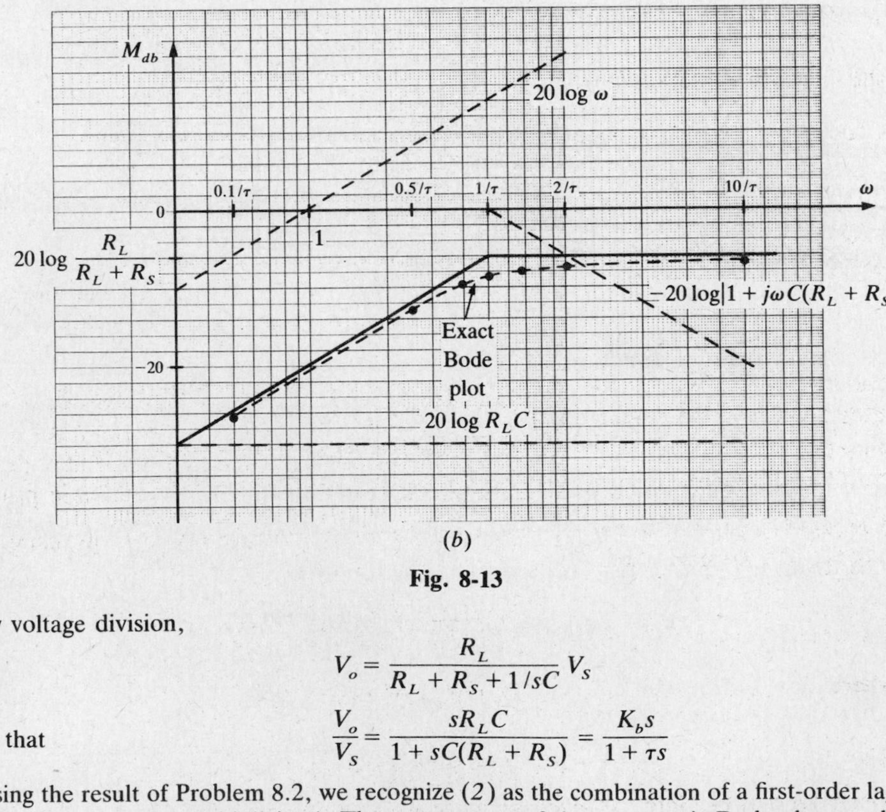

(*b*)

Fig. 8-13

(*a*) By voltage division,

$$V_o = \frac{R_L}{R_L + R_S + 1/sC} V_S \tag{1}$$

so that

$$\frac{V_o}{V_S} = \frac{sR_L C}{1 + sC(R_L + R_S)} = \frac{K_b s}{1 + \tau s} \tag{2}$$

Using the result of Problem 8.2, we recognize (2) as the combination of a first-order lag, a constant gain, and a zero of multiplicity 1. The components of the asymptotic Bode plot are shown dashed in Fig. 8-13(*b*), and the composite is solid. For purposes of illustration, it was assumed that $1/[C(R_L + R_S)] > 1$, which is true in most cases.

(*b*) The correction factors of Table 8-1 lead to the exact Bode plot as drawn in Fig. 8-14(*b*).

8.4 Sketch the asymptotic Bode plot (M_{db} only) associated with the output-to-input voltage ratio of the circuit in Fig. 8-14(a).

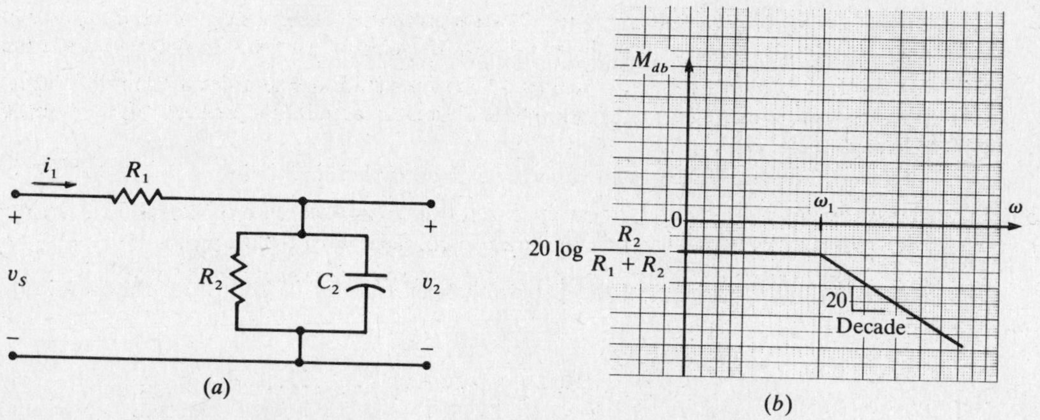

Fig. 8-14

By voltage division,

$$V_2 = \frac{R_2 \| (1/sC_2)}{R_1 + R_2 + 1/sC_2} V_s = \frac{\dfrac{R_2}{sR_2C_2 + 1}}{R_1 + \dfrac{R_2}{sR_2C_2+1}} V_s$$

and the Laplace-domain transfer function is

$$T(s) = \frac{V_2}{V_S} = \frac{R_2/(R_1+R_2)}{s\left(\dfrac{R_1R_2}{R_1+R_2}\right)C + 1} = \frac{K_b}{sR_{eq}C_2 + 1}$$

From $T(s)$, it is apparent that the circuit forms a low-pass filter with low-frequency gain $T(0) = R_2/(R_1 + R_2)$ and a corner frequency at $\omega_1 = 1/\tau_1 = 1/R_{eq}C_2$. Its Bode plot is sketched in Fig. 8-14(b).

8.5 For the amplifier of Fig. 3-6, assume that $C_C \to \infty$, $h_{re} = h_{oe} = 0$, and $R_i = 0$. The bypass capacitor C_E cannot be neglected. Find expressions for (a) the current-gain ratio $A_i(s)$, (b) the current-gain ratio at low frequencies, and (c) the midfrequency current-gain ratio. (d) Determine the low-frequency cutoff point, and sketch the asymptotic Bode plot (M_{db} only).

(a) The small-signal low-frequency equivalent circuit is given in Fig. 8-4. By current division for Laplace-domain quantities,

$$I_b = \frac{R_B}{R_B + h_{ie} + Z_E} I_i \tag{1}$$

where

$$Z_E = R_E \| \frac{1}{sC_E} = \frac{R_E}{sR_EC_E + 1} \tag{2}$$

Also

$$I_L = \frac{-R_C}{R_C + R_L} h_{fe} I_b \tag{3}$$

Substitution of (1) into (3) gives the current-gain ratio as

$$\frac{I_L}{I_i} = \frac{-R_C}{R_C + R_L} \frac{h_{fe}R_B}{R_B + h_{ie} + Z_E} \tag{4}$$

Using (2) in (4) and rearranging leads to the desired current-gain ratio:

$$A_i(s) = \frac{I_L}{I_i} = \frac{\dfrac{-h_{fe}R_CR_B}{(R_C+R_L)(R_E+R_B+h_{ie})}(sR_EC_E+1)}{s\dfrac{R_EC_E(R_B+h_{ie})}{R_E+R_B+h_{ie}}+1} \tag{5}$$

(b) The low-frequency current-gain ratio follows from letting $s \to 0$ in (5):

$$A_i(0) = \lim_{s\to 0}\frac{I_L}{I_i} = \frac{-h_{fe}R_CR_B}{(R_C+R_L)(R_E+R_B+h_{ie})} \tag{6}$$

(c) The midfrequency current-gain ratio is obtained by letting $s \to \infty$ in (5):

$$A_i(\infty) = \lim_{s\to\infty}\frac{I_L}{I_i} = \frac{-h_{fe}R_CR_B}{(R_C+R_L)(R_B+h_{ie})} \tag{7}$$

(d) Inspection of (5) shows that the Laplace-domain transfer function is of the form

$$A_i(s) = K_b\frac{\tau_1 s+1}{\tau_2 s+1}$$

where
$$\omega_1 = \frac{1}{\tau_1} = \frac{1}{R_EC_E} \quad \text{and} \quad \omega_2 = \frac{1}{\tau_2} = \frac{R_E+R_B+h_{ie}}{R_EC_E(R_B+h_{ie})} \tag{8}$$

With ω_1 and ω_2 as given by (8) and with

$$M_{dbL} = 20\log A_i(0) \quad \text{and} \quad M_{dbM} = 20\log A_i(\infty)$$

the Bode plot is identical to that of Fig. 8-5. Since $\omega_2 > \omega_1$, ω_2 is closer to the midfrequency region and thus is the low-frequency cutoff point.

8.6 In the amplifier of Fig. 3-6, $C_C \to \infty$, $R_i = 0$, $R_E = 1\ \text{k}\Omega$, $R_1 = 3.2\ \text{k}\Omega$, $R_2 = 17\ \text{k}\Omega$, $R_L = 10\ \text{k}\Omega$, and $h_{oe} = h_{re} = 0$. The transistors used are characterized by $75 \le h_{fe} \le 100$ and $300 \le h_{ie} \le 1000\ \Omega$. (a) By proper selection of R_C and C_E, design an amplifier with low-frequency cutoff $f_L \le 200$ Hz and high-frequency voltage gain $|A_v| \ge 50$. (b) For the finished design, determine the low-frequency voltage-gain ratio if h_{ie} and h_{fe} have median values.

(a) According to (8.17), the worst-case transistor parameters for high $A_v(\infty)$ are minimum h_{fe} and maximum h_{ie}. Using those parameter values allows us to determine a value for the parallel combination of R_C and R_L:

$$R_{eq} = R_C \| R_L \ge |A_v(\infty)|\frac{h_{ie}}{h_{fe}} = 50\,\frac{1000}{75} = 666.7\ \Omega$$

Then
$$R_C = \frac{R_{eq}R_L}{R_L - R_{eq}} \ge \frac{(666.7)(10{,}000)}{9333.3} = 714.3\ \Omega$$

Now, from (8.20), for $f_L \le 200$ Hz,

$$C_E \ge \frac{h_{ie}+(h_{fe}+1)R_E}{\omega_L R_E h_{ie}} = \frac{300+(101)(1000)}{2\pi(200)(1000)(300)} = 268.7\ \mu\text{F}$$

(b) By (8.16),

$$A_v(0) = -\frac{h_{fe}R_{eq}}{h_{ie}+(h_{fe}+1)R_E} = -\frac{\left(\dfrac{100+75}{2}\right)(666.7)}{\dfrac{300+1000}{2}+\left(\dfrac{100+75}{2}+1\right)(1000)} = -0.654$$

8.7 Let $C_1, C_E \to \infty$ in the capacitor-coupled amplifier of Fig. 6-19. Assume $h_{oe1} = h_{re1} = h_{oe2} = h_{re2} = 0$. Find an expression for the voltage-gain ratio $A_v(s)$.

The first-stage amplifier can be replaced with a Thévenin equivalent, and the second stage represented by its input impedance, as shown in Fig. 8-15. A'_v follows from voltage division and *(6.46)* if R_L, h_{fe}, and h_{ie} are replaced with R_{C1}, h_{fe1}, and h_{ie1}, respectively:

$$A'_v = \frac{R_{eq}}{R_{eq} + R_i}\,\frac{-h_{fe1}R_{C1}}{h_{ie1}} = -\frac{h_{fe1}R_{C1}R_{eq}}{h_{ie1}(R_{eq} + R_i)} \tag{1}$$

where

$$R_{eq} = h_{ie1} \parallel R_{B1} = h_{ie1} \parallel R_{11} \parallel R_{12} = \frac{h_{ie1}R_{11}R_{12}}{h_{ie1}(R_{11} + R_{12}) + R_{11}R_{12}} \tag{2}$$

Z_{o1} is given by *(6.50)* with h_{oe} replaced with R_{C1} (and with $h_{re1} = h_{oe1} = 0$):

$$Z_{o1} = R_{C1} \tag{3}$$

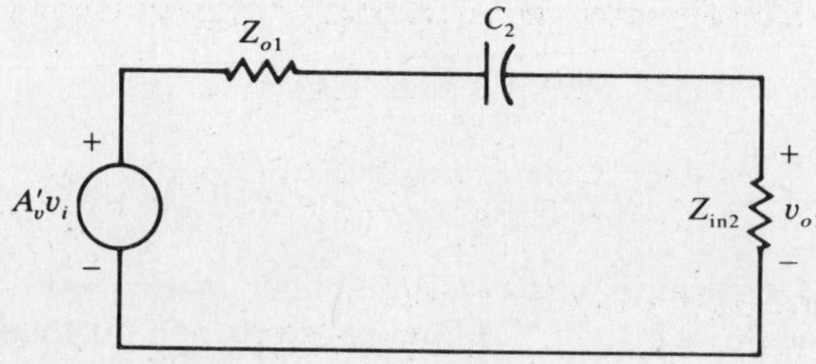

Fig. 8-15

The second-stage input impedance is given by *(6.47)* if h_{ie} is replaced with $h_{ie2} \parallel R_{B2} = h_{ie2} \parallel R_{21} \parallel R_{22}$:

$$Z_{in2} = \frac{h_{ie2}R_{B2}}{h_{ie2} + R_{B2}} = \frac{h_{ie2}R_{21}R_{22}}{h_{ie2}(R_{21} + R_{22}) + R_{21}R_{22}} \tag{4}$$

Now, from *(2)* of Problem 8.3,

$$\frac{V_{o1}}{A'_v V_i} = \frac{sZ_{in2}C_2}{sC_2(Z_{in2} + Z_{o1}) + 1} \tag{5}$$

and rearranging yields the first-stage gain as

$$A_{v1} = \frac{V_{o1}}{V_i} = A'_v\,\frac{sZ_{in2}C_2}{sC_2(Z_{in2} + Z_{o1}) + 1} \tag{6}$$

The second-stage gain follows directly from *(6.46)* if R_L is replaced with R_{C2}:

$$A_{v2} = -\frac{h_{fe2}R_{C2}}{h_{ie2}}$$

Consequently, the overall gain is

$$A_v = A_{v1}A_{v2} = -A'_v\,\frac{sZ_{in}C_2}{sC_2(Z_{in2} + Z_{o1}) + 1}\,\frac{h_{fe2}R_{C2}}{h_{ie2}} \tag{7}$$

Substituting *(1)* into *(7)* and simplifying yield the desired gain:

$$A_v(s) = \frac{h_{fe1}h_{fe2}R_{C1}R_{C2}R_{eq}}{h_{ie1}h_{ie2}(R_{eq} + R_i)}\,\frac{sZ_{in2}C_2}{sC_2(Z_{in2} + Z_{o1}) + 1} \tag{8}$$

8.8 In the cascaded amplifier of Problem 8.7 (Fig. 6-19 with C_1, $C_E \to \infty$), let $h_{ie1} = h_{ie2} = 1500\ \Omega$, $h_{fe1} = h_{fe2} = 40$, $C_2 = 1\ \mu\text{F}$, $R_i = 1\ \text{k}\Omega$, $R_{C1} = 10\ \text{k}\Omega$, $R_{C2} = 20\ \text{k}\Omega$, and $R_{B1} = R_{B2} = 5\ \text{k}\Omega$. Determine (*a*) the low-frequency gain, (*b*) the midfrequency gain, and (*c*) the low-frequency cutoff point.

(a) Letting $s \to 0$ in (8) of Problem 8.7 makes apparent the fact that the low-frequency gain $A_v(0) = 0$.

(b) The midfrequency gain is determined by letting $s \to \infty$ in (8) of Problem 8.7:

From (2), (3) and (4) of Problem 8.7,

$$A_v(\infty) = \lim_{s\to\infty} A_v(s) = \frac{h_{fe1}h_{fe2}R_{C1}R_{C2}R_{eq}}{h_{ie1}h_{ie2}(R_{eq} + R_i)} \frac{Z_{\text{in}2}}{Z_{\text{in}2} + Z_{o1}}$$

$$R_{eq} = \frac{h_{ie1}R_{B1}}{h_{ie1} + R_{B1}} = \frac{(1500)(5000)}{6500} = 1153.8\ \Omega$$

$$Z_{o1} = R_{C1} = 10\ \text{k}\Omega$$

and

$$Z_{\text{in}2} = \frac{h_{ie2}R_{B2}}{h_{ie2} + R_{B2}} = \frac{(1500)(5000)}{6500} = 1153.8\ \Omega$$

Then

$$A_v(\infty) = \frac{(40)(40)(10 \times 10^3)(20 \times 10^3)(1153.8)}{(1500)(1500)(2153.8)} \frac{1153.8}{1153.8 + 10 \times 10^3} = 7881.3$$

(c) The low-frequency cutoff point is computed from the lag term in (8) of Problem 8.7:

$$f_L = \frac{\omega_L}{2\pi} = \frac{1}{2\pi C_2(Z_{\text{in}2} + Z_{o1})} = \frac{1}{2\pi(1 \times 10^{-6})(1153.8 + 10 \times 10^3)} = 14.3\ \text{Hz}$$

8.9 The two coupling capacitors in the CB amplifier of Fig. 6-13 are identical and cannot be neglected. Assume $h_{rb} = h_{ob} = 0$. (a) Find an expression for the voltage-gain ratio V_L/V_S. (b) Find an expression for the midfrequency voltage-gain ratio.

(a) The small-signal low-frequency equivalent circuit is given in Fig. 8-16. Applying Ohm's law in the Laplace domain, we obtain

$$I_S = \frac{V_S}{1/sC_C + R_E h_{ib}/(R_E + h_{ib})} = \frac{sC_C V_S}{sC_C R_E h_{ib}/(R_E + h_{ib}) + 1}$$

Voltage division then gives

$$I_e = \frac{R_E}{h_{ib} + R_E} I_S = \frac{R_E}{h_{ib} + R_E} \frac{sC_C}{sC_C R_E h_{ib}/(R_E + h_{ib}) + 1} V_S \tag{1}$$

By current division at the output,

$$V_L = R_L I_L = -R_L \frac{R_C}{R_C + 1/sC_C + R_L} h_{fb} I_e = -\frac{sh_{fb}R_L R_C C_C I_e}{sC_C(R_L + R_C) + 1} \tag{2}$$

Substituting (1) into (2) and rearranging lead to the desired voltage-gain ratio:

$$A_v(s) = \frac{V_L}{V_S} = -\frac{R_E R_L R_C h_{fb} C_C^2 s^2}{(h_{ib} + R_E)[sC_C R_E h_{ib}/(R_E + h_{ib}) + 1][sC_C(R_L + R_C) + 1]} \tag{3}$$

(b) Letting $s \to \infty$ in (3) leads to the midfrequency gain:

$$A_v(\infty) = -\frac{R_L R_C h_{fb}}{h_{ib}(R_L + R_C)} \tag{4}$$

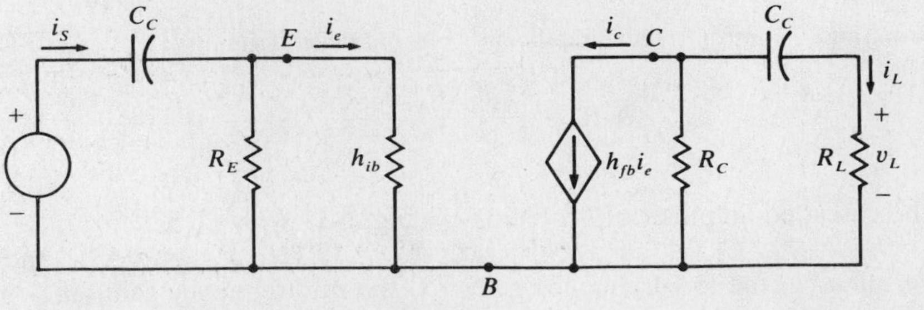

Fig. 8-16

8.10 The two coupling capacitors in the CB amplifier of Fig. 6-13 are identical. Also, $h_{rb} = h_{ob} = 0$. (*a*) Find an expression for the current-gain ratio $A_i(s)$ that is valid at any frequency. (*b*) Find an expression for the midfrequency current-gain ratio.

(*a*) The low-frequency equivalent circuit is displayed in Fig. 8-16. By current division,

$$I_e = \frac{R_E}{h_{ib} + R_E} I_S \tag{1}$$

and

$$I_L = -\frac{R_C}{R_C + 1/sC_C + R_L} h_{fb} I_e = -\frac{sh_{fb} R_L R_C C_C I_e}{sC_C(R_L + R_C) + 1} \tag{2}$$

Substituting (*1*) into (*2*) and dividing by I_S give the desired current-gain ratio:

$$A_i(s) = \frac{I_L}{I_i} = -\frac{sh_{fb} R_L R_C R_E C_C}{(h_{ib} + R_E)[sC_C(R_L + R_C) + 1]} \tag{3}$$

(*b*) The midfrequency current-gain ratio is found by letting $s \to \infty$ in (*3*):

$$A_i(\infty) = -\frac{h_{fb} R_L R_C R_E}{(h_{ib} + R_E)(R_L + R_C)} \tag{4}$$

8.11 On a common set of axes, sketch the asymptotic Bode plots (M_{db} only) for the voltage- and current-gain ratios of the CB amplifier of Fig. 6-13, and then correct them to exact plots. Assume that the coupling capacitors are identical and that, for typical values, $1 \ll C_C(R_E \| h_{ib}) \ll C_C(R_L + R_E)$.

The Laplace-domain transfer functions that serve as bases for Bode plots of the voltage- and current-gain ratios are, respectively, (*3*) of Problem 8.9 and (*3*) of Problem 8.10. Under the given assumptions inspection shows that the two transfer functions share a break frequency at $\omega = 1/[C_C(R_L + R_C)]$ and the voltage-gain transfer function has another at a higher frequency. Moreover, the voltage plot rises at 40 db per decade to its first break point, and the current plot at 20 db per decade. With

$$\omega_{1v} = \omega_{1i} = \omega_{Li} = \frac{1}{C_C(R_L + R_C)} \quad \text{and} \quad \omega_{2v} = \omega_{Lv} = \frac{R_E + h_{ib}}{C_C R_E h_{ib}}$$

the low-frequency asymptotic Bode plots of voltage and current gain are sketched in Fig. 8-17. The given assumption assures a separation of at least a decade between ω_{1v} and ω_{2v} and between $\omega = 1$ and ω_{1v}.

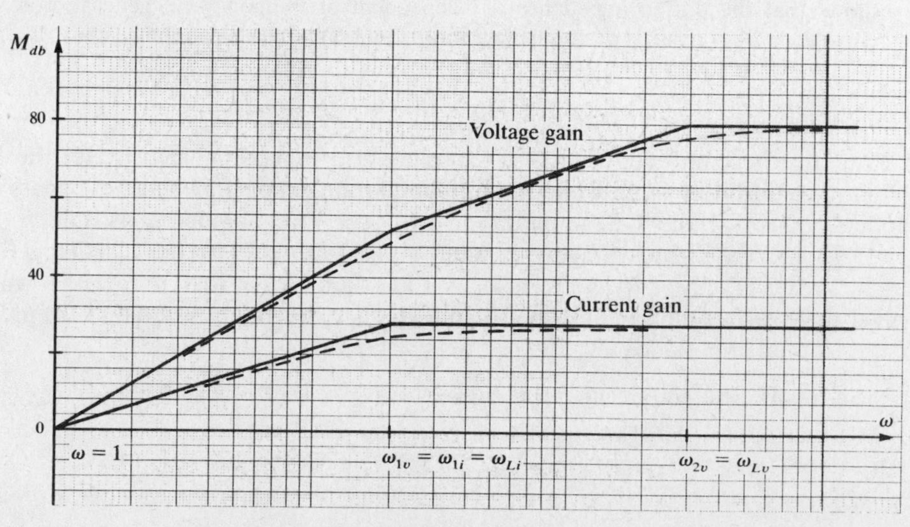

Fig. 8-17

Since the parameter values are not known, the sketches were made under the assumption that $K_b = 1$ in both plots. When values become known, the Bode plots must be shifted upward by

$$20 \log K_{bv} = 20 \log \frac{R_E R_L R_C h_{fb} C_C^2}{h_{ib} + R_E} \quad \text{for the voltage plot}$$

and

$$20 \log K_{bi} = 20 \log \frac{h_{fb} R_L R_C R_E C_C}{h_{ib} + R_E} \quad \text{for the current plot}$$

Correction of the asymptotic plot requires only the application of Table 8-1. The exact plots are shown dashed.

8.12 For the CE amplifier of Fig. 3-6, determine (*a*) Z'_{in}, (*b*) Z_{in}, and (*c*) Z_o if $C_C \to \infty$ but C_E cannot be neglected.

(*a*) The small-signal low-frequency equivalent circuit is given in Fig. 8-4. Using (*8.11*) and (*8.13*), we have

$$Z'_{\text{in}} = \frac{V_i}{I_b} = h_{ie} + (h_{fe} + 1)Z_E = \frac{sh_{ie}R_E C_E + h_{ie} + (h_{fe} + 1)R_E}{sR_E C_E + 1} \tag{1}$$

(*b*)

$$Z_{\text{in}} = R_B \,\|\, Z'_{\text{in}} = \frac{R_B Z'_{\text{in}}}{R_B + Z'_{\text{in}}} \tag{2}$$

Substituting (*1*) into (*2*) and rearranging give

$$Z_{\text{in}} = \frac{R_B[sh_{ie}R_E C_E + h_{ie} + (h_{fe} + 1)R_E]}{sR_E C_E(R_B + h_{ie}) + R_B + h_{ie} + (h_{fe} + 1)R_E} \tag{3}$$

(*c*) With voltage source v_i deactivated (shorted), KVL requires that

$$I_b = \frac{-h_{fe} I_b (Z_E \,\|\, h_{ie})}{h_{ie}}$$

so that

$$\left[1 + \frac{h_{fe} Z_E h_{ie}}{h_{ie}(Z_E + h_{ie})}\right] I_b = 0 \tag{4}$$

Since (*4*) can be satisfied in general only by $I_b = 0$, the output impedance is simply

$$Z_o = R_C \tag{5}$$

In this particular case, (*3*) shows that the input impedance is frequency-dependent, while (*5*) shows that the output impedance is independent of frequency. In general, however, the output impedance does depend on frequency, through a finite-valued coupling capacitor C_C. (See Problem 8.13.)

8.13 To examine the combined effects of coupling and bypass capacitors, let the input coupling capacitor be infinitely large while the output coupling capacitor and the bypass capacitor have practical values in the CE amplifier of Fig. 3-6. For simplicity, assume $h_{re} = h_{oe} = 0$ and $R_B \gg Z'_{\text{in}}$. (*a*) Find the voltage-gain ratio $A_v(s) = v_L / v_i$. (*b*) If $C_E = 200\ \mu\text{F}$, $C_C = 10\ \mu\text{F}$, $R_i = R_E = 100\ \Omega$, $R_C = R_L = 2\ \text{k}\Omega$, $h_{ie} = 1\ \text{k}\Omega$, and $h_{fe} = 100$, determine what parameters control the low-frequency cutoff point and whether it is below 100 Hz. (*c*) Find an expression for the output impedance Z_o.

(*a*) The small-signal equivalent circuit is given in Fig. 8-18. We first define

$$Z_E = R_E \,\|\, \frac{1}{sC_E} = \frac{R_E}{sR_E C_E + 1} \tag{1}$$

Then, by KCL,

$$I_e = I_b + h_{fe} I_b = (h_{fe} + 1) I_b \tag{2}$$

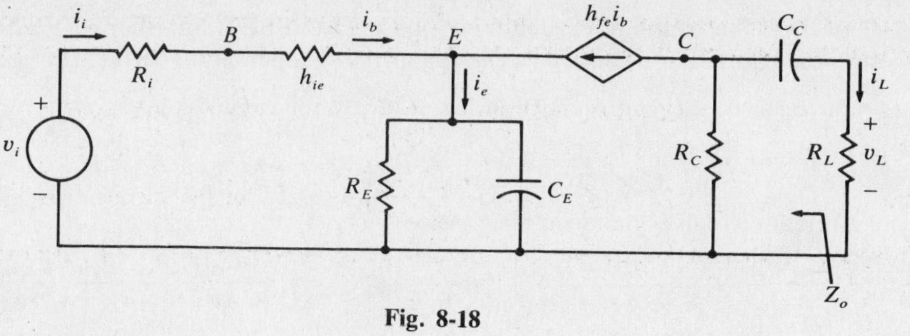

Fig. 8-18

KVL around the input mesh requires that

$$V_i = (R_i + h_{ie})I_b + Z_E I_e \tag{3}$$

Substituting (2) into (3) and solving for I_b yields

$$I_b = \frac{V_i}{R_i + h_{ie} + (h_{fe} + 1)Z_E} \tag{4}$$

Current division at the collector node gives

$$I_L = -\frac{R_C}{R_C + R_L + 1/sC_C} h_{fe} I_b \tag{5}$$

and Ohm's law and (5) yield

$$V_L = R_L I_L = -\frac{R_L R_C}{R_C + R_L + 1/sC_C} h_{fe} I_b \tag{6}$$

Substituting (4) and (1) into (6) and rearranging now lead to the desired voltage-gain ratio:

$$A_v(s) = \frac{V_L}{V_i} = -\frac{\dfrac{h_{fe} R_L R_C C_C}{(h_{fe}+1)R_E + h_{ie} + R_i} s(sR_E C_E + 1)}{[sC_C(R_C + R_L) + 1]\left[s \dfrac{C_E R_E (R_i + h_{ie})}{(h_{fe}+1)R_E + R_i + h_{ie}} + 1\right]} \tag{7}$$

(b) The Laplace-domain transfer function (7) is of the form

$$T(s) = \frac{-K_b s(\tau_2 s + 1)}{(\tau_1 s + 1)(\tau_3 s + 1)}$$

where
$$\omega_1 = \frac{1}{\tau_1} = \frac{1}{C_C(R_C + R_L)} = \frac{1}{(10 \times 10^{-6})(4000)} = 25 \text{ rad/s}$$

$$\omega_2 = \frac{1}{\tau_2} = \frac{1}{R_E C_E} = \frac{1}{(100)(200 \times 10^{-6})} = 50 \text{ rad/s}$$

$$\omega_3 = \frac{1}{\tau_3} = \frac{(h_{fe}+1)R_E + R_i + h_{ie}}{C_E R_E (R_i + h_{ie})} = \frac{(101)(100) + 100 + 1000}{(200 \times 10^{-6})(100)(1100)} = 509.1 \text{ rad/s}$$

Since there is at least a decade of frequency (in which the gain can attenuate from its midfrequency value) between ω_3 and the other (lower) break frequencies, ω_3 must be the low-frequency cutoff ω_L. Then

$$f_L = \frac{\omega_3}{2\pi} = \frac{509.1}{2\pi} = 81.02 \text{ Hz} < 100 \text{ Hz}$$

(c) As in Problem 8.12, $I_b = 0$ if v_i is deactivated; a driving-point source replacing R_L would then see a frequency-dependent output impedance given by

$$Z_o = Z_{dp} = R_C + \frac{1}{sC_C} \tag{8}$$

8.14 Assume that the coupling capacitors in the CS MOSFET amplifier of Fig. 4-18 are identical. Determine the voltage-gain ratio (*a*) for any frequency and (*b*) for midfrequency operation.

(*a*) The equivalent circuit is drawn in Fig. 8-19. By voltage division,

$$V_{gs} = \frac{R_G}{R_G + 1/sC_C} V_i = \frac{sR_G C_C}{sR_G C_C + 1} V_1 \qquad \text{where} \qquad R_G = R_1 \| R_2 = \frac{R_1 R_2}{R_1 + R_2} \tag{1}$$

Current division at the drain node yields

$$I_L = -\frac{R_D \| r_{ds}}{R_D \| r_{ds} + 1/sC_C + R_L} g_m V_{gs} = -\frac{sC_C[R_D r_{ds}/(R_D + r_{ds})]g_m V_{gs}}{sC_C[R_D r_{ds}/(R_D + r_{ds}) + R_L] + 1} \tag{2}$$

from which

$$V_o = R_L I_L = -\frac{sg_m R_D R_L r_{ds} C_C/(R_D + r_{ds})}{sC_C[R_D r_{ds}/(R_D + r_{ds}) + R_L] + 1} V_{gs} \tag{3}$$

Substitution of (*1*) into (*3*) and rearrangement then give

$$A_v(s) = \frac{V_o}{V_i} = -\frac{s^2 g_m R_G R_D R_L r_{ds} C_C^2/(R_D + r_{ds})}{\left[sC_C\left(\dfrac{R_D r_{ds}}{R_D + r_{ds}} + R_L\right) + 1\right][sC_C R_G + 1]} \tag{4}$$

(*b*) Since high-frequency capacitances have not been modeled, the midfrequency gain follows from letting $s \to \infty$ in (*4*):

$$A_{\text{mid}} = A_v(\infty) = -\frac{g_m R_D R_L r_{ds}}{R_D r_{ds} + R_L(R_D + r_{ds})}$$

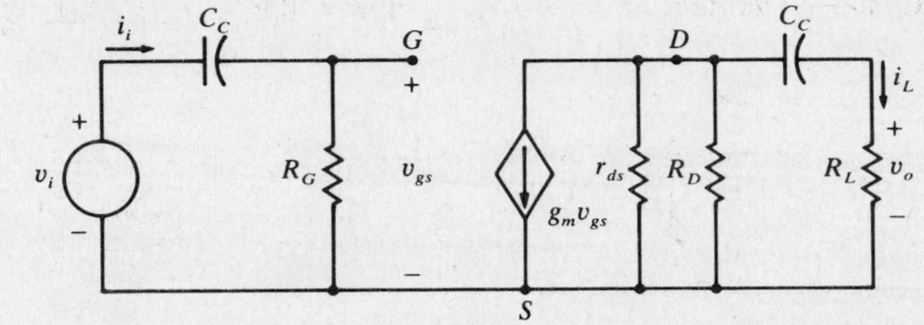

Fig. 8-19

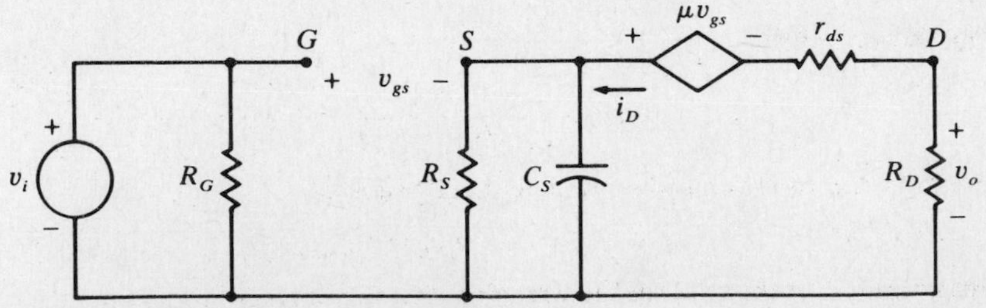

Fig. 8-20

8.15 For the CS JFET amplifier of Fig. 7-2, (*a*) find an expression for the voltage-gain ratio $A_v(s)$ and (*b*) determine the low-frequency cutoff point.

(*a*) The low-frequency equivalent circuit is shown in Fig. 8-20. By KVL,

$$I_d = \frac{\mu V_{gs}}{(R_S \| 1/sC_S) + r_{ds} + R_D} = \frac{\mu(sR_S C_S + 1)V_{gs}}{sC_S R_S(R_D + r_{ds}) + R_S + R_D + r_{ds}} \tag{1}$$

But KVL requires that

$$V_{gs} = V_i - I_D\left(R_S \parallel \frac{1}{sC_S}\right) = V_i - \frac{R_S I_D}{sR_S C_S + 1} \tag{2}$$

Substituting (*1*) into (*2*) and solving for V_{gs} give

$$V_{gs} = \frac{sC_S R_S(R_D + r_{ds}) + R_S + R_D + r_{ds}}{sC_S R_S(R_D + r_{ds}) + R_D + r_{ds} + (\mu + 1)R_S} V_i \tag{3}$$

Now, by Ohm's law and (*1*),

$$V_o = -R_D I_D = -\frac{\mu R_D(sR_S C_S + 1)V_{gs}}{sC_S R_S(R_D + r_{ds}) + R_S + R_D + r_{ds}} \tag{4}$$

Substituting V_{gs} as given by (*3*) into (*4*) and rearranging yield, finally,

$$A_v(s) = \frac{V_o}{V_i} = -\frac{\mu R_D}{R_D + r_{ds} + (\mu + 1)R_S} \frac{sR_S C_S + 1}{s\dfrac{C_S R_S(R_D + r_{ds})}{R_D + r_{ds} + (\mu + 1)R_S} + 1} \tag{5}$$

(*b*) It is apparent that the low-frequency cutoff is the larger of the two break frequencies; from (*5*), it is

$$\omega_L = \frac{R_D + r_{ds} + (\mu + 1)R_S}{C_S R_S(R_D + r_{ds})}$$

8.16 The hybrid-π equivalent circuit for the CE amplifier of Fig. 3-6 with the output shorted is shown in Fig. 8-21. (*a*) Find an expression for the so-called β *cutoff frequency* f_β, which is simply the high-frequency current-gain cutoff point of the transistor with the collector and emitter terminals shorted. (*b*) Evaluate f_β if $r_x = 100\ \Omega$, $r_\pi = 1\ \text{k}\Omega$, $C_\mu = 3$ pF, and $C_\pi = 100$ pF.

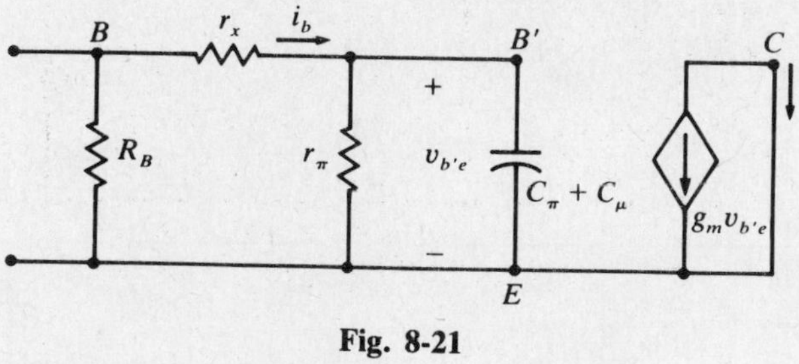

Fig. 8-21

(*a*) Ohm's law gives

$$V_{b'e} = \frac{I_b}{g_\pi + s(C_\pi + C_\mu)} \qquad \text{where} \qquad g_\pi = \frac{1}{r_\pi} \tag{1}$$

But with the collector and emitter terminals shorted,

$$I_L = -g_m V_{b'e} \tag{2}$$

Substituting (*1*) into (*2*) and rearranging give the current-gain ratio

$$\frac{I_L}{I_b} = -\frac{g_m}{g_\pi + s(C_\pi + C_\mu)} = -\frac{g_m r_\pi}{sr_\pi(C_\pi + C_\mu) + 1} \tag{3}$$

From (*3*), the β cutoff frequency is seen to be

$$f_\beta = \frac{\omega_\beta}{2\pi} = \frac{1}{2\pi r_\pi(C_\pi + C_\mu)} \tag{4}$$

(*b*) Substituting the given high-frequency parameters in (*4*) yields

$$f_\beta = \frac{1}{2\pi(1000)(103 \times 10^{-12})} = 1.545\ \text{MHz}$$

8.17 Apply the hybrid-π high-frequency model to the CB amplifier of Fig. 6-13(b): (a) Find an expression for the high-frequency voltage-gain ratio. (b) Describe the high-frequency behavior of the CB amplifier.

(a) Use of the hybrid-π model of Fig. 8-8 results in the high-frequency small-signal equivalent circuit of Fig. 8-22. The coupling capacitors are assumed to be short circuits at high frequency. For typical values, $r_x \ll 1/sC_\pi$, r_π, $1/sC_\mu$ for frequencies near the break frequencies; thus, letting $r_x = 0$ introduces little error (but considerable simplicity).

A Thévenin equivalent can be found for the network to the left of terminal pair a,a'. With $r_x = 0$, current from the dependent source flows only through C, so

$$V_{Th} = -\frac{1}{sC_\mu} g_m V_{b'e} \tag{1}$$

By the method of node voltages,

$$\frac{V_S + V_{b'e}}{R_S} + g_m V_{b'e} + V_{b'e}(sC_\pi + G_E + g_\pi) = 0 \tag{2}$$

Solving (2) for $V_{b'e}$ and substituting the result into (1) yield

$$V_{Th} = \frac{g_m V_S}{sC_\mu[1 + R_S g_m + R_S(sC_\pi + G_E + g_\pi)]} \tag{3}$$

Deactivating (shorting) V_S also shorts E to B'. Consequently, $V_{b'e} = 0$, the dependent current source is open-circuited, and $Z_{Th} = 1/sC_\mu$.

Now, the Thévenin equivalent and voltage division lead to

$$V_L = \frac{R_C \| R_L}{R_C \| R_L + Z_{Th}} V_{Th} \tag{4}$$

Substitution of (3) into (4) and rearrangement give the desired voltage-gain ratio:

$$A_v = \frac{V_L}{V_S} = \frac{g_m R_C \| R_L}{[sC_\mu(R_C \| R_L) + 1][sC_\pi R_S + R_S(g_m + g_\pi + G_E) + 1]} \tag{5}$$

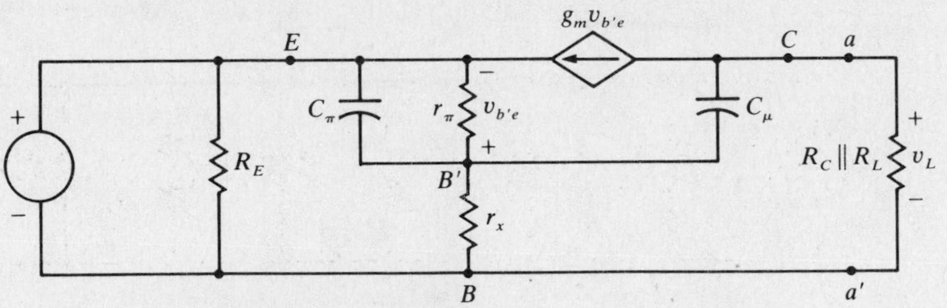

Fig. 8-22

(b) Since (5) involves the upper frequency range, it describes the amplifier as a low-pass (midfrequency) filter with break frequencies at

$$\omega_1 = \frac{1}{C_\mu(R_C \| R_L)} \quad \text{and} \quad \omega_2 = \frac{R_S(g_m + g_\pi + G_E) + 1}{C_\pi R_S} \tag{6}$$

8.18 (a) Apply the results of Section 8.6 to the small-signal equivalent circuit of Fig. 8-9(a) to determine the Miller capacitance. (b) Using the Miller capacitance, draw the associated equivalent circuit and from it find an expression for the high-frequency voltage-gain ratio.

(*a*) First, the gain K_F must be found with capacitor C_μ and load resistor R_L removed. Since

$$V_L = -g_m V_{b'e} R_C$$

the desired gain is

$$K_F = \frac{V_L}{V_{b'e}} = -g_m R_C \tag{1}$$

The Miller capacitance C_M is the input shunt capacitance suggested by (*8.46*):

$$C_M = (1 - K_F)\,\frac{Y_F}{s} = (1 + g_m R_C) C_\mu \tag{2}$$

since comparison of Figs. 8-9(*a*) and 8-12(*a*) shows that C_μ forms a feedback path analogous to Y_F.

(*b*) The output shunt capacitance, as suggested by (*8.48*), must also be determined. Since h_{re} = 0 underlies the hybrid - π model, the reverse voltage-gain ratio $K_R = 0$, hence:

$$Y_o = Y_2 + \left(1 - K_R\right) Y_F \approx Y_2 + Y_F = Y_2 + sC_\mu \tag{3}$$

Comparison of Fig. 8-9(*a*) with Fig. 8-12(*b*) and the use of (*1*) to (*3*) lead to the equivalent circuit of Fig. 8-23. Let

$$C_{eq} = C_M + C_\pi = (1 + g_m R_C) C_\mu + C_\pi$$

Then, by voltage division,

$$V_{b'e} = \frac{r_\pi/(sr_\pi C_{eq} + 1)}{r_x + r_\pi/(sr_\pi C_{eq} + 1)}\, V_s = \frac{r_\pi/(r_x + r_\pi)}{s(r_x \| r_\pi) C_{eq} + 1}\, V_s \tag{4}$$

and by Ohm's law,

$$V_L = -\frac{R_C \| R_L}{s(R_C \| R_L) C_\mu + 1}\, g_m V_{b'e} \tag{5}$$

Substitution of (*4*) into (*5*) and rearrangement yield the desired voltage-gain ratio:

$$A_v(s) = \frac{V_L}{V_s} = -\frac{g_m (R_C \| R_L) r_\pi/(r_x + r_\pi)}{[s(R_C \| R_L) C_\mu + 1][s(r_x \| r_\pi) C_{eq} + 1]}$$

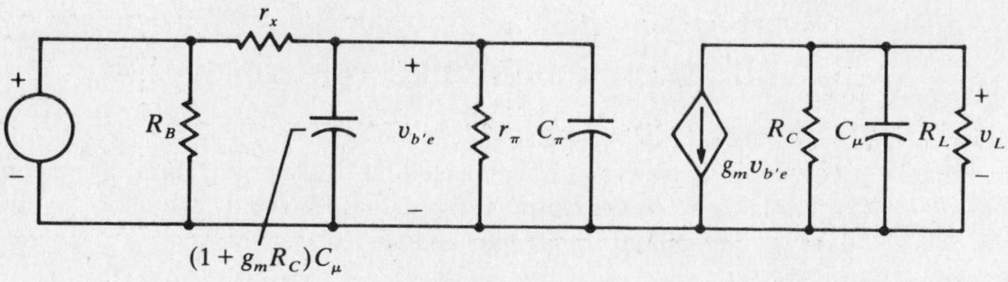

Fig. 8-23

8.19 (*a*) Apply the results of Section 8.6 to the small-signal equivalent circuit of Fig. 8-11 to determine the Miller admittance. (*b*) Utilizing the Miller admittance, draw the high-frequency small-signal equivalent circuit and determine the voltage-gain ratio.

(*a*) With load resistor R_L and feedback capacitor C_{gd} removed from the circuit of Fig. 8-11, the forward gain K_F follows from an application of Ohm's law:

$$K_F = \frac{V_L}{V_{gs}} = -\frac{g_m (r_{ds} \| R_D)}{s(r_{ds} \| R_D) C_{ds} + 1} \tag{1}$$

The Miller admittance suggested by (*8.46*) is

$$Y_M = (1 - K_F)Y_F = \left[1 + \frac{g_m(r_{ds} \| R_D)}{s(r_{ds} \| R_D)C_{ds} + 1}\right] s\, C_{gd} \tag{2}$$

In the frequency range of interest and for typical values of r_{ds}, R_D, and C_{ds}, generally $|s(r_{ds} \| R_D)C_{ds}| \ll 1$; thus, the Miller admittance can be synthesized as a capacitor with value

$$C_M = \frac{Y_M}{s} = [1 + g_m(r_{ds} \| R_D)]C_{gd} \tag{3}$$

(*b*) Since there is no feedback of output voltage to the input network of Fig. 8-11, $K_R = 0$. Hence, the output admittance, as suggested by (*8.48*), is simply

$$(1 - K_R)\, Y_F = Y_F = sC_{gd} \tag{4}$$

The equivalent circuit of Fig. 8-11 can be converted to the form of Fig. 8-12(*b*), as displayed in Fig. 8-24. By Ohm's law,

$$V_L = -\frac{g_m V_{gs}}{s(C_{ds} + C_{gd}) + g_{ds} + G_D + G_L} \tag{5}$$

Since $V_{gs} = V_i$, the required voltage-gain ratio follows as

$$A_v(s) = \frac{V_L}{V_i} = -\frac{g_m}{s(C_{ds} + C_{gd}) + g_{ds} + G_D + G_L} \tag{6}$$

As long as the source resistance is negligible, A_v is independent of C_M. (See Problem 8.23.)

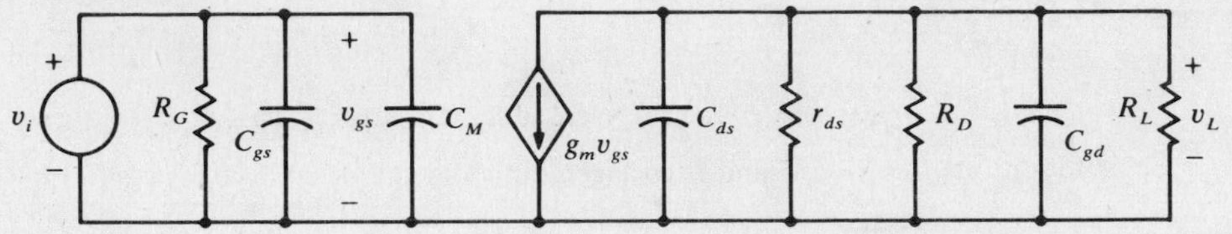

Fig. 8-24

Supplementary Problems

8.20 Show that if two linear networks are connected in cascade to form a new network such that $T(j\omega) = T_1(j\omega)T_2(j\omega)$, then the composite Bode plot is obtained by adding the individual amplitude ratios M_{db1} and M_{db2} and phase angles (ϕ_1 and ϕ_2) associated with $T_1(j\omega)$ and $T_2(j\omega)$ at each frequency.

8.21 Show that (*8.3*) follows from the evaluation of k_1 and k_2 of (*8.2*).

8.22 An amplifier has a Laplace-domain transfer function (voltage-gain ratio) given by

$$A_v(s) = \frac{V_o}{V_i} = \frac{K_s}{(s + 100)(s + 10^5)}$$

(*a*) If an asymptotic Bode plot of $A_v(j\omega)$ is made, over what values of frequency (in the midfrequency range) is the gain constant in amplitude? (*b*) Find the midfrequency gain in decibels if $K_s = 10^8$. (*c*) Within 2 percent accuracy, over what range of frequencies is the exact gain constant?
Ans. (*a*) $100 \le \omega \le 10^5$ rad/s; (*b*) $M_{dbM} = 60$ db; (*c*) $M_{db} \ge 58.8$ db for $500 \le \omega \le 5 \times 10^4$ rad/s

8.23 In Problem 8.19, the gain of the FET amplifier does not depend on the Miller capacitance C_M; however, the situation changes if the source resistance is nonzero. (*a*) Add a source resistance R_i to Fig. 8-24, and find an expression for the voltage-gain ratio. (*b*) Evaluate the gain for $R_i = 0$ and for $R_i = 100\ \Omega$ if $C_{gs} = 3$ pF, $C_{ds} = 1$ pF, $C_{gd} = 2.7$ pF, $r_{ds} = 50$ kΩ, $g_m = 0.016$ S, $R_L = R_D = 2$ kΩ, $R_G = 1$ MΩ, and $f = 50$ MHz.

Ans. (*a*) $$A_v(s) = \frac{-g_m R_G/(R_i + R_G)}{[s(C_{ds} + C_{dg}) + g_{ds} + G_D + G_L][s(R_G \| R_i)(C_{gs} + C_M) + 1]}$$

(*b*) For $R_i = 0$, $A_v = 10.348\underline{|131.53°}$; for $R_i = 100$, $A_v = 3.49\underline{|61.26°}$

8.24 Consider the high-pass filter circuit of Fig. 8-13(*a*). (*a*) Show that as ω becomes large, the amplitude ratio M_{db} actually approaches $20 \log [R_L/(R_L + R_S)]$ as indicated in Fig. 8-13(*b*). (*b*) Show that $|M^2(j\omega_L)|$, where $\omega_L = 1/C(R_L + R_S)$, has the value $\frac{1}{2}|M^2(j\infty)| = \frac{1}{2}[R_L/(R_L + R_S)]^2$.

8.25 In the high-pass filter circuit of Fig. 8-13(*a*), the source impedance $R_S = 5$ kΩ. If the circuit is to have a high-frequency gain of 0.75 and a break or cutoff frequency of 100 rad/s, size R_L and C. *Ans.* $R_L = 15$ kΩ, $C = 0.5\ \mu$F

8.26 In the circuit of Fig. 3-15, replace V_S with a sinusoidal source to give the small-signal circuit of Fig. 8-6. (*a*) If the impedance of the coupling capacitor is not negligible, find the current-gain ratio $A_i(s) = I_L/I_S$. (*b*) Determine the low-frequency cutoff point.
Ans. (*a*) $A_i = h_{fe}R_B/(R_B + h_{ie})$; (*b*) the gain is independent of frequency down to $f = 0$

8.27 Show that the *RC* network of Fig. 8-25 is a high-pass filter. Determine its low-frequency cutoff point.

Ans. $$\frac{V_o}{V_S} = \frac{R_2 + R_3}{R_1 + R_2 + R_3}\,\frac{sR_2R_3C_3/(R_2 + R_3) + 1}{s(R_1 + R_2)R_3C_3/(R_1 + R_2 + R_3) + 1} \qquad \omega_L = \frac{R_1 + R_2 + R_3}{(R_1 + R_2)R_3C_3}$$

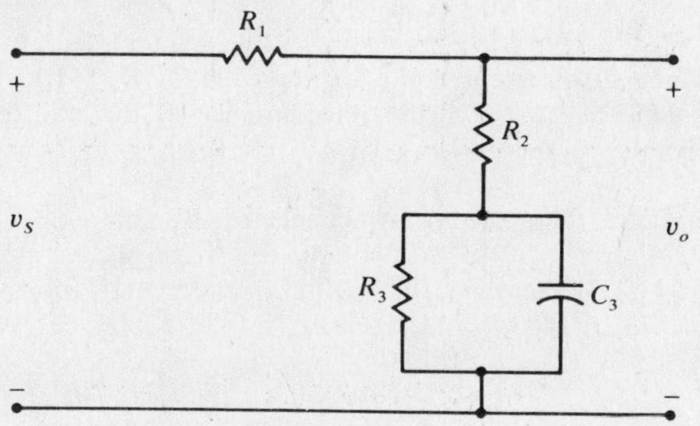

Fig. 8-25

8.28 The amplifier of Fig. 3-6 is modeled for small-signal operation by Fig. 8-4. Let $C_C \to \infty$, $C_E = 100\ \mu$F, $R_E = 100\ \Omega$, $R_C = R_L = 2$ kΩ, $h_{ie} = 200\ \Omega$ and $h_{fe} = 75$. Determine (*a*) the low-frequency voltage gain, (*b*) the midfrequency gain, and (*c*) the low-frequency cutoff point.
Ans. (*a*) −9.62; (*b*) −375; (*c*) 3750 rad/s

8.29 For the amplifier of Fig. 3-6, show that if the source internal impedance R_i is not negligible, but $R_i \ll R_B = R_1 \| R_2$, then the low-frequency cutoff point is given by

$$\omega_L = \frac{(h_{ie} + R_i) + (h_{fe} + 1)R_E}{R_E C_E (h_{ie} + R_i)}$$

8.30 Show that, for the amplifier of Fig. 3-6 as described in Problem 8.5, if $R_B \gg R_E + h_{ie}$, then the current-gain ratio becomes independent of frequency.

8.31 In the amplifier of Fig. 3-6, let $C_C \to \infty$, $C_E = 100\ \mu$F, $R_E = 20$ kΩ, $h_{ie} = 100\ \Omega$, $h_{fe} = 75$, $R_C = R_L = 2$ kΩ, $R_1 = 2$ kΩ, and $R_2 = 20$ kΩ. Determine (*a*) the low-frequency current gain, (*b*) the midfrequency current gain, and (*c*) the low-frequency cutoff point. *Ans.* (*a*) −3.11; (*b*) −35.54; (*c*) 5.71 rad/s

8.32 In the amplifier of Problem 8.6, let $R_i = 500\ \Omega$ and all else remain unchanged. Determine the value of the emitter bypass capacitor required to ensure that $f_L \leq 200$ Hz. Compare your result with that of Problem 8.6 to see that consideration of the source internal impedance allows the use of a smaller bypass capacitor. (*Hint*: See Problem 8.29.) *Ans.* $C_E \geq 101.3\ \mu$F

8.33 In the amplifier of Fig. 3-6, $C_C \to \infty$, $R_i = 500\ \Omega$, $R_E = 30\ \text{k}\Omega$, $R_1 = 3.2\ \text{k}\Omega$, $R_2 = 17\ \text{k}\Omega$, $R_L = 10\ \text{k}\Omega$, $h_{oe} = h_{re} = 0$, $h_{fe} = 100$, and $h_{ie} = 100\ \Omega$. Determine R_C and C_E so that the amplifier has a midfrequency current-gain ratio $|A_i| \geq 30$ with low-frequency cutoff $f_L \geq 20$ Hz. (*Hint*: See Problem 8.5.) *Ans.* $R_C \geq 4517.8\ \Omega$, $C_E \geq 3.13\ \mu$F

8.34 In the CE amplifier of Fig. 3-6, let $C_C \to \infty$, $C_E = 100\ \mu$F, $R_E = 100\ \Omega$, $R_i = 0$, $R_B = 5\ \text{k}\Omega$, $R_C = R_L = 2\ \text{k}\Omega$, $h_{oe} = h_{re} = 0$, $h_{fe} = 75$, and $h_{ie} = 1\ \text{k}\Omega$. The small-signal ac equivalent circuit is given by Fig. 8-4. If a sinusoidal signal $v_i = V_m \sin \omega t$ is impressed (with $\omega = 400$ rad/s), determine (*a*) the phase angle between v_i and i_i, the (*b*) phase shift between input and output voltages, and (*c*) the phase shift between input and output currents.
Ans. (*a*) Current leads voltage by 35.52°; (*b*) output voltage lags input voltage by 128.98°; (*c*) output current lags input current by 180°.

8.35 In the amplifier of Problem 8.13, let $C_E = 200\ \mu$F, $C_C = 10\ \mu$F, $R_E = 50\ \Omega$, $R_C = R_L = 2\ \text{k}\Omega$, $R_i = 100$, $h_{re} = h_{oe} = 0$, $h_{ie} = 1\ \text{k}\Omega$, and $h_{fe} = 50$. (*a*) Sketch the asymptotic Bode plot (M_{db} only) for the voltage-gain ratio. (*b*) Is the 3-db attenuation point below 40 Hz?
Ans. (*a*) $A_v(s) = -0.548s(0.01s + 1)/[0.04s + 1)(0.00301s + 1)]$. The associated Bode plot is given in Fig. 8-26; (*b*) no, because $M_{db}(j\infty) - M_{db}(j80\pi) = 3.79$ db

8.36 In the CE amplifier of Example 8.7, let $g_m = 0.035$ S, $r_\pi = 8\ \text{k}\Omega$, $r_x = 30\ \Omega$, $R_C = R_L = 10\ \text{k}\Omega$, $C_\pi = 10$ pF, and $C_\mu = 2$ pF. (*a*) Determine the high-frequency cutoff point. (*b*) Find the midfrequency gain. *Ans.* (*a*) $f_H = 16.49$ MHz; (*b*) $A_{v\text{mid}} = -174.3$

8.37 In the CB amplifier of Problem 8.17, let $R_S = 100\ \Omega$, $R_E = 1\ \text{k}\Omega$, $R_C = R_L = 10\ \text{k}\Omega$, $C_\mu = 2$ pF, $C_\pi = 40$ pF, $g_m = 0.035$ S, and $r_\pi = 5\ \text{k}\Omega$. Determine (*a*) the midfrequency gain and (*b*) the high-frequency cutoff point. *Ans.* (*a*) $A_{v\text{mid}} = 37.88$; (*b*) $f_H = 15.91$ MHz

8.38 Add a source resistance R_i to the high-frequency small-signal equivalent circuit for the CS amplifier given by Fig. 8-11. Let $C_{gs} = 3$ pF, $C_{ds} = 1$ pF, $C_{gd} = 2.7$ pF, $r_{ds} = 50\ \text{k}\Omega$, $g_m = 0.016$ S, $R_L = R_D = 2\ \text{k}\Omega$, and $R_G = 1\ \text{M}\Omega$. Determine the high-frequency cutoff point (*a*) with $R_i = 0$ and (*b*) with $R_i = 100$. *Ans.* (*a*) $f_H = 43.875$ MHz; (*b*) $f_H = 13.69$ MHz

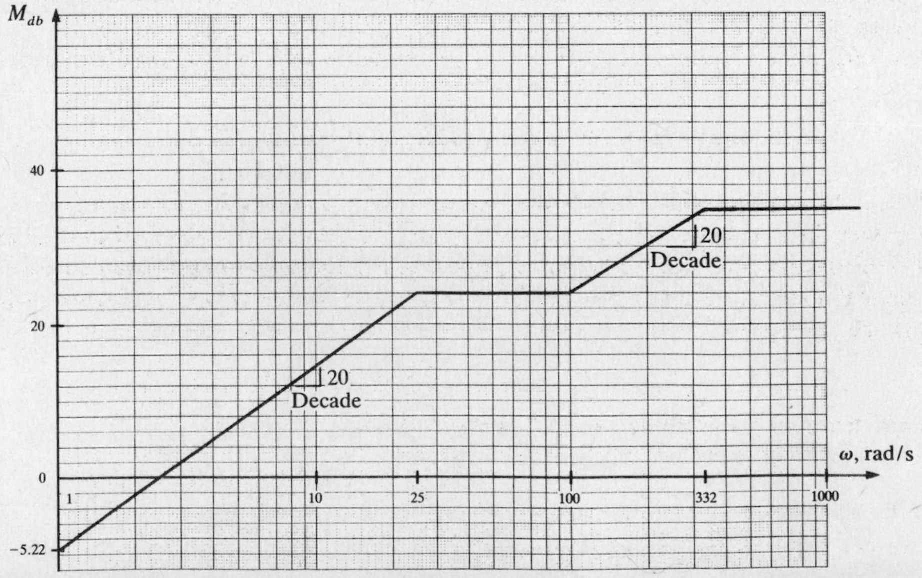

Fig. 8-26

Chapter 9

Power Amplifiers

9.1 AMPLIFIER CLASSIFICATION AND CHARACTERISTICS

In general, a deliberate effort is made to obtain maximum output power from the last stage of a cascaded amplifier. In such a case, this last stage (or amplifier) is called a *power amplifier*. The classification scheme of Table 3-3, based on the percentage of active-region operation, applies to power amplifiers.

Power amplifiers operate with large current and voltage swings that may approach the limit of transistor capability. Further, large signal excursions can introduce *distortion*—nonlinear relationship between input and output signals leading to frequency components in the output signal that are not present in the input signal and which are due to nonlinear terms in the input-output relation (see Problem 9.18). Distortionless operation will be assumed here unless otherwise noted.

Physically larger components are required to handle larger signal swings, and wider base regions are used than in small-signal transistors. Consequently, a smaller percentage of the emitted majority carriers cross the base region, so that values of β are generally smaller in power amplifiers than in small-signal amplifiers.

Midfrequency-range operation is assumed, so coupling and bypass capacitors can be considered to be short circuits.

9.2 POWER AND EFFICIENCY OF AMPLIFIERS

The analysis of amplifiers with regard to power centers on three quantities: the power supplied by the bias sources, the power delivered to the load, and the power dissipated by the transistor. The *supplied power* includes all dc power flowing from bias supplies to the amplifier circuit; however, the power dissipated by the emitter-base network is practically negligible (see Example 9.1). Hence, supplied power is calculated as

$$P_{CC} = V_{CC}\,\frac{1}{T}\int_0^T i_C\,dt = V_{CC}i_{C\text{av}} = V_{CC}I_{CQ} \tag{9.1}$$

where T is the period of the collector current i_C. The right-hand equality holds only if the presence of a signal does not alter the average value of i_C (that is, there is no distortion).

Load power, the average value of the ac power delivered to the amplifier load, is given by

$$P_L = \frac{1}{T}\int_0^T i_L^2 R_L\,dt = \frac{I_{Lm}^2}{2}\,R_L \tag{9.2}$$

The right-hand equality holds only if the load current has zero average value.

The power dissipated by the transistor is the sum of the power delivered to the base-emitter port and that delivered to the collector-emitter port, but the former is negligible compared to the latter. Thus, practical analysis requires only calculation of the *collector power dissipation*, which is found as the difference

$$P_C = P_{CC} - (I^2R\text{ losses}) - P_L \tag{9.3}$$

The *efficiency* of an amplifier is then computed as

$$\eta = \frac{\text{load power}}{\text{supplied power}}\,(100\%) = \frac{\dfrac{1}{T}\displaystyle\int_0^T i_L^2 R\,dt}{V_{CC}i_{C\text{av}}}\,(100\%) \tag{9.4}$$

9.3 RATINGS AND THERMAL CONSIDERATIONS

Effective and safe utilization of power transistors requires primary attention to voltage, current, power, and temperature limitations, which are of only secondary concern in small-signal amplifiers. Manufacturer-furnished transistor specifications include

$i_{C\max}$ ≡ maximum continuous collector current
BV_{CEO} ≡ avalanche breakdown voltage ($v_{CE} > 0$) with $i_B = 0$
$P_{C\max}$ ≡ maximum average power that the transistor can dissipate with case temperature $\leq T_{CO}$
T_{CO} ≡ maximum case temperature without derating from $P_{C\max}$ (typical value is 25°C)
$T_{J\max}$ ≡ maximum allowable junction (collector-base) temperature (80 to 100°C for Ge, 125 to 200°C for Si)

In active-region operation, the reverse-biased collector junction is the greatest source of heat; thus the power dissipation is associated with the collector junction.

Example 9.1 A Si power transistor is biased so that $V_{CEQ} = 14$ V and $I_{CQ} = 100$ mA. At quiescent conditions, determine the power dissipated by the base-emitter junction and the power dissipated by the collector-base junction.

For active-region operation, $V_{BEQ} \approx 0.7$ V and

$$P_B = V_{BEQ} I_{CQ} = (0.7)(0.1) = 0.07 \text{ W}$$

By KVL,

$$V_{CBQ} = V_{CEQ} - V_{BEQ} = 14 - 0.7 = 13.3 \text{ V}$$

so that

$$P_C = V_{CBQ} I_{CQ} = (13.3)(0.1) = 1.33 \text{ W}$$

In this typical example, 97.9 percent of the power dissipated by the transistor is associated with the collector junction.

The manufacturer's specifications can be used to sketch the *dissipation derating curve* of Fig. 9-1, which shows the allowable collector power dissipation P_C as a function of transistor case temperature T_C. The dissipation derating curve follows from the average-value heat-flow model for the BJT in Fig. 9-2, wherein the average value of the collector power dissipation, temperature, and *thermal resistance* are treated, respectively, as electrical analogs of current, voltage, and resistance. The value of the junction-to-case thermal resistance θ_{jc} (measured in degrees Celsius per watt) is given by the manufacturer. A typical value for the case-to-heat-sink thermal resistance θ_{cs} is usually suggested by the transistor manufacturer, but the user must ensure the integrity of the case-to-heat-sink junction. The heat-sink-to-ambient thermal resistance θ_{sa} is specified by the heat-sink manufacturer. The case-to-ambient thermal impedance θ_{ca} is formed by series combination as $\theta_{ca} = \theta_{cs} + \theta_{sa}$.

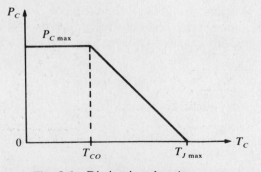

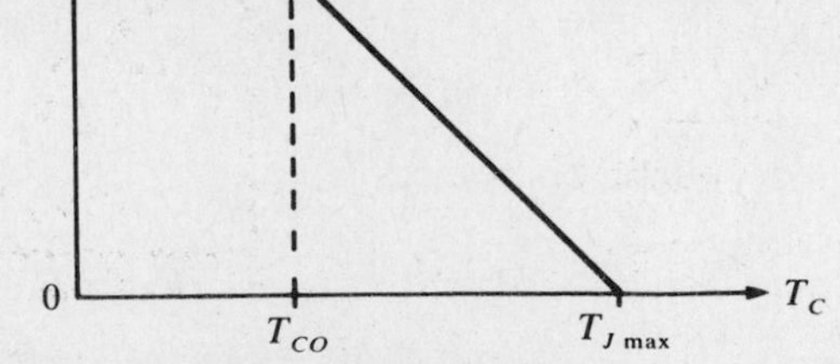

Fig. 9-1 Dissipation derating curve

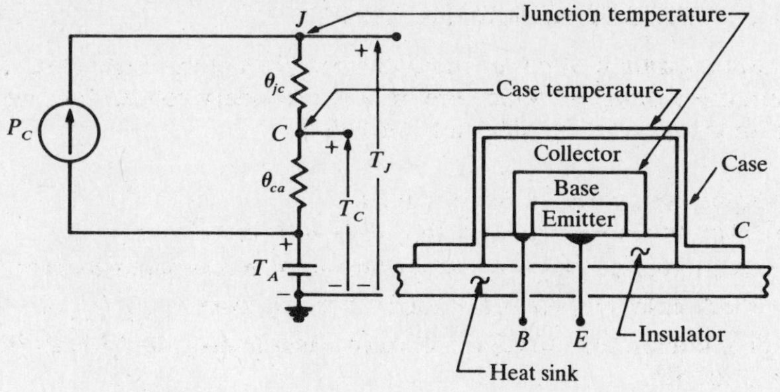

Fig. 9-2 Average-value heat-flow model for BJT

Example 9.2 A power transistor is to operate with $I_{CQ} = 1.25$ A and $V_{CEQ} = 7.5$ V. If $\theta_{jc} = 3°C/W$ and $\theta_{ca} = 10°C/W$, determine the junction temperature if the transistor is operating in a 40°C ambient.

The average power dissipated by the transistor collector is

$$P_C = V_{CEQ}I_{CQ} = (7.5)(1.25) = 9.375 \text{ W}$$

Applying KVL to the analog circuit of Fig. 9-2 yields

$$T_J = P_C(\theta_{jc} + \theta_{ca}) + T_A = (9.375)(3 + 10) + 40 = 161.9°C \tag{9.5}$$

Example 9.3 A power transistor is to dissipate 10 W of power. Let $\theta_{jc} = 2.5°C/W$ and $\theta_{cs} = 0.5°C/W$. The maximum allowable junction temperature is $T_{J\max} = 150°C$. The device must operate in a maximum ambient of 50°C. Three different heat sinks are available, with thermal resistances of 3, 6, and 9°C/W. Select the best heat sink.

Letting $\theta_{ca} = \theta_{cs} + \theta_{sa}$ in (*9.5*) and rearranging yield

$$\theta_{sa} \leq \frac{T_{J\max} - T_A}{P_C} - \theta_{jc} - \theta_{cs} = \frac{150 - 50}{10} - 2.5 - 0.5 = 7°C/W$$

We choose the heat sink with $\theta_{sa} = 6°C/W$, as it is the smaller of the heat sinks that will be adequate.

9.4 CLASS A DIRECT-COUPLED AMPLIFIER

Although its collector characteristics involve considerably larger values of current, the class A direct-coupled power amplifier of Fig. 9-3(*a*) does not differ in operating principle from the small-signal, low-power amplifiers of Chapter 3. Maximum symmetrical swing (Problem 3.23) is universally implemented to yield maximum power output.

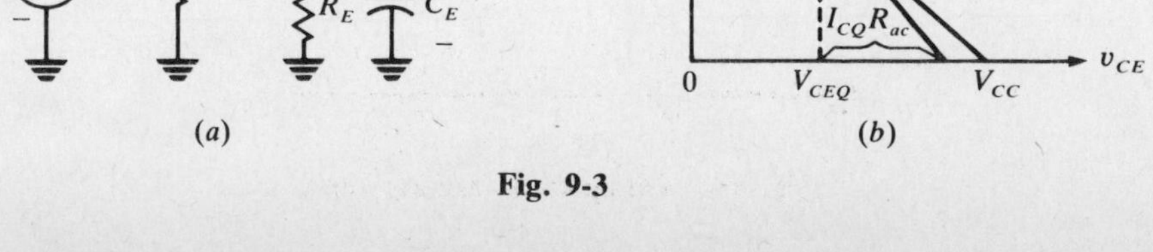

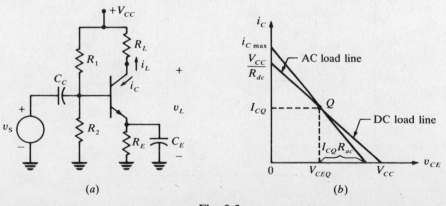

Fig. 9-3

Example 9.4 In the class A power amplifier of Fig. 9-3(*a*), let $R_E = 0.5\ \Omega$, $R_L = 8\ \Omega$, $R_1 = 20\ \text{k}\Omega$, $R_2 = 1\ \text{k}\Omega$, and $V_{CC} = 24$ V. Find (*a*) the supplied power, (*b*) the maximum power to the load, (*c*) the range of collector power dissipation, and (*d*) the maximum efficiency. Assume no distortion.

(*a*) We must find the Q point of Fig. 9-3(*b*) for maximum symmetrical swing. We first note that

$$R_{dc} = R_L + R_E \quad \text{and} \quad R_{ac} = R_L$$

Then, (*3*) of Problem 3.23,

$$I_{CQ} = \frac{V_{CC}}{R_{ac} + R_{dc}} = \frac{V_{CC}}{2R_L + R_E} = \frac{24}{2(8) + 0.5} = 1.454 \text{ A}$$

The delivered power is then determined with (*9.1*):

$$P_{CC} = V_{CC} i_{Cav} = V_{CC} I_{CQ} = (24)(1.454) = 34.896 \text{ W}$$

(*b*) The maximum power is delivered to the load by the maximum signal, so, by (*9.2*),

$$P_{L\max} = \frac{I_{CQ}^2}{2} R_L = \frac{(1.454)^2}{2} 8 = 8.456 \text{ W}$$

(*c*) The collector power dissipation is found by use of (*9.3*). Maximum collector power occurs for $P_L = 0$, so $\max(P_C) = P_{CC} = 34.896$ W. Minimum collector power coincides with $P_L = P_{L\max}$, so $\min(P_C) = P_{CC} - P_{L\max} = 34.896 - 8.456 = 26.44$ W.

(*d*) The amplifier efficiency is at a maximum when maximum power is being delivered to the load:

$$\eta_{\max} = \frac{P_{L\max}}{P_{CC}}(100\%) = \frac{8.456}{34.896}(100\%) = 24.2\%$$

Example 9.4 shows that if bias stability is sacrificed by allowing R_E to go to zero, then $\eta_{\max}$ has 25 percent as an upper limit. In power amplifier applications, such low efficiency means ineffective use of the transistor and presents significant cooling problems; it is the impetus for seeking alternative circuit designs for power amplifiers.

9.5 CLASS A INDUCTOR-COUPLED AMPLIFIER

In the amplifier of Fig. 9-4(*a*), I_{CQ} (dc bias current) can flow through the large inductor, but the inductor presents an infinite impedance to an ac signal. Consequently, the collector is not an ac ground. If the inductor has negligible resistance, the dc load-line equation is

$$i_C = -\frac{1}{R_E} v_{CE} + \frac{V_{CC}}{R_E} \tag{9.6}$$

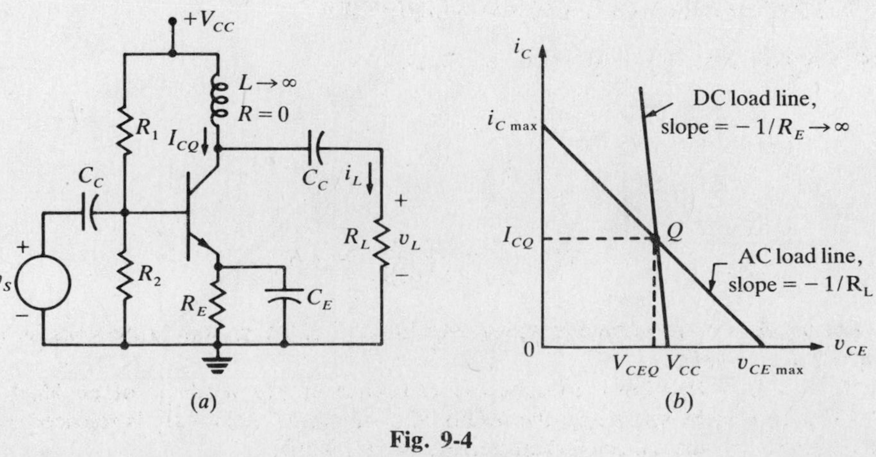

Fig. 9-4

If R_E is selected to be quite small (just large enough to allow stabilized bias design), then the dc load line approaches a vertical line as is illustrated in Fig. 9-4(*b*).

Example 9.5 For the class A inductor-coupled amplifier of Fig. 9-4(*a*), find expressions for I_{CQ} and V_{CEQ} if $(R_E/R_L)^2 \ll 1$ and if $R_E/R_L \ll 1$.

By (*3*) of Problem 3.23,

$$I_{CQ} = \frac{V_{CC}}{R_{ac} + R_{dc}} = \frac{V_{CC}}{R_L + R_E} = \frac{V_{CC}/R_L}{1 + R_E/R_L} \tag{9.7}$$

$$= \frac{V_{CC}/R_L}{1 + R_E/R_L} \frac{1 - R_E/R_L}{1 - R_E/R_L} = \frac{(V_{CC}/R_L)(1 - R_E/R_L)}{1 - (R_E/R_L)^2} \tag{9.8}$$

If $(R_E/R_L)^2 \ll 1$, (*9.8*) becomes

$$I_{CQ} \approx \frac{V_{CC}}{R_L}\left(1 - \frac{R_E}{R_L}\right) \tag{9.9}$$

which, for $R_E/R_L \ll 1$, can be further approximated as

$$I_{CQ} \approx \frac{V_{CC}}{R_L} \tag{9.10}$$

Application of KVL around the collector circuit of Fig. 9-4(*a*) and the use of (*9.8*) yield

$$V_{CEQ} = V_{CC} - I_{CQ}R_E = V_{CC} - \frac{(V_{CC}/R_L)(1 - R_E/R_L)}{1 - (R_E/R_L)^2} R_E \tag{9.11}$$

If $(R_E/R_L)^2 \ll 1$, (*9.11*) reduces to

$$V_{CEQ} \approx V_{CC}\left(1 - \frac{R_E}{R_L}\right) \tag{9.12}$$

For $R_E/R_L \ll 1$, (*9.12*) can be further simplified to

$$V_{CEQ} \approx V_{CC} \tag{9.13}$$

It follows from (*9.10*) and (*9.13*) that if $R_E/R_L \ll 1$ and the Q point is selected at the midpoint of the ac load line (for maximum symmetrical swing), then $i_{C\max} \approx 2V_{CC}/R_L$ and $v_{CE\max} \approx 2V_{CC}$. The maximum efficiency of this inductor-coupled amplifier is 50 percent, which is twice that of the direct-coupled class A amplifier. However, for a fixed power-supply voltage V_{CC} and a particular load R_L, the maximum power that can be delivered to the load is fixed as

$$P_{L\max} = \frac{I_{Lm}^2}{2} R_L = \frac{I_{CQ}^2}{2} R_L \approx \frac{1}{2}\left(\frac{V_{CC}}{R_L}\right)^2 R_L = \frac{V_{CC}^2}{2R_L}$$

If a greater value of P_L is desired, the transformer-coupled amplifier should be considered.

9.6 CLASS A TRANSFORMER-COUPLED AMPLIFIER

The transformer of Fig. 9-5(*a*) is ideal, so that

$$v_L' = av_L \tag{9.14}$$

$$i_C = \frac{1}{a} i_L \tag{9.15}$$

$$R_L' = \frac{v_L'}{i_C} = \frac{av_L}{(1/a)i_L} = a^2 R_L \tag{9.16}$$

Example 9.6 For the class A transformer-coupled amplifier of Fig. 9-5(*a*), find expressions for I_{CQ} and V_{CEQ} if $(R_E/R_L')^2 \ll 1$ and if $R_E/R_L') \ll 1$.

Note that $R_{ac} = a^2 R_L = R_L'$. Further, since the resistance of the windings of an ideal transformer is negligible, $R_{dc} = R_E$. Hence, by symmetry, the results of Example 9.5 hold if R_L is replaced with $R_L' = a^2 R_L$. The load lines are shown on the collector characteristic of Fig. 9-5(*b*).

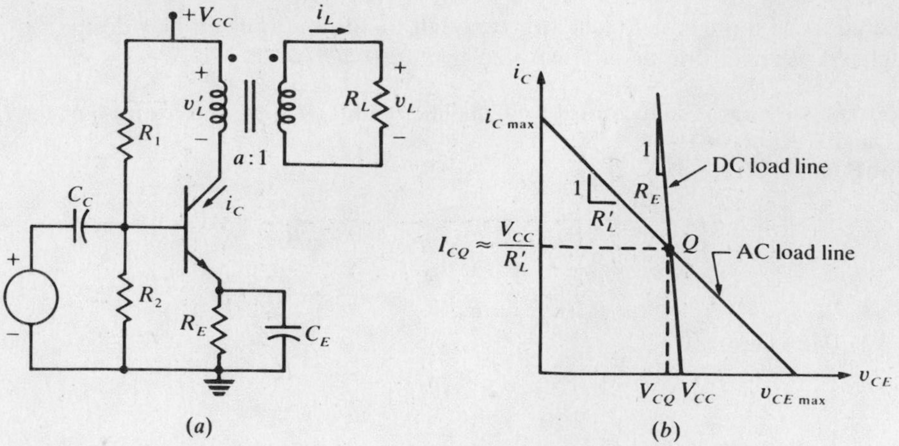

Fig. 9-5

Example 9.7 Find the maximum power that can be delivered to the load R_L of a class A transformer-coupled amplifier, and explain how this can be adjusted to a desired value.

An ideal transformer is a lossless device; hence the power delivered to the primary is identically the power delivered to R_L. Using (*9.15*) and (*9.16*), we have

$$P_L = \frac{I_{Lm}^2}{2} R_L = \frac{(aI_{cm})^2}{2} R_L = \frac{I_{cm}^2}{2} R_L'$$

From Fig. 9-5(*b*), with $R_E \ll R_L'$, for maximum symmetrical swing $I_{cm} \approx I_{CQ} \approx V_{CC}/R_L'$. Therefore,

$$P_{L\max} \approx \frac{I_{CQ}^2}{2} R_L' \approx \frac{V_{CC}^2}{2R_L'} = \frac{V_{CC}^2}{2a^2 R_L} \tag{9.17}$$

From (*9.17*), it is obvious that proper selection of the turns ratio a allows adjustment of the power delivered to R_L.

9.7 PUSH-PULL AMPLIFIERS

The maximum attainable efficiency for a class A transformer-coupled amplifier is shown to be 50 percent in Problem 9.5. In a class B amplifier, the dc collector current I_{CQ} is lower than the peak value of ac current. Consequently, there is less collector dissipation and the efficiency can be as great as 78.54 percent (see Problem 9.11).

The common-emitter push-pull amplifier of Fig. 9-6 has one active transistor and one transistor in cutoff at any instant of time. Owing to the symmetry, we need only study the operation of the active transistor, as diagrammed in Fig. 9-7(*a*).

Example 9.8 Determine the load lines for the half class B push-pull amplifier of Fig. 9-7(*a*).

Since the resistance of the windings of an ideal transformer is negligible, the dc load-line equation is simply $V_{CEQ} = V_{CC}$; the dc load line is then a vertical line, as shown in Fig. 9-7(*b*). When no ac signal is present (quiescent conditions), $i_{B2} = 0$, so that the transistor is cut off. Hence, $I_{CQ} = 0$, and the ac load-line equation follows from (*3.12*) as

$$i_{C2} = \frac{V_{CEQ2}}{R_L'} - \frac{v_{CE2}}{R_L'} = \frac{V_{CC}}{R_L'} - \frac{v_{CE2}}{R_L'} \tag{9.18}$$

When Q_2 is active, (*9.18*) describes a line [on the collector characteristics of Fig. 9-7(*b*)] with vertical intercept $i_{C2} = V_{CC}/R_L'$ and horizontal intercept $v_{CE2} = V_{CC}$; when Q_2 is cut off, $i_{C2} = 0$ and $v_{CE2} = V_{CC}$, giving the horizontal line segment from $v_{CE2} = V_{CC}$ to $v_{CE2} = 2V_{CC}$.

Class A operation of push-pull amplifiers is also possible (see Problem 9.13).

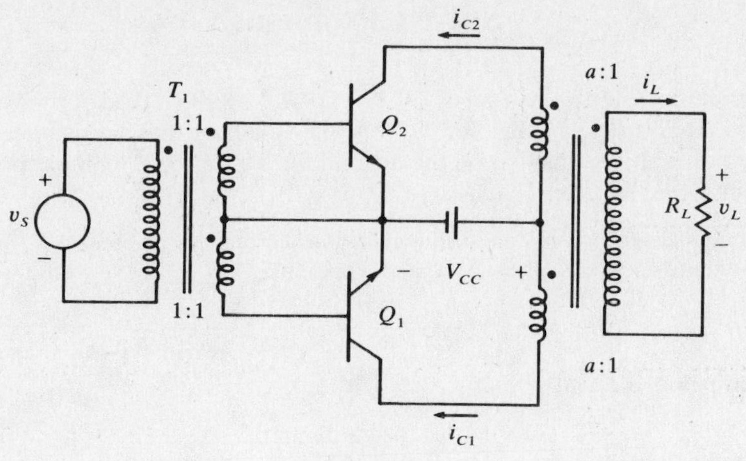

Fig. 9-6

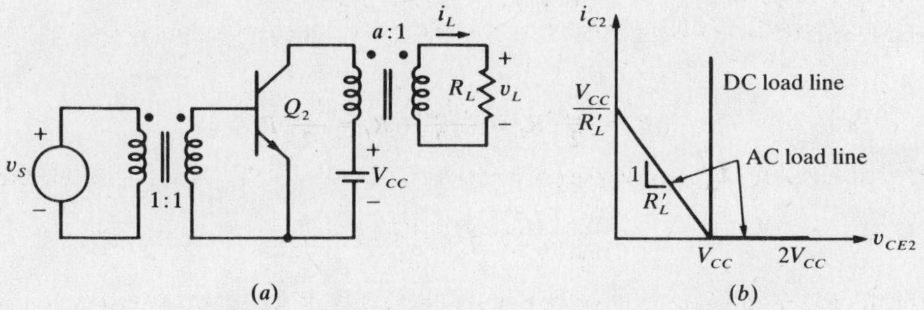

Fig. 9-7

9.8 COMPLEMENTARY-SYMMETRY AMPLIFIERS

By taking advantage of the two possible polarities of BJTs (*npn* and *pnp*), a complementary-symmetry amplifier can be devised to use direct coupling of both the source and the load, eliminating the need for transformers. A basic complementary-symmetry amplifier is illustrated in Fig. 9-8. Each transistor is essentially a class B emitter follower. The load line and power relations are the same as for the class B push-pull amplifier of Section 9.7.

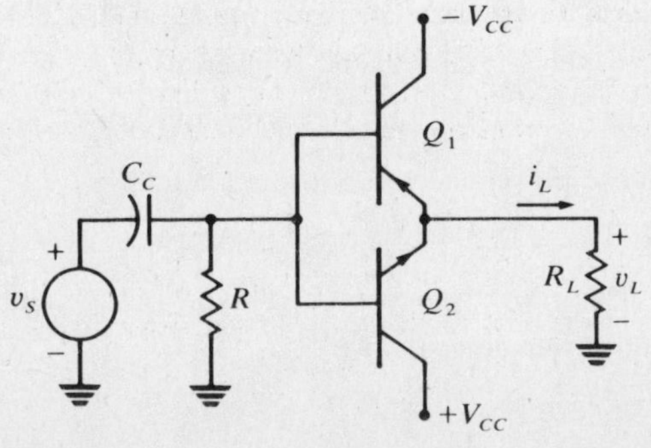

Fig. 9-8

Solved Problems

9.1 In the amplifier of Fig. 3-6(*a*), let $R_E = 470\ \Omega$, $R_C = 800\ \Omega$, and $R_L = 1.2\ \text{k}\Omega$. Assume that the power to R_1 and R_2 is negligible, but that the values of these resistances can be selected to place the Q point at any position on the ac load line. Find the maximum possible efficiency for this amplifier with the given values.

Maximum efficiency occurs with maximum ac load current, so we assume R_1 and R_2 are selected for maximum symmetrical swing. Then, by (*3*) of Problem 3.23,

$$I_{CQ} = \frac{V_{CC}}{R_{ac} + R_{dc}} = \frac{V_{CC}}{R_C \| R_L + R_E + R_C} \tag{1}$$

The delivered power is found with (*1*):

$$P_{CC} = V_{CC} I_{CQ} = \frac{V_{CC}^2}{R_C \| R_L + R_E + R_C}$$

Typically $V_{CE\text{sat}} \approx 0$, and we assume $I_{CBO} = 0$, so that

$$I_{cm} = I_{CQ} = \frac{V_{CC}}{R_C \| R_L + R_E + R_C}$$

Now current division gives

$$I_{Lm} = \frac{R_C}{R_L + R_C} I_{cm} = \frac{R_C V_{CC}}{(R_L + R_C)(R_C \| R_L + R_E + R_C)}$$

The maximum efficiency follows from (*9.2*) and (*9.4*) as

$$\eta_{\max} = \frac{P_{L\max}}{P_{CC}}(100\%) = \frac{\dfrac{1}{2}\left[\dfrac{R_C V_{CC}}{(R_L + R_C)(R_C \| R_L + R_E + R_C)}\right]^2 R_L}{\dfrac{V_{CC}^2}{R_C \| R_L + R_E + R_C}}(100\%)$$

$$= \frac{R_C^2 R_L (100\%)}{2(R_L + R_C)^2 (R_C \| R_L + R_E + R_C)} = \frac{(800)^2(1200)(100\%)}{2(2000)^2(480 + 470 + 800)} = 5.49\%$$

9.2 Add a large emitter bypass capacitor to the circuit of Fig. 3-4(*a*), and let $V_{CC} = 24$ V, $R_C = 5\ \text{k}\Omega$, and $R_E = 200\ \Omega$. Assume that R_1 and R_2 are selected for maximum symmetrical swing but dissipate negligible power. Also, let $V_{CE\text{sat}} \approx 0$ and $I_{CBO} = 0$. (*a*) Find the greatest possible amplitude for the undistorted collector current. Then calculate, for 0, 50, and 100 percent of this maximum undistorted collector current, (*b*) P_{CC}, (*c*) P_L, (*d*) P_C, and (*e*) the amplifier efficiency. The output signal is taken from the common to the collector terminal.

(*a*)
$$I_{cm} = I_{CQ} = \frac{1}{2}\frac{V_{CC}}{R_C + R_E} = \frac{1}{2}\frac{24}{5000 + 200} = 2.308\ \text{mA}$$

Thus, in parts *b* to *e*, calculations are to be made for $I_{cm} = 0$, 1.154, and 2.308 mA.

(*b*) The supplied power is independent of I_{cm} and is given by

$$P_{CC} = V_{CQ} I_{CQ} = (24)(2.308 \times 10^{-3}) = 55.38\ \text{mW}$$

(*c*) The average value of the ac power delivered to the load R_C is

$$P_L = \tfrac{1}{2} I_{cm}^2 R_C = 2500 I_{Cm}^2 \quad \text{W}$$

which, when evaluated at the three values of I_{Cm}, gives 0, 3.33, and 13.32 mW.

(*d*) From (*9.3*), the collector power dissipation is

$$P_C = P_{CC} - I_{CQ}^2 (R_C + R_E) - P_L = 27.7 \times 10^{-3} - P_L \quad \text{W}$$

For the three computed values of P_L, this gives P_C values of 27.7, 24.37, and 14.38 mW.

(*e*) Since $\eta = (P_L / P_{CC})(100\%)$, the three efficiencies are 0, 6.01, and 24.04 percent. All the results of this problem are plotted in Fig. 9-9.

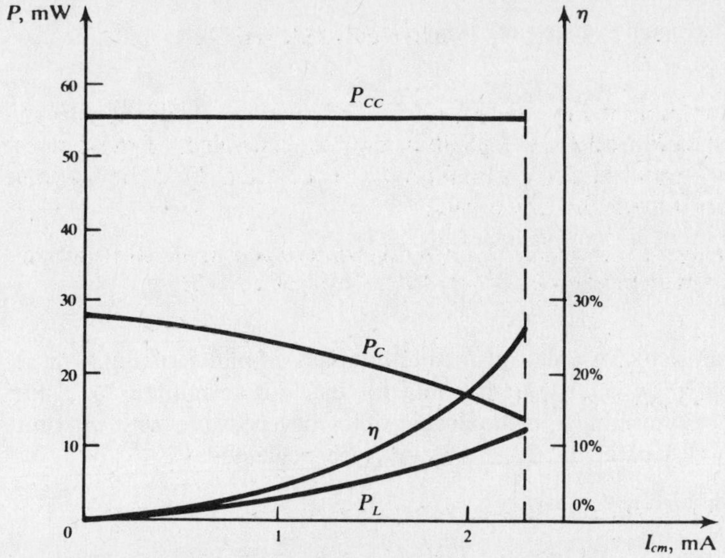

Fig. 9-9

9.3 Transistors are rated according to the power-dissipating capability $P_{C\max}$ of the collector. A plot of $P_{C\max} = v_{CE}i_C$ on the collector characteristics (i_C vs. v_{CE}) is called the *maximum-dissipation hyperbola*. Show that if a transistor is biased for maximum symmetrical swing, then the ac load line is tangent to the maximum-dissipation hyperbola at the Q point.

From the maximum-dissipation equation,

$$i_C = \frac{P_{C\max}}{v_{CE}} \tag{1}$$

We may differentiate (*1*) with respect to v_{CE} and evaluate the result at the Q point to find the slope of the maximum-dissipation hyperbola at that point:

$$\left.\frac{\partial i_C}{\partial v_{CE}}\right|_Q = -\left.\frac{P_{C\max}}{v_{CE}^2}\right|_Q = -\frac{V_{CEQ}I_{CQ}}{V_{CEQ}^2} = -\frac{I_{CQ}}{V_{CEQ}} \tag{2}$$

But for maximum symmetrical swing, $I_{CQ} = V_{CEQ}/R_{ac}$, so (*2*) becomes

$$\left.\frac{\partial i_C}{\partial v_{CE}}\right|_Q = -\frac{V_{CEQ}/R_{ac}}{V_{CEQ}} = -\frac{1}{R_{ac}} \tag{3}$$

From (*3*) it is apparent that the slope of the maximum-dissipation hyperbola is the same as that of the ac load line. Since the line and the curve have identical slopes at a common point (the Q point), they are tangent at that point.

9.4 The transformer-coupled amplifier of Fig. 9-5(*a*) is biased for maximum symmetrical swing. The ideal transformer is coupled to an 8-Ω speaker load R_L which requires an input of 9 W. Let $V_{CC} = 24$ V and $R_E = 0$. (*a*) Select the transformer turns ratio *a*. (*b*) Determine the minimum power rating of the transistor.

(*a*) The average power to R_L must be 9 W; thus,

$$\frac{V_{Lm}^2}{2R_L} = 9 \qquad \text{so} \qquad V_{Lm} = \sqrt{9(2R_L)} = \sqrt{9(2)(8)} = 12 \text{ V}$$

For maximum symmetrical swing with $R_E = 0$, $V'_{Lm} = V_{CEQ} = V_{CC}$. Since the transformer is ideal,

$$a = \frac{V'_{Lm}}{V_{Lm}} = \frac{V_{CC}}{V_{Lm}} = \frac{24}{12} = 2$$

(*b*) For maximum symmetrical swing,

$$I_{CQ} = \frac{V_{CC}}{R_{ac} + R_{dc}} = \frac{V_{CC}}{R_L' + 0} = \frac{V_{CC}}{a^2 R_L} = \frac{24}{(2)^2(8)} = 0.75 \text{ A}$$

and the power supplied is

$$P_{CC} = V_{CC} I_{CQ} = (24)(0.75) = 18 \text{ W}$$

Since there are no I^2R losses, by (*9.3*)

$$P_C = P_{CC} - P_L = 18 - 9 = 9 \text{ W}$$

9.5 For the transistor in the transformer-coupled amplifier of Fig. 9-5(*a*), assume that $V_{CE\text{sat}} = 0$ and $I_{CBO} = 0$. (*a*) Derive a formula for easily determining V_{CEQ} for maximum symmetrical swing. (*b*) Determine the maximum efficiency possible with maximum symmetrical swing if the power dissipated in R_1 and R_2 may be neglected.

(*a*) The dc load-line equation is

$$V_{CC} = v_{CE} + i_E R_E \approx v_{CE} + i_C R_{dc} \tag{1}$$

Using (*3*) of Problem 3.23 in (*1*) and evaluating at the Q point give

$$V_{CC} \approx V_{CEQ} + \frac{V_{CC}}{R_{ac} + R_{dc}} R_{dc} \quad \text{so} \quad V_{CEQ} \approx \frac{V_{CC}}{1 + R_{dc}/R_{ac}} \tag{2}$$

(*b*) The supplied power is

$$P_{CC} = V_{CC} I_{CQ} = V_{CC} \frac{V_{CC}}{R_{ac} + R_{dc}} = \frac{V_{CC}^2}{a^2 R_L + R_E} \tag{3}$$

The maximum power delivered to the load R_L is

$$P_{L\max} = \frac{I_{Lm}^2}{2} R_L = \frac{I_{CQ}^2}{2} a^2 R_L = \frac{1}{2}\left(\frac{V_{CC}}{R_{ac} + R_{dc}}\right)^2 a^2 R_L = \frac{a^2 V_{CC}^2 R_L}{2(a^2 R_L + R_E)^2} \tag{4}$$

Since there are no I^2R losses, the maximum efficiency is given by

$$\eta_{\max} = \frac{P_{L\max}}{P_{CC}}(100\%) = \frac{a^2 V_{CC}^2 R_L/[2(a^2 R_L + R_E)^2]}{V_{CC}^2/(a^2 R_L + R_E)}(100\%) = \frac{a^2 R_L}{2(a^2 R_L + R_E)}(100\%)$$

$$= \frac{1}{2(1 + R_{dc}/R_{ac})}(100\%) = \frac{50}{1 + R_{dc}/R_{ac}}\ \% \tag{5}$$

If $R_{dc} \to 0$, then $\eta_{\max} \to 50$ percent as an upper limit.

9.6 In the transformer-coupled class A amplifier of Fig. 9-5(*a*), $V_{CC} = 15$ V, $R_E = 0.5\ \Omega$, and $R_L = 1.5\ \Omega$. Neglect power losses in R_1 and R_2, but assume their values were selected for maximum symmetrical swing. The resistance R_p of the primary coil of the transformer is 2.5 Ω, and that of the secondary (R_s) is 0.1 Ω. The transformer turns ratio is 5:1. Find (*a*) I_{CQ} and (*b*) P_{CC}, and (*c*) specify the minimum values of $i_{C\max}$, BV_{CEO}, and $P_{C\max}$ for the transistor used in this amplifier. Assume that the transformer leakage reactance is negligible at the frequencies of interest.

(*a*) The secondary winding resistance can be reflected, along with R_L, to the primary side to give

$$R_{ac} = R_p + a^2(R_s + R_L) = 2.5 + (5)^2(0.1 + 1.5) = 42.5\ \Omega$$

$$R_{dc} = R_p + R_E = 2.5 + 0.5 = 3\ \Omega$$

By (*3*) of Problem 3.23,

$$I_{CQ} = \frac{V_{CC}}{R_{ac} + R_{dc}} = \frac{15}{42.5 + 3} = 0.33 \text{ A}$$

(*b*) The supplied power is

$$P_{CC} = V_{CC}I_{CQ} = (15)(0.33) = 4.95 \text{ W}$$

(*c*) Since $R_{dc}/R_{ac} = 3/42.5 \approx 0.07 \ll 1$, the conclusions of Example 9.6 allow us to write $V_{CEQ} \approx V_{CC} =$ 15 V. Thus, we specify

$$i_{C\max} \geq 2I_{CQ} = 2(0.33) = 0.66 \text{ A}$$

and

$$BV_{CEO} \geq 2V_{CC} = 2(15) = 30 \text{ V}$$

The maximum collector dissipation occurs for $P_L = 0$, so we specify

$$P_{C\max} \geq P_{CC} - I_{CQ}^2(R_E + R_p) = 4.95 - (0.33)^2(0.5 + 2.5) = 4.623 \text{ W}$$

9.7 In the transformer-coupled class A amplifier of Fig. 9-5(*a*), $R_L = 1.5\ \Omega$, $V_{CC} = 15$ V, and $\beta = 20$. The maximum delivered load power is 3 W. Neglect R_E, and assume R_1 and R_2 are sized for maximum symmetrical swing but dissipate negligible power. The transformer is ideal. Find (*a*) the quiescent collector current and (*b*) the supplied power; (*c*) choose a transistor for this application by specifying $i_{C\max}$, BV_{CEO}, and $P_{C\max}$; and (*d*) determine the transformer turns ratio.

(*a*) For maximum symmetrical swing, the dc load line should bisect the ac load line; thus, $V_{CEQ} = V_{CC} = 15$ V. For the given resistive load,

$$P_{L\max} = \frac{V_{cem}I_{cm}}{2} = \frac{V_{CEQ}I_{CQ}}{2}$$

so that

$$I_{CQ} = \frac{2P_{L\max}}{V_{CEQ}} = \frac{(2)(3)}{15} = 0.4 \text{ A}$$

(*b*) The supplied power is

$$P_{CC} = V_{CC}I_{CQ} = (15)(0.4) = 6 \text{ W}$$

(*c*) Maximum v_{CE} occurs at cutoff; consequently, we specify

$$BV_{CEO} \geq 2V_{CC} = 2(15) = 30 \text{ V}$$

With $R_{dc}/R_{ac} \ll 1$, (*9.10*) leads to

$$R_{ac} = \frac{V_{CC}}{I_{CQ}} = \frac{15}{0.4} = 37.5\ \Omega$$

and since collector current is a maximum at saturation, we specify

$$i_{C\max} \geq \frac{2V_{CC}}{R_{ac}} = \frac{2(15)}{37.5} = 0.8 \text{ A}$$

With R_E neglected and for an ideal transformer, we specify

$$P_{C\max} \geq V_{CEQ}I_{CQ} = (15)(0.4) = 6 \text{ W}$$

(*d*) The impedance-reflection property of the ideal transformer requires that

$$a^2 = \frac{R_{ac}}{R_L} = \frac{37.5}{1.5} = 25 \qquad \text{so} \qquad a = 5$$

9.8 If the maximum collector power dissipation $P_{C\max}$ is known, show that, for the case of maximum symmetrical swing, the Q point is described by $I_{CQ} = \sqrt{P_{C\max}/R_{ac}}$ and $V_{CEQ} = \sqrt{R_{ac}P_{C\max}}$.

From Problem 9.3, we know that the Q point lies on the maximum-dissipation hyperbola; thus,

$$I_{CQ} = \frac{P_{C\max}}{V_{CEQ}} \tag{1}$$

Moreover, from (*3.11*) evaluated at the Q point,

$$V_{CEQ} = I_{CQ}R_{ac} \tag{2}$$

Substituting (*2*) into (*1*) and rearranging yield

$$I_{CQ} = \frac{P_{C\max}}{I_{CQ}R_{ac}} \quad \text{so that} \quad I_{CQ} = \sqrt{\frac{P_{C\max}}{R_{ac}}} \tag{3}$$

and then

$$V_{CEQ} = \frac{P_{C\max}}{I_{CQ}} = \frac{P_{C\max}}{\sqrt{P_{C\max}/R_{ac}}} = \sqrt{R_{ac}P_{C\max}} \tag{4}$$

9.9 A Si transistor with ratings $P_{C\max} = 6$ W, $BV_{ECO} = 40$ V, and $i_{C\max} = 2$ A is to be used in the amplifier of Fig. 9-4(*a*), where $\beta = 20$, $R_L = 24\ \Omega$, and $R_E \approx 0$. Select V_{CC}, R_1, and R_2 to give maximum symmetrical swing.

Equations (*3*) and (*4*) of Problem 9.8 are equally applicable to an induction-coupled amplifier. Thus,

$$I_{CQ} = \sqrt{\frac{P_{C\max}}{R_{ac}}} = \sqrt{\frac{6}{24}} = 0.5 \text{ A} \quad \text{and} \quad V_{CEQ} = \sqrt{R_L P_{C\max}} = \sqrt{(24)(6)} = 12 \text{ V}$$

For maximum symmetrical swing, we would select $V_{CC} = V_{CEQ} = 12$ V.

Applying KVL around the base-emitter loop gives

$$V_{BB} = \frac{I_{CQ}}{\beta} R_B + V_{BEQ} = \frac{0.5}{20} R_B + 0.7 \tag{1}$$

We *arbitrarily* select $R_B = 400\ \Omega$; then, by (*1*), $V_{BB} = 10.7$ V. Next we simultaneously solve equation set (*3.5*) for R_1 and R_2 and use the results to find

$$R_1 = \frac{R_B}{1 - V_{BB}/V_{CC}} = \frac{400}{1 - 10.7/12} = 3.69 \text{ k}\Omega$$

and

$$R_2 = R_B \frac{V_{CC}}{V_{BB}} = 400 \frac{12}{10.7} = 448.6\ \Omega$$

9.10 Rather than use a bulky center-tapped transformer to provide input signals that are 180° out of phase to a push-pull amplifier, a *phase-inverter* circuit such as that of Fig. 9-10(*a*) can be used. In that circuit, assume that $R_3 = R_4$ and that $h_{oe} = h_{re} = 0$. Use small-signal analysis to show that $v_1/v_2 = -1$, proving that the signals are equal in magnitude and 180° out of phase.

The small-signal equivalent circuit is drawn in Fig. 9-10(*b*). By Ohm's law,

$$v_1 = h_{fe}i_bR_4 \quad \text{and} \quad v_2 = -h_{fe}i_bR_3$$

so that

$$\frac{v_1}{v_2} = \frac{h_{fe}i_bR_4}{-h_{fe}i_bR_3} = -1$$

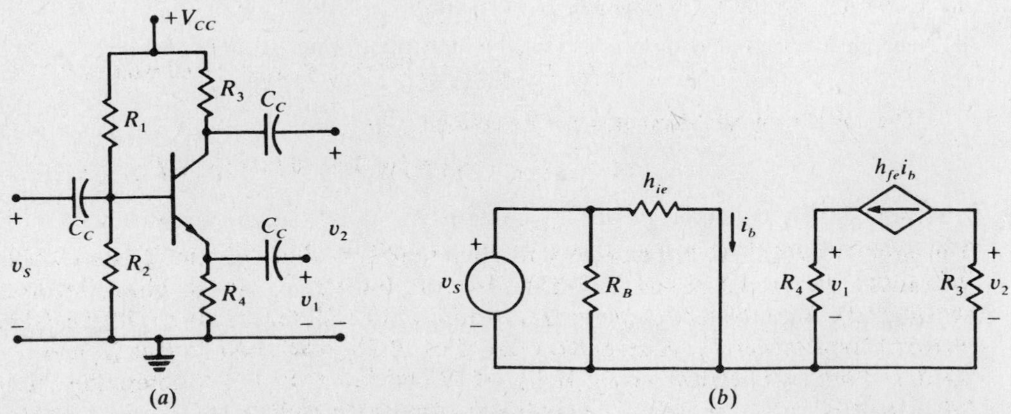

Fig. 9-10

9.11 For the class B push-pull amplifier of Fig. 9-6, find the maximum possible efficiency (which occurs at maximum signal swing) for a sinusoidal input signal.

For a sinusoidal signal $v_s = V_m \sin \omega t$, the current through each transistor is a positive half sine wave such that

$$i_{C1} = I_{cm} \sin \omega t \qquad 0 \le \omega t < \pi$$

$$i_{C2} = -I_{cm} \sin \omega t \qquad \pi \le \omega t < 2\pi$$

Thus, the average power supplied by V_{CC} is found as

$$P_{CC} = \frac{1}{\pi} \int_0^{\pi} V_{CC} I_{cm} \sin \omega t \, d(\omega t) = \frac{V_{CC}}{\pi} I_{cm} \cos \omega t \Big|_{\pi}^{0} = \frac{2}{\pi} V_{CC} I_{cm} \tag{1}$$

For maximum swing, $V_{Lm} = V_{CC}/a$, and for any signal, $I_{Lm} = aI_{cm}$. The power delivered to the load at maximum swing is then

$$P_{L\max} = \frac{V_{Lm}}{\sqrt{2}} \frac{I_{Lm}}{\sqrt{2}} = \frac{V_{CC}}{a\sqrt{2}} \frac{aI_{cm}}{\sqrt{2}} = \frac{V_{CC} I_{cm}}{2} \tag{2}$$

Hence, the maximum possible efficiency is

$$\eta_{\max} = \frac{P_{L\max}}{P_{CC}} (100\%) = \frac{V_{CC} I_{cm}/2}{(2/\pi) V_{CC} I_{cm}} (100\%) = \frac{\pi}{4} (100\%) = 78.54\%$$

9.12 A loudspeaker with an 8-Ω input resistance requiring a power of 0.5 W is to be driven by the push-pull amplifier of Fig. 9-6. V_{CC} is a 9-V transistor battery, and the identical transistors have $V_{CE\text{sat}} = 0.5$ V and $I_{BEO} = 0$. (*a*) Select a suitable turns ratio for the output transformer. (*b*) Find P_{CC} and P_C when 500 mW is being delivered to the speaker.

(*a*) We design for maximum symmetrical swing; then

$$I_{cm} = \frac{V_{CC} - V_{CE\text{sat}}}{a^2 R_L}$$

and the power delivered to the load is

$$P_{L\max} = \frac{I_{cm}^2}{2} a^2 R_L = \frac{1}{2} \left(\frac{V_{CC} - V_{CE\text{sat}}}{a^2 R_L} \right)^2 a^2 R_L = \frac{(V_{CC} - V_{CE\text{sat}})^2}{2a^2 R_L}$$

Solving for a and evaluating give the desired turns ratio:

$$a = \frac{V_{CC} - V_{CE\text{sat}}}{\sqrt{2 R_L P_{L\max}}} = \frac{9 - 0.5}{\sqrt{2(8)(0.5)}} \approx 3 \text{ turns}$$

(*b*) From (*1*) and (*2*) of Problem 9.11,

$$P_{CC} = \frac{2}{\pi} V_{CC} I_{cm} = \frac{2}{\pi} V_{CC} \frac{2P_{L\max}}{V_{CC}} = \frac{4}{\pi} P_{L\max} = \frac{4}{\pi} 0.5 = 0.636 \text{ W}$$

The collector power dissipated per transistor is

$$P_C = \tfrac{1}{2}(P_{CC} - P_L) = \tfrac{1}{2}(0.636 - 0.5) = 68 \text{ mW}$$

9.13 The circuit of Fig. 9-11 is a class A push-pull amplifier. Both Q_1 and Q_2 are biased for class A operation. The input signal is "split" by transformer T_1 (or, a phase inverter could be substituted). Since collector currents i_{C1} and i_{C2} are 180° out of phase, they establish additive fluxes in transformer T_2. Assume that the value of R_B is selected so that Q_1 and Q_2 are biased for maximum symmetrical swing. If $R_L = 5$ Ω and $V_{CC} = 15$ V, (*a*) determine the transformer turns ratio for $P_{L\max} = 10$ W, (*b*) specify the transistor ratings BV_{CEO}, $i_{C\max}$, and $P_{C\max}$, and (*c*) find the maximum possible efficiency of the amplifier.

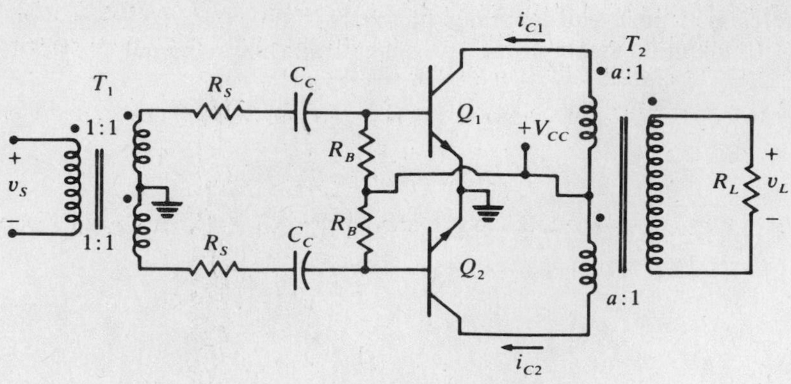

Fig. 9-11

(*a*) The output voltage can be determined from the load power requirement. Since $V_{Lm}^2/2R_L = P_{L\max}$, we have

$$V_{Lm} = \sqrt{2P_{L\max}R_L} = \sqrt{2(10)(5)} = 10 \text{ V}$$

If we assume that the transformer primary resistance is small, the dc load line is nearly vertical; thus,

$$v_{CE\max} \approx 2V_{CC} = 2(15) = 30 \text{ V} \qquad \text{and} \qquad V_{cem} = \frac{v_{CE\max}}{2} = V_{CC} = 15 \text{ V}$$

The transformer turns ratio is then

$$a = \frac{V_{cem}}{V_{Lm}} = \frac{15}{10} = 1.5$$

(*b*) The transistor avalanche breakdown voltage must be greater than or equal to $v_{CE\max}$ of part *a*, so $BV_{CEO} \geq 30$ V. One-half the reflected load current must be supplied by each transistor, so

$$I_{cm} = \tfrac{1}{2}\,\frac{1}{a}\,\frac{V_{Lm}}{R_L} = \frac{1}{2}\,\frac{1}{1.5}\,\frac{10}{5} = 0.667 \text{ A}$$

Since the circuit is designed for maximum symmetrical swing, $I_{CQ} = I_{cm}$; consequently, specify $i_{C\max} = 2I_{CQ} = 2(0.667) = 1.334$ A. The collector power dissipation for a class A amplifier is largest when no signal is present; hence, by (*9.3*),

$$P_{C\max} = P_{CC} = V_{CC}I_{CQ} = (15)(0.667) = 10 \text{ W}$$

(*c*) The power supplied to the entire amplifier is $2P_{CC}$, so the maximum amplifier efficiency is

$$\eta = \frac{P_{L\max}}{2P_{CC}}(100\%) = \frac{10}{20}(100\%) = 50\%$$

9.14 In the complementary-symmetry amplifier of Fig. 9-8, let $R_L = 5\ \Omega$ and $i_{C\max} = 2$ A (for both Q_1 and Q_2). (*a*) Find the minimum value of V_{CC} for maximum power to R_L. (*b*) Find the value of $P_{L\max}$. (*c*) Determine the efficiency when $P_{L\max}$ is being delivered.

(*a*) Since the amplifier is biased for class B operation, each transistor can be considered an independent class B emitter follower. Assume $V_{CE\text{sat}} \approx 0$ and $I_{CEO} = 0$. Then

$$\text{Minimum } V_{CC} = i_{C\max}R_L = (2)(5) = 10 \text{ V}$$

(*b*)

$$P_{L\max} = \frac{i_{C\max}^2}{2}R_L = \frac{(2)^2}{2}(5) = 10 \text{ W}$$

(*c*) From (*1*) of Problem 9.11,

$$P_{CC} = \frac{2}{\pi}V_{CC}\,I_{cm} = \frac{2}{\pi}(10)(2) = 12.73 \text{ W}$$

and

$$\eta = \frac{P_{L\max}}{P_{CC}}(100\%) = \frac{10}{12.73}(100\%) = 78.54\%$$

9.15 For a class B amplifier with negligible I^2R losses (such as that in Fig. 9-6 or 9-8), derive expressions for the value of collector current at which maximum collector dissipation occurs and for the maximum collector power dissipation.

For each transistor, the collector dissipation is one-half the total, or

$$P_C = \frac{1}{2}(P_{CC} - P_L) = \frac{1}{2}\left(\frac{2}{\pi} V_{CC} I_{cm} - \frac{I_{cm}^2}{2} a^2 R_L\right) \tag{1}$$

We differentiate (*1*) with respect to I_{Cm} and equate the result to zero to find the collector current at $P_{C\max}$:

$$\frac{\partial P_C}{\partial I_{cm}} = 0 = \frac{2}{\pi} V_{CC} - I_{cm} a^2 R_L$$

from which

$$I_{cm} = \frac{2V_{CC}}{\pi a^2 R_L} \quad \text{at} \quad P_{C\max} \tag{2}$$

Substitution of (*2*) into (*1*) gives the desired expression for $P_{C\max}$:

$$P_{C\max} = \frac{1}{2}\left[\frac{2}{\pi} V_{CC} \frac{2V_{CC}}{\pi a^2 R_L} - \frac{1}{2}\left(\frac{2V_{CC}}{\pi a^2 R_L}\right)^2 a^2 R_L\right] = \frac{1}{R_L}\left(\frac{V_{CC}}{\pi a}\right)^2 \tag{3}$$

9.16 In the complementary-symmetry amplifier of Fig. 9-12, $R_L = 4\ \Omega$, $R_C = 1\ \Omega$, and $V_{CC} = 10$ V. Find (*a*) the maximum power delivered to the load R_L, (*b*) the power supplied by each dc bias source, and (*c*) the maximum power dissipated by each transistor.

(*a*) When $v_S > 0$, Q_2 conducts and Q_1 is cut off; when $v_S < 0$, Q_1 conducts and Q_2 is cut off. If $V_{CEsat} \approx 0$, then $I_{cm1} = I_{cm2} = V_{CC}/(R_L + R_C)$, and

$$P_{L\max} = \tfrac{1}{2} I_{cm}^2 R_L = \frac{1}{2}\left(\frac{V_{CC}}{R_L + R_C}\right)^2 R_L = \frac{1}{2}\left(\frac{10}{4+1}\right)^2 (4) = 8\ \text{W}$$

(*b*) By (*1*) of Problem 9.11, since each source must supply half the total power P_{CC}, we have

$$\frac{1}{2} P_{CC} = \frac{1}{2}\frac{2}{\pi} V_{CC} I_{cm} = \frac{1}{\pi} V_{CC} \frac{V_{CC}}{R_L + R_C} = \frac{V_{CC}^2}{\pi(R_L + R_C)} = \frac{(10)^2}{\pi(4+1)} = 6.37\ \text{W}$$

(*c*) The maximum collector power dissipation per transistor follows from (*3*) of Problem 9.15 with $a = 1$ and R_L replaced with $R_L + R_C$:

$$P_{C\max} = \frac{1}{R_L + R_C}\left(\frac{V_{CC}}{\pi}\right)^2 = \frac{1}{4+1}\left(\frac{10}{\pi}\right)^2 = 2.026\ \text{W}$$

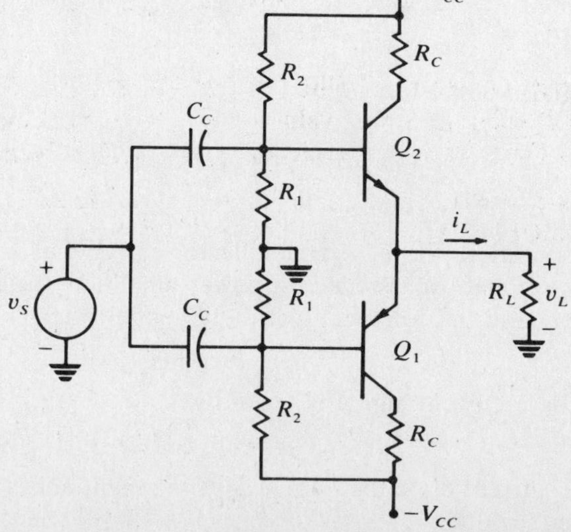

Fig. 9-12

9.17 The complementary-symmetry common-collector amplifier of Fig. 9-13 offers the advantage of using a single-ended (one-polarity, here positive) power supply. Let $V_{CC} = 15$ V, $R_E = 0.25\ \Omega$, $R_L = 10\ \Omega$, $V_{CEsat} \approx 0$, and $I_{CBO} = 0$. The transistors are a *matched-complement pair* (with identical parameters, but with characteristic curves described by complementary voltages and currents). Find (*a*) the peak values of output voltage and current and (*b*) the maximum value of the average power $P_{L\max}$.

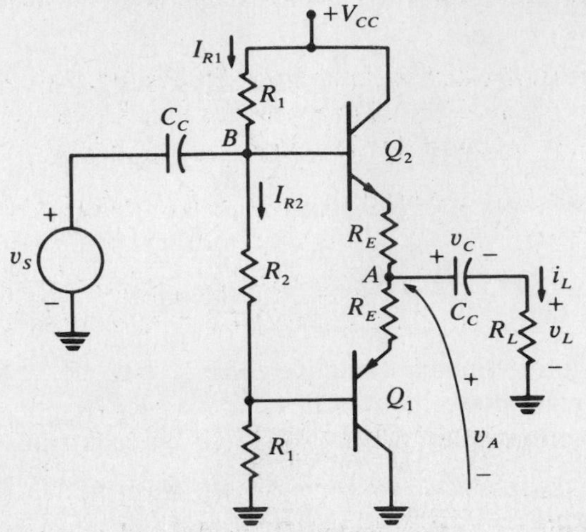

Fig. 9-13

(*a*) From Fig. 9-13, it is apparent that $I_{EQ1} = I_{EQ2}$. Further, $V_{CEQ2} = -V_{CEQ1}$ since the transistors are a matched-complement pair; hence, $V_{AQ} = V_{CC}/2 = v_C$, and the output voltage will exhibit symmetrical swing. Thus, determination of the magnitude of the positive half-cycle of v_L is sufficient. At the point of maximum positive swing of v_S (at which $V_{CE2sat} = 0$ and Q_1 is cut off), voltage division gives

$$V_{Am} = \frac{R_L}{R_E + R_L} V_{CC} \tag{1}$$

Then KVL and (*1*) give

$$V_{Lm} = V_{Am} - v_C = \left(\frac{R_L}{R_E + R_L} - \frac{1}{2}\right) V_{CC} = \left(\frac{10}{10.25} - 0.5\right)(15) = 7.134 \text{ V}$$

and
$$I_{Lm} = \frac{V_{Lm}}{R_L} = \frac{7.134}{10} = 713.4 \text{ mA}$$

(*b*) The maximum load power is

$$P_{L\max} = \frac{I_{Lm}^2}{2} R_L = \frac{(0.7134)^2}{2}(10) = 2.54 \text{ W}$$

9.18 For small-signal operation, the base and collector currents of a BJT are related by $i_c \approx h_{fe} i_b$. However, for large signal excursions in a power amplifier, the relationship between i_c and i_b becomes nonlinear, and the following parabolic model often is used:

$$i_c = h_1 i_b + h_2 i_b^2 \tag{1}$$

Show that if the base current signal is given by

$$i_b = I_{bm} \cos \omega t \tag{2}$$

then the collector current contains both a dc component and a second-harmonic component, as well as the fundamental.

Directly substituting (2) into (1) gives

$$i_c = h_1 I_{bm} \cos \omega t + h_2 I_{bm}^2 \cos^2 \omega t \tag{3}$$

After application of the trigonometric identity $\cos^2 x = \frac{1}{2} + \frac{1}{2}\cos 2x$, the total collector current can be written as

$$i_C = I_{CQ} + i_c = I_{CQ} + \tfrac{1}{2}h_2 I_{bm}^2 + h_1 I_{bm} \cos \omega t + \tfrac{1}{2} h_2 I_{bm}^2 \cos 2\omega t \tag{4}$$

from which the added dc and second-harmonic components are obvious.

9.19 The collector current of a BJT is experimentally observed to swing about a quiescent point I_{CQ} between extremes I_{max} and I_{min}. Find expressions for h_1 and h_2 in (4) of Problem 9.18 in terms of the experimentally measured values of I_{CQ}, I_{min}, and I_{max} if $i_b = I_{bm} \cos \omega t$.

When $\omega t = 0$, $i_c = I_{max}$, and when $\omega t = \pi$, $i_c = I_{min}$. Substituting these values into (4) of Problem 9.18 gives the set of equations

$$I_{max} = I_{CQ} + \tfrac{1}{2}h_2 I_{bm}^2 + h_1 I_{bm} + \tfrac{1}{2}h_2 I_{bm}^2$$
$$I_{min} = I_{CQ} + \tfrac{1}{2}h_2 I_{bm}^2 - h_1 I_{bm} + \tfrac{1}{2}h_2 I_{bm}^2$$

whose simultaneous solution yields

$$h_1 = \frac{I_{max} - I_{min}}{2I_{bm}} \qquad h_2 = \frac{I_{max} + I_{min} - 2I_{CQ}}{2I_{bm}^2}$$

9.20 The *second-harmonic distortion factor* D_2 is defined as the ratio of the magnitude of the second harmonic to that of the fundamental component. Find an expression for the second-harmonic distortion factor of the collector current of a BJT in terms of experimentally measured values I_{max}, I_{min}, and I_{CQ} for an amplifier driven by a base current $i_b = I_{bm} \cos \omega t$.

By the definition of D_2 and (4) of Problem 9.18,

$$D_2 = \left| \frac{\frac{1}{2}h_2 I_{bm}^2}{h_1 I_{bm}} \right| = \left| \frac{h_2 I_{bm}}{2h_1} \right| \tag{1}$$

Substituting the expressions for h_1 and h_2 from Problem 9.19 into (1) leads to

$$D_2 = \left| \frac{\dfrac{I_{max} + I_{min} - 2I_{CQ}}{2I_{bm}^2} I_{bm}}{\dfrac{2(I_{max} - I_{min})}{2I_{bm}}} \right| = \left| \frac{I_{max} + I_{min} - 2I_{CQ}}{2(I_{max} - I_{min})} \right|$$

9.21 A 5-W Si power transistor is to dissipate 5 W in an ambient temperature of 40°C; the device is mounted on a heat sink for which $\theta_{sa} = 10$°C/W. (*a*) Find the junction temperature T_J if the thermal resistance is $\theta_{jc} = 2.4$°C/W. (*b*) What is the maximum ambient temperature in which the device can operate if the maximum allowable junction temperature is 125°C.

(*a*) Application of KVL to the heat-flow model of Fig. 9-2 gives

$$T_J = T_A + P_C(\theta_{jc} + \theta_{ca}) = 40 + 5(2.4 + 10) = 102°\text{C}$$

(*b*)

$$T_A = T_J - P_C(\theta_{jc} + \theta_{ca}) = 125 - 5(2.4 + 10) = 63°\text{C}$$

Supplementary Problems

9.22 Find the value of R_L that will give the maximum possible efficiency to the amplifier of Problem 9.1, and determine what the resulting efficiency is. (*Hint*: Differentiate the efficiency expression of Problem 9.1 with respect to R_L, and equate the result to zero.) *Ans.* $R_L = 626.6\ \Omega$, $\eta_{max} = 6.59\%$

9.23 *Turn-on distortion* occurs at the beginning of an output-signal half-cycle, and *turn-off distortion* at the end; in both cases, distortion exists if actual $v_{ce} = 0$ for $0 < v_{be} < V_{BEQ}$. The amplifier of Fig. 9-5(*a*) is to be operated as a class B amplifier with maximum swing and no *turn-on distortion*. (*a*) If $R_L = 16\ \Omega$, $R_E = 0$, $V_{CC} = 24$ V, $a = 0.5$, and $R_2 = 2$ kΩ, select R_1 to establish class B operation for this Si transistor. Then, for maximum signal swing, calculate (*b*) the collector supplied power, (*c*) the power delivered to the load, and (*d*) the efficiency of the amplifier.
Ans. (*a*) $R_2 = 66.57$ kΩ; (*b*) $P_{CC} = 45.84$ W; (*c*) $P_{L\max} = 36$ W; (*d*) $\eta = 78.54\%$

9.24 For the Si power transistor of Fig. 9-5(*a*), $h_{FE} = \beta = 40$ and $h_{ie} = 25\ \Omega$. Assume that the transformer is ideal and that $V_{CEsat} \approx 0$ and $I_{CBO} = 0$. Let $V_{CC} = 12$ V, $a = 1/6$, $R_E = 0$, and $R_L = 2.5$ kΩ. (*a*) Find the values of R_1 and R_2 that are needed to set $I_{CQ} = 100$ mA. (*b*) Determine the peak values of output voltage and current for class A operation.
Ans. (*a*) $R_1 = 3.809$ kΩ, $R_2 = 4.21$ kΩ; (*b*) $V_{Lm} = 41.64$ V, $I_{Lm} = 16.7$ mA

9.25 For the Si power transistor of Fig. 9-5(*a*), $h_{FE} = \beta = 40$ and $h_{ie} = 25\ \Omega$; the transformer is ideal. Let $R_E = 2\ \Omega$, $R_L = 2.5$ kΩ, $V_{CC} = 12$ V, and $a = 1/6$. (*a*) Select values of R_1 and R_2 to produce β-independent bias while giving maximum symmetrical swing. (*b*) For ideal collector characteristics, determine the maximum load voltage and current for class A operation. (*c*) Find the maximum possible efficiency for this circuit.
Ans. (*a*) $R_1 = 9\ \Omega$, $R_2 = 91.8\ \Omega$; (*b*) $V_{Lm} = 69.96$ V, $I_{Lm} = 28.01$ mA; (*c*) $\eta_{max} = 48.6\%$

9.26 In the class A transformer-coupled amplifier of Fig. 9-5(*a*), $\beta = 30$, $R_E = 2\ \Omega$, $R_1 = 25\ \Omega$, $R_2 = 200\ \Omega$, $R_L = 8\ \Omega$, and $V_{CC} = 15$ V. The transistor is a Si device, and the transformer is ideal. (*a*) Specify the transformer turns ratio for maximum symmetrical swing. (*b*) Specify the minimum ratings for the transistor. (*c*) Determine the maximum possible efficiency (consider the power supplied to the R_1-R_2 bias network).
Ans. (*a*) $a = 2.21$; (*b*) $BV_{CEO} \geq 28.54$ V, $i_{C\max} \geq 0.73$ A, $P_{C\max} \geq 5.21$ W; (*c*) $\eta_{max} = 40.03\%$

9.27 In the inductor-coupled class A power amplifier of Fig. 9-4(*a*), $V_{CC} = 15$ V, $R_1 = 20\ \Omega$, $R_2 = 180\ \Omega$, $R_E = 1\ \Omega$, $R_L = 75\ \Omega$, $R = 2\ \Omega$, $L \to \infty$, and $\beta = 20$. Assume a Si transistor. (*a*) Find equations for the instantaneous values of maximum undistorted v_{CE}, i_C, and i_L if $v_S = V_m \sin \omega t$. (*b*) Specify minimum ratings for the transistor. (*c*) Find the efficiency at the given value of R_L (consider losses in the R_1-R_2 bias network).
Ans. $i_C = 0.421 + 0.183 \sin \omega t$ A, $v_{CE} = 13.737(1 - \sin \omega t)$ V, $i_L = -0.183 \sin \omega t$ A; (*b*) $BV_{CEO} \geq$ 45.31 V, $i_{C\max} \geq 0.604$ A, $P_{C\max} \geqslant 5.78$ W; (*c*) $\eta_{max} = 16.81\%$

9.28 The phase-inverter circuit of Problem 9.10 was shown to yield output voltages v_1 and v_2 that are equal in magnitude and 180° out of phase with no load present. Show that if this circuit is used to replace the input transformer T_1 of Fig. 9-6, then $v_1/v_2 = -1$ if Q_1 and Q_2 are identical, and the amplifier gives class B operation without turn-on distortion. (See Problem 9.23.)

9.29 In the class A push-pull amplifier of Fig. 9-11, let $V_{CC} = 15$ V, $R_L = 20\ \Omega$, and $a = 1/2$. If Q_1 and Q_2 are biased for maximum symmetrical swing, find the maximum possible efficiency. *Ans.* 50%

9.30 In the class B push-pull amplifier of Fig. 9-6, let $R_L = 5\ \Omega$, $a = 2$, and $V_{CC} = 15$ V. (*a*) If, under operating conditions, the power delivered to the load is 4 W, find the amplifier efficiency. (*b*) Specify the minimum power rating for the transistor of this amplifier. *Ans.* (*a*) $\eta = 66.2\%$; (*b*) $P_{C\max} \geq 1.14$ W

9.31 The complementary-symmetry push-pull amplifier of Fig. 9-8 is required to deliver a maximum power of 20 W into a 10-Ω load resistor R_L. Find (*a*) the value of V_{CC} and (*b*) the minimum ratings for the transistors. *Ans.* (*a*) $V_{CC} = 20$ V; (*b*) $i_{C\max} \geq 2$ A, $BV_{CEO} \geq 40$ V, $P_{C\max} \geq 2.026$ W

9.32 To eliminate *crossover distortion* (combined turn-on and turn-off distortion in the transistors) in push-pull amplifiers, the base-emitter junctions are forward-biased so that transistor operation is linear for small values of the input signal. In the complementary-symmetry push-pull amplifier of Fig. 9-12, let $R_1 = 15$ kΩ, and select R_2 so that $V_{BEQ2} = -V_{BEQ1} = 0.7$ V if $V_{CC} = 10$ V. *Ans.* $R_2 = 199.3$ kΩ

9.33 The class B complementary-symmetry push-pull amplifier of Fig. 9-14 uses diodes D_1 and D_2 to offer temperature-insensitive elimination of crossover distortion. Assume that $v_{D1} = v_{D2} = V_{BEQ1} = V_{BEQ2}$ and all devices are Si devices. Let $\beta_1 = \beta_2 = 20$, $V_{CC} = 15$ V, $R_L = 10$ Ω, and $R_1 = 10$ kΩ. (*a*) Find currents i_{D1}, i_{D2}, I_{CQ1}, and I_{CQ2} when $v_S = 0$. (*b*) Determine the maximum collector power dissipation and the associated rms value of the load current.
Ans. (*a*) $i_{D1} = i_{D2} = 1.366$ mA, $I_{CQ1} = I_{CQ2} = 1.301$ mA; (*b*) $P_{C\max} = 2.28$ W, $I_L = 675.4$ mA

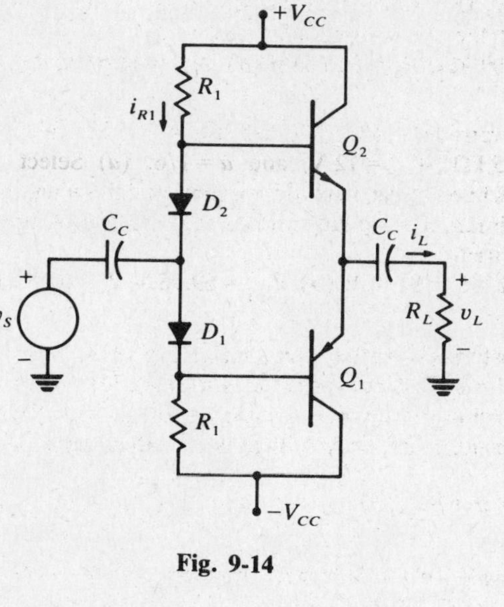

Fig. 9-14

9.34 In the CC complementary-symmetry amplifier of Fig. 9-13, let $\beta_1 = \beta_2 = 30$, $R_E = 0.25$ Ω, and $V_{CC} =$ 15 V. The transistors are Si devices. As a matter of design choice, select values of R_1 and R_2 so that $I_{R1} = 10I_{BQ2}$ (and, thus, $I_{R1} \approx I_{R2}$) and so that $I_{CQ1} = I_{CQ2} = 10$ mA.
Ans. $R_1 = 2.114$ kΩ, $R_2 = 421.5$ Ω

9.35 For the CC complementary-symmetry amplifier of Problem 9.34, find the maximum possible efficiency if $R_L = 10$ Ω. Include the power supplied to the bias network in the calculations. *Ans.* $\eta_{\max} = 73.5\%$

9.36 In the EF amplifier of Problem 3.54, let $i_s = i_b = 50 \cos \omega t$ μA. Find (*a*) the parabolic gain coefficients h_1 and h_2, and (*b*) the second-harmonic distortion factor.
Ans. (*a*) $h_1 = 114$, $h_2 = -2 \times 10^5$ A^{-1}; (*b*) $D_2 = 0.0439$

9.37 Distortion is greater in FET amplifiers than BJT amplifiers, owing to the obviously nonlinear separation of the drain characteristics. For FETs, the parabolic-gain model is written in terms of voltage as $v_{ds} = \mu_1 v_{gs} + \mu_2 v_{gs}^2$. As is done for BJTs in Problems 9.19 and 9.20, determine the gain constants μ_1 and

μ_2 and the second-harmonic distortion factor for FET amplifiers, in terms of the quiescent drain-source voltage V_{DSQ}, the peak value of the gate-source voltage V_{gsm}, and the maximum and minimum values of drain-source voltage excursion V_{max} and V_{min}.

Ans. $\mu_1 = \dfrac{V_{max} - V_{min}}{2V_{gsm}}$ $\quad \mu_2 = \dfrac{V_{max} + V_{min} - 2V_{DSQ}}{2V_{gsm}^2}$ $\quad D_2 = \left| \dfrac{V_{max} + V_{min} - 2V_{DSQ}}{2(V_{max} - V_{min})} \right|$

9.38 For the FET amplifier of Example 4.2, (*a*) find the parabolic gain coefficients μ_1 and μ_2 and (*b*) compute the second-harmonic distortion factor. *Ans.* (*a*) $\mu_1 = 4.35$, $\mu_2 = -0.85\ \text{V}^{-1}$; (*b*) $D_2 = 0.098$

9.39 The second-harmonic distortion factor for the amplifier of Fig. 9-4(*a*) is 5 percent ($D_2 = 0.05$). Determine the error in load power calculation if only the power delivered at the fundamental frequency is considered. [*Hint*: You should be able to show that $P_L = (1 + D_2^2)P_1$, where P_1 is the power delivered to the load at the fundamental frequency.] *Ans.* 0.25% error

9.40 A 5-W Si power transistor is to dissipate 3 W in a particular circuit. The maximum allowable junction temperature is 125°C, the ambient temperature is 40°C, and $\theta_{jc} = 2.4°\text{C/W}$. (*a*) Determine the maximum allowable thermal resistance θ_{ca} between the case and ambient. (*b*) What is the case temperature during operation? *Ans.* (*a*) $\theta_{ca} = 25.93°\text{C/W}$; (*b*) $T_C = 117.8°\text{C}$

Chapter 10

Operational Amplifiers

10.1 INTRODUCTION

The name *operational amplifier* (op amp) was originally given to an amplifier that could be easily modified by external circuitry to perform mathematical operations (addition, scaling, integration, etc.) in analog-computer applications. However, with the advent of solid-state technology, op amps have become highly reliable, miniaturized, temperature-stabilized, and consistently predictable in performance; they now figure as fundamental building blocks in basic amplification and signal conditioning, in active filters, function generators, and switching circuits.

10.2 IDEAL AND PRACTICAL OP AMPS

An op amp amplifies the difference $v_d \equiv v_1 - v_2$ between two input signals (see Fig. 10-1), exhibiting the open-loop voltage gain

$$A_{OL} \equiv \frac{v_o}{v_d} \tag{10.1}$$

In Fig. 10-1, terminal 1 is the *inverting input* (labeled with a minus sign on the actual amplifier); signal v_1 is amplified in magnitude and appears phase-inverted at the output. Terminal 2 is the *noninverting input* (labeled with a plus sign); output due to v_2 is phase-preserved.

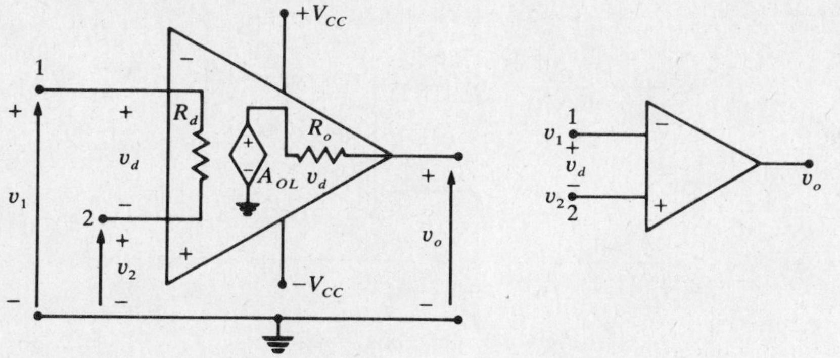

(*a*) Complete representation (*b*) Simplified representation

Fig. 10-1 Operational amplifier

In magnitude, the open-loop voltage gain in op amps ranges from 10^4 to 10^7. The maximum magnitude of the output voltage from an op amp is called its *saturation voltage*; this voltage is approximately 2 V smaller than the power-supply voltage. In other words, the amplifier is linear over the range

$$-(V_{CC} - 2) < v_o < V_{CC} - 2 \qquad \text{V} \tag{10.2}$$

The *ideal* op amp has three essential characteristics which serve as standards for assessing the goodness of a *practical* op amp:

1. The open-loop voltage gain A_{OL} is negatively infinite.
2. The input impedance R_d between terminals 1 and 2 is infinitely large; thus, the input current is zero.
3. The output impedance R_o is zero; consequently, the output voltage is independent of the load.

Figure 10-1(*a*) models the practical characteristics.

Example 10.1 An op amp has saturation voltage $V_{osat} = 10$ V, an open-loop voltage gain of -10^5, and input resistance 100 kΩ. Find (*a*) the value of v_d that will just drive the amplifier to saturation and (*b*) the op amp input current at the onset of saturation.

(*a*) By (*10.1*),

$$v_d = \frac{\pm V_{osat}}{A_{OL}} = \frac{\pm 10}{-10^5} = \pm 0.1 \text{ mV}$$

(*b*) Let i_{in} be the current into terminal 1 of Fig. 10-1(*b*); then

$$i_{in} = \frac{v_d}{R_d} = \frac{\pm 0.1 \times 10^{-3}}{100 \times 10^3} = \pm 1 \text{ nA}$$

In application, a large percentage of negative feedback is used with the operational amplifier, giving a circuit whose characteristics depend almost entirely on circuit elements external to the basic op amp. The error due to treatment of the basic op amp as ideal tends to diminish in the presence of negative feedback.

10.3 INVERTING AMPLIFIER

The *inverting amplifier* of Fig. 10-2 has its noninverting input connected to ground or common. A signal is applied through input resistor R_1, and negative current feedback (see Problem 10.1) is implemented through *feedback resistor* R_F. Output v_o has polarity opposite that of input v_S.

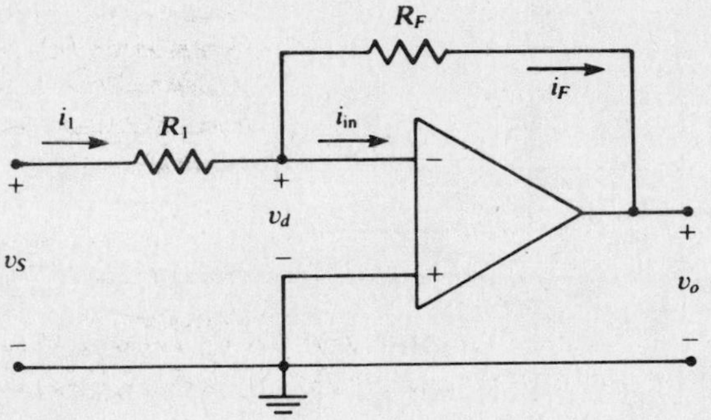

Fig. 10-2 Inverting amplifier

Example 10.2 For the inverting amplifier of Fig. 10-2, find the voltage gain v_o/v_S using (*a*) only characteristic 1 and (*b*) only characteristic 2 of the ideal op amp.

(*a*) By the method of node voltages at the inverting input, the current balance is

$$\frac{v_S - v_d}{R_1} + \frac{v_o - v_d}{R_F} = i_{in} = \frac{v_d}{R_d} \tag{10.3}$$

where R_d is the differential input resistance. By (*10.1*), $v_d = v_o/A_{OL}$ which, when substituted into (*10.3*), gives

$$\frac{v_S - v_o/A_{OL}}{R_1} + \frac{v_o - v_o/A_{OL}}{R_F} = \frac{v_o/R_d}{A_{OL}} \tag{10.4}$$

In the limit as $A_{OL} \rightarrow -\infty$, (*10.4*) becomes

$$\frac{v_S}{R_1} + \frac{v_o}{R_F} = 0 \qquad \text{so that} \qquad A_v \equiv \frac{v_o}{v_S} = -\frac{R_F}{R_1} \tag{10.5}$$

(*b*) If $i_{in} = 0$, then $v_d = i_{in}R_d = 0$, and $i_1 = i_F \equiv i$. The input and feedback-loop equations are, respectively,

$$v_S = iR_1 \qquad \text{and} \qquad v_o = -iR_F$$

whence

$$A_v \equiv \frac{v_o}{v_S} = -\frac{R_F}{R_1} \tag{10.6}$$

in agreement with (*10.5*).

10.4 NONINVERTING AMPLIFIER

The *noninverting amplifier* of Fig. 10-3 is realized by grounding R_1 of Fig. 10-2 and applying the input signal at the noninverting op amp terminal. When v_2 is positive, v_o is positive and current i is positive. Voltage $v_1 = iR_1$ then is applied to the inverting terminal as negative voltage feedback.

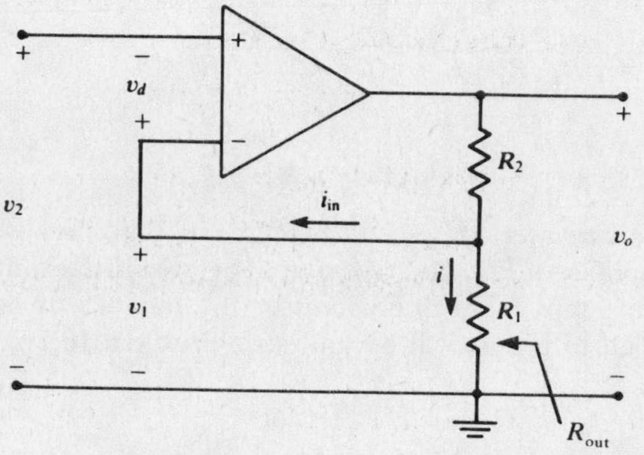

Fig. 10-3 Noninverting amplifier

Example 10.3 For the noninverting amplifier of Fig. 10-3, assume that the current into the inverting terminal of the op amp is zero, so that $v_d \approx 0$ and $v_1 \approx v_2$. Derive an expression for the voltage gain v_o/v_2.

With zero input current to the basic op amp, the currents through R_2 and R_1 must be identical; thus,

$$\frac{v_o - v_1}{R_2} = \frac{v_1}{R_1} \qquad \text{and} \qquad A_v \equiv \frac{v_o}{v_2} \approx \frac{v_o}{v_1} = 1 + \frac{R_2}{R_1} \tag{10.7}$$

10.5 COMMON-MODE REJECTION RATIO

The *common-mode gain* is defined (see Fig. 10-1) as

$$A_{cm} \equiv -\frac{v_o}{v_2} \tag{10.8}$$

where $v_1 = v_2$ by explicit connection. Usually, A_{cm} is much less than unity ($A_{cm} = -0.01$ being typical). Common-mode gain sensitivity is frequently quantized via the *common-mode rejection ratio* (CMRR), defined as

$$\text{CMRR} = \frac{A_{OL}}{A_{cm}} \tag{10.9}$$

and expressed in decibels as

$$\text{CMRR}_{db} = 20 \log \frac{A_{OL}}{A_{cm}} = 20 \log \text{CMRR} \tag{10.10}$$

Typical values for the CMRR range from 100 to 10,000, with corresponding CMRR_{db} values of from 40 to 80 db.

Example 10.4 Find the voltage-gain ratio A_v of the noninverting amplifier of Fig. 10-3 in terms of its CMRR. Assume $v_1 = v_2$ insofar as the common-mode gain is concerned.

The amplifier output voltage is the sum of two components. The first results from amplification of the difference voltage v_d as given by (*10.1*). The second, defined by (*10.8*), is a direct consequence of the common-mode gain. The total output voltage is, then,

$$v_o = A_{OL}v_d - A_{cm}v_2 \tag{10.11}$$

Voltage division (with $i_{in} = 0$) gives

$$v_d = v_1 - v_2 = \frac{R_1}{R_1 + R_2} v_o - v_2 \tag{10.12}$$

and substituting (*10.12*) into (*10.11*) and rearranging give

$$v_o\left(1 - \frac{A_{OL}R_1}{R_1 + R_2}\right) = -(A_{OL} + A_{cm})v_2$$

Then

$$A_v = \frac{v_o}{v_2} = \frac{-(A_{OL} + A_{cm})}{1 - A_{OL}R_1/(R_1 + R_2)} = \frac{-A_{OL}}{1 - A_{OL}R_1/(R_1 + R_2)} - \frac{A_{OL}/\text{CMRR}}{1 - A_{OL}R_1/(R_1 + R_2)} \tag{10.13}$$

10.6 SUMMER AMPLIFIER

The *inverting summer amplifier* (or *inverting adder*) of Fig. 10-4 is formed by adding parallel inputs to the inverting amplifier of Fig. 10-2. Its output is a weighted sum of the inputs, but inverted in polarity. In an ideal op amp, there is no limit to the number of inputs; however, the gain is reduced as inputs are added to a practical op amp (see Problem 10.29).

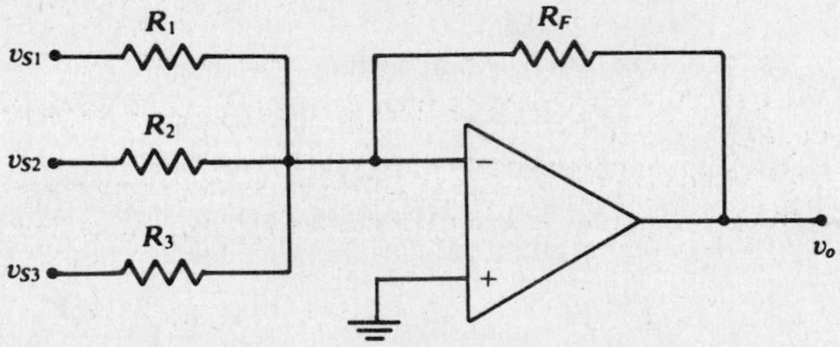

Fig. 10-4 Inverting summer amplifier

Example 10.5 Find an expression for the output of the inverting summer amplifier of Fig. 10-4, assuming the basic op amp is ideal.

We use the principle of superposition. With $v_{S2} = v_{S3} = 0$, the current in R_1 is not affected by the presence of R_2 and R_3, since the inverting node is a virtual ground (see Problem 10.1). Hence, the output voltage due to v_{S1} is, by (*10.5*), $v_{o1} = -(R_F/R_1)v_{S1}$. Similarly, $v_{o2} = -(R_F/R_2)v_{S2}$ and $v_{o3} = -(R_F/R_3)v_{S3}$. Then, by superposition,

$$v_o = v_{o1} + v_{o2} + v_{o3} = -R_F\left(\frac{v_{S1}}{R_1} + \frac{v_{S2}}{R_2} + \frac{v_{S3}}{R_3}\right)$$

10.7 DIFFERENTIATING AMPLIFIER

The introduction of a capacitor into the input path of an op amp leads to time differentiation of the input signal. The circuit of Fig. 10-5 represents the simplest *inverting differentiator* involving an op amp. As such, the circuit finds limited practical use, since high-frequency noise can produce a derivative whose magnitude is comparable to that of the signal. In practice, high-pass filtering is utilized to reduce the effects of noise (see Problem 10.7).

Example 10.6 Find an expression for the output of the inverting differentiator of Fig. 10-5, assuming the basic op amp is ideal.

Since the op amp is ideal, $v_d \approx 0$, and the inverting terminal is a virtual ground. Consequently, v_S appears across capacitor C:

$$i_S = C\frac{dv_S}{dt}$$

But the capacitor current is also the current through R (since $i_{in} = 0$). Hence,

$$v_o = -I_F R = -i_S R = -RC\frac{dv_S}{dt}$$

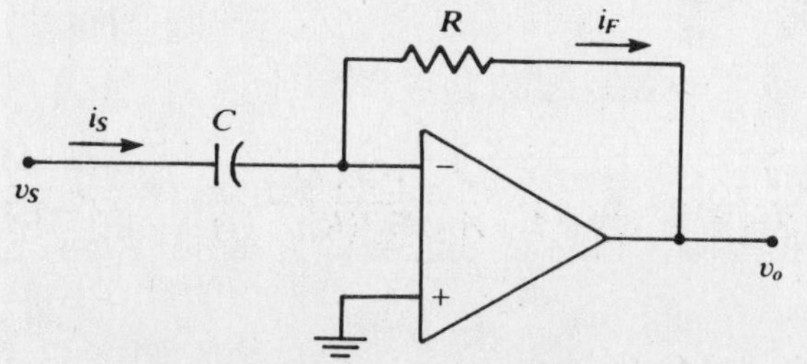

Fig. 10-5 Differentiating amplifier

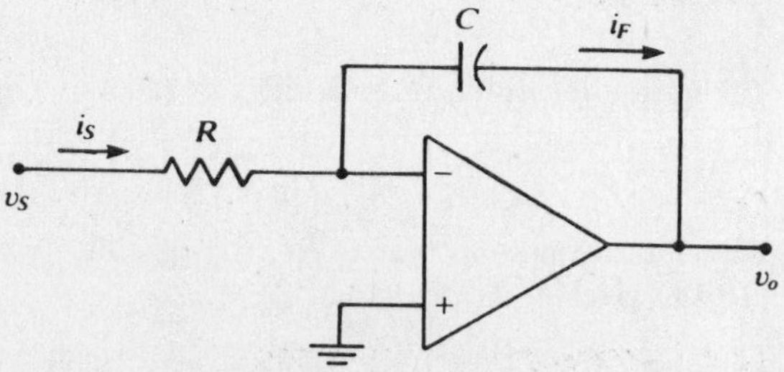

Fig. 10-6 Integrating amplifier

10.8 INTEGRATING AMPLIFIER

The insertion of a capacitor in the feedback path of an op amp results in an output signal that is a time integral of the input signal. A circuit arrangement for a simple *inverting integrator* is given in Fig. 10-6.

Example 10.7 Show that the output of the inverting integrator of Fig. 10-6 actually is the time integral of the input signal, assuming the op amp is ideal.

If the op amp is ideal, the inverting terminal is a virtual ground, and v_S appears across R. Thus, $i_S = v_S/R$. But, with negligible current into the op amp, the current through R must also flow through C. Then

$$v_o = -\frac{1}{C}\int i_F\,dt = -\frac{1}{C}\int i_S\,dt = -\frac{1}{RC}\int v_S\,dt$$

10.9 LOGARITHMIC AMPLIFIER

Analog multiplication can be carried out with a basic circuit like that of Fig. 10-7. Essential to the operation of the logarithmic amplifier is the use of a feedback-loop device that has an exponential terminal characteristic curve; one such device is the semiconductor diode of Chapter 2, which is characterized by

$$i_D = I_o(e^{v_D/\eta V_T} - 1) \approx I_o\,e^{v_D/\eta V_T} \tag{10.14}$$

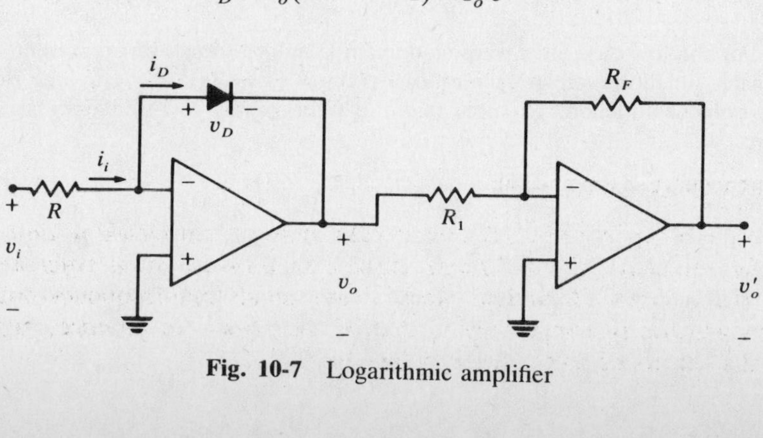

Fig. 10-7 Logarithmic amplifier

A grounded-base BJT can also be utilized, since its emitter current and base-to-emitter voltage are related by

$$i_E = I_S\, e^{v_{BE}/V_T} \tag{10.15}$$

Example 10.8 Determine the condition under which the output voltage v_o is proportional to the logarithm of the input voltage v_i in the circuit of Fig. 10-7.

Since the op amp draws negligible current,

$$i_i = \frac{v_i}{R} = i_D \tag{10.16}$$

Since $v_D = -v_o$, substitution of (*10.16*) into (*10.14*) yields

$$v_i = RI_o\, e^{-v_o/V_T} \tag{10.17}$$

Taking the logarithm of both sides of (*10.17*) leads to

$$\ln v_i = \ln RI_o - \frac{v_o}{V_T} \tag{10.18}$$

Under the condition that $\ln RI_o$ is negligible (which can be accomplished by controlling R so that $RI_o \approx 1$), (*10.18*) gives $v_o \approx -V_T \ln v_i$.

10.10 FILTER APPLICATIONS

The use of op amps in active *RC* filters has increased with the move to integrated circuits. Active filter realizations can eliminate the need for bulky inductors, which do not satisfactorily lend themselves to integrated circuitry. Further, active filters do not necessarily attenuate the signal over the pass band, as do their passive-element counterparts. A simple *inverting, first-order, low-pass filter* using an op amp as the active device is shown in Fig. 10-8(*a*).

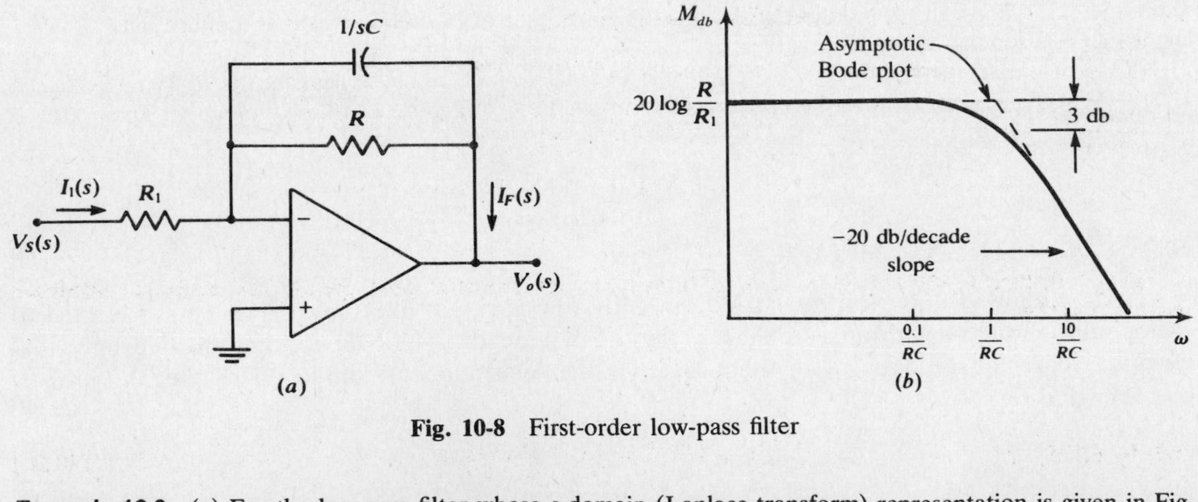

Fig. 10-8 First-order low-pass filter

Example 10.9 (*a*) For the low-pass filter whose *s*-domain (Laplace-transform) representation is given in Fig. 10-8(*a*), find the transfer function (voltage-gain ratio) $A_v(s) = V_o(s)/V_S(s)$. (*b*) Draw the Bode plot (M_{db} only) associated with the transfer function, to show that the filter passes low-frequency signals and attenuates high-frequency signals.

(*a*) The feedback impedance $Z_F(s)$ and the input impedance $Z_1(s)$ are

$$Z_F(s) = \frac{R(1/sC)}{R + (1/sC)} = \frac{R}{sRC + 1} \qquad \text{and} \qquad Z_1(s) = R_1 \tag{10.19}$$

The resistive circuit analysis of Example 10.2 extends directly to the *s* domain; thus,

$$A_v(s) = -\frac{Z_F(s)}{Z_1(s)} = -\frac{R/R_1}{sRC + 1} \tag{10.20}$$

(*b*) Letting $s = j\omega$ in (*10.20*) gives

$$M_{db} \equiv 20 \log |A_v(j\omega)| = 20 \log \frac{R}{R_1} - 20 \log |j\omega RC + 1|$$

A plot of M_{db} is displayed in Fig. 10-8(*b*). The curve is essentially flat below $\omega = 0.1/RC$; thus, all frequencies below $0.1/RC$ are passed with the dc gain R/R_1. A 3-db reduction in gain is experienced at the corner frequency $1/\tau = 1/RC$, and the gain is attenuated by 20 db per decade of frequency change for frequencies greater than $10/RC$.

10.11 FUNCTION GENERATORS AND SIGNAL CONDITIONERS

Frequently in analog system design, the need arises to modify amplifier gain in various ways, to compare signals with a generated reference, or to limit signals depending on their values. Such circuit applications can often be implemented with the high-input-impedance, low-output-impedance and high-gain characteristics of the op amp. The possibilities for op amp circuits are boundless; typically, however, nonlinear elements (such as diodes or transistors) are introduced into negative feedback paths, while linear elements are used in the input branches.

Example 10.10 The signal-conditioning amplifier of Fig. 10-9 changes gain depending upon the polarity of V_S. Find the circuit voltage gain for positive v_S and for negative v_S if diode D_2 is ideal.

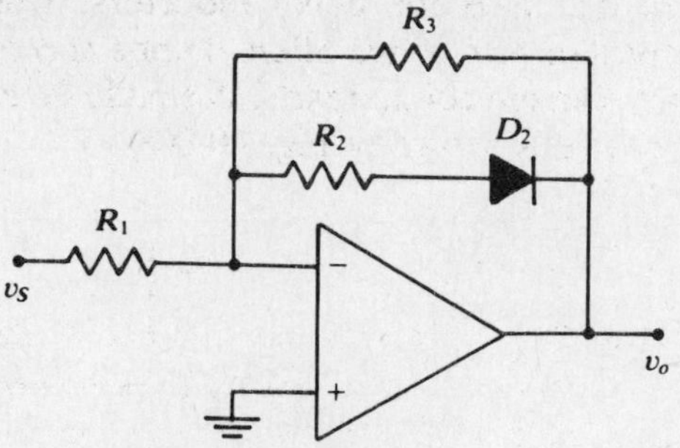

Fig. 10-9

If $v_S > 0$, then $v_o < 0$ and D_2 is forward-biased and appears as a short circuit. The equivalent feedback resistance is then

$$R_{Feq} = \frac{R_2 R_3}{R_2 + R_3}$$

and, by (*10.5*),

$$A_v = -\frac{R_{Feq}}{R_1} = -\frac{R_2 R_3}{R_1(R_2 + R_3)} \tag{10.21}$$

If $v_S < 0$, then $v_o > 0$ and D_2 is reverse-biased and appears as an open circuit. The equivalent feedback resistance is now $R_{Feq} = R_3$, and

$$A_v = -\frac{R_{Feq}}{R_1} = -\frac{R_3}{R_1} \tag{10.22}$$

Solved Problems

10.1 For the inverting amplifier of Fig. 10-2: (*a*) Show that as $A_{OL} \to -\infty$, $v_d \to 0$; thus, the inverting input remains nearly at ground potential (and is called a *virtual ground*). (*b*) Show that the current feedback is actually negative feedback.

(*a*) By KVL around the outer loop,

$$v_S - v_o = i_1 R_1 + i_F R_F \tag{1}$$

Using (*10.1*) in (*1*), rearranging, and taking the limit give

$$\lim_{A_{OL} \to -\infty} v_d = \lim_{A_{OL} \to -\infty} \frac{-i_1 R_1 - i_F R_F + v_S}{A_{OL}} = 0 \tag{2}$$

(*b*) The feedback is negative if i_F counteracts i_1; that is, the two currents must have the same algebraic sign. By two applications of KVL, with $v_d \approx 0$,

$$i_1 = \frac{v_S - v_d}{R_1} \approx \frac{v_S}{R_1} \qquad \text{and} \qquad i_F = \frac{-v_o + v_d}{R_F} \approx \frac{-v_o}{R_F}$$

But in an inverting amplifier, v_o and v_S have opposite signs; therefore, i_1 and i_F have like signs.

10.2 (*a*) Use (*10.4*) to derive an exact formula for the gain of a practical inverting op amp. (*b*) If $R_1 = 1\ \text{k}\Omega$, $R_F = 10\ \text{k}\Omega$, $R_d = 1\ \text{k}\Omega$, and $A_{OL} = -10^4$, evaluate the gain of this inverting amplifier. (*c*) Compare the result of part *b* with the ideal op amp approximation given by (*10.5*).

(*a*) Rearranging (*10.4*) to obtain the voltage-gain ratio gives

$$A_v \equiv \frac{v_o}{v_S} = \frac{A_{OL}}{1 + (R_1/R_F)(1 - A_{OL}) + R_1/R_d}$$

(*b*) Substitution of the given values yields

$$A_v = \frac{-10^4}{1 + (1/10)(1 + 10^4) + 1/1} = -9.979$$

(*c*) From (*10.5*),

$$A_{v\text{ideal}} = -\frac{R_F}{R_1} = -10$$

so the error is

$$\frac{-9.979 - (-10)}{-9.979}(100\%) = -0.21\%$$

Note that R_d and A_{OL} are far removed from the ideal, yet the error is quite small.

10.3 A *differential amplifier* (sometimes called a *subtractor*) responds to the difference between two input signals, removing any identical portions (often a bias or noise) in a process called *common-mode rejection*. Find an expression for v_o in Fig. 10-10 that shows this circuit to be a differential amplifier. Assume an ideal op amp.

Since the current into the ideal op amp is zero, a loop equation gives

$$v_1 = v_{S1} - Ri_1 = v_{S1} - R\frac{v_{S1} - v_o}{R + R_1}$$

By voltage division at the noninverting node,

$$v_2 = \frac{R_1}{R + R_1} v_{S2}$$

In the ideal op amp, $v_d = 0$, so that $v_1 = v_2$, which leads to

$$v_o = \frac{R_1}{R}(v_{S2} - v_{S1})$$

Thus, the output voltage is directly proportional to the difference between the input voltages.

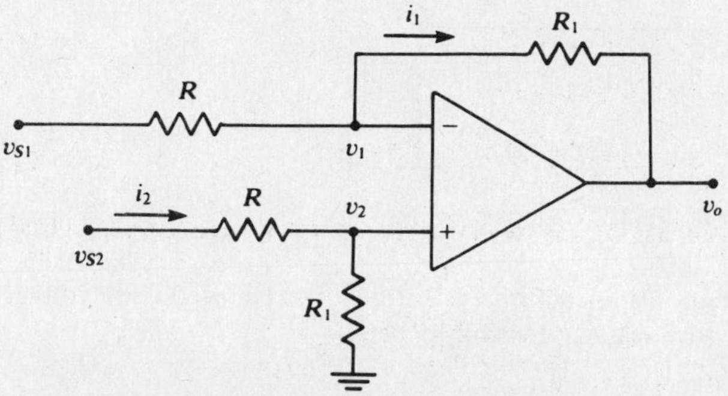

Fig. 10-10 Differential amplifier

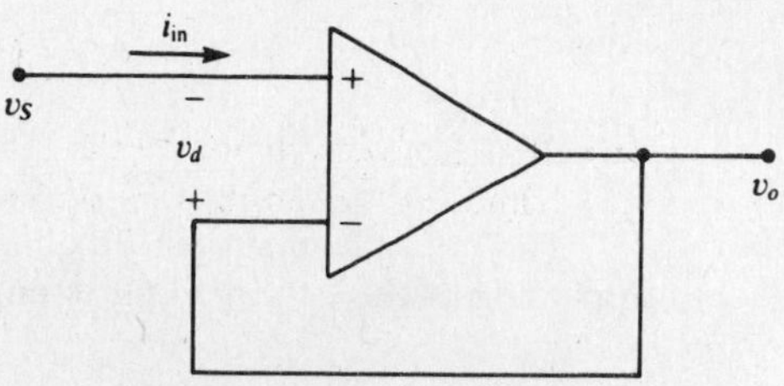

Fig. 10-11 Unity follower

10.4 Find the input impedance Z_1 of the inverting amplifier of Fig. 10-2, assuming the basic op amp is ideal.

Consider v_S a driving-point source. Since the op amp is ideal, the inverting terminal is a virtual ground, and a loop equation at the input leads to

$$v_S = i_1 R_1 + 0 \qquad \text{so that} \qquad Z_1 = \frac{v_S}{i_1} = R_1$$

10.5 The *unity-follower* amplifier of Fig. 10-11 has a voltage gain of 1, and the output is in phase with the input. It also has an extremely high input impedance, leading to its use as an intermediate-stage (*buffer*) amplifier to prevent a small load impedance from loading a source. Assume a practical op amp having $A_{OL} = -10^6$ (a typical value). (*a*) Show that $v_o \approx v_S$. (*b*) Find an expression for the amplifier input impedance, and evaluate it for $R_d = 1\ \text{M}\Omega$ (a typical value).

(*a*) Writing a loop equation and using (*10.1*), we have

$$v_S = v_o - v_d = v_o\left(1 - \frac{1}{A_{OL}}\right)$$

from which $$v_o = \frac{v_S}{1 - 1/A_{OL}} = \frac{v_S}{1 + 10^{-6}} = 0.999999 v_S \approx v_S$$

(*b*) Considering v_S a driving-point source and using (*10.1*), we have

$$v_S = i_{in} R_d + v_o = i_{in} R_d - A_{OL} v_d = i_{in} R_d (1 - A_{OL})$$

and $$Z_{in} = \frac{v_S}{i_{in}} = R_d(1 - A_{OL}) \approx -A_{OL} R_d = -(-10^6)(10^6) = 1\ \text{T}\Omega$$

10.6 Find an expression for the output v_o of the amplifier circuit of Fig. 10-12. Assume an ideal op amp. What mathematical operation does the circuit perform?

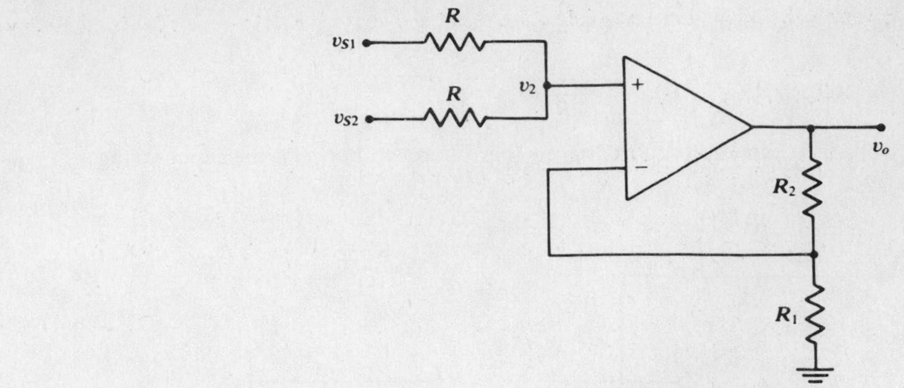

Fig. 10-12

The principle of superposition is applicable to this linear circuit. With $v_{S2} = 0$ (shorted), the voltage appearing at the noninverting terminal is found by voltage division to be

$$v_2 = \frac{R}{R+R} v_{S1} = \frac{v_{S1}}{2} \tag{1}$$

Let v_{o1} be the value of v_o with $v_{S2} = 0$. By the result of Example 10.3 and (*1*),

$$v_{o1} = \left(1 + \frac{R_2}{R_1}\right) v_2 = \left(1 + \frac{R_2}{R_1}\right) \frac{v_{S1}}{2}$$

Similarly, with $v_{S1} = 0$,

$$v_{o2} = \left(1 + \frac{R_2}{R_1}\right) \frac{v_{S2}}{2}$$

By superposition, the total output is then

$$v_o = v_{o1} + v_{o2} = \frac{1}{2}\left(1 + \frac{R_2}{R_1}\right)(v_{S1} + v_{S2})$$

The circuit is a noninverting adder.

10.7 The circuit of Fig. 10-13(*a*) (represented in the *s* domain) is a more practical differentiator than that of Fig. 10-5, because it will attenuate high-frequency noise. (*a*) Find the *s*-domain transfer function relating V_o and V_S. (*b*) Sketch the Bode plot (M_{db} only), and how high-frequency noise effects are reduced. Assume an ideal op amp.

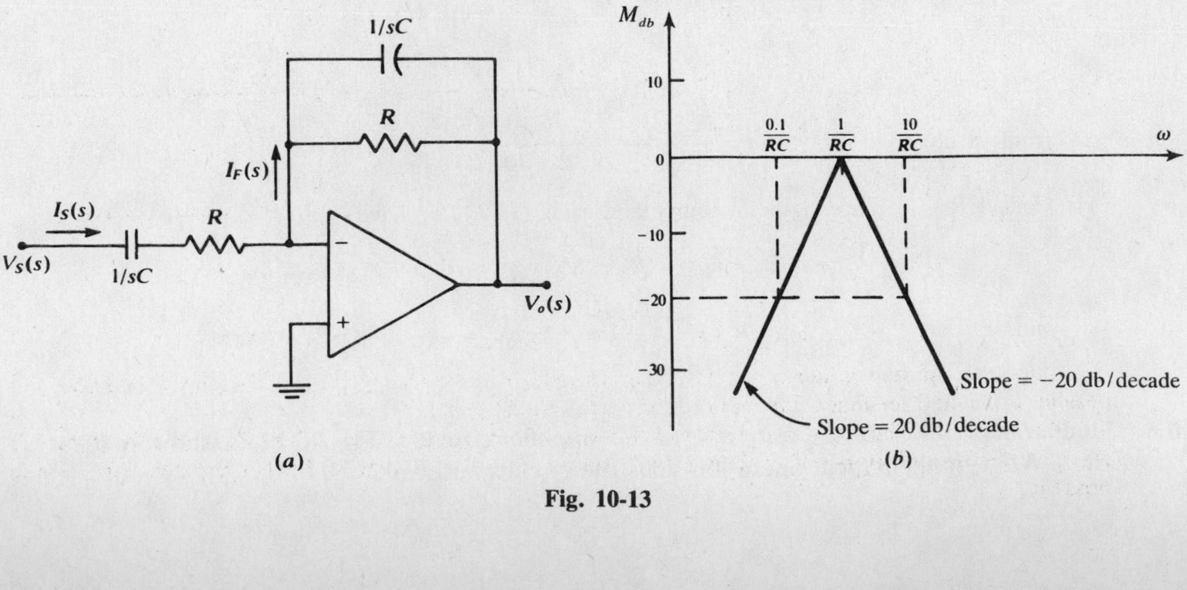

Fig. 10-13

(*a*) In an ideal op amp the inverting terminal is a virtual ground, so $I_S(s) = -I_F(s)$. As in Example 10.9,

$$Z_F(s) = \frac{R}{sRC + 1}$$

Then
$$I_F(s) = \frac{V_o(s)}{Z_F(s)} = \frac{sRC+1}{R} V_o(s)$$

But
$$V_S(s) = I_S(s)Z_{in}(s) = -I_F(s)Z_{in}(s) = -\frac{sRC+1}{R} V_o(s) \frac{sRC+1}{sC}$$

whence
$$A(s) \equiv \frac{V_o(s)}{V_S(s)} = -\frac{sRC}{(sRC+1)^2}$$

(*b*) From the result of part *a*,

$$M_{db} \equiv 20 \log |A(j\omega)| = 20 \log \omega RC - 40 \log |j\omega RC + 1| \approx \begin{cases} 20 \log \omega RC & \text{for } \omega RC \le 1 \\ -20 \log \omega RC & \text{for } \omega RC \ge 1 \end{cases}$$

Figure 10-13(*b*) is a plot of this approximate (asymptotic) expression for M_{db}. For a true differentiator, we would have

$$v_o = K \frac{dv_S}{dt} \qquad \text{or} \qquad V_o = sKV_S$$

which would lead to $M_{db} = 20 \log \omega K$. Thus the practical circuit differentiates only components of the signal whose frequency is less than the break frequency $f_1 \equiv 1/2\pi RC$ Hz. Spectral components above the break frequency including (and especially) noise—will be attenuated; the higher the frequency, the greater the attenuation.

10.8 In analog signal processing, the need often arises to introduce a *level clamp* (linear amplification to a desired output level or value and then no further increase in output level as the input continues to increase). One level-clamp circuit, shown in Fig. 10-14(*a*), uses series Zener diodes in a negative feedback path. Assuming ideal Zeners and op amp, find the relationship between v_o and v_S. Sketch the results on a transfer characteristic.

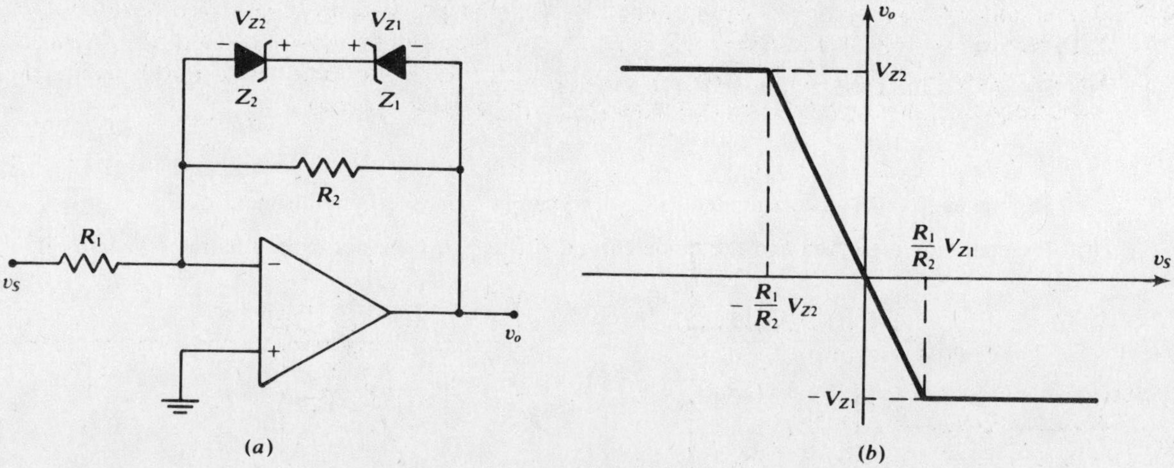

Fig. 10-14

Since the op amp is ideal, the inverting terminal is a virtual ground, and v_o appears across the parallel-connected feedback paths. There are two distinct possibilities:

Case I: $v_S > 0$. For $v_o < 0$, Z_2 is forward-biased and Z_1 reverse-biased. The Zener feedback path is an open circuit until $v_o = -V_{Z1}$; then Z_1 will limit v_o at $-V_{Z1}$ so that no further negative excursion is possible.

Case II: $v_S < 0$. For $v_o > 0$, Z_1 is forward-biased and Z_2 reverse-biased. The Zener feedback path acts as an open circuit until v_o reaches V_{Z2}, at which point Z_2 limits v_o to that value. In summary, for both cases,

$$v_o = \begin{cases} V_{Z2} & \text{for } v_S < -\dfrac{R_1}{R_2} V_{Z2} \\ -\dfrac{R_2}{R_1} v_S & \text{for } -\dfrac{R_1}{R_2} V_{Z2} \le v_S \le \dfrac{R_1}{R_2} V_{Z1} \\ -V_{Z1} & \text{for } v_S > \dfrac{R_1}{R_2} V_{Z1} \end{cases}$$

Figure 10-14(*b*) gives the transfer characteristic.

10.9 The circuit of Fig. 10-15 is an *adjustable-output voltage regulator*. Assume that the basic op amp is ideal. Regulation of the Zener is preserved if $i_Z \ge 0.1 I_Z$ (Section 2.9). (*a*) Find the regulated output v_o in terms of V_Z. (*b*) Given a specific Zener diode and the values of R_S and R_1, over what range of V_S would there be no loss of regulation?

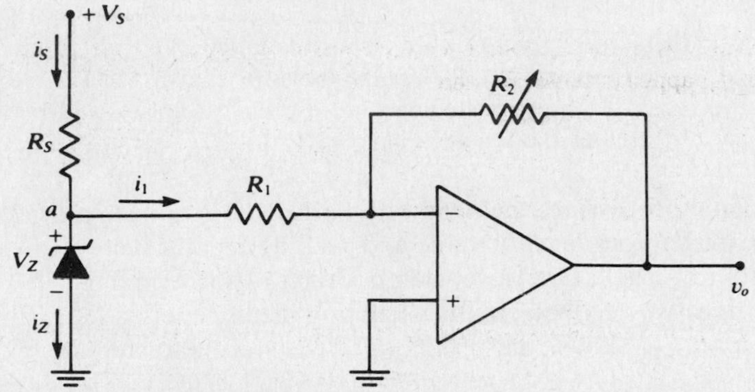

Fig. 10-15

(*a*) Since V_Z is the voltage at node *a*, (*10.5*) gives

$$v_o = -\frac{R_2}{R_1} V_Z$$

So long as $i_Z \ge 0.1 I_Z$, a regulated value of v_o can be achieved by adjustment of R_2.

(*b*) Regulation is preserved and the diode current $i_Z = i_S - i_1$ does not exceed its rated value I_Z if

$$0.1 I_Z \le i_S - i_1 \le I_Z \qquad \text{or} \qquad 0.1 I_Z \le \frac{V_S - V_Z}{R_S} - \frac{V_Z}{R_1} \le I_Z$$

or

$$0.1 I_Z R_S + \left(1 + \frac{R_S}{R_1}\right) V_Z \le V_S \le I_Z R_S + \left(1 + \frac{R_S}{R_1}\right) V_Z$$

10.10 The circuit of Fig. 10-16(*a*) is a *limiter*; it reduces the signal gain to some limiting level rather than imposing the abrupt clamping action of the circuit of Problem 10.8. (*a*) Determine the limiting value V_ℓ of v_o at which the diode D becomes forward-biased, thus establishing a second feedback path through R_3. Assume an ideal op amp and a diode characterized by Fig. 2-2(*a*). (*b*) Determine the relationship between v_o and v_S, and sketch the transfer characteristic.

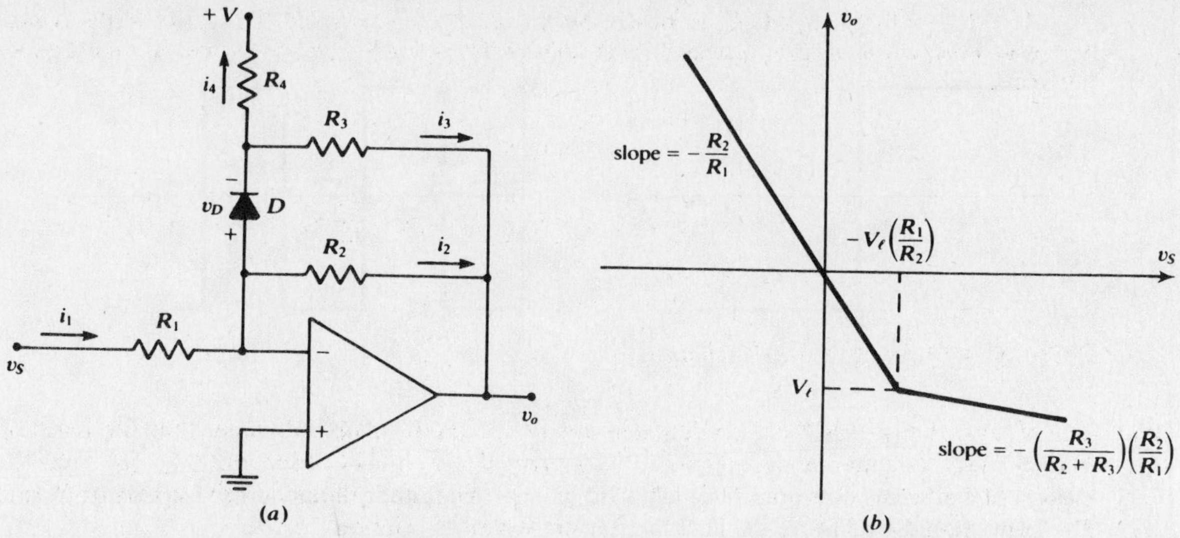

Fig. 10-16

(*a*) The diode voltage v_D is found by writing a loop equation. Since the inverting input is a virtual ground, v_o appears across R_2 and

$$v_D = -v_o - i_3 R_3 = v_o - \frac{V - v_o}{R_3 + R_4} R_3 \tag{1}$$

When $v_D = 0$, $v_o = V_\ell$, and (*1*) gives

$$V_\ell = -\frac{R_3}{R_4} V \tag{2}$$

(*b*) For $v_o > V_\ell$, the diode blocks and R_2 constitutes the only feedback path. Since $i_1 = i_2$,

$$\frac{v_S}{R_1} = -\frac{v_o}{R_2} \tag{3}$$

For $v_o \le V_\ell$, the diode conducts and the parallel combination of R_2 and R_3 forms the feedback path. Since now $i_1 = i_2 + i_3 + i_4$,

$$\frac{v_S}{R_1} = -\left(\frac{v_o}{R_2} + \frac{v_o}{R_3} + \frac{V}{R_4}\right) \tag{4}$$

It follows from (*2*), (*3*), and (*4*) that

$$v_o = \begin{cases} -\dfrac{R_2}{R_1} v_S & \text{for } v_S < \dfrac{R_1 R_3}{R_2 R_4} V \\[2ex] -\dfrac{R_3}{R_2 + R_3}\dfrac{R_2}{R_1} v_S - \dfrac{R_2}{R_2 + R_3}\dfrac{R_3}{R_4} V & \text{for } v_S \ge \dfrac{R_1 R_3}{R_2 R_4} V \end{cases}$$

This transfer characteristic is plotted in Fig. 10-16(*b*).

10.11 What modifications and specifications will change the circuit of Fig. 10-14(*a*) into a 3-V square-wave generator, if $v_S = 0.02 \sin \omega t$ V? Sketch the circuit transfer characteristic and the input and output waveforms.

Modify the circuit by removing R_2, and specify Zener diodes such that $V_{Z1} = V_{Z2} = 3$ V. The transfer characteristic of Fig. 10-14(*b*) will change to that of Fig. 10-17(*a*). The time relationship between v_S and v_o will be that displayed in Fig. 10-17(*b*).

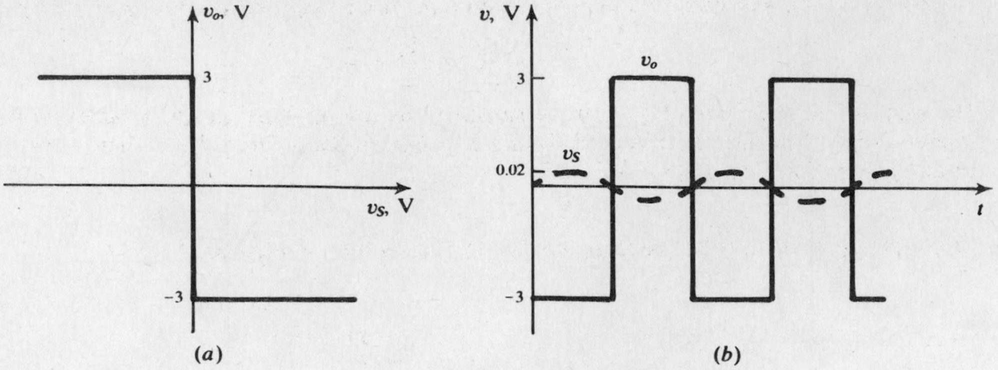

Fig. 10-17

10.12 Design a first-order low-pass filter with dc gain of magnitude 2 and input impedance 5 kΩ. The gain should be flat to 100 Hz.

The filter is shown in Fig. 10-8. For an ideal op amp, Problem 10.4 gives $Z_1 = R_1 = 5$ kΩ. The dc gain is given by (*10.20*) as $A(0) = -R/R_1$, whence $R = 2R_1 = 10$ kΩ. Figure 10-8(*b*) shows that the magnitude of the gain is flat to $\omega = 0.1/RC$, so the capacitor must be sized such that

$$C = \frac{0.1}{2\pi f R} = \frac{0.1}{2\pi(100)(10 \times 10^3)} = 15.9 \text{ nF}$$

10.13 The analog computer utilizes operational amplifiers to solve differential equations. Devise an analog solution for $i(t)$, $t > 0$, in the circuit of Fig. 10-18(*a*). Assume that you have available an inverting integrator with unity gain ($R_1C_1 = 1$), inverting amplifiers, a variable dc source, and a switch.

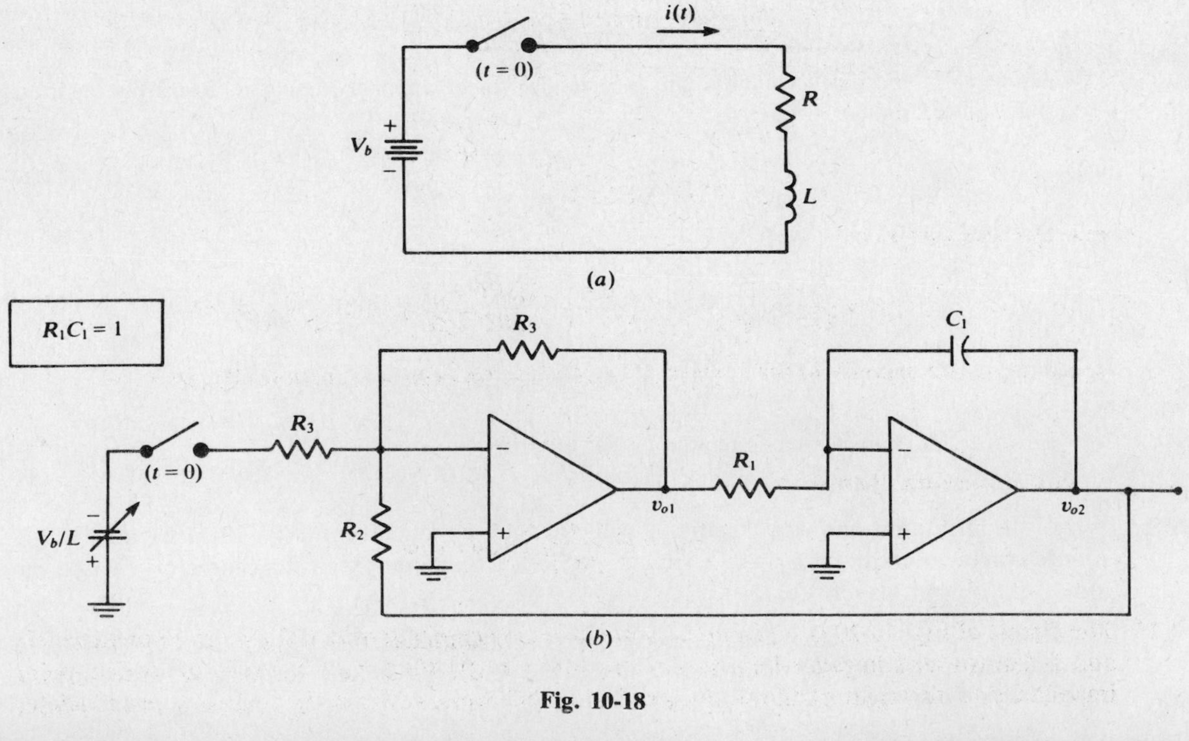

Fig. 10-18

For $t > 0$, the governing differential equation for the circuit of Fig. 10-18(*a*) may be written as

$$-\frac{di}{dt} = -\frac{V_b}{L} + \frac{R}{L}i \tag{1}$$

The sum on the right side of (*1*) can be simulated by the left-hand inverting adder of Fig. 10-18(*b*), where $v_{o1} = -di/dt$ and where R_2 and R_3 are chosen such that $R_3/R_2 = R/L$. Then $v_{o2} = -\int v_{o1}\,dt$ will be an analog of $i(t)$, on a scale of 1 A/V.

10.14 Find the relationship between v_o and v_i in the circuit of Fig. 10-19.

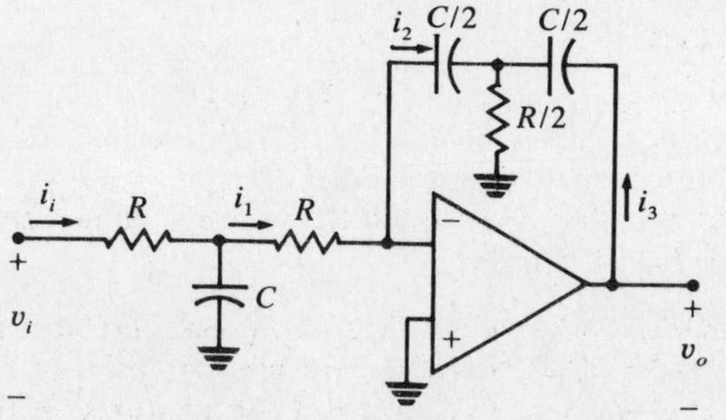

Fig. 10-19

Since the inverting terminal is a virtual ground, the Laplace-domain input current is given by

$$I_i = \frac{V_i}{R + (R \| 1/sC)} = \frac{V_i(sRC+1)}{sR^2C + 2R}$$

With zero current flowing into the op amp inverting terminal, current division yields

$$I_2 = I_1 = \frac{1/sC}{R + 1/sC} I_i = \frac{1}{sRC+1}\,\frac{V_i(sRC+1)}{R(sRC+2)} = \frac{V_i}{R(sRC+2)}$$

Again because the inverting terminal is a virtual ground,

$$I_3 = \frac{V_o}{\dfrac{1}{sC/2} + \dfrac{1}{sC/2}\Big\|\dfrac{R}{2}} = \frac{sC(sRC+4)}{4(sRC+2)}V_o$$

and, by current division,

$$I_2 = \frac{-R/2}{2/sC + R/2} I_3 = \frac{-sRC}{sRC+4}\,\frac{sC(sRC+4)}{4(sRC+2)}V_o = \frac{-s^2RC^2V_o}{4(sRC+2)}$$

Equating the two expressions for I_2 yields a Laplace-domain expression relating V_o and V_i:

$$V_o = -\frac{4}{s^2R^2C^2}V_i \tag{1}$$

or, after inverse transformation,

$$v_o = -\frac{4}{(RC)^2}\int\left(\int v_i\,dt\right)dt$$

10.15 The circuit of Fig. 10-20 is, in essence, a noninverting amplifier with a feedback impedance Z_N and is known as a *negative-impedance converter* (NIC). Find the Thévenin or driving-point impedance to the right of the input terminals, and explain why such a name is appropriate.

At the inverting node, the phasor input current is given by

$$I_i = I_N = \frac{V_i - V_o}{Z_N}$$

so that
$$V_o = V_i - I_i Z_N \tag{1}$$

Since $V_d \approx 0$,
$$I_P = \frac{V_o - V_i}{Z_P} = I_Z = \frac{V_i}{Z}$$

so that
$$V_o = \frac{Z_P}{Z} V_i + V_i \tag{2}$$

If (*1*) and (*2*) are equated and rearranged, they result in

$$Z_{dp} = \frac{V_i}{I_i} = -\frac{Z_N}{Z_P} Z \tag{3}$$

Observe that if $Z_P = Z_N$, then the impedance Z appears to be converted to the negative of its value; hence the name. See Problem 10.16 for another example.

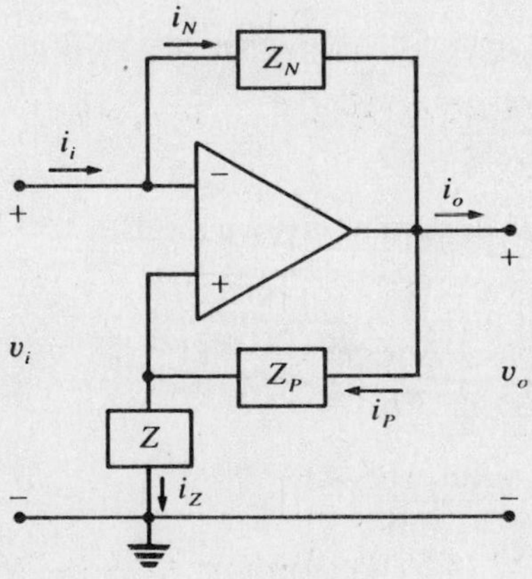

Fig. 10-20

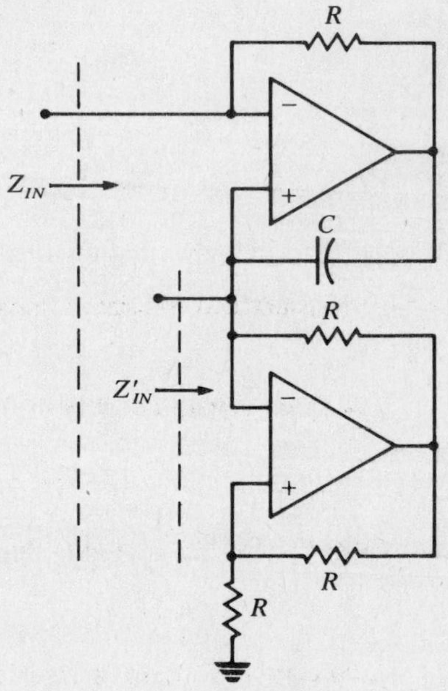

Fig. 10-21

10.16 (*a*) Describe a circuit arrangement that makes use of the NIC of Problem 10.15 and Fig. 10-20, with only resistors and capacitors, to simulate a pure inductor. (*b*) If only four 10-kΩ resistors and a 0.01-μF capacitor are available for use in the circuit, determine the value of L that can be simulated.

(*a*) Consider the circuit of Fig. 10-21. According to (*3*) of Problem 10.15,

$$Z'_{\text{IN}} = \frac{Z_N}{Z_P} Z = -\frac{R}{R} R = -R$$

and
$$Z_{\text{IN}} = -\frac{Z_N}{Z_P} Z = -\frac{R}{1/sC}(-R) = sR^2C \equiv sL_{eq}$$

(*b*) The value of L_{eq} is

$$L_{eq} = R^2C = (10^4)^2(0.01 \times 10^{-6}) = 1\ \text{H}$$

10.17 The logarithmic amplifier of Fig. 10-7 has two undesirable aspects: V_T and I_o are temperature-dependent, and $\ln RI_o$ may not be negligibly small. A circuit that can overcome these shortcomings is presented in Fig. 10-22. Show that if Q_1 and Q_2 are matched transistors, then v_o is truly proportional to $\ln v_S$.

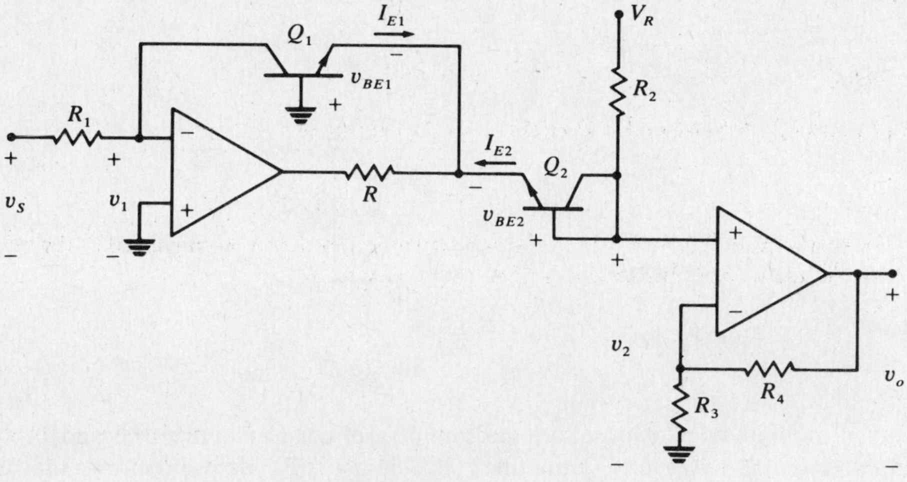

Fig. 10-22

In matched transistors, reverse saturation currents are equal. By KVL, with $v_1 \approx 0$,

$$v_2 = v_{BE2} - v_{BE1} \tag{1}$$

Taking the logarithm of both sides of (*10.15*) leads to

$$v_{BE} = V_T \ln \frac{I_E}{I_S} \tag{2}$$

Now the use of (*2*) in (*1*), with $I_C \approx I_E$, gives

$$v_2 = V_T \ln \frac{I_{E2}}{I_S} - V_T \ln \frac{I_{E1}}{I_S} = -V_T \ln \frac{I_{C1}}{I_{C2}} \tag{3}$$

According to (*1*), v_2 is the difference between two small voltages. Thus, if V_R is several volts in magnitude, then $v_2 \ll V_R$, and

$$I_{C2} \approx I_{E2} = \frac{V_R - v_2}{R_2} \approx \frac{V_R}{R_2} \tag{4}$$

Also, since $v_1 \approx 0$,

$$I_{C1} \approx I_{E1} = \frac{v_S - v_1}{R_1} \approx \frac{v_S}{R_1} \tag{5}$$

Thus, by (*10.7*) along with (*3*) to (*5*),

$$v_o = \frac{R_3 + R_4}{R_3} v_2 = -V_T \frac{R_3 + R_4}{R_3} \ln \frac{I_{C1}}{I_{C2}} = -V_T\left(1 + \frac{R_4}{R_3}\right)\left[\ln v_S - \ln\left(\frac{R_1}{R_2} V_R\right)\right] \tag{6}$$

The selection of $(R_1/R_2)V_R = 1$ forces the last term on the right-hand side of (*6*) to zero. Also, R_3 can be selected with a temperature sensitivity similar to that of V_T, to offset changes in V_T. Further, it is simple to select $R_4/R_3 \gg 1$, so that (*6*) becomes

$$v_o \approx -V_T \frac{R_4}{R_3} \ln v_S$$

10.18 The circuit of Fig. 10-23 is an *exponential* or *inverse log* amplifier. Show that the output v_o is proportional to the inverse logarithm of the input v_i.

Since the input current to the op amp is negligible,

$$i_R \approx i_D = I_o e^{v_D/\eta V_T}$$

But since the inverting terminal is a virtual ground, $v_D = v_i$. Thus,

$$v_o = -i_R R \approx -RI_o e^{v_i/\eta V_T} = -RI_o \ln^{-1} \frac{v_i}{\eta V_T}$$

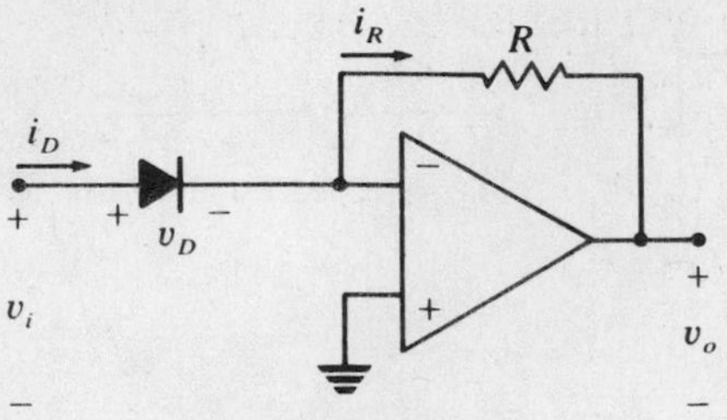

Fig. 10-23

10.19 Having now at your disposal a logarithmic amplifier (Example 10.8 and Problem 10.17) and an exponential (inverse log) amplifier (Problem 10.18), devise a circuit that will multiply two numbers together.

Since $xy = e^{\ln x + \ln y}$, the circuit of Fig. 10-24 is a possible realization.

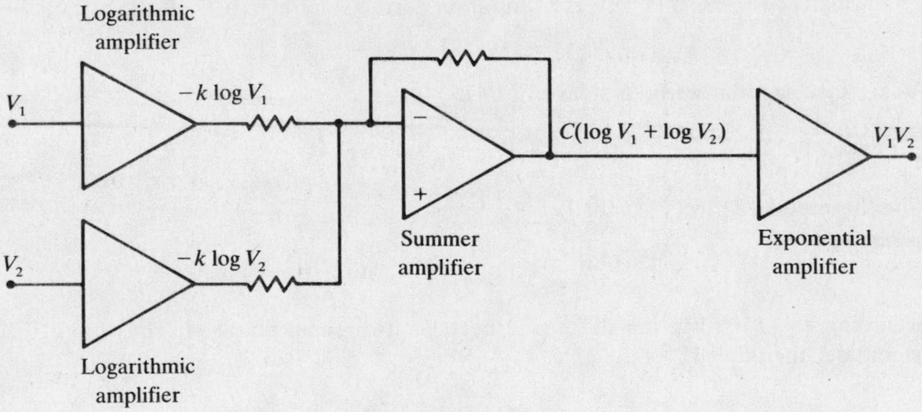

Fig. 10-24

10.20 Two identical passive *RC* low-pass filter sections are to be connected in cascade so as to create a double-pole filter with corner frequency at $1/\tau = 1/RC$. (*a*) Will simple cascade connection of these filters yield the desired transfer function $T(s) = (1/\tau)^2/(s + 1/\tau)^2$? (*b*) If not, how may the desired result be realized?

(*a*) With simple cascading, the overall transfer function would be

$$\frac{V_o}{V_i} = T' = \frac{(1/\tau)^2}{s^2 + 3(1/\tau)s + (1/\tau)^2}$$

which has two distinct negative roots. The desired result is not obtained because the impedance looking into the second stage is not infinite, and thus, the transfer function of the first stage is not simply $(1/\tau)/(s + 1/\tau)$.

(*b*) The desired result can be obtained by adding a unity follower (Fig. 10-11) between stages (see Problem 10.42), as illustrated in Fig. 10-25.

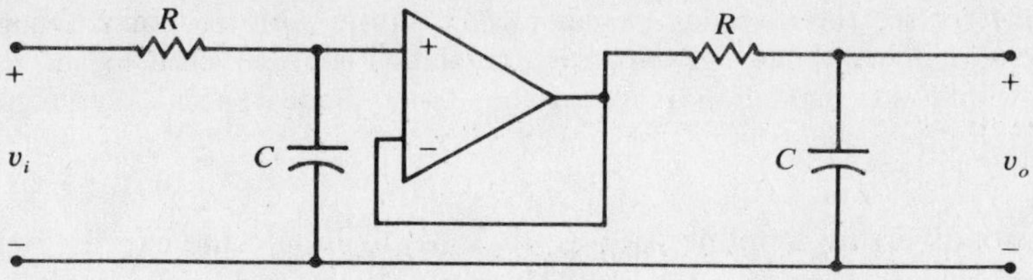

Fig. 10-25

10.21 (*a*) Find the transfer function for the circuit of Fig. 10-26. (*b*) In control theory, there is a compensation network whose transfer function is of the form $(s + 1/\tau_1)/(s + 1/\tau_2)$; it is called a lead-lag network if $1/\tau_1 < 1/\tau_2$, and a lag-lead network if $1/\tau_2 < 1/\tau_1$. Explain how the circuit of Fig. 10-26 may be used as such a compensation network.

(*a*) By extension of (*10.5*),

$$T(s) = \frac{V_o}{V_S} = -\frac{Z_2}{Z_1} = -\frac{\dfrac{R_2}{sR_2C_2 + 1}}{\dfrac{R_1}{sR_1C_1 + 1}} = -\frac{C_1}{C_2}\frac{s + 1/\tau_1}{s + 1/\tau_2} \tag{1}$$

where $\tau_1 = R_1C_1$ and $\tau_2 = R_2C_2$.

(*b*) To obtain unity gain, set $C_1 = C_2$. To obtain a positive transfer function, insert an inverter stage either before or after the circuit. Then, the selection of $R_1 > R_2$ yields $1/\tau_1 < 1/\tau_2$, giving the lead-lag network, and $R_1 < R_2$ results in $1/\tau_2 < 1/\tau_1$, giving the lag-lead network.

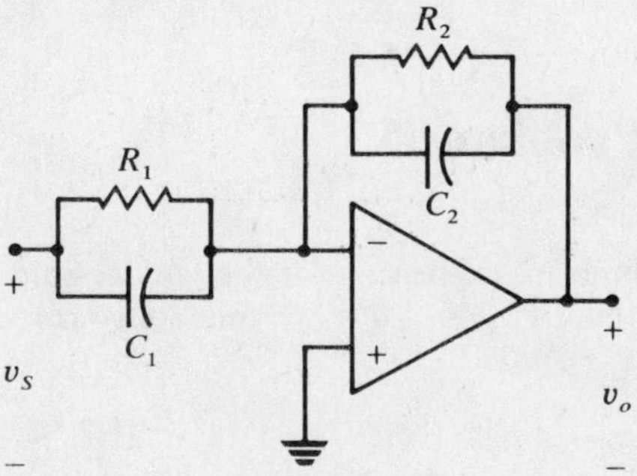

Fig. 10-26

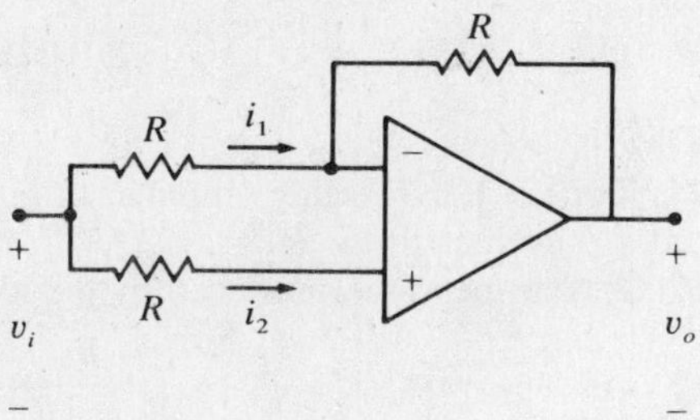

Fig. 10-27

10.22 Show that the transfer function for the op amp circuit of Fig. 10-27 is $v_o/v_i = 1$.

Because the op amp draws negligible current, $i_2 = 0$. Hence, $v_2 = v_i$. However, since $v_d \approx 0$, $v_1 \approx v_2 = v_i$ and

$$i_1 = \frac{v_i - v_1}{R} \approx 0$$

Also, by the method of node voltages,

$$i_1 = \frac{v_i - v_o}{2R} = 0$$

Thus, $v_i = v_o$ and so $v_o/v_i = 1$.

10.23 Use an op amp to design a noninverting voltage source (see Problem 10.9). Determine the conditions under which regulation is maintained in your source.

Simply replace the inverting amplifier of Fig. 10-15 with the noninverting amplifier of Fig. 10-3. Since the op amp draws negligible current, regulation is preserved if V_S and R_S are selected so that i_Z remains within the regulation range of the Zener diode. Specifically, regulation is maintained if $0.1I_Z \le V_S/R_S \le I_Z$.

10.24 For the noninverting amplifier of Fig. 10-3: (*a*) Compare the expressions obtained for voltage gain with common-mode rejection (Example 10.4) and without (in the ideal amplifier of Example 10.3), for $A_{OL} \to -\infty$. (*b*) Show that if CMRR is very large, then it need not be considered in computing the gain.

(*a*) We let $A_{OL} \to -\infty$ in (*10.13*), since that is implicit in Example 10.3:

$$\lim_{A_{OL}\to-\infty} A_v = \lim_{A_{OL}\to-\infty}\left[\frac{-A_{OL}}{1-A_{OL}R_1/(R_1+R_2)} + \frac{-A_{OL}/\mathrm{CMRR}}{1-A_{OL}R_1/(R_1+R_2)}\right]$$

$$= 1 + \frac{R_2}{R_1} + \frac{1}{\mathrm{CMRR}}\left(1+\frac{R_2}{R_1}\right) \qquad (1)$$

Now we can compare (*1*) above with (*10.7*); the difference is the last term on the right-hand side of (*1*) above.

(*b*) Let $\mathrm{CMRR} \to \infty$ in (*1*) above to get

$$\lim_{\substack{\mathrm{CMRR}\to\infty \\ A_{OL}\to-\infty}} A_v = 1 + \frac{R_2}{R_1}$$

which is identical to the ideal case of Example 10.3.

Supplementary Problems

10.25 For the noninverting amplifier of Fig. 10-3, (*a*) find an exact expression for the voltage-gain ratio, and (*b*) evaluate it for $R_1 = 1\ \mathrm{k\Omega}$, $R_2 = 10\ \mathrm{k\Omega}$, $R_d = 1\ \mathrm{k\Omega}$, and $A_{OL} = -10^4$. (*c*) Compare your result in part *b* with the value produced by the ideal expression (*10.7*).

Ans. (*a*) $A_v = \dfrac{R_1 + R_2}{R_1 - \dfrac{R_1R_2}{A_{OL}R_d} - \dfrac{R_1+R_2}{A_{OL}}}$; (*b*) 10.977;

(*c*) $A_{v\mathrm{ideal}} = 11$, for a +0.21% difference

10.26 In the first-order low-pass filter of Example 10.9, $R = 10\ \mathrm{k\Omega}$, $R_1 = 1\ \mathrm{k\Omega}$, and $C = 0.1\ \mu\mathrm{F}$. Find (*a*) the gain for dc signals, (*b*) the break frequency f_1 at which the gain drops off by 3 db, and (*c*) the frequency f_u at which the gain has dropped to unity (called the *unity-gain bandwidth*).
Ans. (*a*) −10; (*b*) 159.2 Hz; (*c*) 1583.6 Hz

10.27 The noninverting amplifier circuit of Fig. 10-3 has an infinite input impedance if the basic op amp is ideal. If the op amp is not ideal, but instead $R_d = 1\ \mathrm{M\Omega}$ and $A_{OL} = -10^6$, find the input impedance. Let $R_2 = 10\ \mathrm{k\Omega}$ and $R_1 = 1\ \mathrm{k\Omega}$. *Ans.* 1 TΩ

10.28 Let $R_1 = R_2 = R_3 = 3R_F$ in the inverting summer amplifier of Fig. 10-4. What mathematical operation does this circuit perform? *Ans.* Gives the negative of the instantaneous average value

10.29 An inverting summer (Fig. 10-4) has n inputs with $R_1 = R_2 = R_3 = \cdots = R_n = R$. Assume that the open-loop basic op amp gain A_{OL} is finite, but that the inverting-terminal input current is negligible.

Derive a relationship that shows how gain magnitude is reduced in the presence of multiple inputs.

Ans. $$A_n \equiv \frac{v_o}{v_{S1} + v_{S2} + \cdots + v_{Sn}} = -\frac{R_F/R}{1 - \dfrac{nR_F}{(R+1)A_{OL}}}$$

For a single input v_{S1}, the gain is A_1. For the same input v_{S1} together with $n-1$ zero inputs $v_{S2} = \cdots = v_{Sn} = 0$, the gain is A_n. But since $A_{OL} < 0$, $|A_n| < |A_1|$ for $n > 1$

10.30 The basic op amp in Fig. 10-28 is ideal. Find v_o and determine what mathematical operation is performed by the amplifier circuit. *Ans.* $v_o = (1 + R_2/R_1)(v_{S2} - v_{S1})$, a subtractor

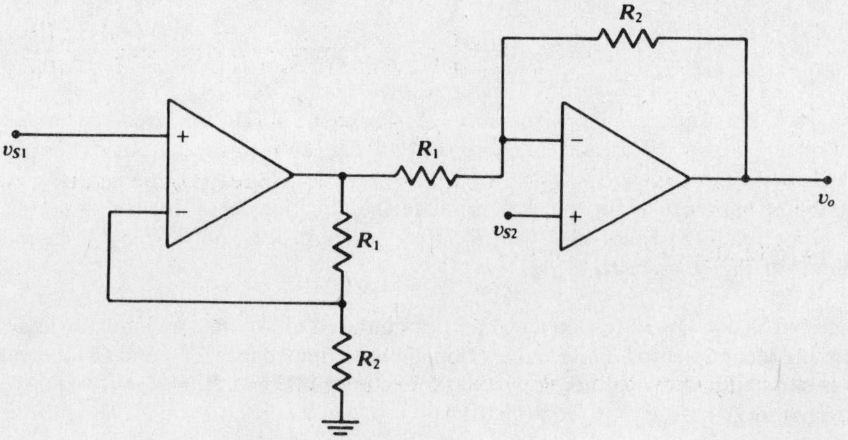

Fig. 10-28

10.31 Describe the transfer characteristic of the level-clamp circuit of Fig. 10-14(*a*) if diode Z_2 is shorted. *Ans.* Let $V_{Z2} = 0$ in Fig. 10-14(*b*)

10.32 Find the gain of the inverting amplifier of Fig. 10-29 if the op amp and diodes are ideal.
Ans. $A_v = -R_2/R_1$ for $v_S > 0$; $A_v = -R_3/R_1$ for $v_S \leq 0$

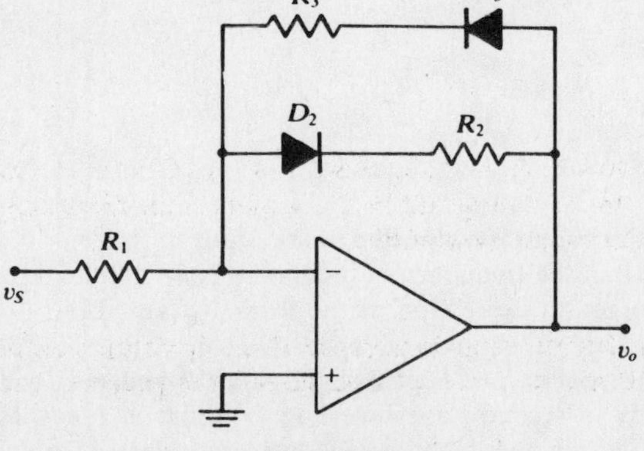

Fig. 10-29

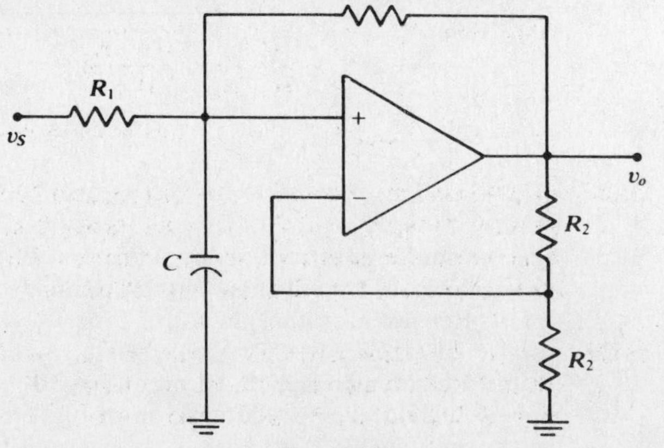

Fig. 10-30

10.33 The op amp in the circuit of Fig. 10-30 is ideal. Find an expression for v_o in terms of v_S, and determine the function of the circuit.
Ans. $v_o = (2/R_1C)\int v_S\,dt$, a noninverting integrator

10.34 If the nonideal op amp of the circuit of Fig. 10-31 has an open-loop gain $A_{OL} = -10^4$, find v_o. *Ans.* $0.9999E_b$

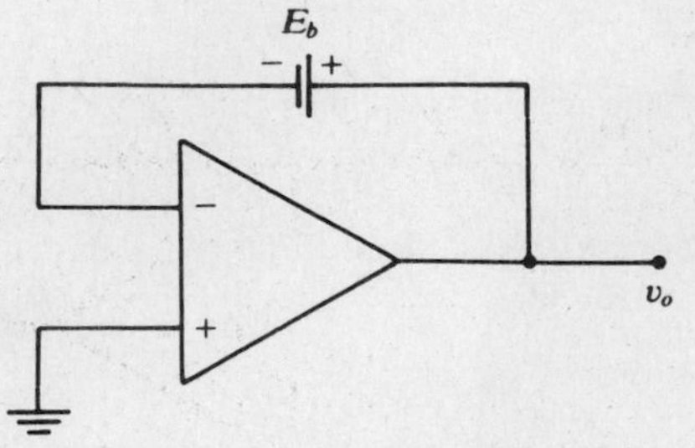

Fig. 10-31

10.35 How can the square-wave generator of Problem 10.11 be used to make a triangular-wave generator? *Ans.* Cascade the integrator of Fig. 10-6 to the output of the square-wave generator

10.36 Describe an op amp circuit that will simulate the equation $3v_1 + 2v_2 + v_3 = v_o$.
Ans. The summer of Fig. 10-4, with $R_F/R_1 = 3$, $R_F/R_2 = 2$, and $R_F/R_3 = 1$, cascaded into the inverting amplifier of Fig. 10-2, with $R_F/R_1 = 1$.

10.37 The circuit of Fig. 10-32 (called a *gyrator*) can be used to simulate an inductor in active RC filter design. Assuming ideal op amps, find (*a*) the *s*-domain input impedance $Z(s)$ and (*b*) the value of the inductance that is simulated if $C = 10$ nF, $R_1 = 20$ kΩ, $R_2 = 100$ kΩ, and $R_3 = R_4 = 10$ kΩ.
Ans. (*a*) $Z(s) = sR_1R_2R_3C/R_4$; (*b*) 2 H

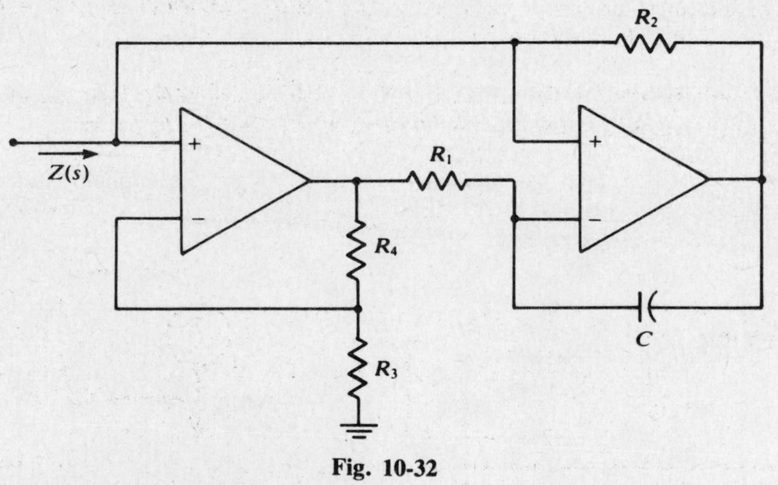

Fig. 10-32

10.38 For the double integrator circuit of Problem 10.14, if the output is connected to the input so that $v_i = v_o$, an oscillator is formed. Show that this claim is so, and that the frequency of oscillation is $f = 1/\pi RC$ Hz. [*Hint*: Replace V_o with V_i in (*1*) of Problem 10.14 to get an expression of the form $V_i f(s) = 0$.]

10.39 In the logarithmic amplifier circuit of Fig. 10-7, v_o must not exceed approximately 0.6 V, or else i_D will not be a good exponential function of v_D. Frequently, a second-stage inverting amplifier is added as shown in Fig. 10-7, so that v_o' is conveniently large. If the second-stage gain is selected to be $A_v = -R_F/R_1 = -1/V_T$, then its output becomes $v_o' = \ln v_i$. In the circuit of Fig. 10-7, v_D is exponential for $0 \le i_D \le 1$ mA, $0 \le v_i \le 10$ V, and $I_o = 100$ pA. Size R, R_F, and R_1 so that v_o' is as given above.
Ans. $R = 10$ MΩ; arbitrarily select $R_1 = 1$ kΩ, and then $R_F = 38.46$ kΩ

10.40 In the logarithmic amplifier of Fig. 10-22, let $v_S = 5$ V, $V_R = 10$ V, $R_1 = 1$ kΩ, $R_2 = 10$ kΩ, $R_3 = 1$ kΩ, and $R_4 = 50$ kΩ. The matched BJTs are operating at 25°C, with $V_T = 0.026$ V. Find (*a*) v_2 and (*b*) v_o (see Problem 10.17). *Ans.* (*a*) −41.8 mV; (*b*) −2.13 V

10.41 Having at your disposal a logarithmic amplifier and an exponential amplifier, devise a circuit that will produce the quotient of two numbers. (*Hint*: $x/y = e^{\ln x - \ln y}$.) *Ans.* See Fig. 10-33

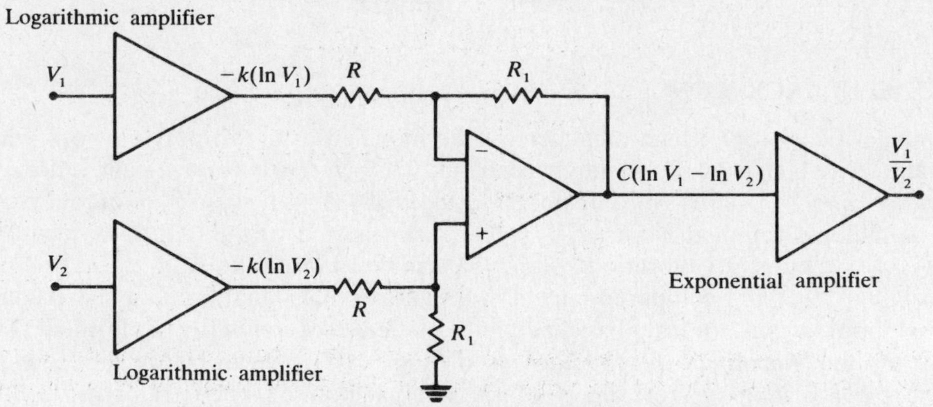

Fig. 10-33

10.42 The unity follower of Fig. 10-11 is the noninverting amplifier of Fig. 10-3 if $R_1 \to \infty$ and $R_2 \to 0$. (*a*) Find the output impedance R_{out} of the noninverting amplifier of Fig. 10-3 subject to the approximation $i_i = 0$. Model the op amp with the practical equivalent circuit of Fig. 10-1(*a*). (*b*) Let $R_1 \to \infty$ and $R_2 \to 0$ in your answer to part *a*, to find the output impedance of the unity follower.
Ans. (*a*) $R_{out} = R_o(R_1 + R_2)/[R_o + R_2 + R_1(1 - A_{OL})]$; (*b*) $R_{out} \approx R_o/(1 - A_{OL})$

10.43 The circuit of Fig. 10-26 is to be used as a high-pass filter having a gain of 0.1 at low frequencies, unity gain at high frequencies, and a gain of 0.707 at 1 krad/s. Arbitrarily select $C_1 = C_2 = 0.1\ \mu F$, and size R_1 and R_2. *Ans.* $R_1 = 100\ k\Omega$, $R_2 = 10\ k\Omega$

10.44 Find the transfer function for the circuit of Fig. 10-34, and explain the use of the circuit.
Ans. $T(s) = 1/(sRC + 1)$, a low-pass filter with zero output impedance

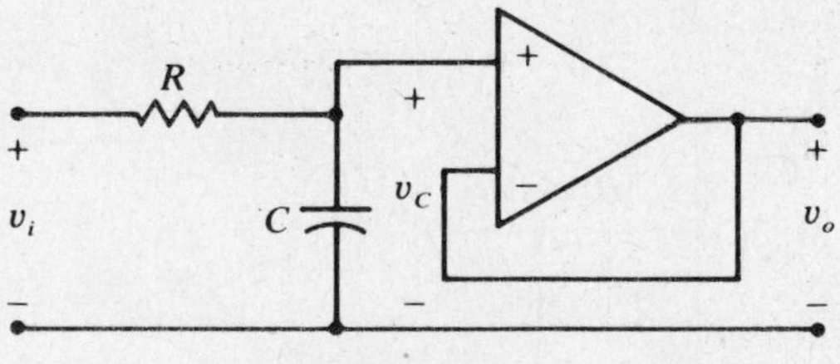

Fig. 10-34

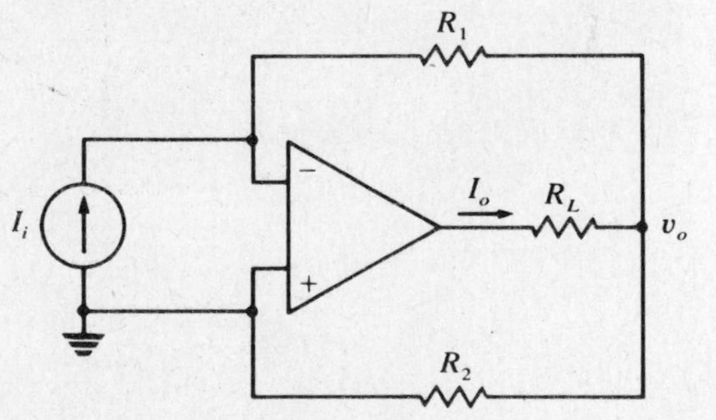

Fig. 10-35

10.45 For the circuit of Fig. 10-35, show that $I_o = -(1 + R_1/R_2)I_i$, so that the circuit is a true current amplifier. (Note that I_o is independent of R_L.)

10.46 If the noninverting terminal of the op amp in Fig. 10-27 is grounded, find the transfer function v_o/v_i. (Compare with Problem 10.22.) *Ans.* $v_o/v_i = -1$

10.47 Devise a method for using the inverting op amp circuit of Fig. 10-2 as a current source.
Ans. Let I_F be the output current; then $i_F = i_1 = v_S/R_1$ regardless of the value of R_F

10.48 A noninverting amplifier with gain $A_v = 21$ is desired. Based on ideal op amp theory, values of $R_1 = 10\ k\Omega$ and $R_2 = 200\ k\Omega$ are selected for the circuit of Fig. 10-3. If the op amp is recognized as nonideal in that $A_{OL} = -10^4$ and $\text{CMRR}_{db} = 40$ db, find the actual gain $A_v = v_o/v_2$.
Ans. $A_v^* = 21.17$

Chapter 11

Feedback Amplifiers

11.1 THE FEEDBACK CONCEPT

Any single- or multiple-stage amplifier can be modeled as a two-port network with a forward transfer ratio A (which may be a current-gain ratio $A_i = i_L/i_S$; a voltage-gain ratio $A_v = v_L/v_S$; a *transresistance ratio* $A_r = v_L/i_S$; or a *transconductance ratio* $A_g = i_L/v_S$). If an output port variable of the basic amplifier is sampled, adjusted by a feedback transfer ratio β (which may be a voltage-gain ratio $\beta_{vv} = v_f/v_a$; a current-gain ratio $\beta_{ii} = i_f/i_L$; a transresistance ratio $\beta_{iv} = v_f/i_L$; or a transconductance ratio $\beta_{vi} = i_L/v_f$) and compared (mixed) with an external signal (from a signal source) to form the driving input signal for the basic amplifier, a *feedback amplifier* is formed. This result is illustrated by the *functional block diagram* of Fig. 11-1(a), which can be converted to the *mathematical block diagram* (or simply *block diagram*) of Fig. 11-1(b). The introduction of a feedback loop reduces the gain of an amplifier; however, feedback can be utilized to alter input and output impedances, increase the range of frequency response, reduce distortion, and desensitize the amplifier gain to parameter changes.

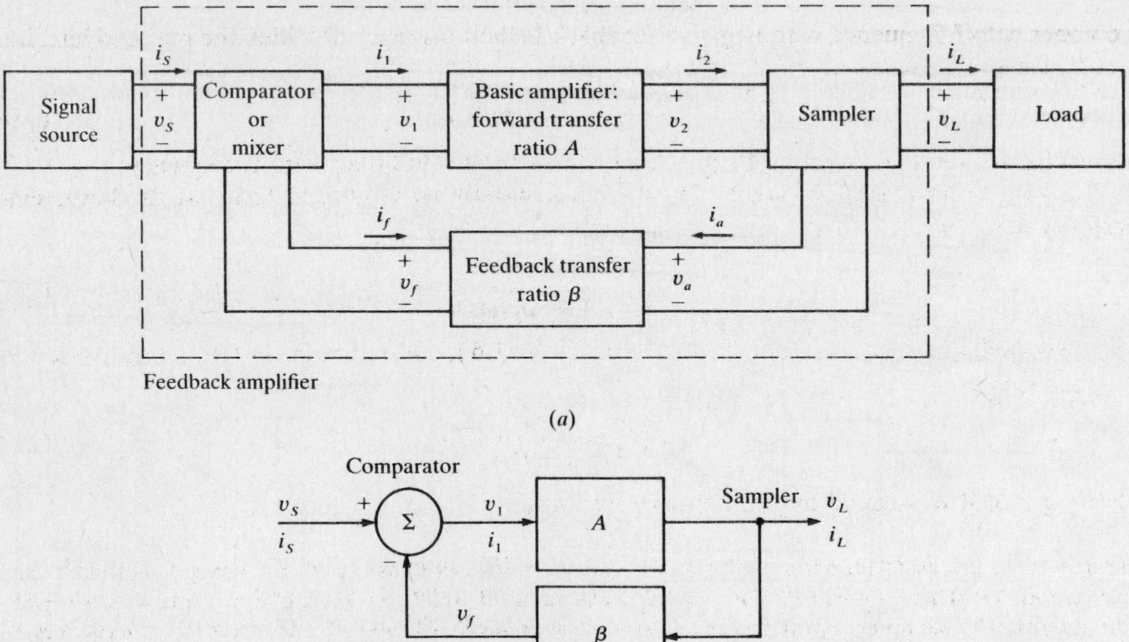

(a)

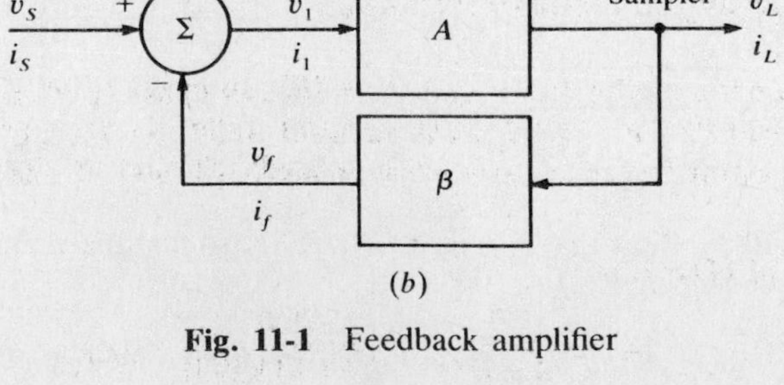

(b)

Fig. 11-1 Feedback amplifier

11.2 EFFECT OF FEEDBACK ON GAIN AND FREQUENCY RESPONSE

For the sake of illustration, let the forward transfer ratio A and feedback transfer ratio β of the amplifier of Fig. 11-1(a) both be voltage-gain ratios; then

$$v_L = A_v v_1 \tag{11.1}$$

and
$$v_1 = v_S - v_f = v_S - \beta_{vv} v_L \tag{11.2}$$

Substituting (*11.2*) into (*11.1*) and rearranging gives the *overall voltage gain* of the feedback amplifier as

$$A_f = A_{vf} \equiv \frac{v_L}{v_S} = \frac{A_v}{1 + \beta_{vv} A_v} \tag{11.3}$$

To ensure *negative feedback* [so that the signal v_f (or i_f) will be subtracted from the signal v_S (or i_S) by the comparator to form v_1 (or i_1)], the signs of A and β must be identical; thus, from (*11.3*) it is apparent that $|A_f| < |A|$. Similar overall transfer ratios for A_{if}, A_{gf}, and A_{rf} are developed in Sections 11.5 to 11.7.

Let an amplifier have the upper cutoff frequency ω_H without feedback; then its gain A can usually be written in terms of the midfrequency gain A_o $(= A_{\text{mid}})$ as

$$A = \frac{A_o}{1 + j\omega/\omega_H} \tag{11.4}$$

From (*11.3*) and (*11.4*), the gain with negative feedback is then

$$A_f = \frac{A}{1 + \beta A} = \frac{\dfrac{A_o}{1 + j\omega/\omega_H}}{1 + \dfrac{\beta A_o}{1 + j\omega/\omega_H}} = \frac{A_o}{1 + \beta A_o + j\omega/\omega_H} \tag{11.5}$$

The upper cutoff frequency with negative feedback is the frequency at which the real and imaginary parts of the denominator of (*11.5*) are equal; thus,

$$\omega_{Hf} \equiv \omega_H(1 + \beta A_o) \tag{11.6}$$

from which we see that negative feedback acts to increase the upper cutoff frequency.

The gain of an amplifier without feedback can usually be written in terms of the lower cutoff frequency as

$$A = \frac{A_o}{1 + j\omega_L/\omega} \tag{11.7}$$

Then, paralleling the process by which we obtained (*11.6*), we find that the lower cutoff frequency with feedback is

$$\omega_{Lf} \equiv \frac{\omega_L}{1 + \beta A_o} \tag{11.8}$$

indicating that it is reduced by the negative feedback.

Example 11.1 An amplifier has a midfrequency gain $A_o = 100$, an upper cutoff frequency $f_H = 40$ kHz, and a lower cutoff frequency $f_L = 20$ Hz. For a feedback transfer ratio $\beta = 0.1$, find (*a*) the overall gain at midfrequency, (*b*) the upper cutoff frequency with negative feedback, and (*c*) the lower cutoff frequency with feedback.

(*a*) From the general form of (*11.3*),

$$A_{fo} = \frac{A_o}{1 + \beta A_o} = \frac{100}{1 + (0.1)(100)} = 9.09$$

(*b*) From (*11.6*),

$$f_{Hf} = f_H(1 + \beta A_o) = (40 \times 10^3)[1 + (0.1)(100)] = 440 \text{ kHz}$$

(*c*) From (*11.8*),

$$f_{Lf} = \frac{f_L}{1 + \beta A_o} = \frac{20}{1 + (0.1)(100)} = 1.82 \text{ Hz}$$

11.3 EFFECT OF FEEDBACK ON INPUT AND OUTPUT IMPEDANCES

For an amplifier with negative feedback, if current i_f is the mixed variable, input impedance is increased. With output voltage v_L as the sampled variable, output impedance is decreased; if output current is sampled, the output impedance is increased. Table 11.1 summarizes these conclusions, which are verified in subsequent sections.

Table 11-1

Sample / Mix	v_L / v_f	v_L / i_f	i_L / v_f	i_L / i_f
R_{of}	$<R_o$	$<R_o$	$>R_o$	$>R_o$
$R_{\text{in}f}$	$>R_{\text{in}}$	$<R_{\text{in}}$	$>R_{\text{in}}$	$<R_{\text{in}}$

11.4 VOLTAGE-SERIES FEEDBACK

The general feedback amplifier model of Fig. 11-2, in which output voltage v_L is the sampled variable and feedback voltage v_f is the mixed variable, is called a *voltage-series* (or *parallel-output, series-input*; or *node-sampling, loop-comparison*) feedback amplifier.

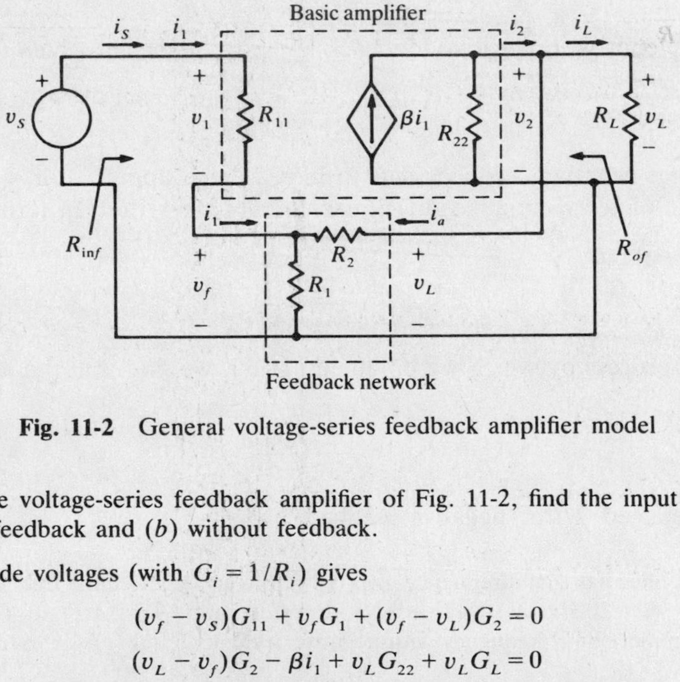

Fig. 11-2 General voltage-series feedback amplifier model

Example 11.2 For the voltage-series feedback amplifier of Fig. 11-2, find the input and output resistances (impedances) (*a*) with feedback and (*b*) without feedback.

(*a*) The method of node voltages (with $G_i = 1/R_i$) gives

$$(v_f - v_S)G_{11} + v_f G_1 + (v_f - v_L)G_2 = 0 \tag{11.9}$$

$$(v_L - v_f)G_2 - \beta i_1 + v_L G_{22} + v_L G_L = 0 \tag{11.10}$$

But

$$i_1 = (v_S - v_f)G_{11} \tag{11.11}$$

Substituting (*11.11*) into (*11.10*) and rearranging yield the following set of linear equations:

$$\begin{bmatrix} G_{11} + G_1 + G_2 & -G_2 \\ -(G_2 - \beta G_{11}) & G_2 + G_{22} + G_L \end{bmatrix} \begin{bmatrix} v_f \\ v_L \end{bmatrix} = \begin{bmatrix} G_{11} v_S \\ \beta G_{11} v_S \end{bmatrix} \tag{11.12}$$

Simultaneous solution of (*11.12*) by means of Cramer's rule gives the feedback voltage v_f as

$$v_f = \frac{\Delta_1}{\Delta} = \frac{\begin{vmatrix} G_{11} v_S & -G_2 \\ \beta G_{11} v_S & G_2 + G_{22} + G_L \end{vmatrix}}{\begin{vmatrix} G_{11} + G_1 + G_2 & -G_2 \\ -(G_2 - \beta G_{11}) & G_2 + G_{22} + G_L \end{vmatrix}} = \frac{v_S G_{11} G}{\Delta} \tag{11.13}$$

where $$G = (1+\beta)G_2 + G_{22} + G_L \tag{11.14}$$

and $$\Delta = G_{11}G + G_1G_2 + (G_1 + G_2)(G_{22} + G_L) \tag{11.15}$$

From Fig. 11-2 and (*11.11*) and (*11.13*), we see that

$$R_{inf} = R_{11} + \frac{v_f}{i_1} = R_{11} + \frac{v_f}{(v_S - v_f)G_{11}} = R_{11} + \frac{G}{\Delta - G_{11}G} \tag{11.16}$$

Substituting (*11.14*) and (*11.15*) into (*11.16*) and rearranging then lead to

$$R_{inf} = R_{11} + \frac{(1+\beta)G_2 + G_{22} + G_L}{G_1G_2 + (G_1 + G_2)(G_{22} + G_L)} \tag{11.17}$$

To determine the output resistance, we deactivate (short) v_S and replace R_L with a driving-point source with polarity matching that of v_L. Then KCL at the output node yields

$$i_{dp} = -\beta i_1 + G_{22}v_{dp} + \frac{v_{dp}}{R_2 + R_{11} \| R_1} \tag{11.18}$$

and current division at the node within the feedback network gives

$$i_1 = \frac{-R_1}{R_{11} + R_1} i_a = \frac{-R_1}{R_{11} + R_1} \frac{v_{dp}}{R_2 + R_{11} \| R_1} \tag{11.19}$$

Substituting (*11.19*) into (*11.18*) leads to an expression for the output conductance:

$$G_{of} = \frac{1}{R_{of}} = \frac{i_{dp}}{v_{dp}} = \frac{1}{R_{22}} + \frac{(1+\beta R_1)/(R_{11} + R_1)}{R_2 + R_{11} \| R_1} = G_{22} + \frac{(1+\beta)G_{11} + G_1}{(G_{11} + G_1)R_2 + 1} \tag{11.20}$$

(*b*) The feedback is deactivated if $R_1 = 1/G_1 = 0$ and $R_2 = 1/G_2 = \infty$; thus, from (*11.17*) and (*11.20*),

$$R_{in} = \lim_{\substack{G_1 \to \infty \\ G_2 \to 0}} R_{inf} = R_{11} \tag{11.21}$$

$$R_o = \lim_{\substack{G_1 \to \infty \\ G_2 \to 0}} \frac{1}{G_{of}} = R_{22} \tag{11.22}$$

Example 11.3 In the voltage-series feedback amplifier of Fig. 11-2, let $G_1 (= 1/R_1) = 1 \times 10^{-3}$ S, $G_2 = 5 \times 10^{-5}$ S, $G_{11} = 0.01$ S, $G_{22} = 1 \times 10^{-6}$ S, $G_L = 0.5 \times 10^{-3}$ S, and $\beta = 75$. Determine the percentage change in (*a*) input resistance R_{in} and (*b*) output resistance R_o due to the addition of feedback.

(*a*) Without feedback, by (*11.21*), $R_{in} = R_{11} = 1/G_{11} = 100\ \Omega$. With feedback, (*11.17*) gives

$$R_{inf} = R_{11} + \frac{(1+\beta)G_2 + G_{22} + G_L}{G_1G_2 + (G_1 + G_2)(G_{22} + G_L)}$$

$$= 100 + \frac{(76)(5 \times 10^{-5}) + 10^{-6} + 0.5 \times 10^{-3}}{(10^{-3})(5 \times 10^{-5}) + (1.05 \times 10^{-3})(501 \times 10^{-6})} = 7.466\ \text{k}\Omega$$

The percentage change is then

$$\%\ \text{increase} = \frac{7466}{100}(100\%) = 7466\%$$

(*b*) By (*11.22*), $R_o = R_{22} = 1/G_{22} = 1\ \text{M}\Omega$. With feedback, from (*11.20*),

$$G_{of} = G_{22} + \frac{(1+\beta)G_{11} + G_1}{(G_{11} + G_1)R_2 + 1} = 10^{-6} + \frac{(76)(0.01) + 10^{-3}}{(0.011)[1/(5 \times 10^{-5})] + 1} = 0.003444\ \text{S}$$

so $$R_{of} = \frac{1}{G_{of}} = 290.36\ \Omega$$

and the percentage change is

$$\%\ \text{decrease} = \frac{99.971 \times 10^4}{1 \times 10^6}(100\%) = 99.97\%$$

The ideal voltage-series feedback amplifier model, shown in block-diagram form in Fig. 11-3, is based on the assumption that the feedback network does not *load* the basic amplifier (i.e., does not alter the forward transfer ratio A).

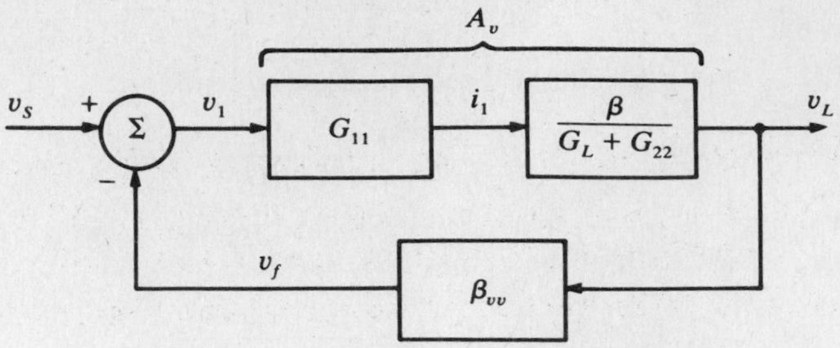

Fig. 11-3 Ideal voltage-series feedback model

Example 11.4 Determine the overall voltage gain of the ideal voltage-series feedback amplifier of Fig. 11-3.

Since the basic amplifier gain $A_v = v_L/v_1$, we have

$$v_L = A_v v_1 = \frac{\beta G_{11}}{G_L + G_{22}} v_1 \tag{11.23}$$

The comparator output is

$$v_1 = v_S - v_f = v_S - \beta_{vv} v_L \tag{11.24}$$

Substituting (*11.24*) into (*11.23*) and rearranging give the desired gain:

$$A_{vf} = \frac{v_L}{v_S} = \frac{A_v}{1 + \beta_{vv} A_v} = \frac{\beta G_{11}}{G_L + G_{22} + \beta_{vv} \beta G_{11}} \tag{11.25}$$

11.5 CURRENT-SERIES FEEDBACK

The general feedback amplifier model of Fig. 11-4, in which output current i_L is the sampled variable and feedback voltage v_f is the mixed variable, is called a *current-series* (or *series-input, series-output*; or *loop-sampling, loop-comparison*) feedback amplifier.

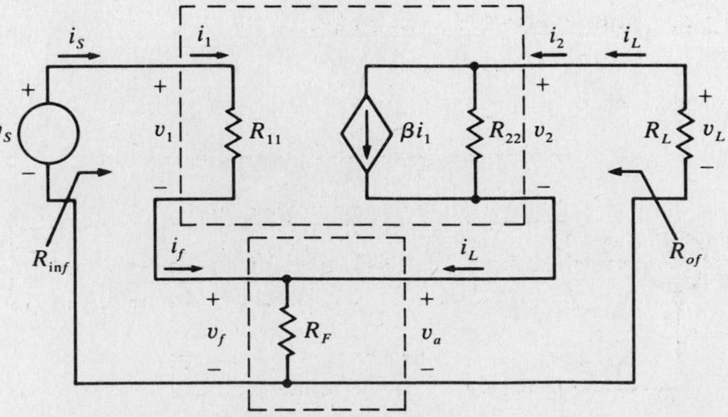

Fig. 11-4 General current-series feedback amplifier model

Example 11.5 For the current-series feedback amplifier of Fig. 11-4, find (*a*) the overall voltage gain and (*b*) the feedback transfer ratio $\beta_{iv} = v_f/i_L$.

(*a*) The method of node voltages requires that

$$(v_f - v_S)G_{11} + v_f G_F - \beta i_1 + (v_f - v_L)G_{22} = 0 \tag{11.26}$$

and

$$\beta i_1 + (v_L - v_f)G_{22} + v_L G_L = 0 \tag{11.27}$$

But

$$i_1 = (v_S - v_f)G_{11} \tag{11.28}$$

Substituting (*11.28*) into (*11.26*) and (*11.27*) and rearranging lead to the linear system of equations

$$\begin{bmatrix} (1+\beta)G_{11} + G_F + G_{22} & -G_{22} \\ -(\beta G_{11} + G_{22}) & (G_{22} + G_L) \end{bmatrix} \begin{bmatrix} v_f \\ v_L \end{bmatrix} = \begin{bmatrix} (1+\beta)G_{11}v_S \\ -\beta G_{11} v_S \end{bmatrix} \tag{11.29}$$

Simultaneous solution of (*11.29*) for the load voltage v_L by Cramer's rule leads to

$$v_L = \frac{\Delta_2}{\Delta} = \frac{\begin{vmatrix} (1+\beta)G_{11} + G_F + G_{22} & (1+\beta)G_{11}v_S \\ -(\beta G_{11} + G_{22}) & -\beta G_{11} v_S \end{vmatrix}}{\begin{vmatrix} (1+\beta)G_{11} + G_F + G_{22} & -G_{22} \\ -(\beta G_{11} + G_{22}) & G_{22} + G_L \end{vmatrix}} = \frac{G_{11}(G_{22} - \beta G_F)v_S}{\Delta} \tag{11.30}$$

where

$$\Delta = G_{22}(G_{11} + G_F) + G_L[(1+\beta)G_{11} + G_{22} + G_F] \tag{11.31}$$

Using (*11.31*) in (*11.30*) and rearranging yield

$$A_{vf} = \frac{v_L}{v_S} = \frac{G_{11}(G_{22} - \beta G_F)}{G_{22}(G_{11} + G_F) + G_L[(1+\beta)G_{11} + G_{22}] + G_L G_F} \tag{11.32}$$

(*b*) Simultaneous solution of (*11.29*) for v_f yields

$$v_f = \frac{\begin{vmatrix} (1+\beta)G_{11}v_S & -G_{22} \\ -\beta G_{11}v_S & G_{22} + G_L \end{vmatrix}}{\Delta} = \frac{G_{11}[(1+\beta)G_L + G_{22}]v_S}{\Delta} \tag{11.33}$$

But

$$\beta_{iv} = \frac{v_f}{i_L} = \frac{v_f}{-v_L/R_L} = -G_L \frac{v_f}{v_L} \tag{11.34}$$

Substituting (*11.30*) and (*11.33*) into (*11.34*) and simplifying thus give

$$\beta_{iv} = \frac{(1+\beta)G_L + G_{22}}{G_L(\beta G_F - G_{22})} \tag{11.35}$$

The ideal current-series feedback amplifier model, shown in block-diagram form in Fig. 11-5, is based on the assumption that the feedback network does not load the basic amplifier. The basic amplifier is, in this case, a transconductance amplifier.

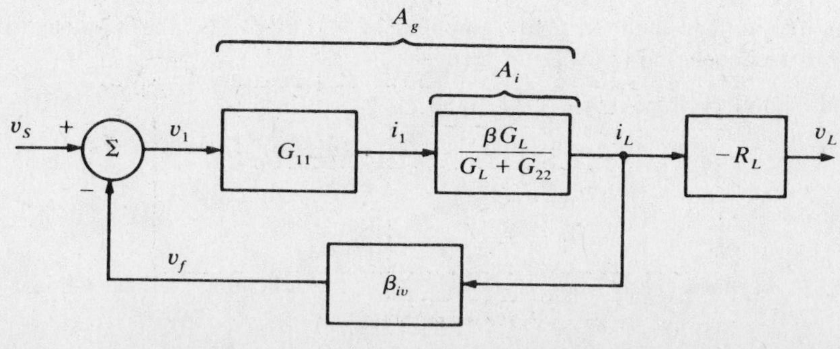

Fig. 11-5 Ideal current-series feedback model

Example 11.6 For the ideal current-series feedback amplifier of Fig. 11-5, find (*a*) the overall transconductance gain and (*b*) the overall voltage gain.

(*a*) The basic amplifier gain $A_g = i_L/v_1$ requires that

$$i_L = A_g v_1 = G_{11} A_i v_1 = \frac{\beta G_L G_{11}}{G_L + G_{22}} v_1 \tag{11.36}$$

From the comparator output,

$$v_1 = v_S - v_f = v_S - \beta_{iv} v_L \tag{11.37}$$

Substitution of (*11.37*) into (*11.36*) allows solution for the overall transconductance gain as

$$A_{gf} = \frac{i_L}{v_S} = \frac{A_g}{1 + \beta_{iv} A_g} = \frac{G_{11} A_i}{1 + \beta_{iv} G_{11} A_i} = \frac{\beta G_L G_{11}}{G_L + G_{22} + \beta_{iv} \beta G_L G_{11}} \tag{11.38}$$

(*b*) Since $v_L = -i_L R_L$,

$$A_{vf} = \frac{v_L}{v_S} = -R_L \frac{i_L}{v_S} = -R_L A_{gf} = \frac{-R_L A_g}{1 + \beta_{iv} A_g} \tag{11.39}$$

11.6 VOLTAGE-SHUNT FEEDBACK

The general feedback amplifier model of Fig. 11-6, with output voltage v_L as the sampled variable and feedback current i_f as the mixed variable, is called a *voltage-shunt* (or *parallel-output, series input*; or *node-sampling, node-comparison*) feedback amplifier.

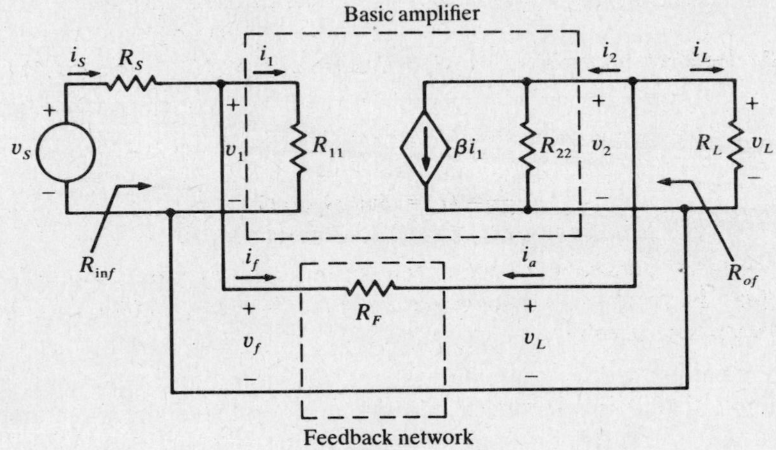

Fig. 11-6 General voltage-shunt feedback amplifier model

Example 11.7 For the voltage-shunt feedback amplifier of Fig. 11-6, find the input and output resistances (impedances) (*a*) with feedback and (*b*) without feedback.

(*a*) By the method of node voltages for $v_1\ (= v_f)$ and v_L,

$$(v_1 - v_S)G_S + v_1 G_{11} + (v_1 - v_L)G_F = 0 \tag{11.40}$$

and

$$\beta i_1 + v_L G_{22} + (v_L - v_1)G_F + v_L G_L = 0 \tag{11.41}$$

But

$$i_1 = G_{11} v_1 \tag{11.42}$$

We substitute (*11.42*) into (*11.41*) and rearrange to get the following system of linear equations:

$$\begin{bmatrix} G_F + G_S + G_{11} & -G_F \\ -(G_F - \beta G_{11}) & G_{22} + G_F + G_L \end{bmatrix} \begin{bmatrix} v_1 \\ v_L \end{bmatrix} = \begin{bmatrix} G_S v_S \\ 0 \end{bmatrix} \tag{11.43}$$

Simultaneous solution of (*11.43*) by Cramer's rule gives v_1 as

$$v_1 = \frac{\Delta_1}{\Delta} = \frac{\begin{vmatrix} G_S v_S & -G_F \\ 0 & G_{22} + G_F + G_L \end{vmatrix}}{\begin{vmatrix} G_F + G_S + G_{11} & -G_F \\ -(G - \beta G_{11}) & G_{22} + G_F + G_L \end{vmatrix}} = \frac{G_S(G_{22} + G_F + G_L)}{\Delta} \tag{11.44}$$

where
$$\Delta = G_{11}[(1+\beta)G_F + G_{22} + G_L] + G_S(G_{22} + G_F) + G_L(G_S + G_F) \tag{11.45}$$

Utilizing $i_S = G_S(v_S - v_1)$ and (*11.44*), we find the input resistance as

$$R_{inf} = \frac{v_i}{i_S} = \frac{v_1}{G_S(v_S - v_1)} = \frac{G_{22} + G_F + G_L}{\Delta - G_S(G_{22} + G_F + G_L)} \tag{11.46}$$

Substituting (*11.45*) into (*11.46*) and rearranging, we obtain

$$R_{inf} = \frac{G_{22} + G_F + G_L}{G_{11}[(1+\beta)G_F + G_{22} + G_L] + G_L G_F} \tag{11.47}$$

If we replace R_L with a driving-point source and short v_S, then we have

$$i_{dp} = \beta i_1 + \frac{v_{dp}}{R_{22}} + \frac{v_{dp}}{R_F + (R_S \| R_{11})} = \beta i_1 + G_{22} v_{dp} + \frac{(G_{11} + G_S) v_{dp}}{1 + R_F(G_{11} + G_S)} \tag{11.48}$$

But
$$i_1 = \frac{v_{dp}}{R_F + (R_S \| R_{11})} \frac{R_S}{R_S + R_{11}} = \frac{G_{11} v_{dp}}{1 + R_F(G_{11} + G_S)} \tag{11.49}$$

Using (*11.49*) in (*11.48*) and solving for the output conductance now give

$$G_{of} = \frac{1}{R_{of}} = \frac{i_{dp}}{v_{dp}} = G_{22} + \frac{(1+\beta)G_{11} + G_S}{1 + R_F(G_{11} + G_S)} \tag{11.50}$$

(*b*) To deactivate the feedback network, we let $G_F = 1/R_F = 0$; then (*11.47*) and (*11.50*) yield

$$R_{in} = \lim_{G_F \to 0} R_{inf} = \frac{1}{G_{11}} = R_{11} \tag{11.51}$$

$$\frac{1}{R_o} = G_o = \lim_{G_F \to 0} G_{of} = G_{22} \tag{11.52}$$

The ideal voltage-shunt feedback amplifier, shown in block-diagram form in Fig. 11-7, is based on the assumption that the feedback network does not load the basic amplifier. In this case, the basic amplifier is a transresistance amplifier.

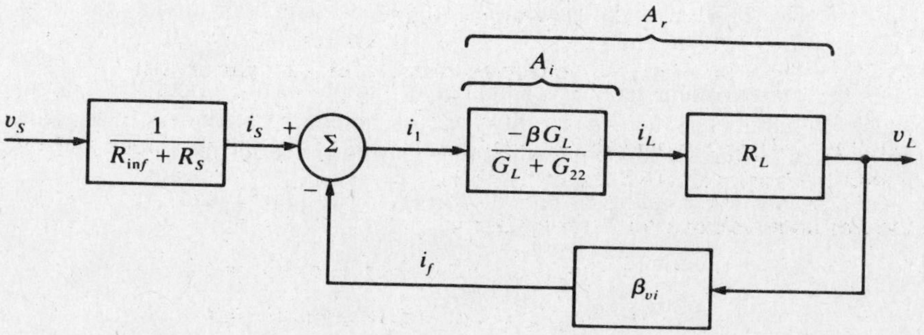

Fig. 11-7 Ideal-voltage-shunt feedback model

Example 11.8 For the ideal voltage-shunt amplifier of Fig. 11-7, find (*a*) the overall voltage gain and (*b*) the overall current gain.

(*a*) The basic amplifier gain provides that

$$v_L = A_r i_1 = R_L A_i i_1 = -\frac{\beta}{G_L + G_{22}} i_1 \tag{11.53}$$

and from the comparator output,

$$i_1 = i_S - i_f = \frac{v_S}{R_{\text{in}f} + R_S} - \beta_{vi} v_L \tag{11.54}$$

Substituting (*11.54*) into (*11.53*) and rearranging give

$$A_{vf} = \frac{v_L}{v_S} = \frac{A_r/(R_{\text{in}f} + R_S)}{1 + \beta_{vi} A_r} = \frac{R_L A_i/(R_{\text{in}f} + R_S)}{1 + \beta_{vi} R_L A_i} = \frac{-\beta/(R_{\text{in}f} + R_S)}{G_L + G_{22} - \beta_{vi}\beta} \tag{11.55}$$

(*b*) The overall current gain follows as

$$A_{if} = \frac{i_L}{i_S} = \frac{v_L/R_L}{v_S/(R_{\text{in}f} + R_S)} = A_{vf}\frac{R_{\text{in}f} + R_S}{R_L} = \frac{A_r/R_L}{1 + \beta_{vi} A_r} \tag{11.56}$$

11.7 CURRENT-SHUNT FEEDBACK

The general feedback amplifier of Fig. 11-8, in which output current i_L is the sampled variable and feedback current i_f is the mixed variable, is called a *current-shunt* (or *parallel-output*, *parallel-input*; or *node-sampling*, *node-comparison*) feedback amplifier.

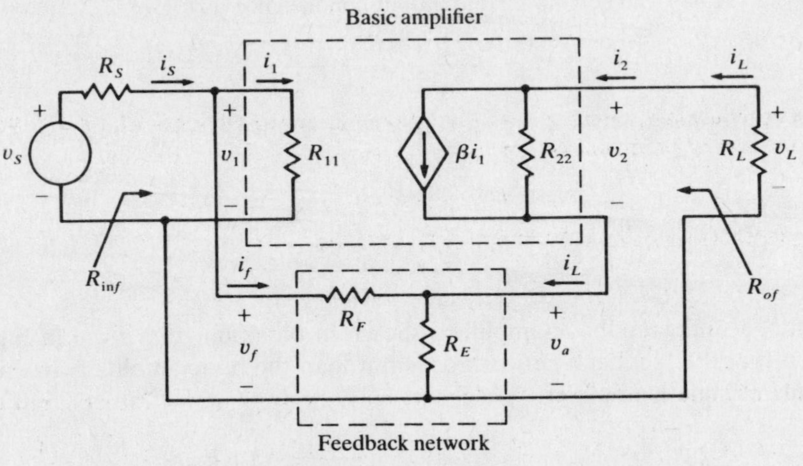

Fig. 11-8 General current-shunt feedback amplifier model

Example 11.9 For the current-shunt feedback amplifier of Fig. 11-8, (*a*) model the feedback network as a two-port network using *h* parameters, and (*b*) modify the basic amplifier parameters β, R_{11}, and R_{22} to account for any loading introduced by the feedback network, thus leaving the latter ideal.

(*a*) Using the feedback-network variables of Fig. 11-8 in (*1.14*) to (*1.19*) leads to

$$v_a = h_{11} i_L + h_{12} v_f \tag{11.57}$$

$$i_f = h_{21} i_L + h_{22} v_f \equiv i_f' + h_{22} v_f \tag{11.58}$$

where $$h_{11} = \left.\frac{v_a}{i_L}\right|_{v_f=0} \qquad h_{12} = \left.\frac{v_a}{v_f}\right|_{i_L=0} \qquad h_{21} = \left.\frac{i_f}{i_L}\right|_{v_f=0} \qquad h_{22} = \left.\frac{i_f}{v_f}\right|_{i_L=0} \tag{11.59}$$

For the particular feedback network at hand

$$h_{11} = R_E \| R_F = \frac{R_E R_F}{R_E + R_F} \qquad h_{12} = \frac{R_E}{R_E + R_F} \qquad h_{21} = \frac{-R_E}{R_E + R_F} \qquad h_{22} = \frac{1}{R_E + R_F}$$

(*b*) Since $v_1 = v_f$ in Fig. 11-8, admittance h_{22} of (*11.58*) is connected in parallel with R_{11}; thus, an equivalent resistance,

$$R'_{11} = R_{11} \,\|\, \frac{1}{h_{22}} = \frac{R_{11}}{1 + h_{22}R_{11}} \tag{11.60}$$

can be formed, leaving only a controlled current source within the left port of the feedback network, as shown in Fig. 11-9.

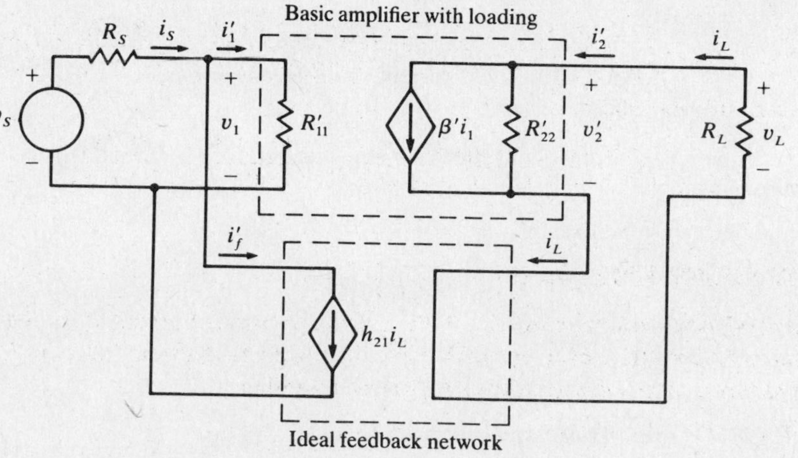

Fig. 11-9

A Thévenin equivalent circuit can be found for the network to the left of R_L. With v_S deactivated (shorted) and a driving-point source replacing R_L, KCL yields

$$i_{dp} = \beta i_1 + G_{22}[v_{dp} - (i_{dp}h_{11} + h_{12}v_f)] \tag{11.61}$$

But, by current division at the node of the input network,

$$i_1 = \frac{-1/h_{22}}{1/h_{22} + R_{11}}\, h_{21} i_L = \frac{-h_{21}}{1 + h_{22}R_{11}}\, i_L = \frac{-h_{21}}{1 + h_{22}R_{11}}\, i_{dp} \tag{11.62}$$

Also, $$v_f = R_{11} i_1 \tag{11.63}$$

Substituting (*11.62*) and (*11.63*) into (*11.61*) and rearranging yield the Thévenin resistance:

$$R_{Th} = \frac{1}{Y_N} = R'_{22} = \frac{v_{dp}}{i_{dp}} = R_{22}\left(1 + \frac{\beta h_{21}}{1 + h_{22}R_{11}}\right) + h_{11} - \frac{G_{22}h_{12}h_{21}R_{11}}{1 + h_{22}R_{11}} \tag{11.64}$$

The Thévenin voltage is found by allowing $R_L \to \infty$:

$$V_{Th} = -v_L = (\beta R_{22} + h_{12}R_{11}) i_1 \tag{11.65}$$

The Norton current that would flow if R_L were zero is found by use of (*11.64*) and (*11.65*):

$$I_N = \frac{V_{Th}}{R_{Th}} = \frac{(\beta R_{22} + h_{12}R_{11})(1 + h_{22}R_{11}) i_1}{R_{22}(1 + h_{22}R_{11} + \beta h_{21}) + h_{11}(1 + h_{22}R_{11}) = G_{22}h_{12}h_{21}R_{11}} = \beta' i_1 \tag{11.66}$$

With the elements of the Norton equivalent circuit as given by (*11.64*) and (*11.66*), the output network of Fig. 11-8 can be redrawn as in Fig. 11-9. The ideal feedback network merely multiplies current i_L by h_{21} and feeds the result to the input node for mixing with the input signal i_S.

The ideal current-shunt feedback amplifier model shown in block-diagram form in Fig. 11-10 is based on the assumption that the feedback network does not load the basic amplifier; however, if the loading were not negligible, the procedure of Example 11.9 could be utilized to replace β with β' and G_{22} with $G'_{22} = 1/R'_{22}$. The basic amplifier in this case is a current amplifier.

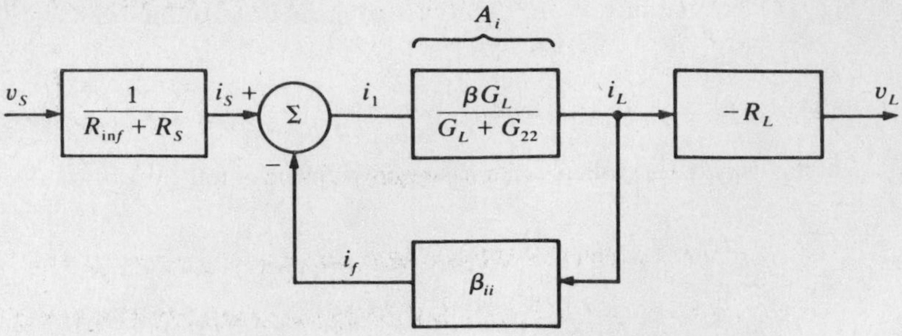

Fig. 11-10 Ideal current-shunt feedback model

Example 11.10 For the ideal current-shunt feedback amplifier of Fig. 11-10, find (*a*) the overall current gain and (*b*) the overall voltage gain.

(*a*) The basic amplifier gain relationship is

$$i_L = A_i i_1 = \frac{\beta G_L}{G_L + G_{22}} i_1 \qquad (11.67)$$

and the comparator output gives

$$i_1 = i_S - i_f = i_S - \beta_{ii} i_L \qquad (11.68)$$

We substitute (*11.67*) into (*11.68*) and rearrange to find

$$A_{if} = \frac{i_L}{i_S} = \frac{A_i}{1 + \beta_{ii} A_i} = \frac{\beta G_L}{G_L + G_{22} + \beta_{ii} G_L} \qquad (11.69)$$

(*b*) The overall voltage gain follows as

$$A_{vf} = \frac{v_L}{v_S} = \frac{-i_L R_L}{i_S(R_{inf} + R_S)} = \frac{-A_{if} R_L}{R_{inf} + R_S} = \frac{-\beta}{(G_L + G_{22} + \beta_{ii} G_L)(R_{inf} + R_S)} \qquad (11.70)$$

Solved Problems

11.1 A one-pole amplifier ($f_L = 0$) with midfrequency gain $A_{vo} = -1000$ has an upper cutoff frequency (3-db point) of 20 kHz. Negative voltage-series feedback is to be implemented to increase the upper cutoff frequency to 100 kHz. Determine (*a*) the new midfrequency gain and (*b*) the necessary feedback transfer ratio.

(*a*) The addition of feedback to an amplifier reduces the gain while increasing the bandwidth. If $f_L \ll f_H$ and $f_{Lf} \ll f_{Hf}$, then the *gain-bandwidth product* $\text{GBW} \equiv A(f_H - f_L) \approx Af_H$ remains approximately constant ($\text{GBW} \approx \text{GBW}_f$). Thus, the new midfrequency overall gain is

$$A_{vfo} = \frac{\text{GBW}}{f_{Hf}} = \frac{A_{vo} f_H}{f_{Hf}} = \frac{(-1000)(20 \times 10^3)}{100 \times 10^3} = -200$$

(*b*) For $\omega \ll \omega_H$, (*11.5*) applied to voltage-series feedback gives

$$A_{vfo} = \frac{A_{vo}}{1 + \beta_{vv} A_{vo}} \qquad (1)$$

from which

$$\beta_{vv} = \frac{1}{A_{vfo}} - \frac{1}{A_{vo}} = \frac{1}{-200} - \frac{1}{-1000} = -0.004$$

11.2 To see that voltage-series feedback is characterized by an increase in input resistance and a decrease in output resistance for all types of amplifiers, replace the current-source model within the basic amplifier of Fig. 11-2 with a voltage-source model, to obtain the voltage-series feedback amplifier of Fig. 11-11. Find the input and output resistances (*a*) with feedback and (*b*) without feedback.

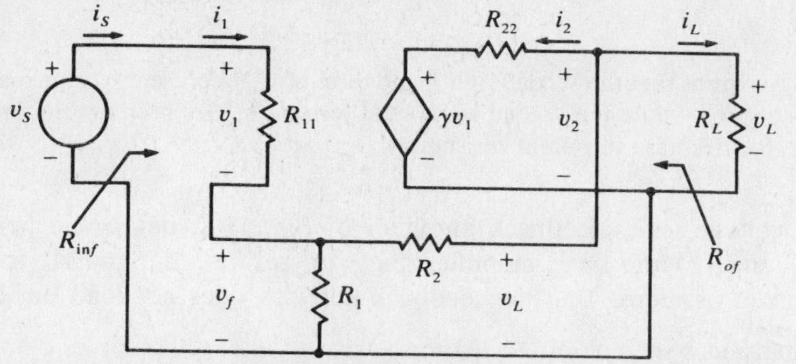

Fig. 11-11

(*a*) Application of the method of node voltages to Fig. 11-11 gives

$$(v_f - v_S)G_{11} + v_f G_1 + (v_f - v_L)G_2 = 0 \tag{1}$$

$$(v_L - v_f)G_2 + (v_L - \gamma v_1)G_{22} + v_L G_L = 0 \tag{2}$$

But

$$v_1 = v_S - v_f \tag{3}$$

Substitution of (*3*) into (*2*) and rearrangement result in the linear system of equations

$$\begin{bmatrix} G_{11} + G_1 + G_2 & -G_2 \\ -(G_2 - \gamma G_{22}) & G_2 + G_{22} + G_L \end{bmatrix} \begin{bmatrix} v_f \\ v_L \end{bmatrix} = \begin{bmatrix} G_{11} v_S \\ \gamma G_{22} v_S \end{bmatrix} \tag{4}$$

Simultaneous solution of (*4*) by Cramer's rule gives

$$v_f = \frac{\Delta_1}{\Delta} = \frac{\begin{vmatrix} G_{11} v_S & -G_2 \\ \gamma G_{22} v_S & G_2 + G_{22} + G_L \end{vmatrix}}{\begin{vmatrix} G_{11} + G_1 + G_2 & -G_2 \\ -(G_2 - \gamma G_{22}) & G_2 + G_{22} + G_L \end{vmatrix}} = \frac{G_{11}(G_2 + G_{22} + G_L) + \gamma G_{2m} G_{22}}{\Delta} v_S$$

where

$$\Delta = (G_{11} + G_1)(G_2 + G_{22} + G_L) + G_2[(1 + \gamma)G_{22} + G_L] \tag{5}$$

Now,

$$i_1 = (v_S - v_f)G_{11}$$

so that

$$R_{inf} = R_{11} + \frac{v_f}{i_1} = R_{11} + \frac{v_f}{(v_S - v_f)G_{11}} \tag{6}$$

Substitution of (*5*) into (*6*) and simplification give

$$R_{inf} = R_{11} + \frac{G_{11}(G_2 + G_{22} + G_L) + \gamma G_2 G_{22}/G_{11}}{G_1(G_2 + G_{22} + G_L) + G_{22}(G_2 + G_L)} \tag{7}$$

To find the output resistance (Thévenin resistance), we deactivate (short) v_S and replace R_L with a driving-point source. Then, by KCL,

$$i_{dp} = \frac{v_{dp}}{R_2 + R_1 \| R_{11}} + \frac{v_{dp} - \gamma v_1}{R_{22}} \tag{8}$$

By voltage division,

$$v_1 = -\frac{R_1 \| R_{11}}{R_1 \| R_{11} + R_2} v_{dp} \tag{9}$$

Substituting (*9*) into (*8*), converting to conductances, and solving for the output conductance give

$$G_{of} = \frac{1}{R_{of}} = \frac{i_{dp}}{v_{dp}} = G_{22} + \frac{(G_1 + G_{11} + \gamma G_{22})G_2}{(G_1 + G_{11} + G_{22})G_{22}} \tag{10}$$

(*b*) To deactivate the feedback network, let $R_1 (= 1/G_1) = 0$ and $R_2 (= 1/G_2) = \infty$. Then the input and output resistances without feedback are found from (*7*) and (*10*) to be

$$R_{\text{in}} = \lim_{\substack{G_1 \to \infty \\ G_2 \to 0}} R_{\text{inf}} = R_{11}$$

$$R_o = \lim_{\substack{G_1 \to \infty \\ G_2 \to 0}} \frac{1}{G_{of}} = \frac{1}{G_{22}} = R_{22}$$

It is apparent that the second right-hand term of (*7*) represents the increase in input resistance due to feedback, while the second right-hand term of (*10*) represents the increase in output conductance (or decrease in output resistance).

11.3 For the voltage-series feedback amplifier of Fig. 11-2, determine (*a*) the feedback transfer ratio β_{vv} and (*b*) the basic amplifier gain A_v. (*c*) Use β_{vv} and A_v to determine the overall voltage gain, assuming that the feedback network does not load or alter A_v.

(*a*) Simultaneous solution of (*11.12*) for v_L gives

$$v_L = \frac{\Delta_2}{\Delta} = \frac{\begin{vmatrix} G_{11} + G_1 + G_2 & G_{11}v_S \\ -(G_2 - \beta G_{11}) & \beta G_{11} v_S \end{vmatrix}}{\Delta} = \frac{G_{11}[(1+\beta)G_2 + \beta G_1]v_S}{\Delta} \tag{1}$$

Then dividing (*11.13*) by (*1*) gives

$$\beta_{vv} = \frac{v_f}{v_L} = \frac{(1+\beta)G_2 + G_{22} + G_L}{(1+\beta)G_2 + \beta G_1} \tag{2}$$

(*b*) With the feedback network deactivated ($R_1 = 0$, $R_2 = \infty$),

$$i_1 = G_{11} v_1 \tag{3}$$

and
$$v_2 = \beta i_1 (R_{22} \| R_L) = \frac{\beta i_1}{G_{22} + G_L} \tag{4}$$

Substitution of (*3*) into (*4*) then allows solution for the basic amplifier voltage gain:

$$A_v = \frac{v_2}{v_1} = \frac{\beta G_{11}}{G_{22} + G_L} \tag{5}$$

(*c*) Since the feedback network does not alter A_v, (*11.25*) yields

$$A_{vf} = \frac{A_v}{1 + \beta_{vv} A_v} = \frac{\beta G_{11}/(G_{22} + G_L)}{1 + \dfrac{(1+\beta)G_2 + G_{22} + G_L}{(1+\beta)G_2 + \beta G_1} \dfrac{\beta G_{11}}{G_{22} + G_L}}$$

$$= \frac{\beta G_{11}[(1+\beta)G_2 + \beta G_1]}{(G_{22} + G_L)[(1+\beta)G_2 + \beta G_1] + \beta G_{11}[(1+\beta)G_2 + G_{22} + G_L]} \tag{6}$$

11.4 If distortion is introduced into the output voltage of a high-gain amplifier or, as commonly occurs, into the output(power) stage of a cascaded amplifier, then feedback can be used to reduce the amplitude of the distortion. A model for a voltage-series feedback amplifier, with distortion introduced as voltage v_D, is shown in Fig. 11-12, where the gain A_v may be the product of the gains of cascaded stages as long as v_1 and v_2 are in phase. Find the output v_L in terms of v_S and v_D, to show that the amplitude of the distortion voltage is reduced if $\beta_{vv} A_v \gg 1$.

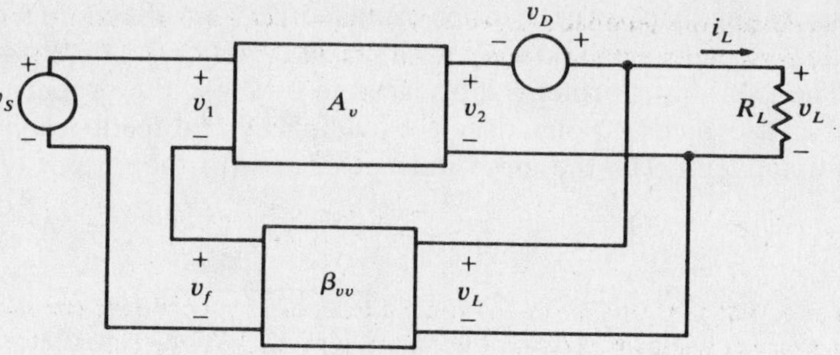

Fig. 11-12

KVL around the input loop requires that

$$v_1 = v_S - v_f = v_S - \beta_{vv} v_L \qquad (1)$$

The gain relationship for the basic amplifier and an application of KVL around the output loop then lead to

$$v_1 = \frac{v_2}{A_v} = \frac{v_L - v_D}{A_v} \qquad (2)$$

Equating (*1*) and (*2*) and solving for v_L give

$$v_L = \frac{A_v}{1 + \beta_{vv} A_v} v_S + \frac{v_D}{1 + \beta_{vv} A_v} \qquad (3)$$

Equation (*3*) shows that the magnitude of the distortion voltage is reduced significantly if $\beta_{vv} A_v \gg 1$.

11.5 Using the voltage-series feedback amplifier of Fig. 11-11, show that if loading by the feedback network is negligible, then (*a*) $R_{inf} = R_{in}(1 + \beta_{vv} A_v)$ and (*b*) $R_{of} = R_o/(1 + \beta_{vv} A'_v)$, where, for the specific circuit at hand, $R_{in} = R_{11}$, $R_o = R_{22}$, $A_v = \gamma R_L/(R_L + R_{22})$, $A'_v = \gamma$, and $\beta_{vv} = R_1/(R_1 + R_2)$.

(*a*) By KVL around the input loop of Fig. 11-11,

$$v_S = i_1 R_{11} + v_f = i_1 R_{11} + \frac{R_1}{R_1 + R_2} v_L = i_1 R_{11} + \beta_{vv} v_L \qquad (1)$$

Neglecting feedback-network loading, voltage division applied to the output loop gives

$$v_L = \frac{R_L}{R_L + R_{22}} \gamma v_1 = \frac{R_L}{R_L + R_{22}} \gamma (R_{11} i_1) = A_v R_{11} i_1 \qquad (2)$$

where $A_v = \gamma R_L/(R_L + R_{22})$ is the gain of the basic amplifier with feedback deactivated. Substituting (*2*) into (*1*) and rearranging give

$$v_S = i_1 R_{11} + \beta_{vv}(A_v R_{11} i_1) = i_1 R_{11}(1 + \beta_{vv} A_v)$$

Hence,

$$R_{inf} = \frac{v_S}{i_1} = \frac{v_S}{i_S} = R_{11}(1 + \beta_{vv} A_v) = R_{in}(1 + \beta_{vv} A_v) \qquad (3)$$

where $R_{in} = R_{11}$ is the input resistance with feedback deactivated.

(*b*) With v_S deactivated (shorted) and R_L replaced with a driving-point source, KCL applied at the node of the output network requires that

$$i_{dp} = \frac{v_{dp} - \gamma v_1}{R_{22}} \qquad (4)$$

But with $v_S = 0$,

$$v_1 = -v_F = -\beta_{vv} v_{dp} \qquad (5)$$

Substituting (*5*) into (*4*) and rearranging yield

$$R_{of} = \frac{v_{dp}}{i_{dp}} = \frac{R_{22}}{1 + \beta_{vv}\gamma} = \frac{R_o}{1 + \beta_{vv} A'_v} \qquad (6)$$

11.6 Voltage-series feedback is used in the cascaded FET amplifiers of Fig. 11-13(a) and the FETs are identical. Let $R_2 = 190\ \text{k}\Omega$, $R_1 = 10\ \text{k}\Omega$, $R_D = R_L = 2\ \text{k}\Omega$, $R_G = 1\ \text{M}\Omega$, $r_{ds} = 30\ \text{k}\Omega$, and $\mu = 60$. The ideal transformer is introduced to establish the proper polarity for v_f for negative feedback at the mixing point. Neglect loading by the feedback network, and find (a) the overall voltage gain, (b) the input resistance, and (c) the output resistance.

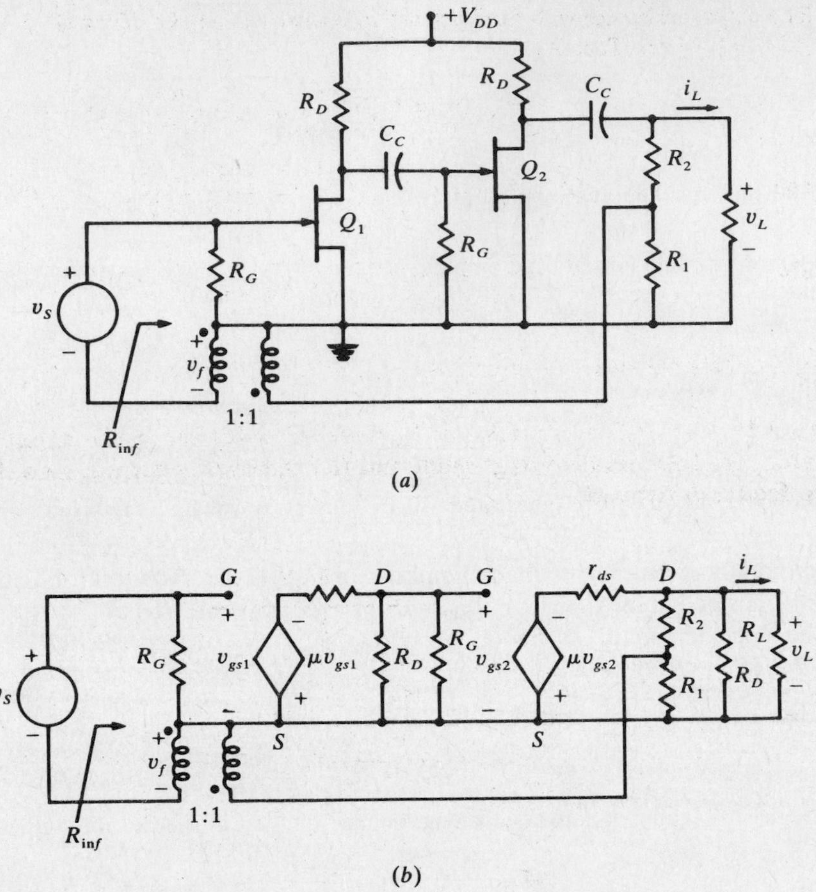

Fig. 11-13

(a) The small-signal equivalent circuit is displayed in Fig. 11-13(b). Since $R_G \| R_D \approx R_D$ and $R_G \| R_L \| R_D \approx R_L \| R_D$, the voltage gain without feedback is

$$A_v = A_{v1}A_{v2} \approx \frac{-\mu R_D}{R_D + r_{ds}} \, \frac{-\mu R_D \| R_L}{R_D \| R_L + r_{ds}} = \frac{\mu^2 R_D^2 R_L}{(R_D + r_{ds})[R_D R_L + r_{ds}(R_D + R_L)]}$$

$$= \frac{(60)^2(2000)^2(2000)}{(32 \times 10^3)[(2000)(2000) + 30 \times 10^3(4000)]} = 7.26$$

Voltage division gives the feedback transfer ratio as

$$\beta_{vv} = \frac{R_1}{R_1 + R_2} = \frac{10 \times 10^3}{(10 + 190) \times 10^3} = 0.05$$

With feedback loading neglected, (*11.25*) gives the overall voltage gain as

$$A_{vf} = \frac{A_v}{1 + \beta_{vv} A_v} = \frac{7.26}{1 + (0.05)(7.26)} = 5.33$$

(*b*) Without feedback ($R_1 = 0$, $R_2 = \infty$), $R_{in} = R_G = 1\ \text{M}\Omega$. Then by (*3*) of Problem 11.5,

$$R_{inf} = R_{in}(1 + \beta_{vv}A_v) = (1 \times 10^6)[1 + (0.05)(7.26)] = 1.363\ \text{M}\Omega$$

(*c*) Without feedback, the output resistance is

$$R_o = r_{ds} \parallel R_D = \frac{r_{ds}R_D}{r_{ds} + R_D} = \frac{(30 \times 10^3)(2 \times 10^3)}{(30 + 2) \times 10^3} = 1.875\ \text{k}\Omega$$

The output resistance with feedback active is given by (*6*) of Problem 11.5, where A_v' is the voltage gain with $i_L = 0$. That is,

$$V_{Th} = -\mu(A_{v1}v_1)\frac{R_D}{R_D + r_{ds}} = -\mu\frac{-\mu R_D}{R_D + r_{ds}}\frac{R_D}{R_D + r_{ds}}v_1$$

and
$$A_v' = \frac{V_{Th}}{v_1} = \left(\frac{\mu R_D}{R_D + r_{ds}}\right)^2 = \left[\frac{(60)(2000)}{(2 + 30) \times 10^3}\right]^2 = 14.06$$

Then,
$$R_{of} = \frac{R_o}{1 + \beta_{vv}A_v'} = \frac{1875}{1 + (0.05)(14.06)} = 1101\ \Omega$$

11.7 Voltage-series feedback is used in the cascaded BJT amplifier of Fig. 11-14(*a*). Let $R_E = 100\ \Omega$, $R_B = 500\ \text{k}\Omega$, $R_C = 2\ \text{k}\Omega$, $R_L = 3\ \text{k}\Omega$, $R_F = 15\ \text{k}\Omega$, $h_{ie1} = 100\ \Omega$, $h_{ie2} = 250\ \Omega$, $h_{fe1} = 75$, $h_{fe2} = 90$, and $h_{re1} = h_{re2} = h_{oe1} = h_{oe2} = 0$. Find (*a*) the input resistance, (*b*) the output resistance, (*c*) the overall voltage gain, and (*d*) the overall current gain. Neglect any loading by the feedback network.

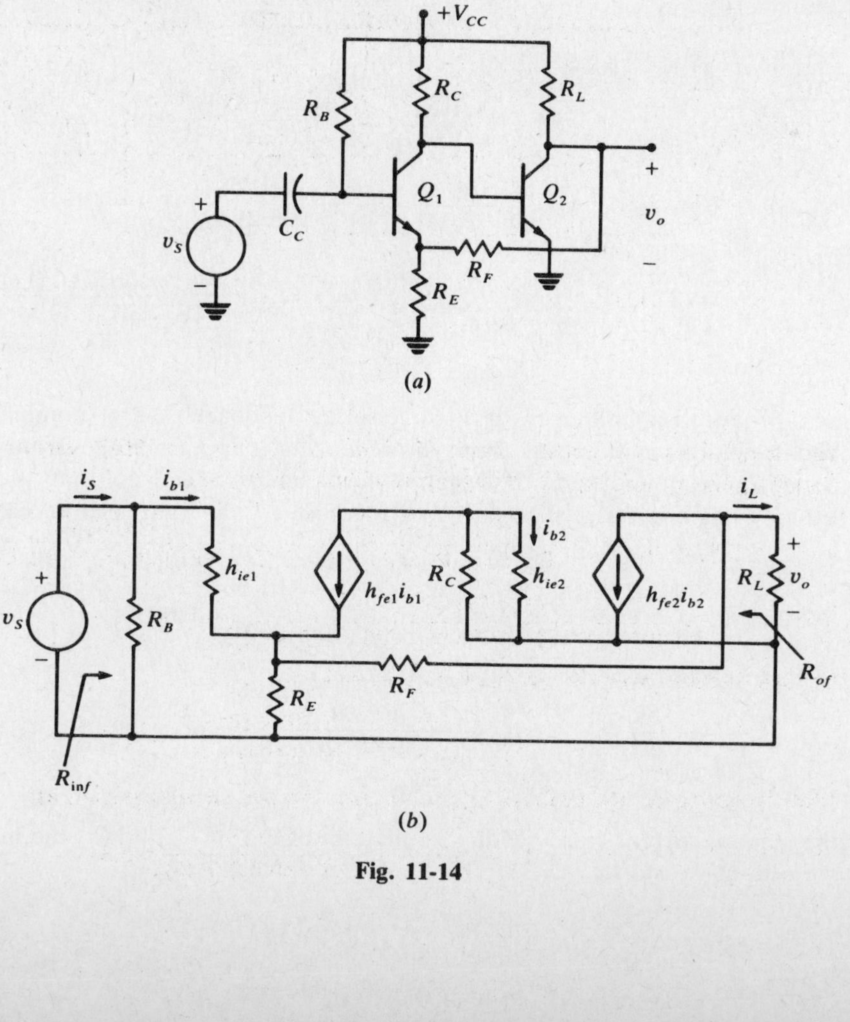

Fig. 11-14

(*a*) The small-signal equivalent circuit is drawn in Fig. 11-14(*b*), from which, without feedback, $R_{in} = R_{11} = h_{ie1}$ and $R_o = 1/h_{oe2} = \infty$. With no feedback and with R_L open-circuited, the controlled current source βi_1 of Fig. 11-6 becomes

$$\beta i_1 = \beta i_{b1} = -h_{fe2} i_{b2} = -h_{fe2}(-h_{fe1} i_{b1}) \frac{R_C}{R_C + h_{ie2}}$$

so that
$$\beta = \frac{h_{fe1} h_{fe2} R_C}{R_C + h_{ie2}} = \frac{(75)(90)(2000)}{2100} = 6428.6$$

Then according to (*11.17*),

$$R_{inf} = R_{11} + \frac{(1+\beta)G_2 + G_{22} + G_L}{G_1 G_2 + (G_1 + G_2)(G_{22} + G_L)}$$

$$= 100 + \frac{(6429.6)(0.0667 \times 10^{-3}) + 0 + 3.333 \times 10^{-4}}{(0.01)(0.0667 \times 10^{-3}) + (0.010067)(3.333 \times 10^{-4})} = 106.79\ \text{k}\Omega$$

(*b*) From (*11.20*),

$$G_{of} = G_{22} + \frac{(1+\beta)G_{11} + G_1}{(G_{11} + G_1)R_2 + 1} = 0 + \frac{(6429.6)(0.01) + 0.01}{(0.02)(15 \times 10^3) + 1} = 0.2136$$

so
$$R_{of} = \frac{1}{G_{of}} = \frac{1}{0.2136} = 4.68\ \Omega$$

(*c*) By voltage division,

$$\beta_{vv} = \frac{R_E}{R_E + R_F} = \frac{100}{15{,}100} = 0.00662$$

and from (*5*) of Problem 11.3,

$$A_v = \frac{\beta G_{11}}{G_{22} + G_L} = \frac{(6428.6)(0.01)}{0 + 3.333 \times 10^{-4}} = 192{,}877$$

Then (*11.25*) gives
$$A_{vf} = \frac{A_v}{1 + \beta_{vv} A_v} = \frac{192{,}858}{1 + (0.00662)(192{,}858)} = 150.94$$

(*d*)
$$A_{if} = \frac{i_L}{i_S} = \frac{v_o/R_L}{v_S/R_{inf}} = A_{vf}\frac{R_{inf}}{R_L} = 150.94\,\frac{106.79 \times 10^3}{3 \times 10^3} = 5372.9$$

11.8 A mass-produced amplifier with voltage-series feedback has a nominal open-loop (no-feedback) voltage gain A_v of 2000; however, its gain can range from 1200 to 2800 as component parameters vary. If negative feedback is added so that the nominal overall (closed-loop) gain is reduced to 50, what range of overall gain is to be expected?

Solving (*11.25*) for the feedback transfer ratio gives

$$\beta_{vv} = \frac{1}{A_{vf}} - \frac{1}{A_v} = \frac{1}{50} - \frac{1}{2000} = 0.0195$$

And, the range to be expected is, again from (*11.25*),

$$\frac{1200}{1 + (0.0195)(1200)} \le A_{vf} \le \frac{2800}{1 + (0.0195)(2800)} \qquad \text{or} \qquad 49.18 \le A_{vf} \le 50.36$$

11.9 For the general current-series feedback amplifier of Fig. 11-4, find (*a*) the input resistance and (*b*) the output resistance.

(*a*) From Fig. 11-4, with $i_1 = G_{11}(v_S - v_f)$,

$$R_{\text{in}f} = R_{11} + \frac{v_f}{i_1} = R_{11} + \frac{v_f}{G_{11}(v_S - v_f)} \tag{1}$$

Substituting for v_f from (*11.33*) and simplifying then give

$$R_{\text{in}f} = R_{11} + \frac{(1+\beta)G_L + G_{22}}{G_{22}G_F + G_L G_{22} + G_L G_F} \tag{2}$$

(*b*) To find R_{of}, we deactivate (short) v_S and replace R_L with a driving-point source. Then the method of node voltages gives

$$-\beta i_1 + (v_f - v_{dp})G_{22} + v_f G_F + v_f G_{11} = 0 \tag{3}$$

But now $i_1 = -v_f G_{11}$ so that, from (*3*),

$$v_f = \frac{G_{22} v_{dp}}{(1+\beta)G_{11} + G_{22} + G_F} \tag{4}$$

KCL applied at the node of the output network requires

$$i_{dp} = (v_{dp} - v_f)G_{22} + \beta i_1 = (v_{dp} - v_f)G_{22} - \beta v_f G_{11} \tag{5}$$

The use of (*4*) in (*5*) and rearrangement yield

$$R_{of} = \frac{v_{dp}}{i_{dp}} = \frac{(1+\beta)G_{11} + G_{22} + G_F}{G_{22}(G_{11} + G_F)} \tag{6}$$

11.10 For the general voltage-series feedback amplifier of Fig. 11-2, (*a*) model the feedback network as a two-port network using *h* parameters and (*b*) modify the basic amplifier parameters β, R_{11}, and R_{22} to account for any loading introduced by the feedback network, leaving it ideal.

(*a*) In terms of the feedback network variables of Fig. 11-2, (*1.14*) to (*1.19*) lead to

$$v_f = h_{11} i_1 + h_{12} v_L \tag{1}$$

$$i_a = h_{21} i_1 + h_{22} v_L \tag{2}$$

where $$h_{11} = \frac{v_f}{i_1}\bigg|_{v_L=0} \qquad h_{12} = \frac{v_f}{v_L}\bigg|_{i_1=0} \qquad h_{21} = \frac{i_a}{i_1}\bigg|_{v_L=0} \qquad h_{22} = \frac{i_a}{v_L}\bigg|_{i_1=0} \tag{3}$$

For the particular feedback network at hand,

$$h_{11} = R_1 \| R_2 = \frac{R_1 R_2}{R_1 + R_2} \qquad h_{12} = \frac{R_1}{R_1 + R_2} \qquad h_{21} = -\frac{R_1}{R_1 + R_2} \qquad h_{22} = \frac{1}{R_1 + R_2}$$

(*b*) The series combination of R_{11} and h_{11} can be added to give

$$R'_{11} = R_{11} + h_{11}$$

The parallel combination of R_{22} and h_{22} can be replaced with

$$R'_{22} = R_{22} \,\Big\|\, \frac{1}{h_{22}} = \frac{R_{22}}{h_{22}R_{22} + 1}$$

And an equivalent current source $\beta' i_1$ can be formed as

$$\beta' = \beta - h_{21}$$

These new parameters result in the equivalent circuit of Fig. 11-15, where the feedback network is ideal.

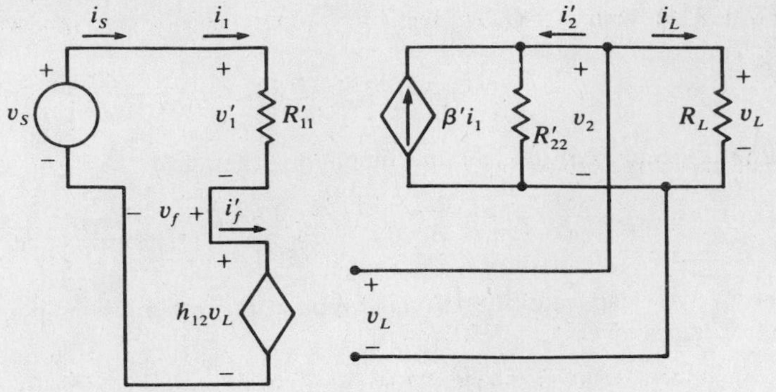

Fig. 11-15

11.11 Develop a means to account for the loading effects of the feedback network of the current-series feedback amplifier by (*a*) modeling the feedback network as a two-port network using z parameters and (*b*) modifying the basic amplifier parameters β, R_{11}, and R_{22} of Fig. 11-4 to account for any loading introduced by the feedback circuitry.

(*a*) In terms of the feedback network variables of Fig. 11-4, (*1.8*) to (*1.13*) lead to

$$v_f = z_{11}i_1 + z_{12}i_L \equiv z_{11}i_1 + v_f' \qquad (1)$$

$$v_a = z_{21}i_1 + z_{22}i_L \qquad (2)$$

where $$z_{11} = \left.\frac{v_f}{i_1}\right|_{i_L=0} \qquad z_{12} = \left.\frac{v_f}{i_L}\right|_{i_1=0} \qquad z_{21} = \left.\frac{v_a}{i_1}\right|_{i_L=0} \qquad z_{22} = \left.\frac{v_a}{i_L}\right|_{i_1=0} \qquad (3)$$

For the particular feedback network at hand,

$$z_{11} = z_{12} = z_{21} = z_{22} = R_F$$

(*b*) Impedance (resistance) z_{11} can immediately be combined with R_{11} of Fig. 11-4 to form

$$R_{11}' = R_{11} + z_{11} \qquad (4)$$

leaving only the current-controlled voltage source $v_f' = z_{12}i_L$ in the feedback network.

A Norton equivalent circuit can be formed for the network to the left of R_L. With R_L replaced with a driving-point source and v_S deactivated (shorted),

$$i_1 = \frac{-z_{12}i_L}{R_{11}'} \qquad (5)$$

Applying KCL at the upper node of the output network gives, with (*2*),

$$i_{dp} = \beta i_1 + \frac{v_{dp} - v_a}{R_{22}} = \beta i_1 + \frac{1}{R_{22}}(v_{dp} - z_{21}i_1 + z_{22}i_{dp}) \qquad (6)$$

Substituting (*5*) into (*6*) and solving for the driving-point impedance give

$$R_{Th} = \frac{1}{Y_N} = R_{22}\left(1 + \frac{\beta z_{12}}{R_{11}'}\right) + z_{22} - \frac{z_{12}z_{21}}{R_{11}'} = R_{22}' \qquad (7)$$

If the output were open-circuited (R_L removed), we would have $i_L = 0$ and $i_1 = v_S/R_{11}'$. Hence, the Thévenin voltage (with polarity opposite that of v_L) is found from (*2*) as

$$V_{Th} = \beta i_1 R_{22} - v_a = \beta i_1 R_{22} - z_{12}i_1 = (\beta R_{22} - z_{12})\frac{v_S}{R_{11}'} \qquad (8)$$

It is convenient to determine the input resistance. By KVL applied at the input and output networks,

$$v_S = R_{11}'i_1 + z_{12}i_L \qquad (9)$$

$$V_{Th} = (R_{Th} + R_L)i_L \qquad (10)$$

Substituting (8) into (10), solving for i_L, and substituting the result in (9) give the input impedance as

$$R_{\text{in}f} = \frac{v_S}{i_1} = \frac{R'_{11}(R_{Th} + R_L) + z_{12}(\beta R_{22} - z_{12})}{R_{Th} + R_L} \tag{11}$$

The Norton source follows as

$$I_N = \beta' i_1 = \frac{V_{Th}}{R_{Th}} = \frac{(\beta R_{22} - z_{21})v_S}{R_{Th}R'_{11}} = \frac{(\beta R_{22} - z_{21})R_{\text{in}f}i_1}{R'_{11}R'_{22}} \tag{12}$$

so

$$\beta' = \frac{(\beta R_{22} - z_{21})R_{\text{in}f}}{R'_{11}R'_{22}} \tag{13}$$

Thus, the current-series feedback amplifier of Fig. 11-4 can be redrawn as shown in Fig. 11-16. The ideal feedback network simply multiplies load current i_L by the transresistance gain and feeds the result (v'_f) to the input loop for mixing with the input signal v_S.

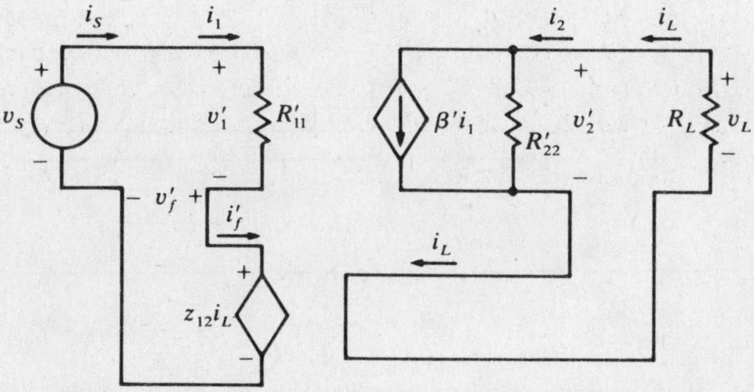

Fig. 11-16

11.12 Figure 11.17(a) displays a current-series feedback amplifier. Let $h_{fe1} = h_{fe2} = h_{fe3} = 50$, $h_{ie1} = h_{ie2} = h_{ie3} = 200\ \Omega$, $R_{C1} = R_{C2} = R_L = 2\ \text{k}\Omega$, $R_{E1} = 50\ \Omega$, $R_{E3} = 150\ \Omega$, and $R_{13} = 100\ \Omega$. Assume $h_{re1} = h_{re2} = h_{re3} = h_{oe1} = h_{oe2} = h_{oe3} = 0$. For this amplifier, find (a) R_{of}, (b) $R_{\text{in}f}$, (c) A_{vf}, and (d) A_{if}.

(a) The small-signal equivalent circuit is shown in Fig. 11-17(b). By (3) of Problem 11.11, the z parameters are

$$z_{11} = \frac{v_f}{i_{b1}}\bigg|_{i_L=0} = R_{E1} \| (R_{13} + R_{E3}) = \frac{(50)(250)}{300} = 41.67\ \Omega$$

$$z_{12} = \frac{v_f}{i_L}\bigg|_{i_{b1}=0} = \frac{R_{E1}R_{E3}}{R_{E1} + R_{13} + R_{E3}} = \frac{(50)(150)}{300} = 25\ \Omega$$

$$z_{21} = \frac{v_a}{i_1}\bigg|_{i_L=0} = \frac{R_{E1}R_{E3}}{R_{E1} + R_{13} + R_{E3}} = z_{12} = 25\ \Omega$$

$$z_{22} = \frac{v_a}{i_L}\bigg|_{i_1=0} = R_{E3} \| (R_{13} + R_{E1}) = \frac{(150)(50)}{200} = 75\ \Omega$$

The basic amplifier parameters of Fig. 11-4 are needed. With the feedback network of Fig. 11-17(b) deactivated, so that $v_f = v_a = 0$, it is apparent that

$$R_{11} = R_{\text{in}} = h_{ie1} = 200\ \Omega$$

Also, with v_S shorted and R_L replaced with a driving-point source,

$$R_{22} = R_o = \frac{1}{h_{oe3}} = \infty$$

From (7) of Problem 11.11 with $R_{22} = \infty$,

$$R_{of} = R_{Th} = R'_{22} = \infty$$

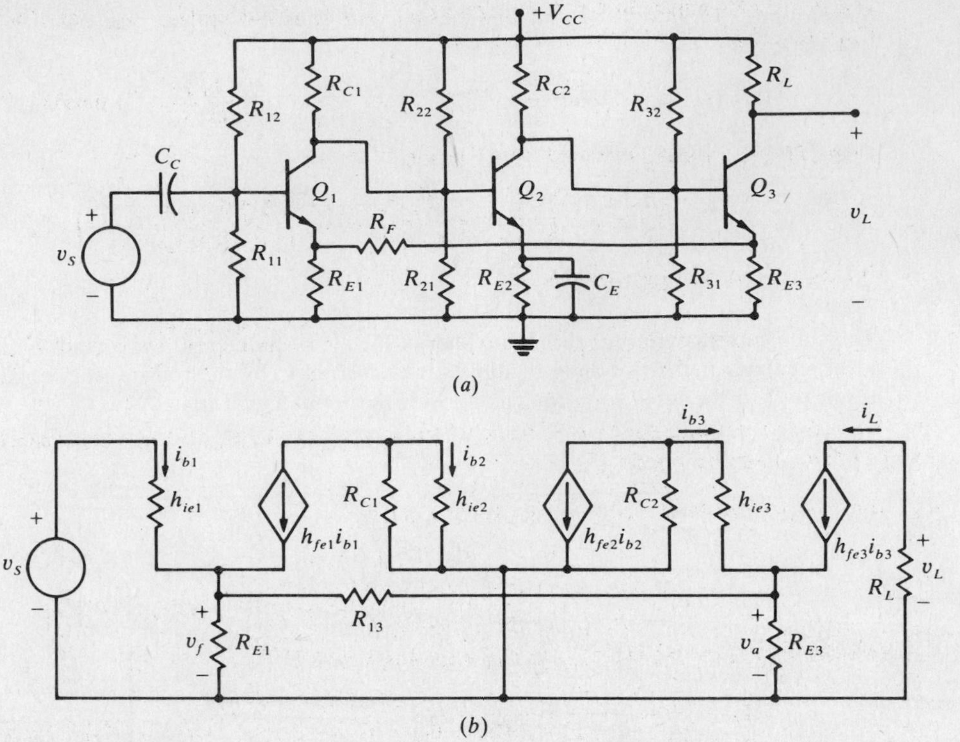

Fig. 11-17

(*b*) With the feedback network still deactivated, current division gives

$$i_{b2} = \frac{R_{C1}}{R_{C1} + h_{ie2}} h_{fe1} i_{b1} \quad \text{and} \quad i_{b3} = \frac{R_{C2}}{R_{C2} + h_{ie3}} h_{fe2} i_{b2} \tag{1}$$

so that $$i_L = h_{fe3} i_{b3} = \frac{R_{C1} R_{C2} h_{fe1} h_{fe2} h_{fe3}}{(R_{C1} + h_{ie2})(R_{C2} + h_{ie3})} i_{b1}$$

whence $$\beta = \frac{i_L}{i_{b1}} = \frac{R_{C1} R_{C2} h_{fe1} h_{fe2} h_{fe3}}{(R_{C1} + h_{ie2})(R_{C2} + h_{ie3})} = \frac{(2000)^2 (50)^3}{(2200)^2} = 103.3 \times 10^3$$

Now, substituting R_{Th} as given by (*7*) of Problem 11.11 into (*11*) of Problem 11.11 gives

$$R_{\text{inf}} = \frac{R'_{11}\left[R_{22}(1 + \beta z_{12} / R'_{11}) + z_{22} + z_{12} z_{21} / R'_{11} + R_L\right] + z_{12}(\beta R_{22} - z_{12})}{R_{22}(1 + \beta z_{12} / R'_{11}) + z_{22} + z_{12} z_{21} / R'_{11} + R_L} \tag{2}$$

which, in the limit as $R_{22} \to \infty$, yields

$$R_{\text{inf}} = R'_{11} + \frac{\beta R'_{11} z_{12}}{R'_{11} + \beta z_{12}} = (h_{ie1} + z_{11}) + \frac{\beta (h_{ie1} + z_{11}) z_{12}}{h_{ie1} + z_{11} + \beta z_{12}} = (200 + 41.67) + (103.3 \times 10^3)(25)$$

$$= 2.582\ \text{M}\Omega$$

$$R_{\text{inf}} = 200 + 41.67 + \frac{(103.3 \times 10^3)(200 + 41.67)(25)}{200 + 41.67 + (103.3 \times 10^3)(25)} = 483.34\Omega$$

(*c*) We use (*7*) of Problem 11.11 to rewrite (*13*) of Problem 11.11 as

$$\beta' = \frac{(\beta R_{22} - z_{21}) R_{\text{inf}}}{[R_{22}(1 + \beta z_{12}/R'_{11}) + z_{22} - z_{12} z_{21}/R'_{11}] R'_{11}}$$

which, in the limit as $R_{22} \to \infty$, becomes

$$\beta' = \frac{\beta R_{\text{inf}}}{R'_{11} + \beta z_{12}} = \frac{\beta R_{\text{inf}}}{h_{ie1} + z_{11} + \beta z_{12}} = \frac{(103.3 \times 10^3)(483.34)}{200 + 41.67 + (103.3 \times 10^3)(25)} = 19.33$$

The transconductance gain is found from Fig. 11-5 by replacing G_{11}, G_{22}, and β with $G'_{11} = 1/R'_{11}$, $G'_{22} = 0$, and β', respectively. The result is

$$A'_g = \frac{\beta' G'_{11} G_L}{G_L + G'_{22}} = \beta' G'_{11} = \frac{\beta'}{h_{ie1} + z_{11}} = \frac{19.33}{241.67} = 0.0452$$

Then (*11.39*) with A_g replaced by A'_g yields

$$A_{vf} = \frac{-R_L A'_g}{1 + \beta_{iv} A'_g} = \frac{-R_L A'_g}{1 + z_{12} A'_g} = \frac{-(2000)(0.0452)}{1 + (25)(0.0452)} = -42.44$$

(*d*) The overall current gain follows as

$$A_{if} = \frac{i_L}{i_S} = \frac{-v_L/R_L}{v_S/R_{\text{inf}}} = -A_{vf}\frac{R_{\text{inf}}}{R_L} = -\frac{(-42.44)(483.34)}{2 \times 10^3} = 10.26^3$$

11.13 For the general voltage-shunt feedback amplifier of Fig. 11-6, find (*a*) the overall voltage gain and (*b*) the overall current gain.

(*a*) Building on the work of Example 11.7, we have

$$v_L = \frac{\Delta_2}{\Delta} = \frac{\begin{vmatrix} (G_F + G_S + G_{11}) & G_S v_S \\ -(G_F - \beta G_{11}) & 0 \end{vmatrix}}{\Delta} = \frac{G_S(G_F - \beta G_{11})}{\Delta} v_S \tag{1}$$

so that, with Δ as in (*11.45*), the overall voltage gain is

$$A_{vf} = \frac{v_L}{v_S} = \frac{G_S(G_F - \beta G_{11})}{G_{11}[(1+\beta)G_F + G_{22} + G_L] + G_S(G_{22} + G_F) + G_L(G_S + G_F)} \tag{2}$$

(*b*) The overall current gain is found as

$$A_{if} = \frac{i_L}{i_1} = \frac{v_L/R_L}{v_S/(R_{\text{inf}} + R_S)} = A_{vf}\frac{(R_{\text{inf}} + R_S)}{R_L} \tag{3}$$

Substituting (*2*) and (*11.47*) into (*3*) then yields

$$A_{if} = \frac{G_L G_S (G_F - \beta G_{11})(G_{22} + G_F + G_L + R_S G)}{G[G + G_S(G_{22} + G_F + G_L)]}$$

where $$G = G_{11}[(1+\beta)G_F + G_{22} + G_L] + G_L G_F \tag{4}$$

11.14 The loading effects of the feedback network in the voltage-shunt feedback amplifier can be accounted for by modifying the basic amplifier through the use of *short-circuit admittance parameters or y parameters,* defined in terms of the variables of Fig. 11-6 to model the feedback network as a two-port network:

$$i_a = y_{11} v_L + y_{12} v_f \tag{1}$$

$$i_f = y_{21} v_L + y_{22} v_f \tag{2}$$

where $$y_{11} = \left.\frac{i_a}{v_L}\right|_{v_f=0} \qquad y_{12} = \left.\frac{i_a}{v_f}\right|_{v_L=0} \qquad y_{21} = \left.\frac{i_f}{v_L}\right|_{v_f=0} \qquad y_{22} = \left.\frac{i_f}{v_f}\right|_{v_L=0} \tag{3}$$

Draw the modified voltage-shunt feedback amplifier that results when the loading of the feedback network of Fig. 11-6 is transferred to the basic amplifier through use of the *y* parameters, leaving an ideal feedback network.

Equations (*1*) and (*2*) each suggest a current source in parallel with an admittance. Admittance y_{11} of (*1*) can be combined with R_{11}, and admittance y_{22} of (*2*) can be combined with R_{22}, to give

$$R'_{11} = R_{11} \parallel \frac{1}{y_{11}} = \frac{R_{11}}{1 + y_{11} R_{11}} \tag{4}$$

$$R'_{22} = R_{22} \parallel \frac{1}{y_{22}} = \frac{R_{22}}{1 + y_{22}R_{22}} \tag{5}$$

Further, the current-controlled current source of Fig. 11-6 can be combined with the suggested voltage-controlled current source of (2) to obtain

$$\beta' i_1 = \beta i_1 + y_{12}v_f = \beta i_1 + y_{12}R_{11}i_1$$

so that

$$\beta' = \beta + y_{12}R_{11} \tag{6}$$

The network of Fig. 11-18 results from the use of (4) to (6).

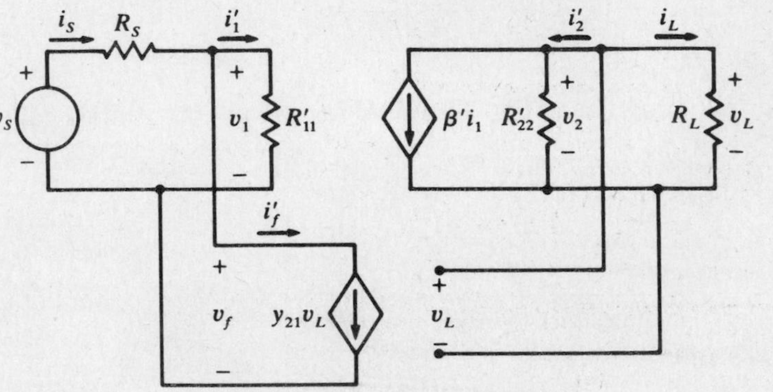

Fig. 11-18

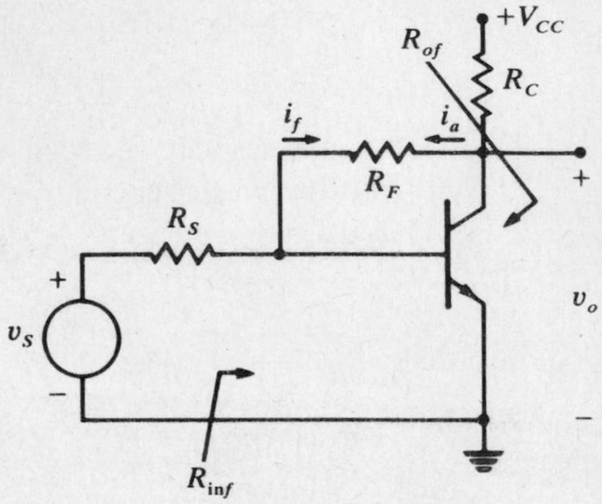

Fig. 11-19

11.15 Voltage-shunt feedback is used in the BJT amplifier of Fig. 11-19. Let $R_C = 4\ \text{k}\Omega$, $R_F = 40\ \text{k}\Omega$, $R_S = 10\ \text{k}\Omega$, $h_{ie} = 1100\ \Omega$, $h_{fe} = 50$, and $h_{re} = h_{oe} = 0$. Using the exact model of Fig. 11-6, find (a) R_{inf}, (b) R_{of}, (c) A_{vf}, and (d) A_{if}.

(a) By (11.47) with $G_{22} = h_{oe} = 0$, $G_L = G_C$, and $G_{11} = 1/h_{ie}$,

$$R_{inf} = \frac{h_{oe} + G_F + G_C}{\dfrac{1}{h_{ie}}[(1 + h_{fe})G_F + h_{oe} + G_C] + G_C G_F}$$

$$= \frac{\dfrac{1}{40 \times 10^3} + \dfrac{1}{4 \times 10^3}}{\dfrac{1}{1100}\left(\dfrac{51}{40 \times 10^3} + \dfrac{1}{4 \times 10^3}\right) + \dfrac{1}{(4 \times 10^3)(40 \times 10^3)}} = 197.47\ \Omega$$

(b) From (11.50),

$$G_{of} = h_{oe} + \frac{(1 + h_{fe})/h_{ie} + 1/R_S}{1 + R_F(1/h_{ie} + 1/R_S)} = \frac{51/1100 + 1/10^4}{1 + 40 \times 10^3(1/1100 + 1/10^4)} = 0.001123$$

and

$$R_{of} = \frac{1}{G_{of}} = \frac{1}{0.001123} = 890.2\ \Omega$$

(c) From (2) of Problem 11.13,

$$A_{vf} = \frac{G_S(G_F - h_{fe}/h_{ie})}{(1/h_{ie})[(1 + h_{fe})G_F + h_{oe} + G_C] + G_S(h_{oe} + G_F) + G_C(G_S + G_F)}$$

$$= \frac{(1 \times 10^{-4})(2.5 \times 10^{-5} - 50/1100)}{9.09 \times 10^{-4}[(51)(2.5 \times 10^{-5}) + 2.5 \times 10^{-4}] + (1 \times 10^{-4})(2.5 \times 10^{-5}) + (2.5 \times 10^{-4})(1 \times 10^{-4} + 2.5 \times 10^{-5})}$$

$$= -3.2$$

(*d*) By (*3*) of Problem 11.13,

$$A_{if} = A_{vf}\frac{(R_{\text{inf}} + R_S)}{R_C} = \frac{(-3.2)(197.47 + 10 \times 10^3)}{4 \times 10^3} = -8.16$$

11.16 Voltage shunt feedback is used in the cascaded BJT amplifier of Fig. 11-20(*a*). The transistors are identical, with $h_{re} = h_{oe} = 0$, $h_{fe} = 100$, and $h_{ie} = 2\ \text{k}\Omega$. Let $R_S = 100\ \Omega$, $R_E = 4\ \text{k}\Omega$, $R_F = 3\ \text{k}\Omega$, and $R_{C2} = 6\ \text{k}\Omega$. Find (*a*) R_{inf}, (*b*) R_{of}, and (*c*) A_{vf}.

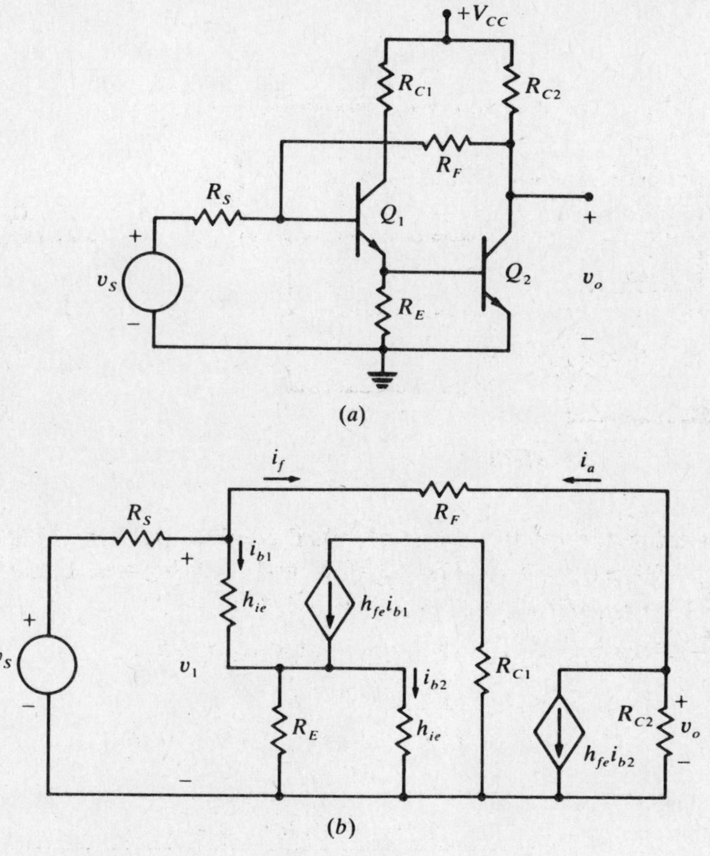

Fig. 11-20

(*a*) The small-signal equivalent circuit is displayed in Fig. 11-20(*b*). With the feedback network deactivated (R_F removed), $\beta i_{b1} = h_{fe} i_{b2}$. But by current division,

$$i_{b2} = \frac{R_E}{R_E + h_{ie}}(h_{fe} + 1)i_{b1}$$

so

$$\beta = \frac{h_{fe}(h_{fe} + 1)R_E}{R_E + h_{ie}} = \frac{(100)(101)(4000)}{6000} = 6733.3$$

We deactivate (short) v_S and replace R_{C2} with a driving-point source to see that $R_o = R_{22} = \infty$.

The input impedance (resistance) with the feedback network deactivated can be found by applying KVL around the input loop:

$$v_S = i_{b1}h_{ie} + (h_{fe} + 1)i_{b1}(R_E \| h_{ie})$$

so

$$R_{11} = R_{\text{in}} = \frac{v_S}{i_{b1}} = h_{ie} + \frac{(h_{fe} + 1)\,R_E\,h_{ie}}{R_E + h_{ie}} = 2000 + \frac{(101)(4000)(2000)}{6000} = 136.67\ \text{k}\Omega$$

Then, from (*11.47*),

$$R_{\text{in}f} = \frac{G_{22} + G_F + G_{C2}}{G_{11}[(1+\beta)G_F + G_{22} + G_{C2}] + G_{C2}G_F}$$

$$= \frac{1/3000 + 1/6000}{\dfrac{1}{136{,}670}\left(\dfrac{6734.3}{3000} + \dfrac{1}{6000}\right) + \dfrac{1}{(6000)(3000)}} = 30.34\ \Omega$$

(*b*) From (*11.50*),

$$G_{of} = G_{22} + \frac{(1+\beta)G_{11} + G_S}{1 + R_F(G_{11} + G_S)} = \frac{6734.3/136{,}670 + 1/100}{1 + 3000(1/136{,}670 + 1/100)} = 0.001911\ \text{S}$$

and $$R_{of} = \frac{1}{G_{of}} = \frac{1}{0.001911} = 523.3\ \Omega$$

(*c*) From (*3*) of Problem 11.14,

$$y_{11} = \left.\frac{i_a}{v_o}\right|_{v_1=0} = \frac{1}{R_F} = \frac{1}{3000} = 3.333 \times 10^{-4}\ \text{S}$$

$$y_{12} = \left.\frac{i_a}{v_1}\right|_{v_0=0} = -\frac{1}{R_F} = -3.333 \times 10^{-4}\ \text{S}$$

$$y_{21} = \left.\frac{i_f}{v_o}\right|_{v_1=0} = -\frac{1}{R_F} = -3.333 \times 10^{-4}\ \text{S}$$

$$y_{22} = \left.\frac{i_f}{v_1}\right|_{v_0=0} = \frac{1}{R_F} = 3.333 \times 10^{-4}\ \text{S}$$

Thus, by (*5*) and (*6*) of Problem 11.14,

$$R'_{22} = \lim_{R_{22}\to\infty} \frac{R_{22}}{1 + y_{22}R_{22}} = \frac{1}{y_{22}} = R_F = 3\ \text{k}\Omega$$

$$\beta' = \beta + y_{12}R_{11} = 6733.3 + (-3.333 \times 10^{-4})(136.67 \times 10^3) = 6687.7$$

From Fig. 11-7 with β and G_{22} replaced by β' and G'_{22}, respectively, and with $G_L = G_{C2}$, the transresistance gain is

$$A_r = \frac{-\beta'}{G_{C2} + G'_{22}} = \frac{-6687.7}{1/6000 + 1/3000} = -13.375 \times 10^6$$

The feedback transfer ratio is $\beta_{vi} = y_{21} = -1/R_F$; hence, by (*11.55*),

$$A_{vf} = \frac{A_r/(R_{\text{in}f} + R_S)}{1 + \beta_{vi}A_r} = \frac{-13.375 \times 10^6/(130.09)}{1 + (-3.333 \times 10^{-4})(-13.375 \times 10^6)} = -23.05$$

11.17 For the general current-shunt feedback amplifier of Fig. 11-8, find (*a*) $R_{\text{in}f}$ and (*b*) R_{of}.

(*a*) For the model of Fig. 11-9, in which the feedback network loading is accounted for within the basic amplifier, KCL at the node of the input network yields

$$i_S = h_{21}i_L + \frac{v_S}{R'_{11}} \tag{1}$$

But $$i_L = \frac{V_{Th}}{R_{Th} + R_L} \tag{2}$$

where V_{Th} is given by (*11.65*), and R_{Th} by (*11.64*). Substituting (*11.65*) into (*2*) with $i_1 = v_S/R_{11}$ and using the result in (*1*) give

$$i_S = h_{21}\frac{\beta R_{22} + h_{12}R_{11}}{R_{Th} + R_L}\frac{v_S}{R_{11}} + \frac{v_S}{R'_{11}} \tag{3}$$

from which

$$R_{\text{in}f} = \frac{v_S}{i_S} = \frac{R_{11}R'_{11}(R_{Th} + R_L)}{h_{21}R'_{11}(\beta R_{22} + h_{12}R_{11}) + R_{11}(R_{Th} + R_L)} \tag{4}$$

(*b*) If R_L were replaced with a driving-point source, we would have

$$R_{of} = R'_{22} = R_{Th}$$

11.18 Current-shunt feedback is used in the cascaded BJT amplifier of Fig. 11-21(*a*). Let $R_{E2} = R_S = 1\ \text{k}\Omega$, $R_{C1} = 12\ \text{k}\Omega$, $R_L = 4\ \text{k}\Omega$, $R_F = 20\ \text{k}\Omega$, $h_{fe1} = h_{fe2} = 100$, $h_{ie1} = h_{ie2} = 2\ \text{k}\Omega$, and $h_{re1} = h_{re2} = h_{oe1} = h_{oe2} = 0$. Find (*a*) R_{of}, (*b*) $R_{\text{in}f}$, (*c*) A_{if}, and (*d*) A_{vf}.

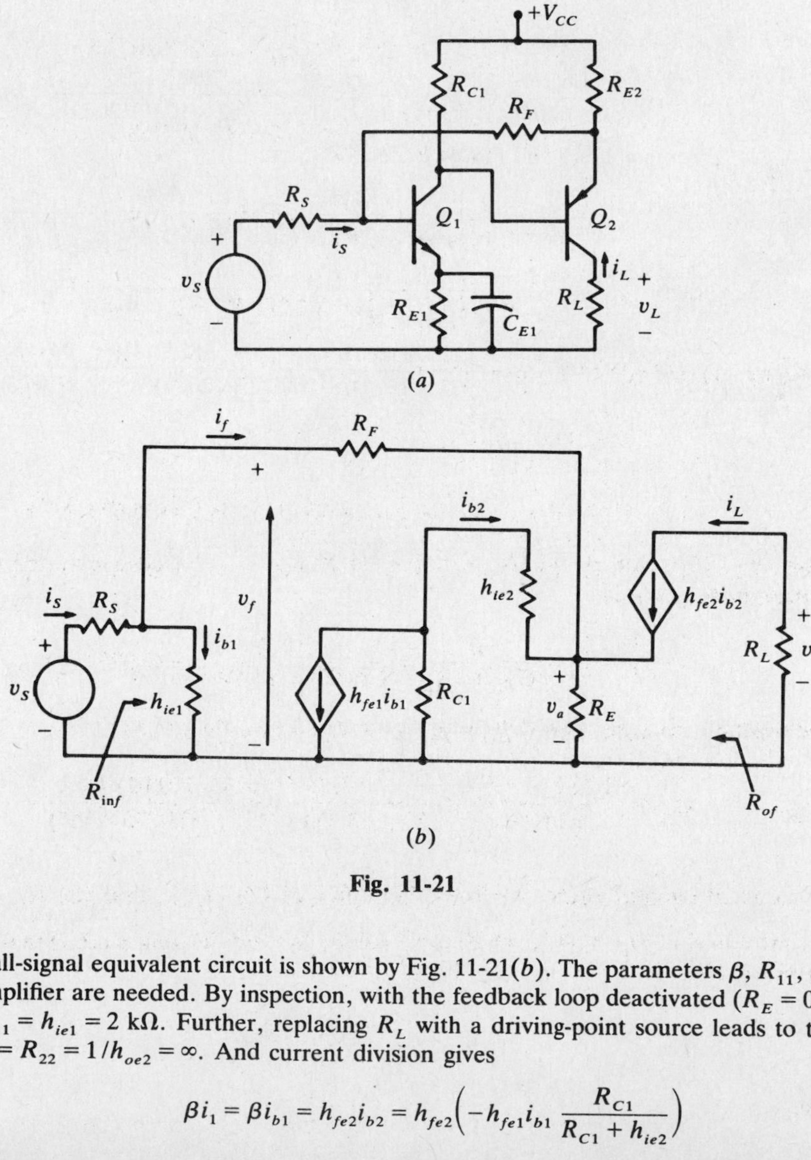

Fig. 11-21

(a) The small-signal equivalent circuit is shown by Fig. 11-21(*b*). The parameters β, R_{11}, and R_{22} of the basic amplifier are needed. By inspection, with the feedback loop deactivated ($R_E = 0$ and $R_F = \infty$), $R_{\text{in}} = R_{11} = h_{ie1} = 2\ \text{k}\Omega$. Further, replacing R_L with a driving-point source leads to the conclusion that $R_o = R_{22} = 1/h_{oe2} = \infty$. And current division gives

$$\beta i_1 = \beta i_{b1} = h_{fe2}i_{b2} = h_{fe2}\left(-h_{fe1}i_{b1}\frac{R_{C1}}{R_{C1} + h_{ie2}}\right)$$

so that
$$\beta = \frac{-h_{fe1}h_{fe2}R_{C1}}{R_{C1}+h_{ie2}} = -\frac{(100)^2(12{,}000)}{14{,}000} = -8571.4$$

The h parameters of the feedback network are determined by use of (*11.59*). We neglect i_{b2} since typically $(h_{fe2}+1)i_{b2} \approx h_{fe2}i_{b2}$; then,

$$h_{11} \approx \left.\frac{v_a}{i_L}\right|_{v_f=0} = R_E \parallel R_F = \frac{(1000)(20{,}000)}{21{,}000} = 952.4\ \Omega$$

$$h_{12} \approx \left.\frac{v_a}{v_L}\right|_{i_L=0} = \frac{R_E}{R_F+R_E} = \frac{1000}{21{,}000} = 0.04762$$

$$h_{21} \approx \left.\frac{i_f}{i_L}\right|_{v_f=0} = \frac{-R_E}{R_F+R_E} = \frac{-1000}{21{,}000} = -0.04762$$

$$h_{22} \approx \left.\frac{i_f}{v_f}\right|_{i_L=0} = \frac{1}{R_F+R_E} = \frac{1}{21{,}000} = 4.762\times 10^{-5}\ \text{S}$$

Since $R_{22}=\infty$, (*11.64*) gives, regardless of the values of the h parameters,

$$R_{of} = R_{Th} = R'_{22} = \infty$$

(*b*) Direct use of (*11.60*) yields

$$R'_{11} = \frac{R_{11}}{1+h_{22}R_{11}} = \frac{h_{ie1}}{1+h_{22}h_{ie1}} = \frac{2000}{1+(4.762\times 10^{-5})(2000)} = 1826.1\ \Omega$$

By (*4*) of Problem 11.17 and (*11.64*), since $R_{22}=\infty$,

$$R_{inf} = \lim_{R_{22}\to\infty} \frac{R_{11}R'_{11}(R_{Th}/R_{22}+R_L/R_{22})}{h_{21}R'_{11}(\beta + h_{12}R_{11}/R_{22}) + R_{11}(R_{Th}/R_{22}+R_L/R_{22})}$$

$$= \frac{R_{11}R'_{11}(1+R_{11}h_{22}+\beta h_{21})}{h_{21}R'_{11}\beta(1+R_{11}h_{22}) + R_{11}(1+R_{11}h_{22}+\beta h_{21})}$$

$$= \frac{(2000)(1826.1)[1+(2000)(4.672\times 10^{-5})+(-8571.4)(-0.04762)]}{(-0.04762)(1826.1)(-8571.4)[1+(2000)(4.672\times 10^{-5})]+2000[1+(2000)(4.672\times 10^{-5})+(-8571.4)(-0.04762)]}$$

$$= 27.03\ \Omega$$

(*c*) From Fig. 11-10 with $G_{22} = 1/R_{22} = 0$,

$$A_i = \frac{\beta G_L}{G_L + G'_{22}} = \beta = -8571.4 \qquad \text{and} \qquad \beta_{ii} = h_{21} = -0.04762$$

Then from (*11.69*),

$$A_{if} = \frac{A_i}{1+\beta_{ii}A_i} = \frac{-8571.4}{1+(-0.04762)(-8571.4)} = -20.95$$

(*d*) The overall voltage gain is found by direct use of (*11.70*):

$$A_{vf} = \frac{-A_{if}R_L}{R_{inf}+R_S} = \frac{-(-20.95)(4000)}{1027.03} = 81.59$$

Supplementary Problems

11.19 The gain-bandwidth product GBW of an amplifier remains essentially constant with the addition of feedback (see Problem 11.1). Illustrate that fact for the amplifier of Example 11.1, for which $GBW \approx GBW_f = 3.999 \times 10^6$ Hz.

11.20 The CE amplifier of Fig. 3-6, as analyzed in Problem 8.13, has a low-frequency cutoff f_L of 81.02 Hz. The amplifier is to be used for frequencies down to the low end of the audio range (20 Hz), and a transformer is to be added, as in Problem 11.6, to introduce voltage-series feedback. (*a*) Specify a feedback gain that will reduce the low-frequency cutoff point to 20 Hz. (Assume the transformer turns ratio is 1:1.) (*b*) Find the resulting midfrequency voltage gain.
Ans. (*a*) $\beta_{vv} = -0.03356$; (*b*) $A_{vof} = -12.44$

11.21 For the voltage-source amplifier with voltage-series feedback of Fig. 11-11, determine the overall voltage gain (*a*) with feedback and (*b*) without feedback.

$$Ans.\quad (a)\ A_{vf} = \frac{G_{11}G_2 + \gamma G_{22}(G_1 + G_2)}{(G_{11} + G_1)(G_2 + G_{22} + G_L) + G_2[(1+\gamma)G_{22} + G_L]};\quad (b)\ A_v = \frac{\gamma G_{22}}{G_{22} + G_L}$$

11.22 For the general voltage-series feedback amplifier of Fig. 11-2, let $G_1 = 1 \times 10^{-3}$ S, $G_2 = 0.05 \times 10^{-3}$ S, $G_{11} = 0.01$ S, $G_{22} \approx 0$, $G_L = 0.5 \times 10^{-3}$ S, and $\beta = 75$. Determine (*a*) A_v, (*b*) A_{vf}, (*c*) A_i, and (*d*) A_{if}. *Ans.* (*a*) $A_v = 1500$; (*b*) $A_{vf} = 18.094$; (*c*) $A_i = 75$; (*d*) $A_{if} = 67.67$

11.23 Determine the error in overall voltage gain that results from neglecting the effects of feedback loading in the voltage-series feedback amplifier of Problem 11.22 (by treating the feedback network as a true voltage divider). *Ans.* Approximate overall voltage gain is 20.72; error = 14.51%

11.24 Show that for the ideal voltage-series feedback amplifier model of Fig. 11-3, (*a*) $R_{inf} = R_{in}(1 + \beta_{vv}A_i')$, where $A_i' = A_iR_L/R_{11} = \beta R_{22}R_L/R_{11}(R_{22} + R_L)$, and (*b*) $R_{of} = R_o/(1 + \beta_{vv}A_i'')$, where $A_i'' = \beta R_{22}/R_{11}$.

11.25 For the cascaded FET amplifier of Fig. 11-13 with parameter values given in Problem 11.6, (*a*) find the values of the parameters R_{11}, R_{22}, and γ of the general voltage-source amplifier of Fig. 11-11 with voltage-series feedback, and (*b*) use those values to determine the overall voltage gain.
Ans. (*a*) $R_{11} = R_G = 1$ MΩ, $R_{22} = r_{ds} \| R_D = 1.875$ kΩ, $\gamma = \mu^2 R_D/(R_D + r_{ds})^2 = 14.06$; (*b*) $A_{vf} = 5.33$

11.26 Show that if $\beta \gg 1$, $\beta_{vv}A_v/\beta \ll 1$, and $G_{22} \to 0$ for the current-source amplifier with current-series feedback of Fig. 11-4, then the following relationships hold:

$$A_{vf} = \frac{A_v}{1 + \beta_{vv}A_v} = \frac{A_v}{1 - G_L\beta_{iv}A_v} \qquad R_{inf} = R_{in}(1 + \beta_{vv}A_v) \qquad R_{of} = R_o(1 + \beta_{vv}A_v) \to \infty$$

11.27 Let $R_B = 200$ kΩ, $R_L = 2$ kΩ, $R_E = 100$ Ω, $h_{ie} = 100$ Ω, $h_{fe} = 60$, $h_{re} = 0$, and $h_{oe} = 5 \times 10^{-5}$ S for the BJT amplifier of Fig. 11-22. Find (*a*) the exact feedback transfer ratio β_{iv}, (*b*) the overall voltage gain, (*c*) the input resistance, and (*d*) the output resistance.
Ans. (*a*) $\beta_{iv} = 101.84$, (*b*) $A_{vf} = -19.289$, (*c*) $R_{inf} = 5.47$ kΩ, (*d*) $R_{of} = 620$ kΩ

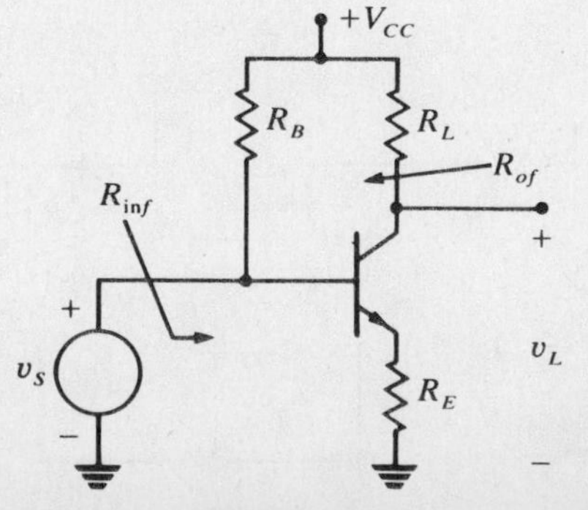

Fig. 11-22

11.28 For the cascaded voltage-series feedback amplifier of Fig. 11-14(*a*), (*a*) transfer the loading effects of the feedback network to the basic amplifier (determine R'_{11}, R'_{22}, and β') so that the model of Fig. 11-15 is applicable. (*b*) Find the overall voltage gain. Also calculate (*c*) $R_{\text{in}f}$, (*d*) R_{of}, and (*e*) the overall current gain. Use the parameter values of Problem 11.7.
Ans. (*a*) $R'_{11} = 199.34\ \Omega$, $R'_{22} = 15.1\ \text{k}\Omega$, $\beta' = 6428.59$; (*b*) $A_{vf} = 150.77$; (*c*) $R_{\text{in}f} = 106.79\ \text{k}\Omega$; (*d*) $R_{of} = 4.68\ \Omega$; (*e*) $A_{if} = 5366.91$

11.29 The *h*-parameter small-signal equivalent circuit model of the BJT includes ideal voltage-series feedback through controlled source $h_{re}v_{ce}$. For the simple CE amplifier of Fig. 6-5(*a*), find expressions for the basic amplifier voltage gain A_v and feedback transfer ratio β_{vv} of Fig. 11-3. Write the overall voltage gain, showing that positive feedback exists.

Ans. $A_v = \dfrac{-h_{fe}}{h_{ie}(G_L + h_{oe})}$ $\qquad \beta_{vv} = h_{re} \qquad A_{vf} = \dfrac{A_v}{1 - h_{re}h_{fe}/h_{ie}(G_L + h_{oe})}$

where the minus sign in the denominator of A_{vf} indicates positive feedback

11.30 For the BJT amplifier of Problem 11.27: (*a*) Transfer the loading effects of the feedback network to the basic amplifier so that the circuit of Fig. 11-16 is applicable (find R'_{11}, R'_{22}, and β'). Then find (*b*) R_{of}, (*c*) $R_{\text{in}f}$, (*d*) A_{vf}, and (*e*) A_{if}.
Ans. (*a*) $R'_{11} = 200\ \Omega$, $R'_{22} = 620.05\ \text{k}\Omega$, $\beta' = 54.47$; (*b*) $R_{of} = 620.05\ \text{k}\Omega$; (*c*) $R_{\text{in}f} = 5629.41$; (*d*) $A_{vf} = -19.29$; (*e*) $A_{if} = 54.29$

11.31 Rework Problem 11.12 with $h_{oe3} = 5 \times 10^{-5}$ S and all else unchanged, to see that it is usually reasonable to neglect h_{oe}.
Ans. (*a*) $R_{of} = 213.7\ \text{M}\Omega$; (*b*) $R_{\text{in}f} = 483.3\ \Omega$; (*c*) $A_{vf} = -51.61$; (*d*) $A_{if} = 12.47$

11.32 For the BJT amplifier of Fig. 11-19, (*a*) determine the parameters of the voltage-shunt feedback model as defined in Fig. 11-18 and (*b*) find the overall current gain by use of (*11.56*), if $R_C = 4\ \text{k}\Omega$, $R_F = 40\ \text{k}\Omega$, $R_S = 10\ \text{k}\Omega$, $h_{ie} = 1100\ \Omega$, $h_{fe} = 50$, and $h_{re} = h_{oe} = 0$.
Ans. (*a*) $R'_{11} = h_{ie}R_F/(h_{ie} + R_F)$, $R'_{22} = R_F$, $y_{21} = -1/R_F$, $\beta' = h_{fe} - h_{ie}/R_F$; (*b*) $A_{if} = -8.19$

11.33 Show that for $\beta \to \infty$ (approximately realizable with a large odd number of cascaded BJT stages), the voltage-shunt feedback amplifier of Fig. 11-6 displays the characteristics of an operational amplifier in that $A_{vf} \approx -R_F/R_S$ and $R_{of} \approx 0$.

11.34 The MOSFET amplifier of Fig. 4-7(*a*) includes voltage-shunt feedback if a gate-to-source resistor R_G is added. Let $R_F = 5\ \text{M}\Omega$, $R_L = 14\ \text{k}\Omega$, $r_{ds} = 40\ \text{k}\Omega$, $R_G = 10\ \text{M}\Omega$, and $g_m = 1$ mS. (The amplifier was analyzed in Problem 7.2, without R_G.) Using the model of Fig. 11-18 with feedback loading accounted for in the basic amplifier, find (*a*) $R_{\text{in}f}$, (*b*) R_{of}, (*c*) A_{vf}, and (*d*) A_{if}.
Ans. (*a*) $R_{\text{in}f} = 431.5\ \text{k}\Omega$; (*b*) $R_{of} = 1.446\ \text{k}\Omega$; (*c*) $A_{vf} = -11.05$; (*d*) $A_{if} = -340.6$

11.35 Current-shunt feedback is used in the cascaded JFET amplifier of Fig. 11-23. Let $R_S = 500\ \Omega$, $R_{G1} = 20\ \text{k}\Omega$, $r_{ds} = 30\ \text{k}\Omega$, $g_m = 1 \times 10^{-1}$ S, $R_{D1} = R_{D2} = 3\ \text{k}\Omega$, $R_F = 50\ \text{k}\Omega$, $R_{G2} = 1\ \text{M}\Omega$, $R_{S22} = 50\ \Omega$, and $R_L = 4\ \text{k}\Omega$. Determine (*a*) R_{of}, (*b*)$R_{\text{in}f}$, (*c*) A_{if}, and (*d*) A_{vf}.
Ans. (*a*) $R_{of} = 369.45\ \text{M}\Omega$; (*b*) $R_{\text{in}f} = 11.117\ \text{k}\Omega$; (*c*) $A_{if} = -1000.95$; (*d*) $A_{vf} = 344.65$

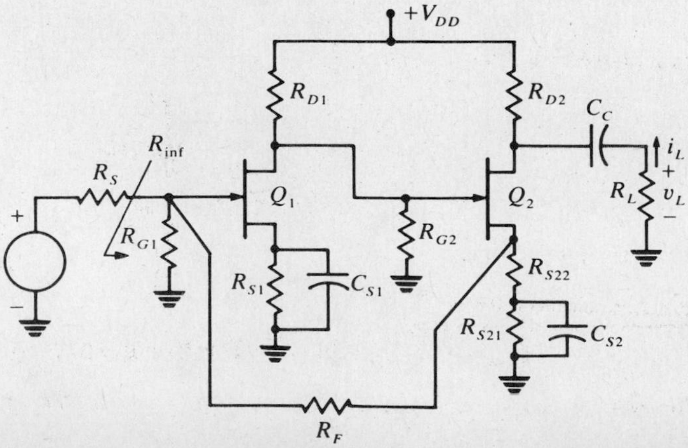

Fig. 11-23

Chapter 12

Switching and Logic Circuits

12.1 INTRODUCTION

The switching of an active device (BJT or FET) is a nonlinear operation that places the device in one of two *basic states* for controlled periods of time:

1. An *ON state* corresponding to a condition of heavy conduction
2. An *OFF state* corresponding to a condition of light conduction

In practical electronic switches, the ON state consists of a range of output voltages $0 < v_o \le V_L$ (called *low logic*), the OFF state consists of the range $V_L < V_H \le v_o <$ power-supply voltage (called *high logic*); V_L is called the *low-logic limit*, and V_H the *high-logic threshold*. Subsequent circuitry must be capable of discriminating effectively between V_L and V_H; for good design, the indeterminate range between V_L and V_H is made as large as possible.

Although transition times between states are important in establishing limits of operation for high-speed switching circuits, zero switching times are assumed in studying the principles involved.

12.2 SWITCHING MODELS FOR THE BJT

Practical switching models for the BJT are derived by considering the boundaries between the saturation, active, and cutoff regions of Fig. 3-3(*c*) to be straight line segments, as indicated in Fig. 12-1. The typical manufacturer's specification sheet gives the value of $V_{CE\text{sat}}$ along with the corresponding values of $i_C = I_{C\text{sat}}$ and $i_B = I_{B\text{sat}}$. The value of I_{CEO} and the associated value of $v_{CE} = V_{CEO} = V_{CE\text{cutoff}}$ are also given.

For the case of cutoff (the OFF state), $i_B = 0$, and the BJT is modeled as a collector-to-emitter cutoff resistance

$$R_{CO} = \frac{V_{CEO}}{I_{CEO}} \tag{12.1}$$

as shown in Fig. 12-2(*a*). Saturation operation (the ON state) can be modeled by describing the base-to-emitter junction as a piecewise-linear diode [see Section 2.5 and Fig. 2-10(*b*)] and the collector-to-emitter path as a saturation resistor

$$R_{\text{sat}} = \frac{V_{CE\text{sat}}}{I_{C\text{sat}}} \tag{12.2}$$

as shown in Fig. 12-2(*b*). The ideal diodes of Fig. 12-2 are reversed for a *pnp* device.

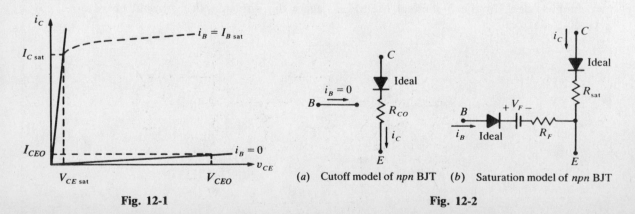

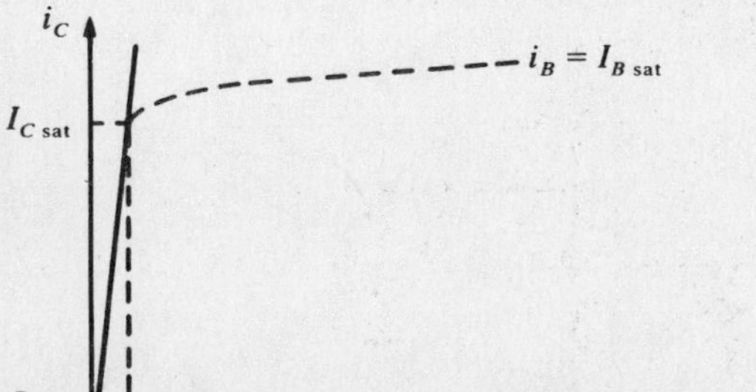

Fig. 12-1

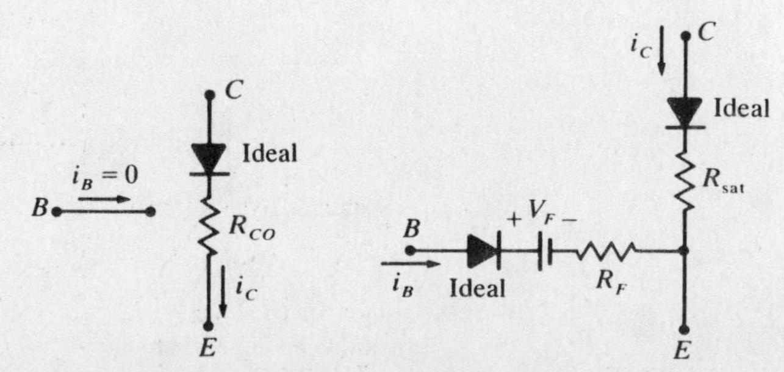

(*a*) Cutoff model of *npn* BJT (*b*) Saturation model of *npn* BJT

Fig. 12-2

Frequently, negligible error is introduced if the assumptions are made that $V_{CEsat} = I_{CEO} R_F - V_F = 0$, which leads directly to the ideal BJT switching models of Fig. 12-3.

Example 12.1 The manufacturer's specification sheet for a particular Si *npn* BJT shows $I_{CEO} = 20\ \mu A$ at test condition $V_{CB} = 25$ V, $V_{CEsat} = 0.2$ V at test conditions $I_C = 15$ mA and $I_B = 100\ \mu A$, and $V_{BE} = 1$ V at test conditions $I_C = 15$ mA and $I_B = .100\ \mu A$. Determine the switching-model parameters of Fig. 12-2.

The cutoff resistance follows directly from (*12.1*):

$$R_{CO} = \frac{V_{CEO}}{I_{CEO}} = \frac{25}{20 \times 10^{-6}} = 1.25\ \text{M}\Omega$$

For the typical value $V_F = 0.5$ V, KVL applied to Fig. 12-2(*b*) requires that

$$R_F = \frac{V_{BE} - V_F}{I_{CEO}} = \frac{1 - 0.5}{100 \times 10^{-6}} = 5\ \text{k}\Omega$$

Then, by (*12.2*),

$$R_{sat} = \frac{V_{CEsat}}{I_{Csat}} = \frac{0.2}{15 \times 10^{-3}} = 13.33\ \Omega$$

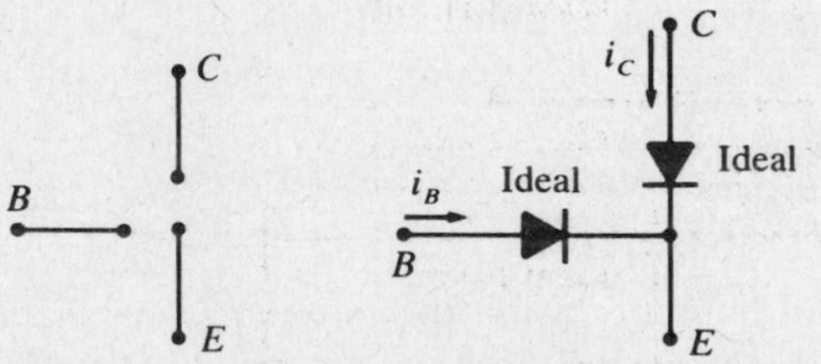

(*a*) Ideal cutoff model of *npn* BJT (*b*) Ideal saturation model of *npn* BJT

Fig. 12-3

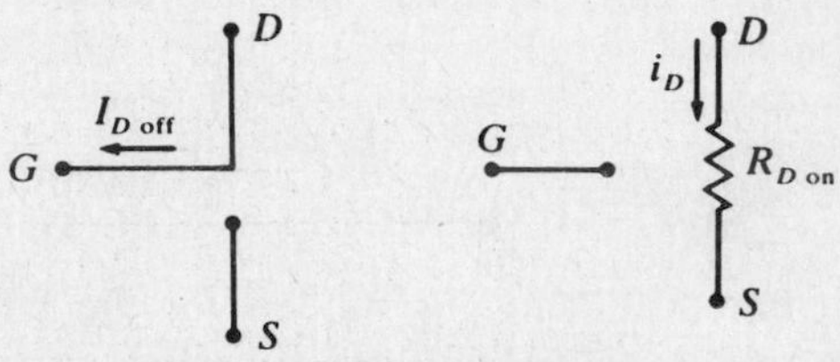

(*a*) OFF-biased *n*-channel JFET (*b*) ON-biased *n*-channel JFET

Fig. 12-4

12.3 SWITCHING MODELS FOR THE FET

The manufacturer's specification sheet for the FET gives a value for R_{Don}, the ohmic resistance of the conducting channel of the ON-biased FET (analogous to saturation for a BJT). Further, the *n*-channel JFET requires negative v_{GS} for the OFF-biased condition (directly analogous to cutoff for the BJT), which results in a small current I_{Doff} flowing from the drain to the gate (the *p*-channel JFET requires positive v_{GS} for OFF bias, so that I_{Doff} flows from gate to drain). For a MOSFET, this gate (or drain) current is negligible.

The implied switching models for the *n*-channel JFET are given in Fig. 12-4; the directions of I_{Doff} and i_D would be reversed if the device were a *p*-channel JFET. Figure 12-5 displays the switching models for the *n*-channel MOSFET; again the direction of i_D would be reversed for a *p*-channel device.

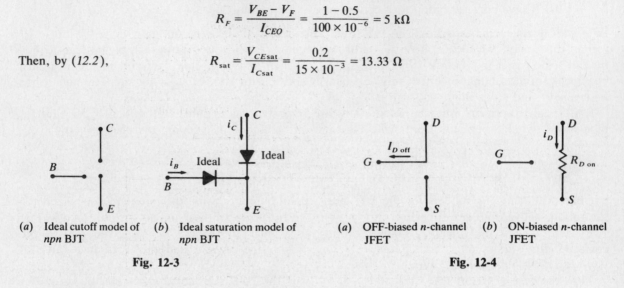

(*a*) OFF-biased *n*-channel MOSFET (*b*) ON-biased *n*-channel MOSFET

Fig. 12-5

12.4 DIGITAL LOGIC AND BOOLEAN ALGEBRA

Digital circuits make use of ON-OFF devices to implement the operations of a system of logic called *Boolean algebra*. (When applied in circuitry, the system is sometimes known as *digital logic*.) Its statements may take the form of algebraic expressions, logic block diagrams, or truth tables, as well as circuits.

The theorems of Boolean algebra are listed in Table 12-1 in *algebraic expression* form;

$A, B, C \equiv$ logic variables $\cdot \equiv$ AND (intersection of sets)

$0 \equiv$ null (or zero) set $+ \equiv$ OR (union of sets)

$1 \equiv$ universal set $\bar{A} \equiv$ negation (or complement) of A

Table 12-1 Boolean Algebra Theorems

Number	Theorem	Name
1	$A + B = B + A$ $A \cdot B = B \cdot A$	commutative law
2	$(A + B) + C = A + (B + C)$ $(A \cdot B) \cdot C = A \cdot (B \cdot C)$	associative law
3	$A \cdot (B + C) = A \cdot B + A \cdot C$ $A + (B \cdot C) = (A + B) \cdot (A + C)$	distributive law
4	$A + A = A$ $A \cdot A = A$	identity law
5	$\bar{\bar{A}} = A$	negation law
6	$A + A \cdot B = A$ $A \cdot (A + B) = A$	redundancy law
7	$0 + A = A$ $1 \cdot A = A$ $1 + A = 1$ $0 \cdot A = 0$	Boolean postulates
8	$\bar{A} + A = 1$ $\bar{A} \cdot A = 0$	
9	$A + \bar{A} \cdot B = A + B$ $A \cdot (\bar{A} + B) = A \cdot B$	
10	$\overline{A + B} = \bar{A} \cdot \bar{B}$ $\overline{A \cdot B} = \bar{A} + \bar{B}$	DeMorgan's laws

12.5 DIGITAL LOGIC BLOCK DIAGRAMS

The elements of digital circuits are called *digital logic gates*, and their symbols are shown in Fig. 12-6. Circuit diagrams that include these symbols are called *logic block diagrams*.

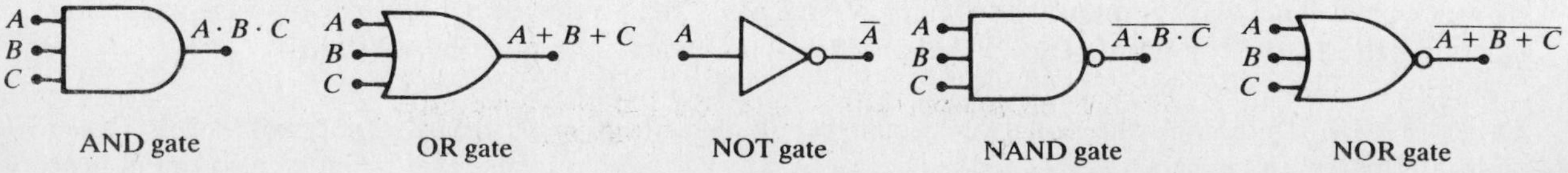

Fig. 12-6

Example 12.2 Write an algebraic expression for X, the overall output of the *logic block diagram* of Fig. 12-7, and then use the theorems of Table 12-1 to simplify it.

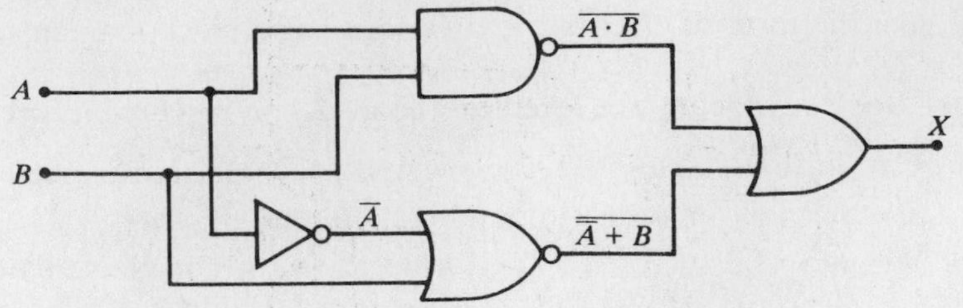

Fig. 12-7

With the outputs of the intermediate logic gates as indicated in Fig. 12-7, the overall output must be

$$X = \overline{A\cdot B} + \overline{\bar{A} + B}$$

Application of DeMorgan's laws and Theorems 9 and 4 yield

$$X = \bar{A} + (\bar{B} + \bar{\bar{A}}\cdot \bar{B}) = \bar{A} + \bar{B}$$

which, according to DeMorgan's laws, can also be written as

$$X = \overline{A\cdot B} \tag{12.3}$$

A *truth table* is constructed by considering all possible combinations of inputs to a logic circuit and finding the corresponding outputs, realizing that, in a digital logic circuit, a variable can take on only one of two discrete values—0 and 1.

Example 12.3 It is apparent from (*12.3*) that the entire logic network of Fig. 12-7 can be replaced with a single two-input NAND gate. Construct a truth table for the network to verify this observation.

A truth table for Fig. 12-7 need only contain all pairs of values for the logic variables A and B and the resulting values of X; however, it is good practice to augment the basic truth table with columns showing intermediate values, as in the following table:

A	B	$\overline{A\cdot B}$	$\overline{\bar{A}+B}$	X
0	0	1	0	1
0	1	1	0	1
1	0	1	1	1
1	1	0	0	0

Obviously, the output column is identical to the $\overline{A\cdot B}$ column and could be replaced by it.

12.6 BISTABLE MULTIVIBRATORS

The *bistable multivibrator* (BMV) or *flip-flop* (FF) is a switching circuit with two stable states. The circuit can be triggered (changed) from either state to the other by applying an input voltage via a suitable trigger circuit; it thus can be used as a basic memory element for digital logic. A simple BMV is shown in Fig. 12-8, where either $v_{o1} \geq V_H$ ($= 1$, or *high logic*) and $v_{o2} \leq V_L$ ($= 0$, or *low logic*) or vice versa. The two momentary switches R and S set the values of v_{o1} and v_{o2} by *base triggering* (shorting the transistor base). (See Problem 12.5 for a practical triggering method.)

Example 12.4 Determine the voltages v_{o1} and v_{o2} in the circuit of Fig. 12-8 (*a*) if switch S is closed momentarily and (*b*) if switch R is closed momentarily. Let $V_{CC} = 5$ V (a value that is commonly used in logic circuits), $R_1 = R_4 = 1$ kΩ, $R_2 = R_3 = 20$ kΩ, $h_{FE\min} = 50$ (see Problem 12.1) for both Q_1 and Q_2, and $V_{CE\text{sat}} =$ 0.2 V for both Q_1 and Q_2. The transistors are Si devices.

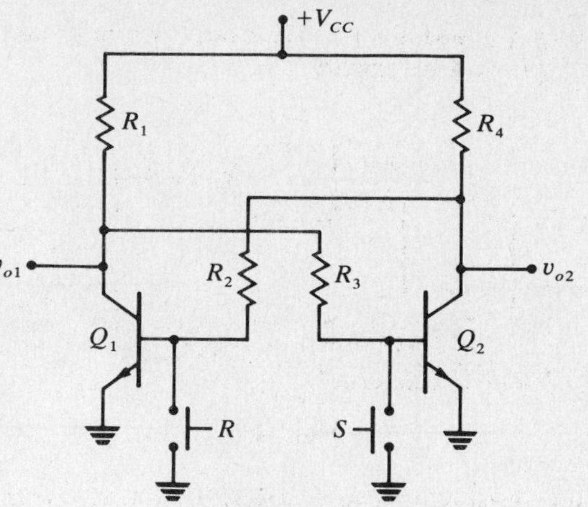

Fig. 12-8

(a) If S is closed, then $V_{BE2} \approx 0$ and Q_2 is OFF; however, Q_1 is forward-biased, and KVL requires that

$$I_{B1} = \frac{V_{CC} - V_{BE1}}{R_2 + R_4} = \frac{5 - 0.7}{21{,}000} \approx 205\ \mu\text{A}$$

Hence, $$v_{o2} = V_{CC} - I_{B1}R_4 = 5 - (205 \times 10^{-6})(1000) \approx 4.8\ \text{V}$$

and since Q_1 is driven into saturation,

$$v_{o1} = V_{CE\text{sat}} = 0.2\ \text{V}$$

(b) If R is closed, then by symmetry $v_{o1} = 4.8$ V and $v_{o2} = 0.2$ V.

Solved Problems

12.1 As a BJT approaches saturation, the collector-base junction (reverse-biased for active-region operation) becomes forward-biased, reducing the width of the collector-base depletion region. Consequently, a lower percentage of majority carriers are swept into the collector, and both α and $\beta = h_{FE} = \alpha/(1 - \alpha)$ are decreased. The manufacturer's specification sheet for switching transistors gives a value $h_{FE\text{min}}$, measured at or near saturation, that is valid for relating collector and base currents for saturation operation. For the circuit of Fig. 12-9, $V_{CC} = 15$ V and the transistor is to be modeled with Fig. 12-3(b). (a) Determine the minimum value of $i_B = I_{B\text{sat}}$ if $h_{FE\text{min}} = 30$ and $R_L = 1$ kΩ. (b) If $h_{FE\text{min}} = 30$ and $i_B = 100$ μA, find the maximum value of R_L that will ensure saturation operation.

(a) The model of Fig. 12-3(b) is based on the assumption that $V_{CE\text{sat}} = 0$; hence, for the saturation condition,

$$I_{C\text{sat}} = \frac{V_{CC}}{R_L} = \frac{15}{1 \times 10^3} = 15\ \text{mA}$$

and the minimum base current is

$$I_{B\text{sat}} = \frac{I_{C\text{sat}}}{h_{FE\text{min}}} = \frac{15 \times 10^{-3}}{30} = 500\ \mu\text{A}$$

(b) The collector current at saturation is

$$I_{C\text{sat}} = h_{FE\text{min}} I_{B\text{sat}} = (30)(100 \times 10^{-6}) = 3\ \text{mA}$$

whence $$R_L \leq \frac{V_{CC}}{I_{C\text{sat}}} = \frac{15}{3 \times 10^{-3}} = 5\ \text{k}\Omega$$

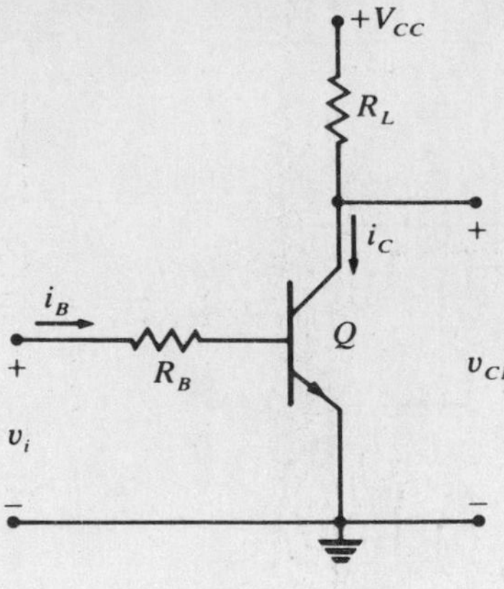

Fig. 12-9

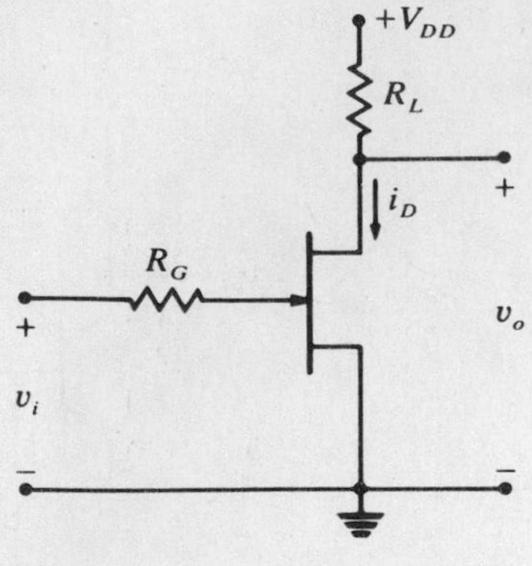

Fig. 12-10

12.2 For the JFET switch of Fig. 12-10, $V_{DD} = 10$ V, $R_L = 2$ kΩ, $I_{Doff} = 0.3$ nA, and $R_{Don} = 30\ \Omega$. Determine v_o (*a*) if $v_i = 0$ and (*b*) if $v_i = -5$ V (OFF-biased).

(*a*) Using the model of Fig. 12-4(*a*) and voltage division, we have

$$v_o = \frac{R_{Don}}{R_{Don} + R_L} V_{DD} = \frac{30}{2030} 10 = 0.148 \text{ V}$$

(*b*) Using the model of Fig. 12-4(*b*) and KVL, we get

$$v_o = V_{DD} - I_{Doff} R_L = 10 - (0.3 \times 10^{-9})(2000) \approx 10 \text{ V}$$

12.3 Show that the circuit of Fig. 12-9 could be used as an inverter or NOT gate in a digital logic circuit.

If $v_i\ (= V_H > 0)$ is large enough to saturate the transistor, then $v_o = v_{CE} = V_{CEsat} \approx 0$. If $v_i\ (= V_L < V_{BEQ} \approx 0)$ is small enough so that the transistor is cut off, then $v_o = v_{CE} \approx V_{CC}$. The truth table clearly shows NOT logic:

v_i		v_o	
Actual	Logic	Actual	Logic
V_H	1	0	0
V_L	0	V_{CC}	1

12.4 By use of a truth table, show that the circuit of Fig. 12-11 is (*a*) a NOR gate if v_{o1} is taken as the output and (*b*) an OR gate if v_{o2} is the output.

(*a*) If $v_1 \geqq V_H$, Q_1 is saturated and $v_{o1} = V_{CEsat}$. If $v_2 \geqq V_H$, Q_2 is saturated and $v_{o1} = V_{CEsat}$. Also, if Q_1 and Q_2 are both saturated, $v_{o1} = V_{CEsat}$. Otherwise, $v_{o1} \approx V_{CC}$. A truth table shows NOR logic for v_{o1} as output:

v_1		v_2		v_{o1}		v_{o2}	
Actual	Logic	Actual	Logic	Actual	Logic	Actual	Logic
V_H	1	V_H	1	V_{CEsat}	0	V_{CC}	1
V_L	0	V_L	0	V_{CC}	1	V_{CEsat}	0
V_H	1	V_L	0	V_{CEsat}	0	V_{CC}	1
V_L	0	V_H	1	V_{CEsat}	0	V_{CC}	1

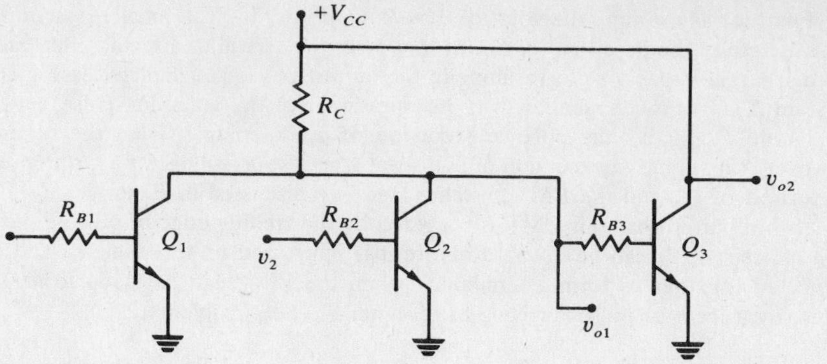

Fig. 12-11

(b) Since $v_{o1} = V_{CEsat}$ is below the level that will render Q_3 conducting, and $v_{o1} = V_{CC}$ will drive Q_3 to saturation if R_{B3} is properly sized, the Q_3 stage is simply an inverter (a NOT gate). Thus, output v_{o2} is the *logic complement* (in which 0s and 1s are interchanged) of v_{o1}, as the truth table shows, and the overall logic is that of an OR gate.

12.5 The BMV of Fig. 12-8 is shown with momentary switches R and S; however, in practical applications, triggering must be done electronically. One possibility would be to impress a short-duration low-logic signal on the upper contact connection of either R or S to trigger the switching of state. But if a negative voltage v_{BE} of short duration were placed across the base-emitter junction of the triggered transistor, the resulting reverse charge flow would accelerate the transistor turnoff process. Such a negative trigger voltage can be realized by use of an RC circuit (differentiator action) as shown in Fig. 12-12. Qualitatively discuss the reset process that occurs if Q_2 is operating in saturation when a rectangular pulse v_R arrives at terminal R in Fig. 12-12.

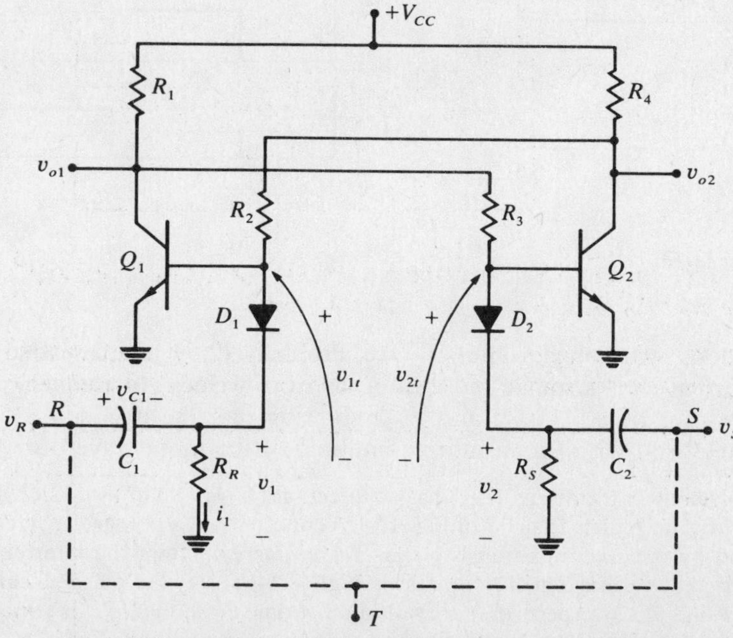

Fig. 12-12

First, let us examine the voltage v_1 ($\approx R_R C_1\, dv_R/dt$ if RC is small), assuming that negligible loading is presented by the base loop of Q_1. In that case, the arrival of a rectangular pulse at terminal R charges and discharges capacitor C_1, resulting in the output voltage v_1 depicted in Fig. 12-13 (provided the time constant $R_R C_1$ is much smaller than the duration τ of the incoming pulse).

Diode D_1 blocks the positive excursion of v_1, keeping it from the base-emitter junction of Q_1. However, the negative excursion of v_1 forward-biases D_1, whence $v_{BE1} < 0$ momentarily; this results in the turnoff of Q_1 and the BMV switches states as discussed in Example 12.4.

It is apparent that this BMV is triggered by the trailing edge of pulse v_R; if response to the leading edge is desired, v_R can be inverted before it is impressed on terminal R. And if terminals R and S are connected together to form a terminal T, then the BMV can be made to *toggle*, or alternately switch states, by a train of pulses arriving at terminal T. (See Problem 12.6.)

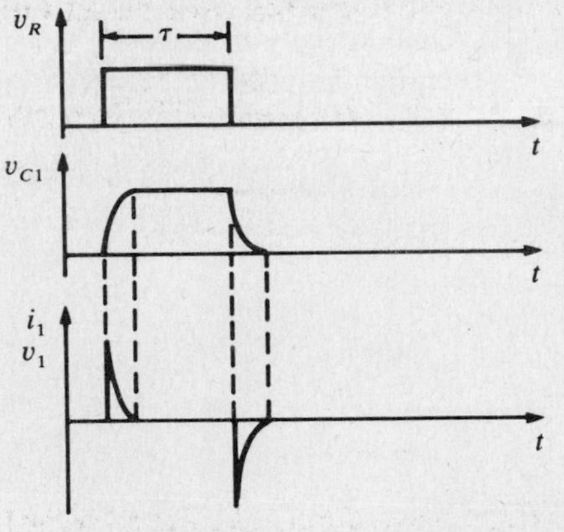

Fig. 12-13

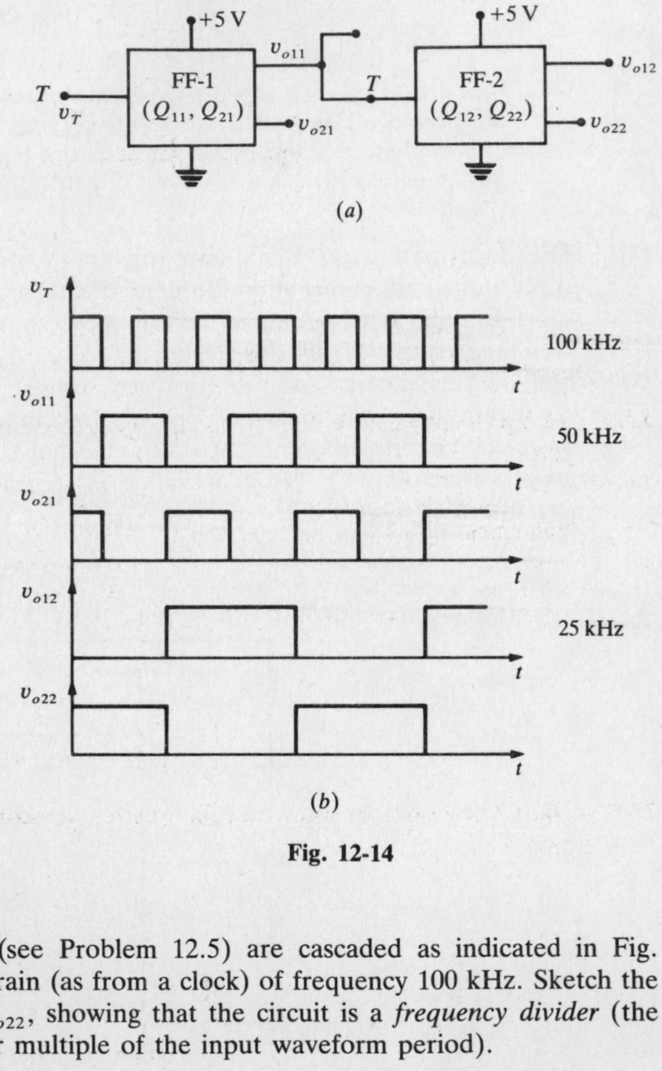

Fig. 12-14

12.6 Two flip-flops with toggle input T (see Problem 12.5) are cascaded as indicated in Fig. 12-14(*a*). Input v_T is a square pulse train (as from a clock) of frequency 100 kHz. Sketch the waveforms of v_{o11}, v_{o21}, v_{o12}, and v_{o22}, showing that the circuit is a *frequency divider* (the output waveform period is an integer multiple of the input waveform period).

Waveform v_1 (or v_2) of Fig. 12-13 has no effect on Q_1 (or Q_2) of Fig. 12-12 if transistor Q_1 (or Q_2) is OFF, even with terminals R and S of Fig. 12-12 connected to form toggle T; thus, switching action must be triggered by the saturated transistor. A *timing diagram* shows the interrelationship of signals in a switching circuit as a function of time; that in Fig. 12-14(*b*) is based on the assumption that Q_{11} and Q_{12} (Q_1 of FF-1 and FF-2, respectively) are saturated at time $t = 0$; then Q_{11} is switched to cutoff by the first trailing edge of v_T; and then Q_{12} is switched off by the first trailing edge of v_{o11}. Note the relationship among the waveform periods.

12.7 Generate the truth table for the *diode logic* (DL) circuit of Fig. 12-15, showing that the circuit is an AND gate.

If either or both of v_1 and v_2 are logic low ($\leq V_L$), then v_o is logic low ($\approx$0.7 V). If both v_1 and v_2 are logic high ($\geq V_H$), then $v_o \geq V_H - 0.7$. The truth table shows the circuit to be an AND gate:

v_1	v_2	v_o
0	0	0
0	1	0
1	0	0
1	1	1

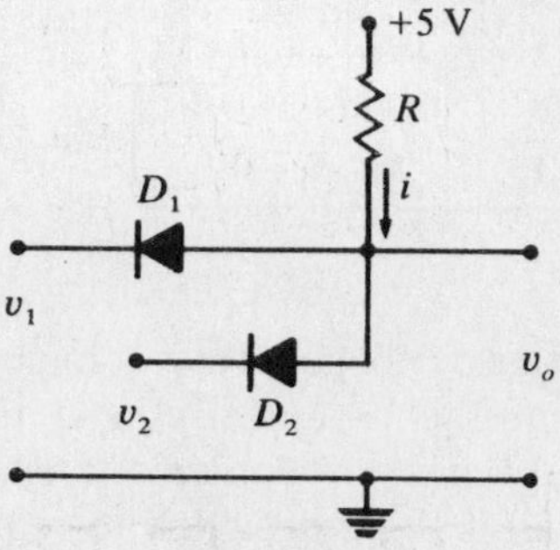

Fig. 12-15

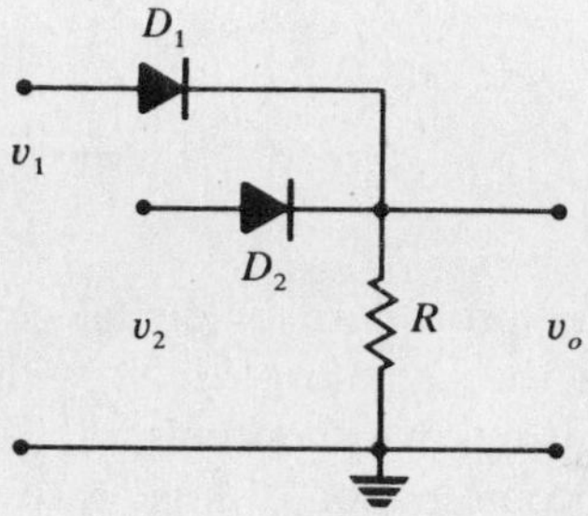

Fig. 12-16

12.8 Set up the truth table for the diode logic circuit of Fig. 12-16, showing that the circuit is an OR gate.

If either or both of v_1 and v_2 are logic high ($\geq V_H$), then $v_o \geq V_H - 0.7$. Otherwise, $v_o = 0$. The truth table shows OR logic:

v_1	v_2	v_o
0	0	0
0	1	1
1	0	1
1	1	1

12.9 Use a truth table to show that the *diode transistor logic* (DTL) circuit of Fig. 12-17 is a NAND gate.

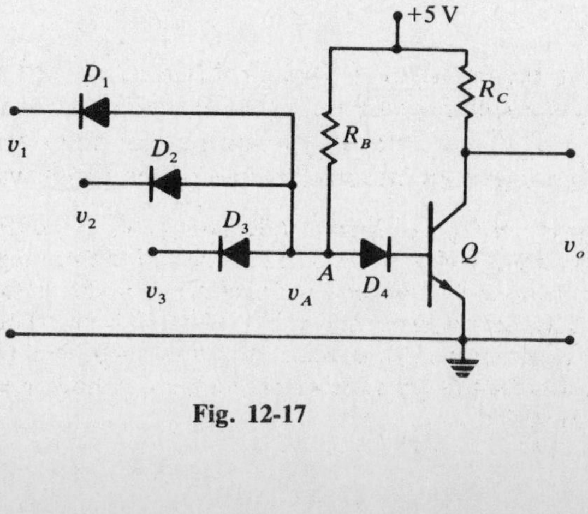

Fig. 12-17

From Problem 12.7, we see that the signal at node A is the output of a three-input AND gate. Further, the transistor circuit to the right of node A forms an inverter [diode D_4 ensures that the transistor is not turned ON when $v_A \approx 0.7$ V (low logic)]. We extend the truth table of Problem 12.7 to include three inputs and the logic complement of v_A, showing the overall logic to be that of a NAND gate:

v_1	v_2	v_3	v_A	$v_o = \bar{v}_A$
0	0	0	0	1
0	0	1	0	1
0	1	0	0	1
0	1	1	0	1
1	0	0	0	1
1	0	1	0	1
1	1	0	0	1
1	1	1	1	0

12.10 A common procedure in digital circuit design is to generate a logic statement or *function* describing the required circuit without regard to complexity, and then to manipulate the function using the theorems of Table 12.1 until an equivalent is found that requires a minimum combination of logic gates. A circuit is to be designed to realize the following logic:

1. If A, B, and C are all present (= 1, true), then the process is correct (= 1, true).
2. If A, B, and C are all absent (= 0, false), then the process is correct (= 1, true).
3. If B is present (= 1, true), then the process is correct (= 1, true).
4. Otherwise the process is incorrect (= 0, false).

(*a*) Express the process state as a digital logic function, (*b*) manipulate the function to simplify it (and thus require fewer logic gates), and (*c*) draw a logic block diagram of the original and simplified systems.

(*a*) We simply form an OR expression encompassing all combinations that lead to a correct process (an output of 1). From statements 1 to 3,

$$X = A \cdot B \cdot C + \bar{A} \cdot \bar{B} \cdot \bar{C} + B \tag{1}$$

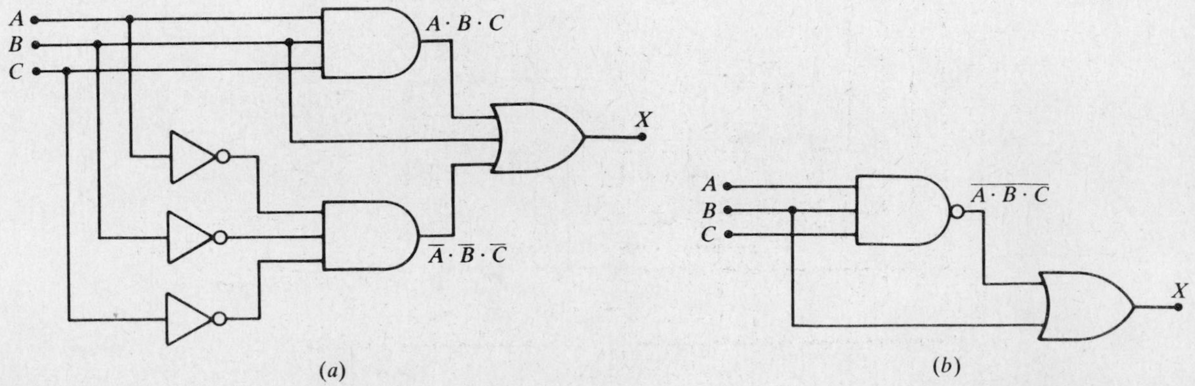

Fig. 12-18

(*b*) Operating on the second term of (*1*) with the first of DeMorgan's laws gives

$$X = A \cdot B \cdot C + \overline{A + B + C} + B \qquad (2)$$

Applying the redundancy law to (*2*) yields

$$X = (B + A \cdot B \cdot C) + \overline{A + B + C} = B + \overline{A + B + C} \qquad (3)$$

(*c*) A logic block diagram for the original system, as described by (*1*), is shown in Fig. 12-18(*a*), where six logic gates are required. Equation (*3*) can be realized with only two logic gates, as illustrated by Fig. 12-18(*b*).

12.11 A tedious but systematic approach to the construction of the logic function describing a process is to generate a table of all combinations of variables, identifying all *intersections* (AND combinations that are true). The logic function is then the *union* (OR combination) of all the intersections. (*a*) Use this systematic approach to generate a logic function for the process of Problem 12.10, and (*b*) draw a logic block diagram of the system.

(*a*) All possible combinations of input variables, as well as the intersections that satisfy statements 1 to 3 of Problem 12.10, are shown in the following table:

A	B	C	Intersections
0	0	0	$\bar{A} \cdot \bar{B} \cdot \bar{C}$
0	0	1	
0	1	0	$\bar{A} \cdot B \cdot \bar{C}$
0	1	1	$\bar{A} \cdot B \cdot C$
1	0	0	
1	0	1	
1	1	0	$A \cdot B \cdot \bar{C}$
1	1	1	$A \cdot B \cdot C$

The union of the intersections is then

$$X = \bar{A} \cdot \bar{B} \cdot \bar{C} + \bar{A} \cdot B \cdot \bar{C} + \bar{A} \cdot B \cdot C + A \cdot B \cdot \bar{C} + A \cdot B \cdot C \qquad (1)$$

(*b*) The logic block diagram suggested by (*1*) is displayed in Fig. 12-19, where it is assumed that an OR gate with no more than four inputs is available (as a practical limit).

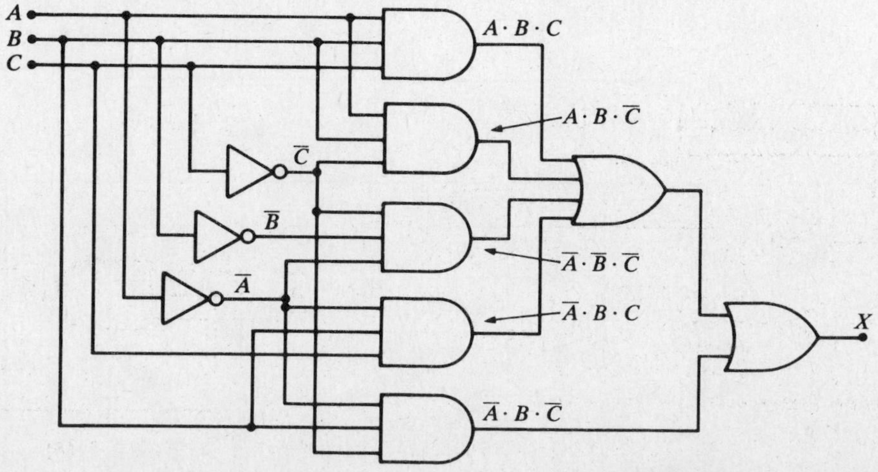

Fig. 12-19

12.12 Illustrate, for the case of three variables, that NAND gates can be interconnected to form the equivalent of an OR gate.

The desired logic function is

$$X = A + B + C \tag{1}$$

Applying the negation and the first of DeMorgan's laws to (*1*) yields

$$X = \overline{\overline{A + B + C}} = \overline{\bar{A} \cdot \bar{B} \cdot \bar{C}} \tag{2}$$

Now (*2*) could be synthesized by a NAND gate if the complements of the three variables were available as inputs. But by the identity law,

$$\overline{A \cdot A} = \bar{A} \tag{3}$$

An interpretation of (*3*) is that if A is the input to both gates of a two-input NAND gate, then its output is $\bar{A}$. Thus, (*1*) is realized by the logic block diagram of Fig. 12-20.

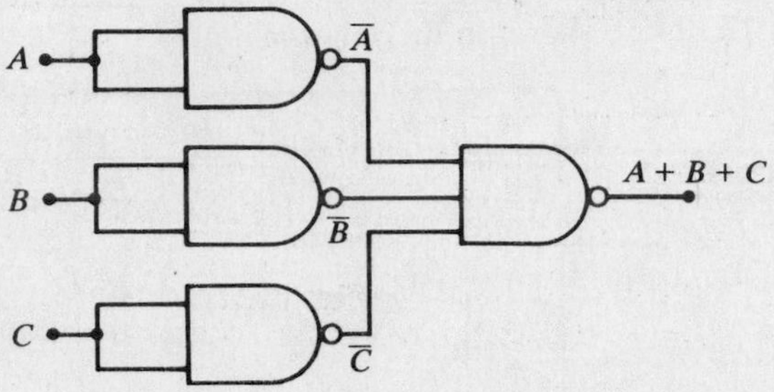

Fig. 12-20

12.13 Construct truth tables for the logic functions (*a*) $f_1 = A \cdot \bar{B} + A \cdot B$ and (*b*) $f_2 = \bar{A} \cdot B \cdot C + \bar{A} \cdot \bar{B}$.

(*a*) The truth table for f_1, showing the values of the individual terms, is:

A	B	$A \cdot \bar{B}$	$A \cdot B$	f_1
0	0	0	0	0
0	1	0	0	0
1	0	1	0	1
1	1	0	1	1

(*b*) The truth table for f_2 is as follows:

A	B	C	$\bar{A} \cdot B \cdot C$	$\bar{A} \cdot \bar{B}$	f_2
0	0	0	0	1	1
0	0	1	0	1	1
0	1	0	0	0	0
0	1	1	1	0	1
1	0	0	0	0	0
1	0	1	0	0	0
1	1	0	0	0	0
1	1	1	0	0	0

12.14 Realize the logic function $f = A\cdot B + \bar{A}\cdot B$ using only NOR logic gates.

The negation law and DeMorgan's laws give

$$\bar{f} = \overline{\overline{\overline{A\cdot B + \bar{A}\cdot B}}} = \overline{\overline{(\overline{A\cdot B})\cdot(\overline{\bar{A}\cdot B})}} = \overline{\overline{(\bar{A}+\bar{B})\cdot(A+\bar{B})}} = \overline{\overline{(\bar{A}+\bar{B})}+\overline{(A+\bar{B})}} \qquad (1)$$

And, by the negation and identity laws,

$$f = \bar{\bar{f}} = \overline{\bar{f}+\bar{f}} \qquad (2)$$

Use of (*1*) and (*2*) leads to the logic block diagram of Fig. 12-21. Note, however, that $A\cdot B + \bar{A}\cdot B = B$, so f could be realized without any NOR gates (or other gates), by straight connection to signal B.

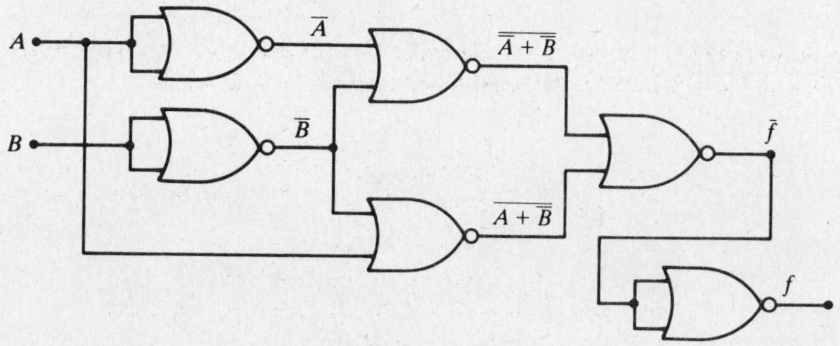

Fig. 12-21

Supplementary Problems

12.15 Rework Problem 12.1 using the model of Fig. 12-2 with $V_F = 0.5$ V, $R_{sat} = 15\ \Omega$, and $R_F = 3$ kΩ. *Ans.* (*a*) $I_{Bsat} = 492.6\ \mu$A; (*b*) $R_L \le 4.985$ kΩ

12.16 Show that the collector power dissipation for BJTs operating in the switching mode is significantly less than that for operation in the active region by calculating the collector power dissipated by the circuit of Fig. 12-9 for (*a*) cutoff, (*b*) saturation, and (*c*) $V_{CEQ} = 7.5$ V. Assume $R_L = 1$ kΩ, $V_{CC} = 15$ V, and that the transistor is characterized by $R_{sat} = 15\ \Omega$, $V_{CEsat} = 0.2$ V, and $R_{CO} = 1$ MΩ.
Ans. (*a*) 225 μW; (*b*) 3.25 mW; (*c*) 56.25 mW

12.17 Use a truth table to show that the circuit of Fig. 12-22 is (*a*) a NAND gate if v_{o1} is taken as the output and (*b*) an AND gate if v_{o2} is the output.

12.18 The *transistor transistor logic* (TTL or T^2L) circuit of Fig. 12-23 applies the input signal directly to transistor terminals (no current-limiting resistors). If either or both of v_1 and v_2 is logic low, Q_1 is driven to saturation. Draw a truth table for this circuit showing that output v_{o1} produces AND logic and output v_{o2} produces NAND logic.

12.19 TTL logic gates are usually applied as NAND gates. Other gate types are formed by appropriate use of inverters in conjunction with NAND gates. Construct a logic block diagram using the two-input TTL NAND gate of Fig. 12-23 in conjunction with inverters (NOT gates) to form an OR gate.
Ans. See Fig. 12-24.

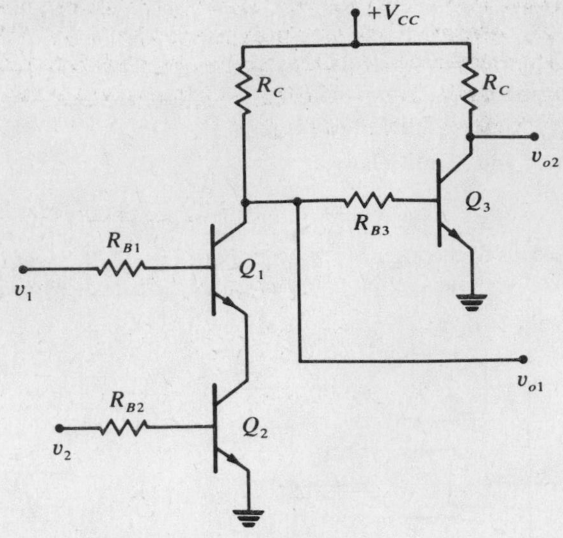

Fig. 12-22

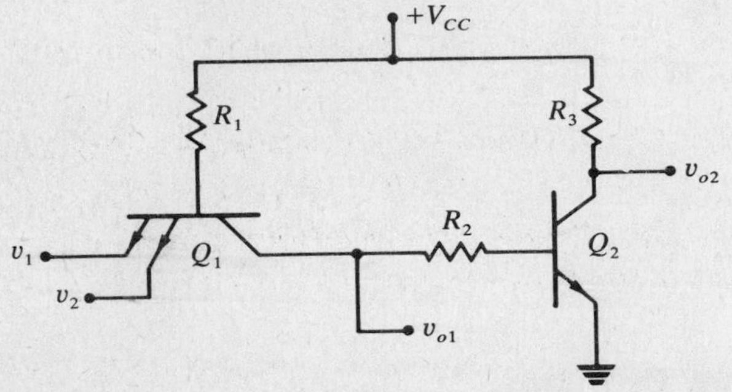

Fig. 12-23

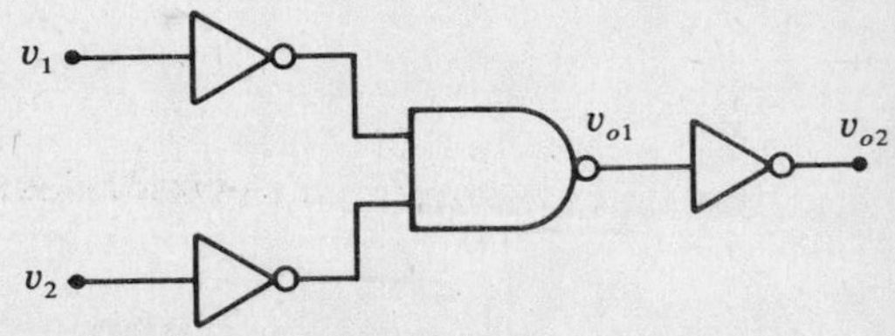

Fig. 12-24

12.20 *Diode transistor logic* (DTL) combines diodes and transistors to form logic gates. Use a truth table to show that the DTL circuit of Fig. 12-25 acts as a NOR gate.

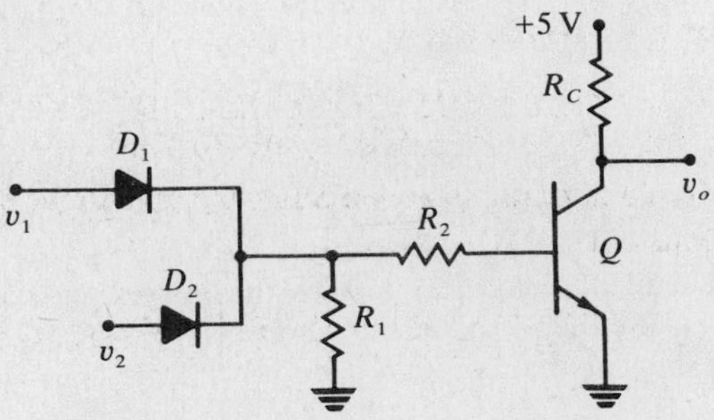

Fig. 12-25

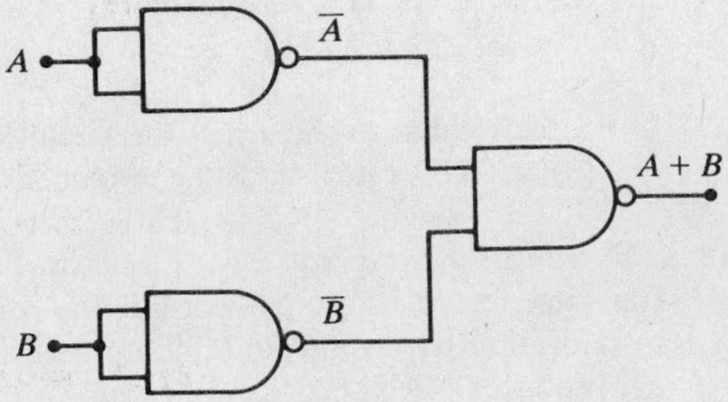

Fig. 12-26

12.21 Use DeMorgan's laws to find a combination of NAND gates that is identical in function to a two-input OR gate. *Ans.* See Fig. 12-26.

12.22 Show that the logic function $X = \bar{A}\cdot\bar{B} + \bar{A}\cdot B + A\cdot B$ can be reduced to $\bar{A} + B$.

12.23 If $X = \bar{A}\cdot B + A\cdot\bar{B}$, show that $\bar{X} = A\cdot B + \bar{A}\cdot\bar{B}$.

12.24 Use the theorems of Table 12-1 to show that the following are identities:

$$A\cdot B\cdot C + A\cdot B\cdot\bar{C} = A\cdot B$$

$$(A + B)\cdot(\bar{A} + C) = A\cdot C + \bar{A}\cdot B$$

$$(A + C)\cdot(\bar{A} + B) = A\cdot B + \bar{A}\cdot C$$

12.25 Use NAND gates only to construct a logic block diagram with output $X = A + \bar{B}\cdot C$.
Ans. See Fig. 12-27.

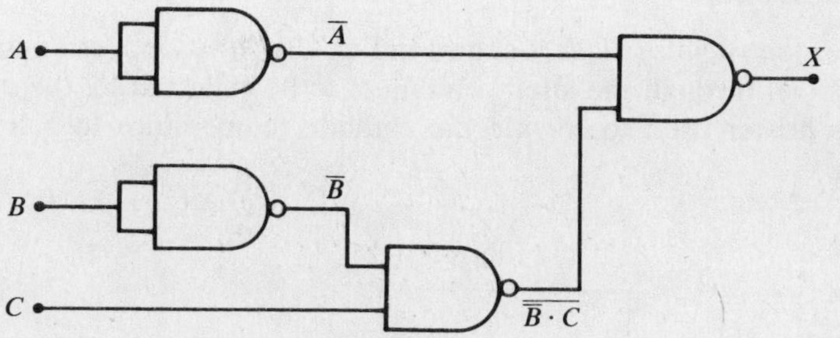

Fig. 12-27

12.26 Evaluate the Boolean expressions (*a*) $f_1 = \bar{A}\cdot\bar{B} + A\cdot B + \bar{B} + C$ and (*b*) $f_2 = A\cdot B\cdot\bar{C} + \bar{A}\cdot B + A\cdot\bar{B}$ for $A = 1$, $B = 0$, and $C = 1$. *Ans.* (*a*) $f_1 = 1$; (*b*) $f_2 = 1$

Chapter 13

Vacuum Tubes

13.1 INTRODUCTION

A *vacuum tube* is an evacuated enclosure containing (1) a *cathode* that emits electrons, with a *heater* acting to enhance the process; (2) an *anode* or *plate* that attracts the emitted electrons when operated at a positive potential relative to the cathode; and usually (3) one or more intermediate electrodes (called *grids*) that modify the emission-attraction process.

13.2 VACUUM DIODES

The symbol for the *vacuum diode* is shown in Fig. 13-1(*a*). Conduction is by electrons that escape the *cathode* and travel through the internal vacuum to be collected by the *anode*. The *filament* is an electric resistance heater used to elevate the cathode temperature to a level at which thermionic emission occurs.

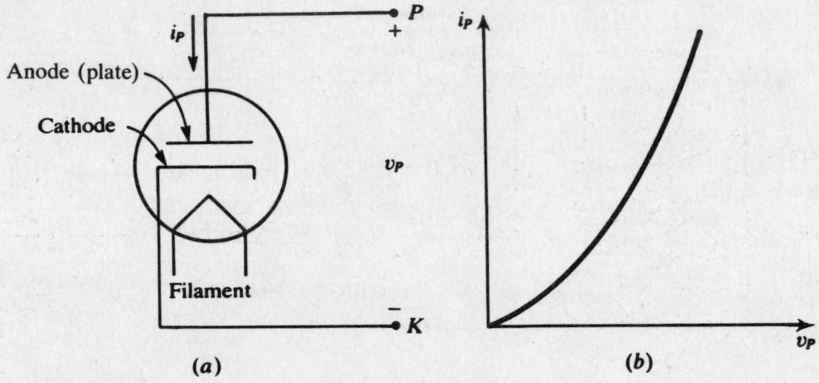

Fig. 13-1

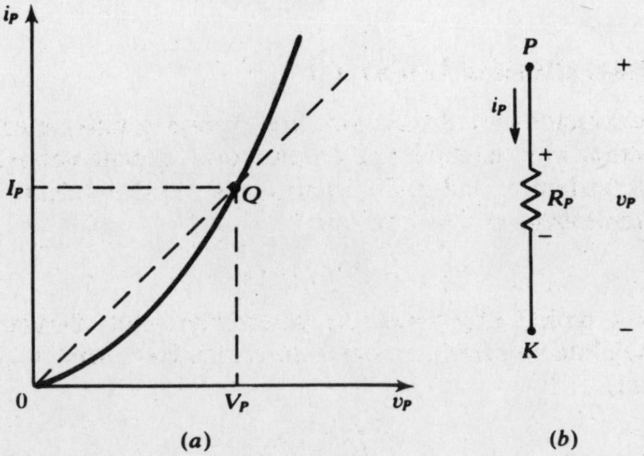

Fig. 13-2

A typical i_P-versus-v_P characteristic curve (the *plate characteristic*) is depicted in Fig. 13-1(b). This characteristic is described by the *Childs-Langmuir three-halves-power law*:

$$i_P = \kappa v_P^{3/2} \tag{13.1}$$

where κ is the *perveance* (a constant that depends upon the mechanical design of the tube).

Like the semiconductor diode, the vacuum diode is a near-unilateral conductor. However, the forward voltage drop of the vacuum diode is typically several volts, which renders the ideal-diode approximation unjustifiable. Most commonly, the vacuum diode is modeled in forward conduction as a resistance $R_P = V_P/I_P$, obtained as the slope of a straight-line approximation of the plate characteristic (Fig. 13-2). A reverse-biased vacuum diode acts like an open circuit.

13.3 VACUUM TRIODE CONSTRUCTION AND SYMBOLS

The single grid of the *vacuum triode* is called the *control grid*; it is made of small-diameter wire and inserted between the plate and cathode as suggested in Fig. 13-3(a). The mesh of the grid is sufficiently coarse so as not to impede current flow from plate to cathode through collision of electrons with the grid wire; moreover, the grid is placed physically close to the cathode so that its electric field can exert considerable control over electron emission from the cathode surface. The symbols for the total instantaneous currents and voltages of the triode are shown in Fig. 13-3(b); component, average, rms, and maximum values are symbolized as in Table 3-1.

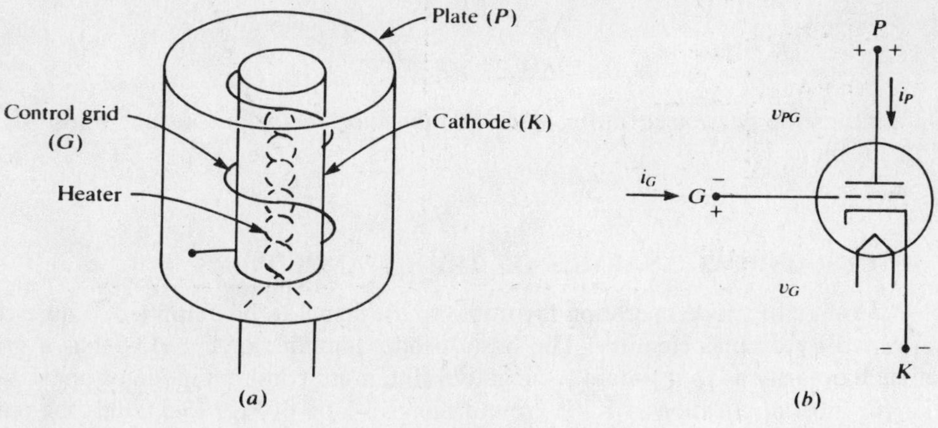

Fig. 13-3

13.4 TRIODE TERMINAL CHARACTERISTICS

The voltage-current characteristics of the triode are experimentally determined with the cathode sharing a common connection with the input and output ports. If plate voltage v_P and grid voltage v_G are taken as independent variables, and grid current i_G as the dependent variable, then the *input characteristics* (or *grid characteristics*) have the form

$$i_G = f_1(v_P, v_G) \tag{13.2}$$

of which Fig. 13-4(a) is a typical experimentally determined plot. Similarly, with v_P and v_G as independent variables, the plate current i_P becomes the dependent variable of the *output characteristics* (or *plate characteristics*)

$$i_P = f_2(v_P, v_G) \tag{13.3}$$

of which a typical plot is displayed in Fig. 13-4(b).

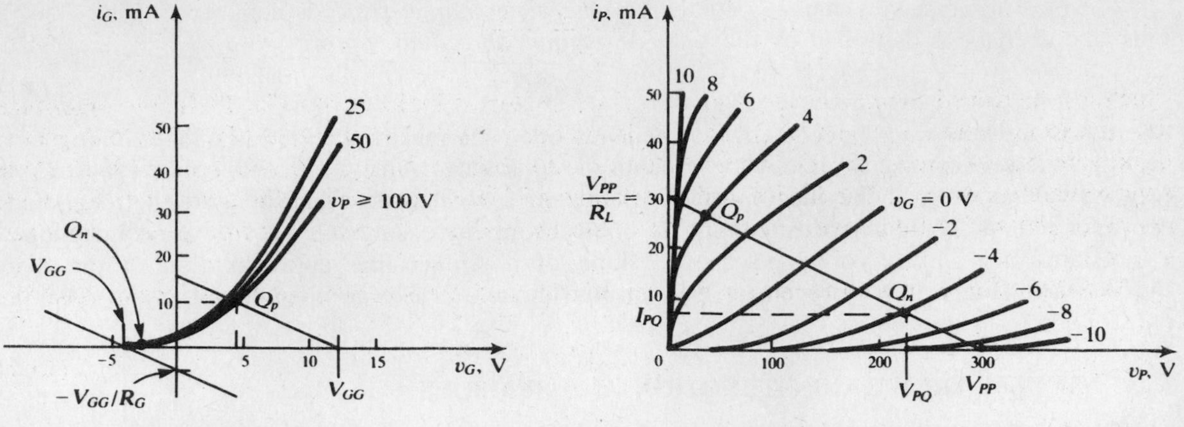

(*a*) Grid characteristics (*b*) Plate characteristics

Fig. 13-4

The triode input characteristics of Fig. 13-4(*a*) show that operation with a positive grid voltage results in flow of grid current; however, with a negative grid voltage (the common application), negligible grid current flows and the plate characteristics are reasonably approximated by a three-halves-power relationship (Section 13.2) involving a linear combination of plate and grid voltages:

$$i_P = \kappa(v_P + \mu v_G)^{3/2} \tag{13.4}$$

where κ again denotes the perveance and μ is the *amplification factor*, a constant whose significance is elucidated in Problem 13.3.

13.5 BIAS AND GRAPHICAL ANALYSIS OF TRIODE AMPLIFIERS

To establish a range of triode operation favorable to the signal to be amplified, a quiescent point must be determined by dc bias circuitry. The basic triode amplifier of Fig. 13-5 has a grid power supply V_{GG} of such polarity as to maintain v_G negative (the more common mode of operation). With no input signal ($v_S = 0$), application of KVL around the grid loop of Fig. 13-5 yields the equation of the *grid bias line*,

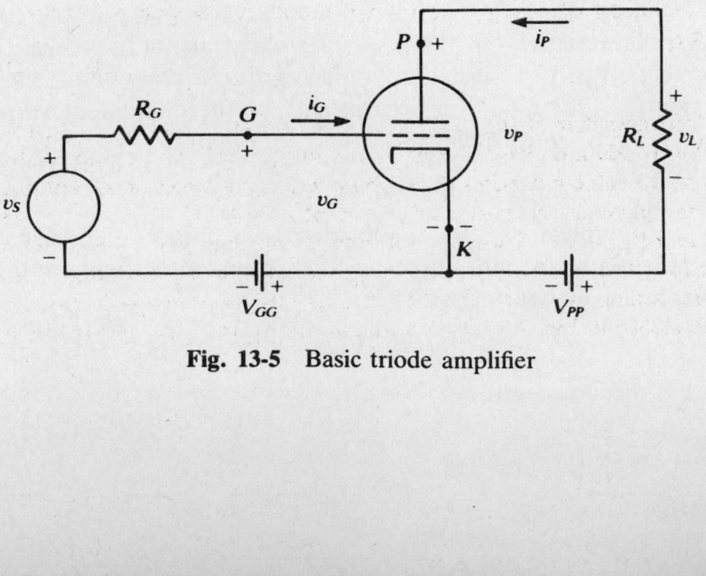

Fig. 13-5 Basic triode amplifier

$$i_G = -\frac{V_{GG}}{R_G} - \frac{v_G}{R_G} \tag{13.5}$$

which can be solved simultaneously with (*13.2*) or plotted as indicated on Fig. 13-4(*a*) to determine the quiescent values I_{GQ} and V_{GQ}. If V_{GG} is of the polarity indicated in Fig. 13-5, the grid is negatively biased, giving the Q point labeled Q_n. At that point, $I_{GQ} \approx 0$ and $V_{GQ} \approx -V_{GG}$; these approximate solutions suffice in the case of negative grid bias. However, if the polarity of V_{GG} were reversed, the grid would have a positive bias, and the quiescent point Q_p would give $I_{GQ} > 0$ and $V_{GQ} < V_{GG}$.

Voltage summation around the plate circuit of Fig. 13-5 leads to the equation of the *dc load line*

$$i_P = \frac{V_{PP}}{R_L} - \frac{v_P}{R_L} \tag{13.6}$$

which, when plotted on the plate characteristics of Fig. 13-4(*b*), yields the quiescent values V_{PQ} and I_{PQ} at its intersection with the curve $v_G = V_{GQ}$.

Example 13.1 In the triode amplifier of Fig. 13-5, $V_{GG} = 4$ V, $V_{PP} = 300$ V, $R_L = 10$ kΩ, and $R_G = 2$ kΩ. The plate characteristics for the triode are given by Fig. 13-4(*b*). (*a*) Draw the dc load line; then determine the quiescent values (*b*) I_{GQ}, (*c*) V_{GQ}, (*d*) I_{PQ}, and (*e*) V_{PQ}.

(*a*) For the given values, the dc load line (*13.6*) has the i_P intercept

$$\frac{V_{PP}}{R_L} = \frac{300}{10 \times 10^3} = 30 \text{ mA}$$

and the v_P intercept $V_{PP} = 300$ V. These intercepts have been utilized to draw the dc load line on the plate characteristics of Fig. 13-4(*b*).

(*b*) Since the polarity of V_{GG} is such that v_G is negative, negligible grid current will flow ($I_{GQ} \approx 0$).

(*c*) For negligible grid current, (*13.5*) evaluated at the Q point yields $V_{GQ} = -V_{GG} = -4$ V.

(*d*) The quiescent plate current is read as the projection of Q_n onto the i_P axis of Fig. 13-4(*b*) and is $I_{PQ} = 8$ mA.

(*e*) Projection of Q_n onto the v_P axis of Fig. 13-4(*b*) gives $V_{PQ} = 220$ V.

The application of a time-varying signal v_S to the amplifier of Fig. 13-5 results in a grid voltage with a time-varying component,

$$v_G = V_{GQ} + v_g$$

It is usual practice to ensure that $v_G \le 0$ by proper selection of the combination of bias and signal. Then $i_G = 0$, and the operating point must move along the dc load line from the Q point in accordance with the variation of v_g, giving instantaneous values of v_P and i_P that simultaneously satisfy (*13.3*) and (*13.6*).

Example 13.2 The triode amplifier of Fig. 13-5 has V_{GG}, V_{PP}, R_G, and R_L as given in Example 13.1. If the plate characteristics of the triode are given by Fig. 13-6 and $v_S = 2 \sin \omega t$ V, graphically find v_P and i_P.

The dc load line, with the same intercepts as in Example 13.1, is superimposed on the characteristics of Fig. 13-6; however, because the plate characteristics are different from those of Example 13.1, the quiescent values are now $I_{PQ} = 11.3$ mA and $V_{PQ} = 186$ V. Then a time axis on which to plot $v_G = -4 + 2 \sin \omega t$ V is constructed perpendicular to the dc load line at the Q point. Time axes for i_P and v_P are also constructed as shown, and values of i_P and v_P corresponding to particular values of $v_G(t)$ are found by projecting through the dc load line, for one cycle of v_G. The result, in Fig. 13-6, shows that v_P varies from 152 to 218 V and i_P ranges from 8.1 to 14.7 mA.

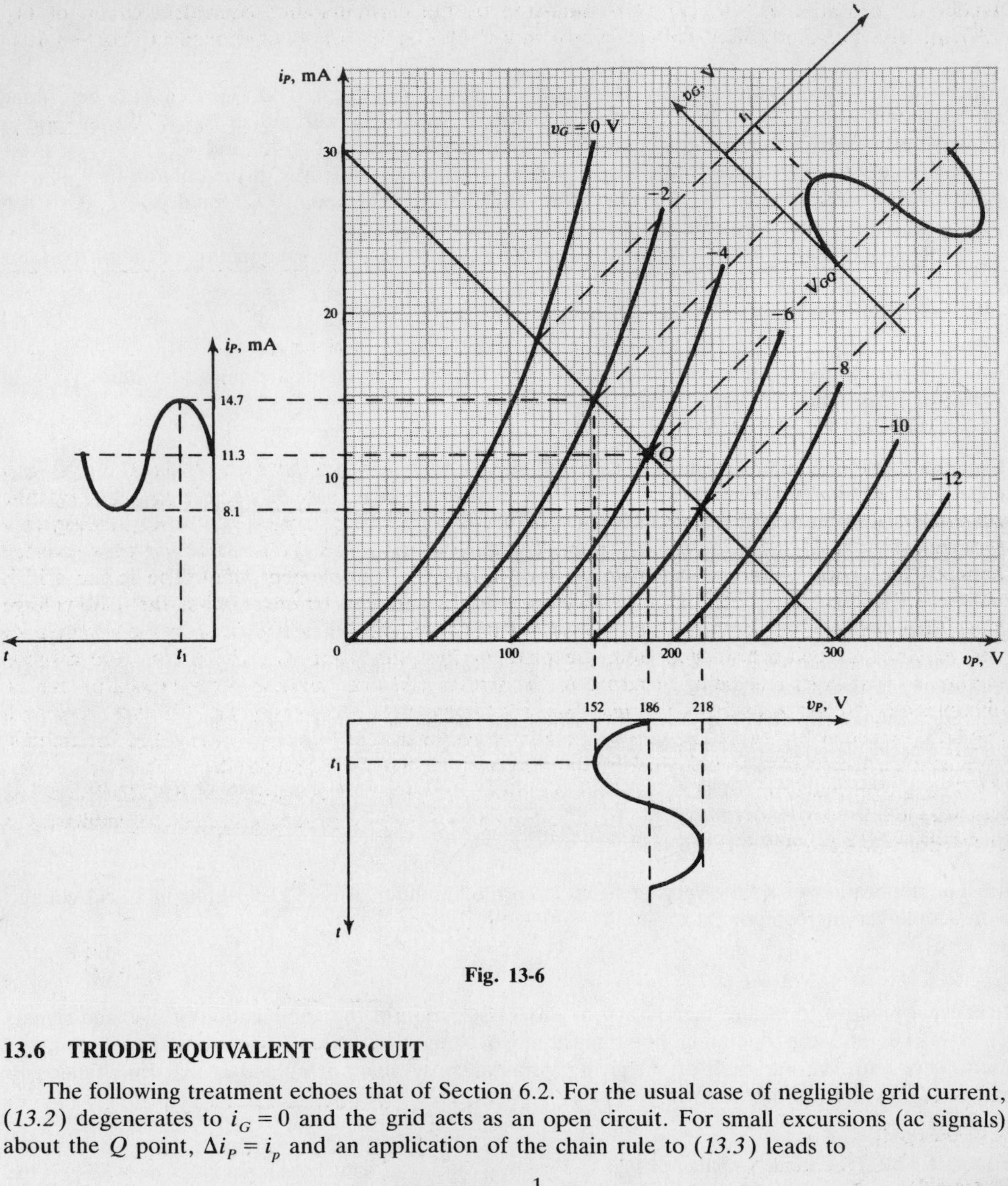

Fig. 13-6

13.6 TRIODE EQUIVALENT CIRCUIT

The following treatment echoes that of Section 6.2. For the usual case of negligible grid current, (*13.2*) degenerates to $i_G = 0$ and the grid acts as an open circuit. For small excursions (ac signals) about the Q point, $\Delta i_P = i_p$ and an application of the chain rule to (*13.3*) leads to

$$i_p = \Delta i_P \approx di_P = \frac{1}{r_p} v_p + g_m v_g \qquad (13.7)$$

where we have defined

$$\text{Plate resistance} \qquad r_p \equiv \left.\frac{\partial v_P}{\partial i_P}\right|_Q \approx \left.\frac{\Delta v_P}{\Delta i_P}\right|_Q \qquad (13.8)$$

$$\text{Transconductance} \qquad g_m \equiv \left.\frac{\partial i_P}{\partial v_G}\right|_Q \simeq \left.\frac{\Delta i_P}{\Delta v_G}\right|_Q \qquad (13.9)$$

Under the condition $i_G = 0$, (*13.7*) is simulated by the current-source equivalent circuit of Fig. 13-7(*a*). The frequently used voltage-source model of Fig. 13-7(*b*) is developed in Problem 13.3.

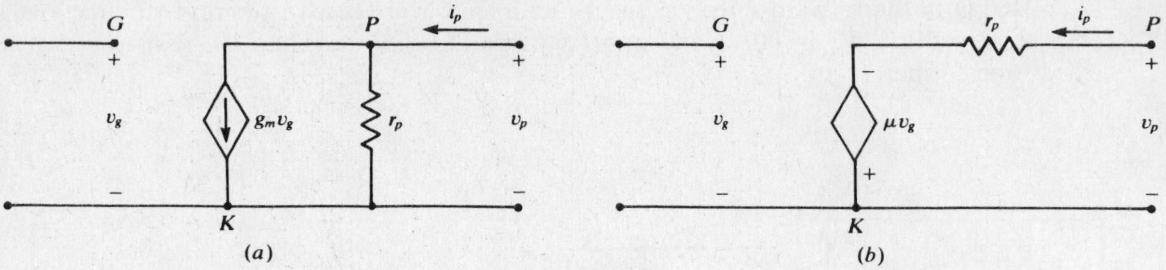

Fig. 13-7 Triode small-signal equivalent circuits

13.7 MULTIPLE-CONTROL-GRID TUBES

Interelectrode capacitances limit the usefulness of the triode at high frequencies; the plate-grid capacitance C_{pg} leads to the greatest detriment in performance because its effect is magnified by tube parameters and because it constitutes undesired coupling between input and output, leading to a reduction in control of the output. A marked reduction in C_{pg} results if a *screen grid* is inserted between the control grid and the plate, to form a *tetrode* (four-element *tube*). The screen grid is maintained at a positive potential v_{G2} with respect to the cathode. However, when the plate voltage is less than the screen-grid voltage ($v_P < v_{G2}$), the plate characteristics exhibit an erratic nonlinearity due to electrons being dislodged from the plate by incident electrons arriving from the cathode (*secondary emission*) and being attracted to the screen grid. This undesirable nonlinearity can be eliminated by adding a third grid (a *suppressor grid*) between the screen grid and the plate, to form a *pentode* (five-element *tube*). The suppressor grid is maintained at the same potential as the cathode by electrical connection, as indicated by the schematic of the vacuum pentode in Fig. 13-8.

The grid characteristics of a pentode are similar to those of a triode [see Fig. 13-4(*a*)]. The experimentally determined plate characteristics are typically like those in Fig. 13-9. (A similarity to JFET drain characteristics is apparent.)

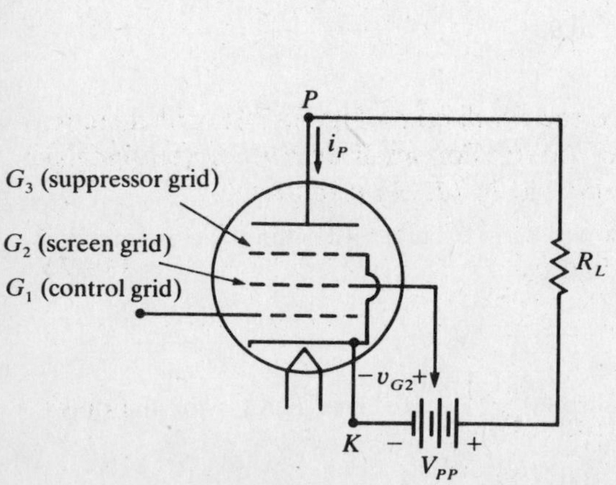

Fig. 13-8

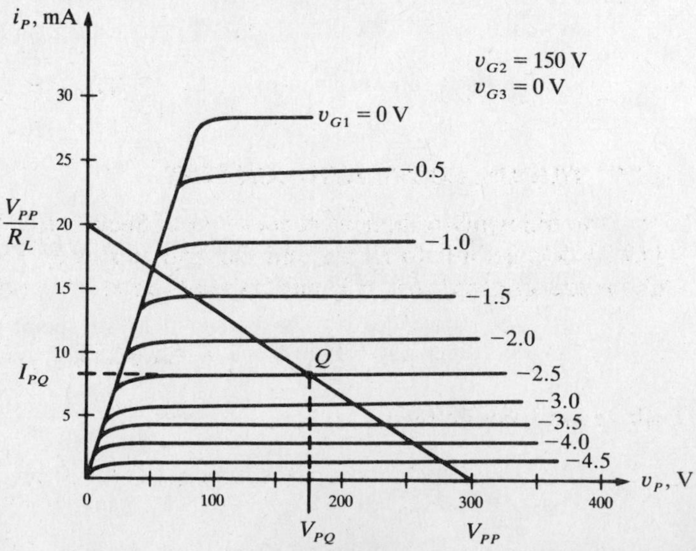

Fig. 13-9

Solved Problems

13.1 The vacuum diodes in the rectifier circuit of Fig. 13-10(*a*) are identical and can be modeled by $R_F = 400\ \Omega$ in the forward direction and by an infinite resistance in the reverse direction. If $C = 0$, $R_L = 5\ \text{k}\Omega$, $V_{Lo} = 110$ V, and $v_S = 120\sqrt{2} \sin 120\pi t$ V, calculate the turns ratio of the ideal transformer.

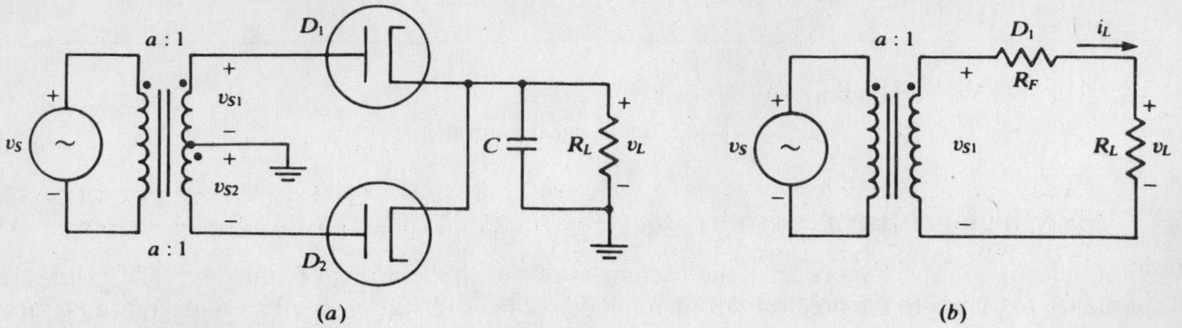

Fig. 13-10

Owing to the symmetry of the problem, it is necessary to analyze only one leg of the rectifier circuit, as indicated in Fig. 13-10(*b*). Since the output is a rectified sine wave, for which (from Problem 2.48) $V_{Lo} = 0.637V_{Lm}$, during the positive half cycle of v_S the load voltage can be written as

$$v_L(t) = \frac{110}{0.637} \sin 120\pi t = 172.96 \sin 120\pi t \qquad \text{V}$$

[$v_L(t)$ is zero during the negative half cycle of v_S]. By Ohm's law,

$$i_L(t) = \frac{v_L(t)}{R_L} = \frac{172.96}{5000} \sin 120\pi t = 0.03459 \sin 120\pi t \qquad \text{A}$$

Also, $$v_{D1}(t) = R_F i_L(t) = (400)(0.03459 \sin 120\pi t) = 13.84 \sin 120\pi t \qquad \text{V}$$

so that $$v_{S1}(t) = v_{D1}(t) + v_L(t) = 186.8 \sin 120\pi t \qquad \text{V}$$

and the turns ratio is given by

$$a = \frac{v_S}{v_{S1}} = \frac{120\sqrt{2}}{186.8} = 0.908$$

13.2 For a triode with plate characteristics given by Fig. 13-6, find (*a*) the preveance κ and (*b*) the amplification factor μ.

(*a*) The perveance can be evaluated at any point on the $v_G = 0$ curve. Choosing the point with coordinates $i_P = 15$ mA and $v_P = 100$ V, we have, from (*13.4*),

$$\kappa = \frac{i_P}{v_P^{3/2}} = \frac{15 \times 10^{-3}}{100^{3/2}} = 15\ \mu\text{A/V}^{3/2}$$

(*b*) The amplification factor is most easily evaluated along the v_P axis. From (*13.4*), for the point $i_P = 0$, $v_P = 100$ V, $v_G = -4$ V, we obtain

$$\mu = -\frac{v_P}{v_G} = -\frac{100}{-4} = 25$$

13.3 Use the current-source small-signal triode model of Fig. 13-7(a) to derive the voltage-source model of Fig. 13-7(b).

We need to find the Thévenin equivalent for the circuit to the left of the output terminals in Fig. 13-7(a). If the independent source is deactivated, then $v_g = 0$; thus, $g_m v_g = 0$, and the dependent current source acts as an open circuit. The Thévenin resistance is then $R_{Th} = r_p$. The open-circuit voltage appearing at the output terminals is

$$v_{Th} = -g_m v_g r_p \equiv -\mu v_g$$

where $\mu \equiv g_m r_p$ is the *amplification factor*. Proper series arrangement of v_{Th} and R_{Th} gives the circuit of Fig. 13-7(b).

13.4 The amplifier of Example 13.1 has plate current

$$i_P = I_P + i_p = 8 + \cos \omega t \qquad \text{mA}$$

Determine (a) the power delivered by the plate supply voltage V_{PP}, (b) the average power delivered to the load R_L, and (c) the average power dissipated by the plate of the triode. (d) If the tube has a plate rating of 2 W, is it being properly applied?

(a) The power supplied by the source V_{PP} is found by integration over a period of the ac waveform:

$$P_{PP} = \frac{1}{T}\int_0^T V_{PP} i_P \, dt = V_{PP} I_P = (300)(8 \times 10^{-3}) = 2.4 \text{ W}$$

(b)
$$P_L = \frac{1}{T}\int_0^T i_P^2 R_L \, dt = R_L(I_P^2 + I_p^2) = 10 \times 10^3\left[(8 \times 10^{-3})^2 + \left(\frac{1 \times 10^{-3}}{\sqrt{2}}\right)^2\right] = 0.645 \text{ W}$$

(c) The average power dissipated by the plate is

$$P_P = P_{PP} - P_L = 2.4 - 0.645 = 1.755 \text{ W}$$

(d) The tube is not properly applied. If the signal is removed (so that $i_p = 0$), then the plate dissipation increases to $P_P = P_{PP} = 2.4$ W, which exceeds the power rating.

13.5 For the amplifier of Example 13.2, (a) use (*13.8*) to evaluate the plate resistance and (b) use (*13.9*) to find the transconductance.

(a)
$$r_p \approx \left.\frac{\Delta v_P}{\Delta i_P}\right|_{v_G = -4} = \frac{218 - 152}{(14.7 - 8.1) \times 10^{-3}} = 10 \text{ k}\Omega$$

(b)
$$g_m \approx \left.\frac{\Delta i_P}{\Delta v_G}\right|_{v_P = 186} = \frac{(14.7 - 8.1) \times 10^{-3}}{-2 - (-6)} = 1.65 \text{ mS}$$

13.6 Find an expression for the voltage gain $A_v = v_p/v_g$ of the basic triode amplifier of Fig. 13-5, using an ac equivalent circuit.

The equivalent circuit of Fig. 13-7(b) is applicable if R_L is connected from P to K. Then, by voltage division in the plate circuit,

$$v_p = \frac{R_L}{R_L + r_p}(-\mu v_g) \qquad \text{so} \qquad A_v = \frac{v_p}{v_g} = \frac{-\mu R_L}{R_L + r_p}$$

13.7 The *plate efficiency* of a vacuum-tube amplifier is defined as the ratio of ac signal power delivered to the load to plate supply power, or $P_{L\text{ac}}/P_{PP}$. (a) Calculate the plate efficiency of the amplifier of Problem 13.4. (b) What is the maximum possible plate efficiency for this amplifier without changing the Q point or clipping the signal?

(*a*) $$\eta = \frac{P_{L\,\mathrm{ac}}}{P_{PP}}(100\%) = \frac{I_p^2 R_L}{V_{PP} I_P}(100\%) = \frac{(10^{-3}/\sqrt{2})^2(10 \times 10^3)}{2.4}(100\%) = 0.208\%$$

(*b*) Ideally, the input signal could be increased until i_p swings ± 8 mA; thus,

$$\eta_{\max} = \left(\frac{8}{1}\right)^2 (0.208\%) = 13.31\%$$

13.8 The triode amplifier of Fig. 13-11 utilizes *cathode bias* to eliminate the need for a grid power supply. The very large resistance R_G provides a path to ground for stray charge collected by the grid; this current is so small, however, that the voltage drop across R_G is negligible. It follows that the grid is maintained at a negative bias, so

$$v_G = -R_K i_P \qquad (1)$$

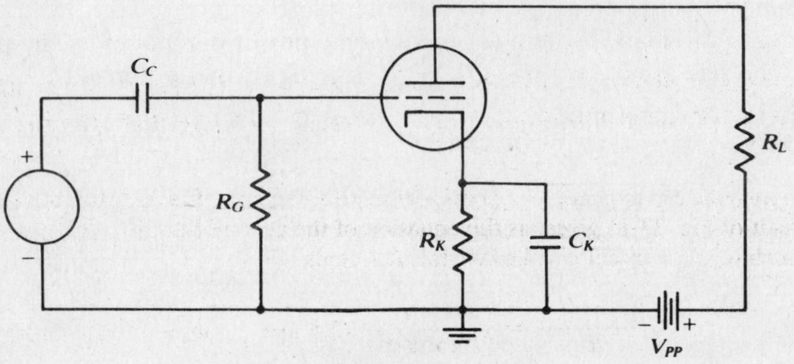

Fig. 13-11

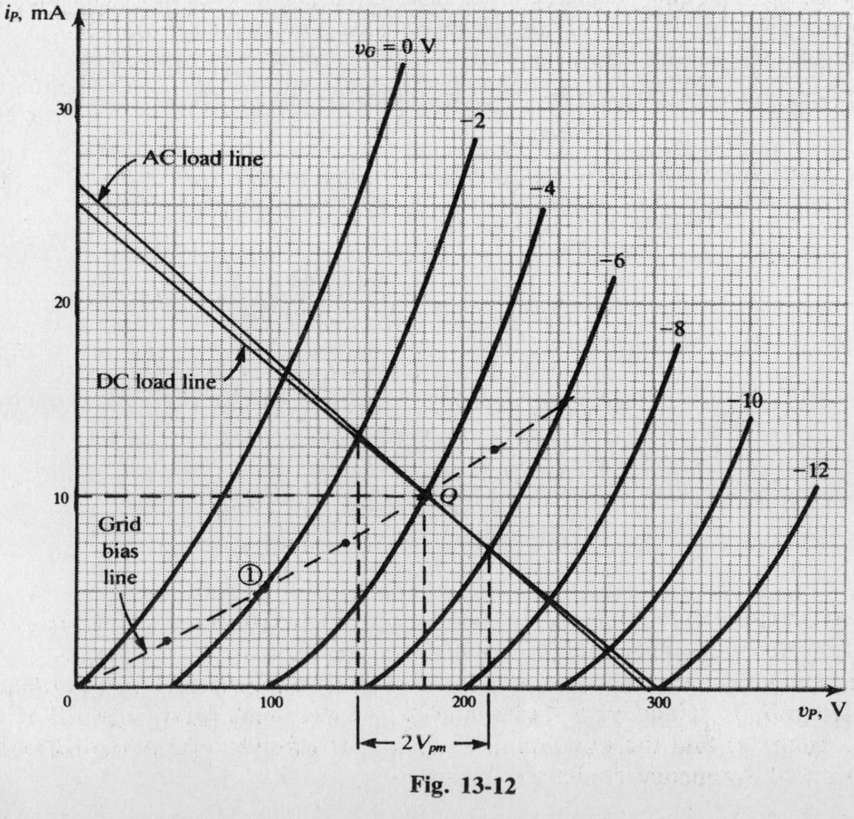

Fig. 13-12

A plot of (*1*) on the plate characteristics is called the *grid bias line*, and its intersection with the dc load line determines the Q point. Let $R_L = 11.6\ \text{k}\Omega$, $R_K = 400\ \Omega$, $R_G = 1\ \text{M}\Omega$, and $V_{PP} = 300$ V. If the plate characteristics of the triode are given by Fig. 13-12, (*a*) draw the dc load line, (*b*) sketch the grid bias line, and (*c*) determine the Q-point quantities.

(*a*) The dc load line has horizontal intercept $V_{PP} = 300$ V and vertical intercept

$$\frac{V_{PP}}{R_{\text{dc}}} = \frac{V_{PP}}{R_L + R_K} = \frac{300}{(11.6 + 0.4) \times 10^3} = 25\ \text{mA}$$

as shown on the plate characteristics of Fig. 13-12.

(*b*) Points for the plot of (*1*) are found by selecting values of i_P and calculating the corresponding values of v_G. For example, if $i_P = 5$ mA, then $v_G = -400(5 \times 10^{-3}) = -2$ V, which plots as point 1 of the dashed grid bias line in Fig. 13-12. Note that this is not a straight line.

(*c*) From the intersection of the grid bias line with the dc load line, $I_{PQ} = 10$ mA, $V_{PQ} = 180$ V, and $V_{GQ} = -4$ V.

13.9 In the amplifier of Problem 13.8, let $v_S = 2 \cos \omega t$ V. (*a*) Draw the ac load line on Fig. 13-12. (*b*) Graphically determine the voltage gain. (*c*) Calculate the voltage gain using small-signal analysis.

(*a*) If capacitor C_K appears as a short circuit to ac signals, then application of KVL around the plate circuit of Fig. 13-11 gives, as the equation of the ac load line, $V_{PP} + V_{GQ} = i_P R_L + v_P$. Thus, the ac load line has vertical and horizontal intercepts

$$\frac{V_{PP} + V_{GQ}}{R_L} = \frac{300 - 4}{11.6 \times 10^3} = 25.5\ \text{mA} \qquad \text{and} \qquad V_{PP} + V_{GQ} = 296\ \text{V}$$

as shown on Fig. 13-12.

(*b*) We have $v_g = v_S$; thus, as v_g swings ± 2 V along the ac load line from the Q point in Fig. 13-12, v_p swings a total of $2V_{pm} = 213 - 145 = 68$ V as shown. The voltage gain is then

$$A_v = -\frac{2V_{pm}}{2V_{gm}} = -\frac{68}{4} = -17$$

where the minus sign is included to account for the phase reversal between v_p and v_g.

(*c*) Applying (*13.8*) and (*13.9*) at the Q point of Fig. 13-12 yields

$$r_p = \left.\frac{\Delta v_P}{\Delta i_P}\right|_{v_G = -4} = \frac{202 - 168}{(15 - 8) \times 10^{-3}} = 4.86\ \text{k}\Omega$$

$$g_m = \left.\frac{\Delta i_P}{\Delta v_G}\right|_{v_p = 180} = \frac{(15.5 - 6.5) \times 10^{-3}}{-3 - (-5)} = 4.5\ \text{mS}$$

Then, $\mu \equiv g_m r_p = 21.87$, and Problem 13.6, yields

$$A_v = -\frac{\mu R_L}{R_L + r_p} = -\frac{(21.87)(11.6 \times 10^3)}{(11.6 + 4.86) \times 10^3} = -15.41$$

13.10 The input admittance to a triode modeled by the small-signal equivalent circuit of Fig. 13-7(*b*) is obviously zero; however, there are interelectrode capacitances that must be considered for high-frequency operation. Add these interelectrode capacitances (grid-cathode capacitance C_{gk}; plate-grid, C_{pg}; and plate-cathode, C_{pk}) to the small-signal equivalent circuit of Fig. 13-7(*b*). Then (*a*) find the input admittance Y_{in}, (*b*) find the output admittance Y_o, and (*c*) develop a high-frequency model for the triode.

(*a*) With the interelectrode capacitances in position, the small-signal equivalent circuit is given by Fig. 13-13. The input admittance is

$$Y_{\text{in}} = \frac{I_S}{V_S} = \frac{I_1 + I_2}{V_S} \tag{1}$$

But
$$I_1 = \frac{V_S}{1/sC_{gk}} = sC_{gk}V_S \tag{2}$$

and
$$I_2 = \frac{V_S - V_o}{1/sC_{pg}} = sC_{pg}(V_S - V_o) \tag{3}$$

Substituting (*2*) and (*3*) into (*1*) and rearranging give

$$Y_{\text{in}} = s\left[C_{gk} + \left(1 - \frac{V_o}{V_S}\right)C_{pg}\right] \tag{4}$$

Now, from the result of Problem 13.6,

$$\frac{V_o}{V_S} = -\frac{\mu R_L}{R_L + r_p} \tag{5}$$

so (*4*) becomes

$$Y_{\text{in}} = s\left[C_{gk} + \left(1 + \frac{\mu R_L}{R_L + r_p}\right)C_{pg}\right] \tag{6}$$

(*b*) The output admittance is

$$Y_o = -\frac{I_L}{V_o} = -\frac{I_2 - I_p - I_{pk}}{V_o} \tag{7}$$

and
$$I_{pk} = sC_{pk}V_o \tag{8}$$

Let Y'_o be the output admittance that would exist if the capacitances were negligible; then

$$I_p = Y'_oV_o \tag{9}$$

so that
$$Y_o = s\left[\left(1 + \frac{R_L + r_p}{\mu R_L}\right)C_{pg} + C_{pk}\right] + Y'_o \tag{10}$$

(*c*) From (6) and (10) we see that high-frequency triode operation can be modeled by Fig. 13-7(*b*) with a capacitor $C_{\text{in}} = C_{gk} + [1 + R_L/(R_L + r_p)]C_{pg}$ connected from the grid to the cathode, and a capacitor $C_o = [1 + (R_L + r_p)/\mu R_L]C_{pg} + C_{pk}$ connected from the plate to the cathode.

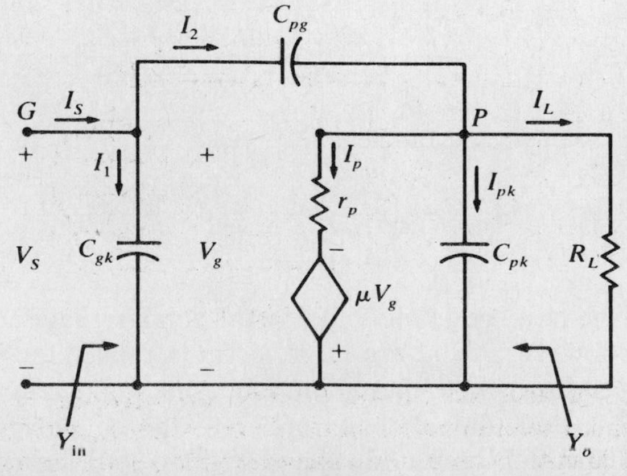

Fig. 13-13

13.11 In the pentode amplifier of Fig. 13-8, assume that the control grid is biased for $V_{G1Q} = -2.5$ V and the screen grid is biased for $V_{G2Q} = 150$ V. Let $V_{PP} = 300$ V and $R_L = 15$ kΩ. The pentode is characterized by Fig. 13-9. (*a*) Draw the dc load line. Then determine the quiescent values (*b*) I_{G1Q}, (*c*) I_{PQ}, and (*d*) V_{PQ}.

(*a*) For the given values the dc load line is described by (*13.6*) and has i_P intercept

$$\frac{V_{PP}}{R_L} = \frac{300}{15 \times 10^3} = 20 \text{ mA}$$

and v_P intercept $V_{PP} = 300$ V. These intercepts have been used to draw the dc load line of Fig. 13-9.

(*b*) If $v_{G1} < 0$, then the control grid draws negligible current and $I_{G1Q} \approx 0$.

(*c*) The quiescent plate current is read, from the projection of Q onto the i_P axis, as $I_{PQ} = 8.13$ mA.

(*d*) The quiescent plate voltage is read from the projection of Q onto the v_P axis and is $V_{PQ} = 175$ V.

13.12 Since the plate characteristics of a pentode can be described by (*13.3*), the small-signal equivalent circuits of Fig. 13-7, with parameters determined by (*13.8*) and (*13.9*), are applicable. For the pentode amplifier of Problem 13.11, (*a*) find the parameters of the small-signal equivalent circuit and (*b*) calculate the voltage gain.

(*a*) Evaluating (*13.8*) and (*13.9*) at the Q point of Fig. 13-9 gives

$$r_p \approx \left.\frac{\Delta v_P}{\Delta i_P}\right|_{v_{G1}=-2.5} = \frac{250 - 100}{(8.44 - 8.125) \times 10^{-3}} = 476.2 \text{ k}\Omega$$

and

$$g_m \approx \left.\frac{\Delta i_P}{\Delta v_{G1}}\right|_{v_P=175} = \frac{(10.94 - 5.94) \times 10^{-3}}{-2 - (-3)} = 5 \text{ mS}$$

(*b*) Using the small-signal equivalent circuit of Fig. 13-7(*a*), we have

$$A_v = \frac{v_p}{v_g} = -g_m(r_p \| R_L) = \frac{-g_m r_p R_L}{r_p + R_L} = \frac{-(5 \times 10^{-3})(476.2 \times 10^3)(15 \times 10^3)}{(476.2 + 15) \times 10^3} = -72.71$$

Supplementary Problems

13.13 The vacuum diode of Fig. 13-14 has plate characteristic $i_P = 0.08 v_P^{3/2}$ mA. If $E_{PP} = 50$ V and $R = 2.5$ kΩ, calculate the value of the plate current. *Ans.* 10 mA

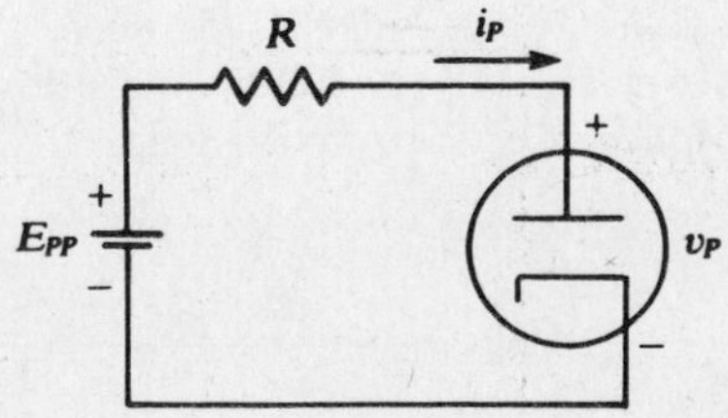

Fig. 13-14

13.14 On a common set of axes sketch the forward characteristics of a vacuum diode described by $i_P = 1.00 v_P^{3/2}$ mA and a semiconductor diode described by $i_D = 10^{-5} e^{40 v_D}$ mA, for values of forward current from 0 to 100 mA. What major difference between the devices is apparent from your sketch? *Ans.* $(i, v_P, v_D) = (0, 0, 0)$, $(20, 7.4, 0.362)$, $(60, 15.3, 0.39)$, $(100, 21.5, 0.403)$

13.15 Analogous to the dynamic resistance defined in (*2.5*), the dynamic resistance of a vacuum diode is

$$r_p \equiv \left.\frac{dv_P}{di_P}\right|_Q = \frac{V_P}{I_P}$$

If a vacuum diode has the static plate characteristic $i_P = 0.08v_P^{3/2}$ mA and is operating with a quiescent plate current $I_P = 15$ mA, calculate (*a*) the quiescent plate voltage V_P and (*b*) the dynamic resistance r_p. *Ans.* (*a*) 32.76 V; (*b*) 1456 Ω

13.16 Suppose the amplifier of Problem 13.4 has plate resistance $r_p = 20$ kΩ and $v_S = 1 \cos \omega t$ V. Find its amplification factor μ using the small-signal voltage-source model of Fig. 13-7(*b*). *Ans.* 30

13.17 Suppose the bypass capacitor C_K is removed from the amplifier of Fig. 13-11. Find (*a*) an expression for the voltage gain and (*b*) the percentage deviation of the voltage gain from the result of Problem 13.9. *Ans.* $A_v = -\mu R_L/[R_L + r_p + (\mu + 1)R_K]$; (*b*) 35.7% decrease

13.18 Two triodes are parallel-connected plate to plate, grid to grid, and cathode to cathode. Find the equivalent amplification factor μ_{eq} and plate resistance r_{peq} for the combination. *Ans.* $\mu_{eq} = (\mu_1 r_{p2} + \mu_2 r_{p1})/(r_{p1} + r_{p2})$, $r_{peq} = r_{p1}r_{p2}/(r_{p1} + r_{p2})$

13.19 The circuit of Fig. 13-15 is a *cathode follower*, so called because v_o is in phase with v_S and nearly equal to it in magnitude. Find a voltage-source equivalent circuit of the form of Fig. 13-7(*b*) that models the cathode follower. *Ans.* See Fig. 13-16.

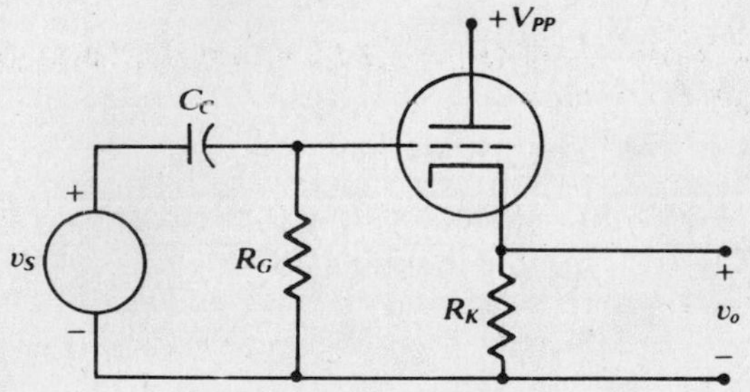

Fig. 13-15

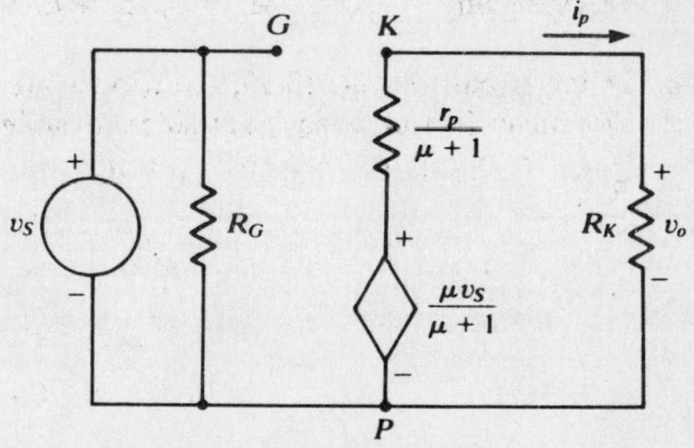

Fig. 13-16

13.20 For the cathode follower of Fig. 13-15, $r_p = 5$ kΩ, $\mu = 25$, and $R_K = 15$ kΩ. (*a*) Use the equivalent circuit of Fig. 13-16 to find a formula for the voltage gain. (*b*) Evaluate the voltage gain. *Ans.* (*a*) $A_v = \mu R_K/[r_p + (\mu + 1)R_K]$; (*b*) 0.95

13.21 The cathode follower is frequently used as a final-stage amplifier to effect an impedance match with a low-impedance load for maximum power transfer. In such a case, the load (resistor R_L) is capacitor-coupled to the right of R_K in Fig. 13-16. Find an expression for the internal impedance (output impedance) of the cathode follower as seen by the load. *Ans.* $R_o = R_K r_p/[r_p + (\mu + 1)R_K]$

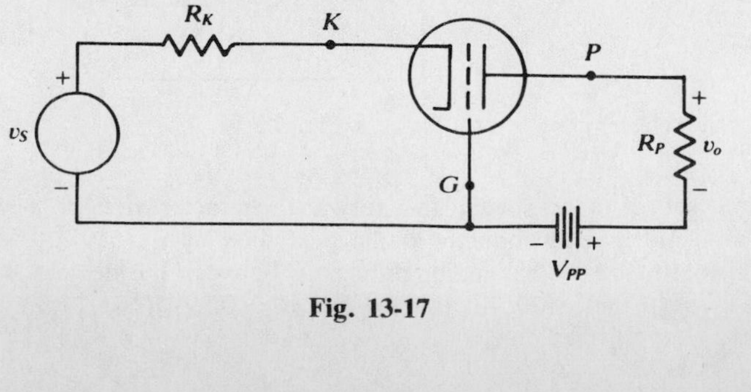

Fig. 13-17

13.22 The amplifier of Fig. 13-17 is a *common-grid* amplifier. By finding a Thévenin equivalent for the network to the right of G,K and another for the network to the left of R_P, verify that the small-signal circuit of Fig. 13-18 is valid. Then, (*a*) find an expression for the voltage gain; (*b*) evaluate the voltage gain for the typical values $\mu = 20$, $r_p = 5\ \text{k}\Omega$, $R_K = 1\ \text{k}\Omega$, and $R_P = 15\ \text{k}\Omega$; (*c*) find the input resistance R_{in}; and (*d*) find the output resistance R_o.
Ans. (*a*) $A_v = (\mu + 1)R_P/[R_P + r_p + (\mu + 1)R_K]$; (*b*) 7.7; (*c*) $R_{in} = 1.95\ \text{k}\Omega$; (*d*) $R_o = 26\ \text{k}\Omega$

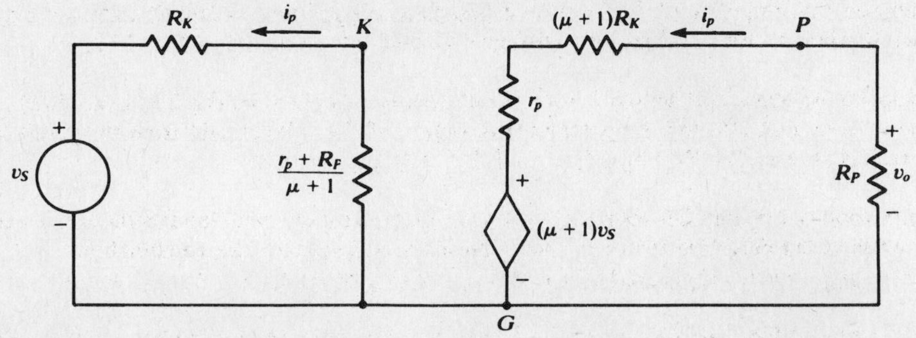

Fig. 13-18

13.23 For the pentode amplifier of Problem 13.11, let $v_{G1} = V_{G1Q} + 0.5 \sin \omega t$ V. Graphically determine the voltage gain. *Ans.* $A_v = v_{g1}/v_p \approx 75$ V

13.24 In the circuit of Fig. 13-19, the triodes are identical, $R_G \approx \infty$, and $(R_L + r_p)/(\mu + 1) \ll R_K$. Show that the circuit is a difference amplifier, meaning that $v_o = f(v_1 - v_2)$.
Ans. $v_o = \mu R_L(v_1 - v_2)/(2R_L + 2r_p)$

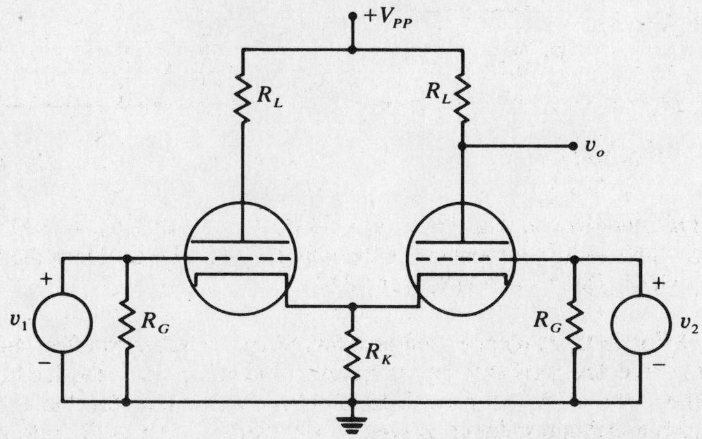

Fig. 13-19

Index

Index